Benchmark Papers in Electrical Engineering and Computer Science

SERIES EDITOR: John B. Thomas
Princeton University

Published Volumes and Volumes in Preparation

SYSTEM SENSITIVITY ANALYSIS / José B. Cruz, Jr.
RANDOM PROCESSES: Multiplicity Theory and Canonical Decompositions / Anthony Ephremides and John B. Thomas
ALGEBRAIC CODING THEORY: History and Development / Ian F. Blake
DIGITAL FILTERS: Design, Theory and Application / Bede Liu
COMPUTER AIDED DESIGN / S. W. Director
NATURAL LANGUAGE QUESTION-ANSWERING SYSTEMS / Robert Chien
CIRCUIT THEORY: / M. E. Van Valkenburg
NONLINEAR SYSTEMS STABILITY / J. K. Aggarwal
DATA COMPRESSION / Lee Davisson
MODELING AND PHYSICAL FOUNDATIONS OF RANDOM PROCESSES / Robert Lugannani
NONPARAMETRIC PATTERN CLASSIFICATION / T. J. Wagner
FUNDAMENTALS OF DATA COMMUNICATION / Lewis E. Franks
SEMICONDUCTOR DEVICE ANALYSIS / James F. Gibbons
PATTERN RECOGNITION: Introduction and Foundations / Jack Sklansky
INTERSYMBOL INTERFERENCE / Kung Yao
BANDLIMITED SIGNAL EXPANSION / Kung Yao
FORMAL LANGUAGES / Jack W. Carlyle and Sheila A. Greibach
N-PORT NETWORK THEORY / Nhan Levan
METHODS OF ENVIRONMENTAL MODELING / R. H. Pantell and D. Daetz
NONLINEAR SYSTEMS: Processing of Random Signals / A. H. Haddad
RANDOM PROCESSES—LINEAR ESTIMATION, VOLUMES I AND II / Thomas Kailath
CLASSIC PAPERS IN CONTROL / George J. Thaler
SEMICONDUCTOR SURFACE PHENOMENA / Peter Mark

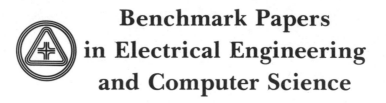

Benchmark Papers in Electrical Engineering and Computer Science

A *BENCHMARK*™ Books Series

RANDOM PROCESSES

v. 1

Multiplicity Theory and Canonical Decompositions

Edited by
ANTHONY EPHREMIDES and JOHN B. THOMAS
University of Maryland
Princeton University

Dowden, Hutchinson & Ross, Inc.
Stroudsburg, Pennsylvania

Library of Congress Catalog Card Number: 72–96190
ISBN: 0–87933–022–8

Library of Congress Cataloging in Publication Data

Ephremides, Anthony, comp.
Random processes: multiplicity theory and canonical decompositions.

(Benchmark papers in electrical engineering and computer science)
Includes bibliographical references.
1. Statistical communication theory. 2. Stochastic processes. I. Thomas, John Bowman, 1925- joint comp. II. Title.
TK5102.5.E62 621.38'043 72-96190
ISBN 0-87933-022-8

Manufactured in the United States of America.

Exclusive distributor outside the United States and Canada: John Wiley & Sons, Inc.

Acknowledgments and Permissions

ACKNOWLEDGMENTS

University of Minnesota, Department of Statistics—*Technical Report 49* and the University of North Carolina Press—*The University of North Carolina Monograph Series in Probability and Statistics*
"On the Connection between Multiplicity Theory and O. Hanner's Time Domain Analysis of Weakly Stationary Stochastic Processes"

PERMISSIONS

The following papers have been reprinted with the permission of the authors and copyright owners.

American Society of Mechanical Engineers—*Journal of Basic Engineering*
"New Results in Linear Filtering and Prediction Theory"

Dunod/Gauthier-Villars—*Annales de Science de l'Ecole Normale Supérieure*
"Sur une classe de courbes de l'espace de Hilbert et sur une équation intégrale non-linéaire"

Institute of Electrical and Electronics Engineers
IEEE Transactions on Automatic Control
"An Innovations Approach to Least-Squares Estimation. Part I: Linear Filtering in Additive White Noise"
"An Innovations Approach to Least-Squares Estimation. Part II: Linear Smoothing in Additive White Noise"
IEEE Transactions on Information Theory
"Nonsingular Detection and Likelihood Ratio for Random Signals in White Gaussian Noise"
Proceedings of the IEEE
"The Innovations Approach to Detection and Estimation Theory"
Proceedings of the I.R.E.
"A Simplified Derivation of Linear Least Square Smoothing and Prediction Theory"

Institute of Mathematical Statistics—*Annals of Mathematical Statistics*
"Radon–Nikodym Derivatives of Gaussian Measures"

Mathematical Institute, Nagoya University—*Nagoya Mathematical Journal*
"On Multivariate Wide-Sense Markov Processes"

Princeton University Press—*The First Samuel Stanley Wilks Lecture at Princeton University, March 17, 1970*
"Structural and Statistical Problems for a Class of Stochastic Processes"

Royal Swedish Academy of Science—*Arkiv för Matematik*
"Deterministic and Non-deterministic Stationary Random Processes"
"On the Structure of Purely Non-deterministic Stochastic Processes"

Society for Industrial and Applied Mathematics—*Theory of Probability and Its Applications*
"Analytic Random Processes"
"Multiplicity and Representation Theory of Purely Non-deterministic Stochastic Processes"
"Stochastic Processes as Curves in Hilbert Space"
"On the Canonical Hida–Cramér Representation for Random Processes"

University of California Press
Proceedings of the Fourth Berkeley Symposium on Mathematical Statistics and Probability
"On Some Classes of Nonstationary Stochastic Processes"
Proceedings of the Fifth Berkeley Sumposium on Mathematical Statistics and Probability
"A Contribution to the Multiplicity Theory of Stochastic Processes"

University of Kyoto—*Memoirs of the College of Science*
"Canonical Representations of Gaussian Processes and Their Applications"

Series Editor's Preface

The "Benchmark Papers in Electrical Engineering and Computer Science" series is aimed at sifting, organizing, and making readily accessible to the reader the vast literature that has accumulated on both subjects. Although the series is not intended as a complete substitute for a study of this literature, it will serve at least three major critical purposes. First, it provides a practical point of entry into a given area of research. Each volume offers an expert's selection of the critical papers on a given topic as well as his views on its structure, development, and present status. Second, the series provides a convenient and time-saving means for study in areas related to but not contiguous with one's principal interests. Last, but by no means least, the series allows the collection, in a particularly compact and convenient form, of the major works on which present research activities and interests are based.

Each volume in the series has been collected, organized, and edited by an authority in the area to which it pertains. In order to present a unified view of the area, the volume editor has prepared an introduction to the subject, has included his comments on each article, and has provided a subject index to facilitate access to the papers.

We believe that this series will provide a manageable working library of the most important technical articles in electrical engineering and computer science. We hope that it will be equally valuable to students, teachers, and researchers.

This volume, *Random Processes: Multiplicity Theory and Canonical Decompositions*, has been edited jointly by me and a former student, Professor A. Ephremides of the University of Maryland. We have selected twenty papers which treat the decomposition and representation of a second-order random process in terms of uncorrelated white-noise components.

In addition to the insight which such a decomposition provides into the structure of random processes, it forms the theoretical foundation and justification of the "prewhitening" technique widely used in detection, estimation, and filtering of signals in noise.

John B. Thomas

Contents

Contents by Author

Introduction

In many applied fields such as the engineering disciplines and, in particular, in the areas of estimation, communication, and control, the sources of interference in signal processing are numerous and diverse both in origin and in magnitude. In the construction of mathematical models for most systems, these interferences are standardly portrayed as random processes.

This common practice has produced a growing interest on the part of engineers and other applications-oriented workers in the various aspects of the theory of random processes and in the field of statistics in general. A major portion of this current interest and research is oriented toward the physical modeling of random processes and toward their representation in terms of simple components.

The mathematical sophistication required to comprehend the properties of random processes and the role they play in the corresponding applications is considerable; this tends to create a gap between mathematically oriented engineers, with limited interest in design and development, and applications-oriented engineers, who tend to dismiss some of the mathematical tools available to them as too complicated and of questionable usefulness.

One purpose of the present collection of papers, drawn from both engineering and mathematical fields, is to contribute to the narrowing of this gap by demonstrating the fundamental relationship that exists between a selected set of commonly accepted landmark results and ideas in engineering and an appropriate set of involved mathematical concepts.

One of the major statistical tools borrowed by engineers in the modeling of problems in the areas of detection, estimation, control, and optimization was the adoption of the mean-square error as a criterion of performance. Originally, in dealing with discrete-time systems, the use of this criterion amounted to a direct application of techniques of the analysis of variance. Gradually, though, with the introduction of time-varying, continuous-time systems, the need for more advanced approaches was created. Moreover, linear data processing became popular for reasons of easy design implementation and of tractability in the analysis of the mathematical models.

These two general concepts created the area of linear, least-squares processing that has been a major branch of the theories of statistical communication and of automatic control.

The first models used in this area arose from N. Wiener's ideas on time-series analysis in both the time and frequency domains (Wiener, 1949). Some of the results include the classical Wiener and matched filters (North, 1943). At the same time, in order for these rather rudimentary processing schemes to be valid, a number of restrictive assumptions were made on the statistical nature of the signal and noise waveforms, such as stationarity, independence, and "whiteness." These assumptions generated questions concerning their necessity and led naturally to the need for a more careful study of the quantities involved in the mathematical models.

A few general conclusions that emerged from this study concern the nature of the mathematical counterparts of the essential concepts of linearity and of the mean-square error.

First, the establishment of the mean-square error as a performance criterion necessarily restricts the class of permissible random processes $x(t)$ to include those processes for which this error can be defined, namely the second-order processes (i.e., $E[x(t)]^2 < \infty$, for every t in an interval of interest). Furthermore, the requirement of linear processing allows considerable flexibility in the distribution properties of the permissible random processes. It imposes restrictions only on the second-order moments of the process. In addition, it provides powerful geometric tools for the analysis of the appropriate models, such as Hilbert-space environment, curve properties, projections, etc.

Finally, the analysis of the possible error magnitudes introduces an interesting dichotomy in the body of the random processes under consideration. On the one hand there are those processes whose future can be unambiguously determined from their pasts, and on the other hand there are the ones without this property. The first class consists of the deterministic random processes, and the second of the nondeterministic ones. A special subclass of the latter consists of the purely nondeterministic (pnd) processes; that is, the ones that do not have a deterministic component and are therefore entirely unpredictable, as for example the Brownian motion process. The notion of determinism accepts an intuitive, physical interpretation; its abstract mathematical counterpart is related to the analyticity property, which was also studied by Wiener.

The eventual understanding of these notions and the familiarity with the structure of random processes gave rise to brilliant engineering schemes in data processing, such as the prewhitening technique of Bode and Shannon (1950: paper 1 of this volume), the Kalman–Bucy filtering model (Kalman and Bucy, 1961: paper 8), the shaping-filter approaches for nonwhite noise, and the innovations approach by Kailath and others (Kailath, 1968, 1970: papers 13, 16; Kailath and Frost, 1968: paper 14).

These engineering developments have emerged from a process of continuous interaction with the field of mathematical statistics, where similar parallel developments have occurred. A theoretical development that pertains to the idea of linear, least-squares filtering is the canonical decomposition of a random process, known also as the Cramér–Hida multiplicity representation (Cramér, 1960, 1961: papers 5 and 7; Hida, 1960: paper 6). This decomposition provides a linear, geometric description of a second-order random process, and it very clearly embodies the notion of determinism. Its precise form depends, as expected, only on wide-sense, second-order moment properties of the process, and in particular on the form of the autocovariance function $R(t,s) = E\left[x(t)x(s)\right]$. Furthermore, the decomposition can be expressed in terms immediately translatable into standard engineering language, such as white-noise inputs, causal filters, and a very powerful block-diagram description. The theory associated with this representation is essentially a time-domain spectral analysis of a wide class of random processes.

The story of this canonical decomposition began in the late thirties and early forties when, following Wiener's ideas, Wold (1938) and Karhunen (1947) obtained a representation for a regular (i.e., pnd), wide-sense stationary random process $x(t)$ in terms of a white-noise generating process $n(t)$, and an invertible, causal, linear filter with impulse-response function $h(t,\tau)$. It was shown that such a process could be obtained from $n(t)$ as in Fig. 1a. Furthermore, it was shown that $n(t)$ could be recovered from $x(t)$ by a similar inverse operation, as in Fig. 1b. Following further investigations and generalizations of this representation by a number of authors such as Hanner (1950: paper 2), Lévy (1956: paper 3) and others, both Hida and Cramér, working independently, derived the complete version of the canonical decomposition for an arbitrary, second-order, left mean-square continuous random process. They showed that any such process may be obtained by causally filtering a set of M uncorrelated white-noise components, and by adding to the result a deterministic component, as shown in Fig. 2a. The number M is determined uniquely by the form of the autocovariance function of the process and is called the "multiplicity" of the process. Each of the generators of the process can be recovered from the process itself by a similar set of inverse operations, as in Fig. 2b.

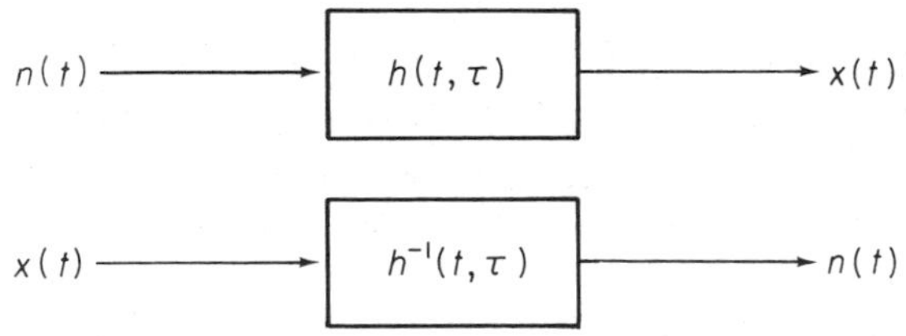

Figure 1. The prewhitening problem: (a) shaping of white noise by linear filtering; (b) the inverse or prewhitening filter.

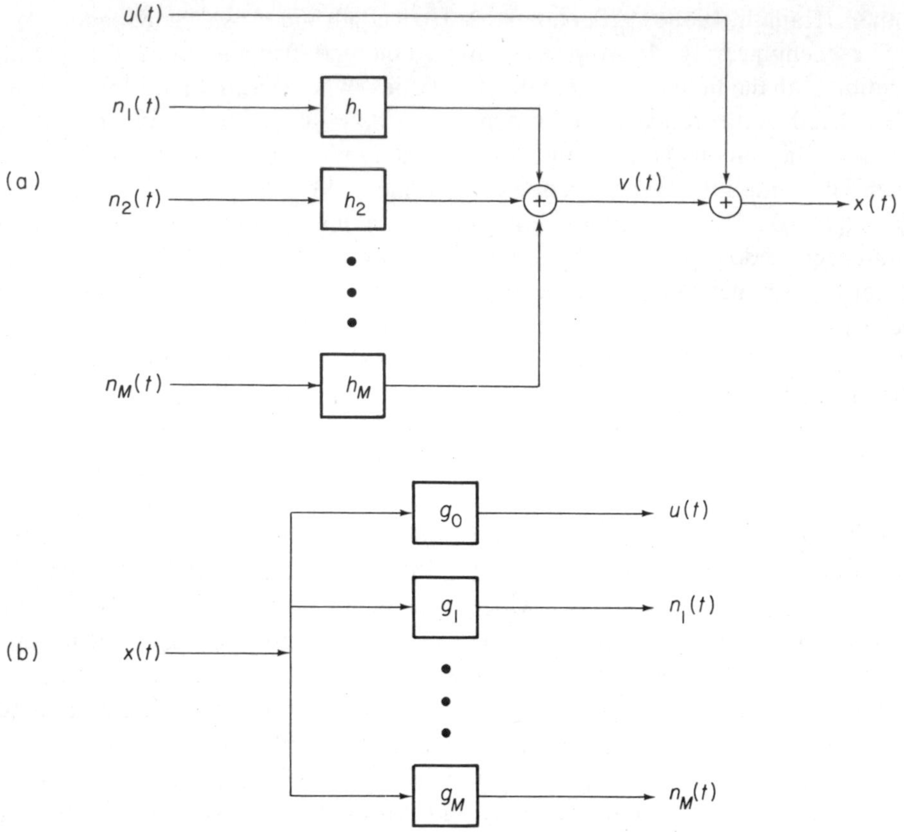

Figure 2. A block diagram of the canonical representation of a random process: (a) the canonical composition; (b) the canonical decomposition.

The canonical decomposition can of course be described in more sophisticated and rigorous mathematical terms. The block-diagram interpretation, however, demonstrates the engineering relevance of the representation. In recent years further work by Cramér (1964, 1965, 1971: papers 11, 12, and 17), Kallianpur and Mandrekar (1965,1970: papers 10 and 9), Ivkovich and Rozanov (1971: paper 20) has added elaborate new results to the theory, increasing the knowledge of the linear structure of random processes and offering new tools for the study of statistical inference.

Of special significance to optimum data processing is the possibility of prewhitening the incoming observation process without loss of information. Such a transformation constitutes an orthogonalization procedure that maps the observed data onto an equivalent set of data of a definite known structure. Although the lack of hardware without a finite cutoff frequency precludes the actual physical implementation of this scheme, the theoretical possibility alone provides a number of advantages. First, the least-squares estimate of a signal corrupted by additive white noise can assume a simple, compact expression that facilitates further analysis. Second, the problem of nonsingular detection of such a signal acquires a definite answer by the determination of the existence of the likelihood ratio.

Bode and Shannon (1950) proposed such a transformation for noncausal processing. Since then the extensive use of the idea in various forms has resulted in modifications and improvements of the original technique. For example, the use of prewhitening in the matched filter analysis allows a solution to the colored-noise problem. The Kalman–Bucy filtering model, although of wider significance and applicability, implicitly assumes (at least in its linear version) the possibility of invertible prewhitening of the state and observation processes. The shaping-filter models for the processing of signals in colored noise are again implicitly concerned with the information-lossless prewhitening of the observation process. Finally, the innovations approach to detection and estimation makes direct use of the existence of such a transformation.

It should be observed that the engineering concern for the existence of an invertible, causal, linear filter for the prewhitening of a random process $x(t)$ can find its answer in the context of the canonical decomposition. The answer is positive if and only if the process $x(t)$ is pnd and has multiplicity equal to one. It is known, for example, that discrete-time processes, wide-sense stationary processes, and wide-sense Markov processes do have unit multiplicity. Most of the more general results, however, are expressed in indirect and unverifiable terms, and as such they are at present of little practical use to the engineer.

The theory of multiplicity, as well as the outlined engineering ideas associated with it, aims primarily at problems of a linear nature. These problems have not been resolved in their entirety, and several aspects of them remain quite obscure. There is a considerable current interest in nonlinear methods of data processing. In such cases the required tool is in general a complete statistical description of the random process through its joint distributions of all orders. Needless to say, such nonlinear approaches are still at an early state of development. Linear analysis on the other hand is at a comparatively advanced state. However, as demonstrated by the papers appearing in this volume, as well as by numerous recent investigations along the lines of the theory of multiplicity and its applications, in addition to a number of linear problems in which there is still vigorous interest, several nonlinear problems are peripherally fusing into this theory and broadening its scope.

The literature on the subject of the present volume is rich in contributions from both the engineering and the mathematical statistics fields. For this reason the selection of papers has been a difficult task. The principal criteria that were used were the significance and quality of the contributions, their relevance to the major topic of the collection, and their role in the historical evolution of the theory and its application. The reference list of each selected paper contains additional outstanding contributions that could not be included here. An effort was made throughout to present a coherent sequence of results that covers the entire field of the canonical decomposition and that enhances the aspect of interaction between engineering considerations and the underlying mathematical ideas.

The papers are presented in their entirety, and the order in which they appear is basically chronological, but slightly perturbed to achieve conceptual continuity. Some of the papers are focused on topics outside the main subject of multiplicity. The highlight comments preceding the papers point out their relationships to the central topic and provide additional background information.

We have been extraordinarily fortunate in that both Professor Cramér and Professor Hida have contributed their own introductory comments on multiplicity theory for inclusion in this volume. Their remarks follow this introduction. We are most grateful to them for their interest and for the additional insights afforded by these comments.

We believe that this collection of outstanding contributions constitutes reference material useful to both the theoretician and to the applications-oriented reader.

References

Karhunen, K. (1947). Uber Lineare Methoden in der Wahrscheinlichkeitsrechnung. *Ann. Acad. Sci. Fennicae, Ser. AI,* No. 37, 1–79.

North, D. O. (1943). Analysis of factors which determine signal–noise discrimination in pulsed carrier systems. RCA Tech. Rept. PTR-6C.

Wiener, N. (1949). "Extrapolation, Interpolation, and Smoothing of Stationary Time Series." M.I.T. Press, Cambridge, Mass.

Wold, H. (1938). "A Study in the Analysis of Stationary Time Series." Almquist and Wiksell, Uppsala, Sweden.

Comments on Multiplicity Theory

HARALD CRAMÉR

In order to deal successfully with problems of statistical inference connected with stochastic processes, it is of fundamental importance to have recourse to a convenient type of analytical representation for the particular class of processes under consideration. Such a representation should express in mathematical form some essential features of the random mechanism assumed to generate the process. It will then lend itself in a natural way to a discussion of the various statistical questions arising in the applications of the process. As examples of such types of representations we may refer to the spectral representation for stationary processes, the Karhunen–Loève representation in terms of eigenfunctions and eigenvalues of a symmetrical kernel, etc.

Multiplicity theory, as we know it today, is concerned with a very broad class of stochastic processes, and has developed out of a systematic study of one particularly important type of analytical representation, namely, the representation in terms of past innovations. Consider the simple case of a real-valued stochastic process $x(t)$ with $Ex(t) = 0$ and $Ex^2(t) < \infty$, the continuous time parameter t varying over $(-\infty,\infty)$. When such a process is encountered in an applied problem, it will often appear realistic to regard the random variable $x(t)$ associated with a given instant t as the cumulative effect produced at this time by a stream of random impulses or "innovations" acting throughout the past. Let us assume that the impulse $dz(u)$ occurs within the infinitesimal time element $(u, u + du)$, and that a unit impulse acting at time u produces the effect $g(t,u)$ at time $t > u$. If effects are additive, we should then be led to expect a representation of the form

$$x(t) = \int_{-\infty}^{t} g(t,u)\, dz(u)$$

with some appropriate definition of the stochastic integral involved.

It was soon found that, whenever an innovations representation of this form did exist, it could render useful service in various applied problems connected with the $x(t)$ process. As an important example we may mention the problem of estimating an unknown signal from an $x(t)$ message containing the signal corrupted by random noise. Accordingly, several early works in this field were concerned with the problem of deciding whether a given process $x(t)$ will admit an innovations representation of the form just given. It was easy to show by means of examples that this is by no means always so, but it seemed difficult to find a general criterion. Also, it was easy to see that a representation of this type, even when it does exist, will not be unique. It thus seemed desirable to find, among the various possible representations, one which might reasonably be regarded as canonical.

The definite turning point was reached when it became clear that a slight generalization of the concept of an innovations representation would be sufficient to make it apply to a very broad class of processes. In fact, a representation of the form

$$x(t) = \sum_{n=1}^{N} \int_{-\infty}^{t} g_n(t,u)\, dz_n(u)$$

can be shown to hold under fairly general conditions, and may be regarded as showing how the random variable $x(t)$ of the process is built up as the joint effect of a stream of N-dimensional random innovations $(dz_1(u), \ldots, dz_N(u))$ acting throughout the past. The random innovation element $dz_n(u)$ is then associated with the nonrandom response function $g_n(t,u)$. It follows from the papers reproduced in this volume that a representation of this form does hold in very general cases. Also, under appropriate conditions, the number N of terms required in the second term of the above equation is uniquely determined by the $x(t)$ process. This characteristic number N, which may be finite or infinite, is called the multiplicity of the $x(t)$ process.

Analogous innovations representations hold in more general cases, such as complex-valued processes, and vector processes of various degrees of generality. By "multiplicity theory" we may understand the theory of this type of representation.

The natural approach to multiplicity theory is by way of the geometry of a separable Hilbert space.

The basic proposition of multiplicity theory states that, under appropriate conditions, a stochastic process admits a representation in terms of past innovations. This proposition can be directly expressed in Hilbert-space terminology, and will then follow from a well-known classical theorem on a separable Hilbert space H. In order to state and prove this theorem, only the most elementary geometric properties of the space H are required. Advanced Hilbert-space concepts, such as self-adjoint transformations, self-reproducing kernels, etc., do not even have to be mentioned. The statement of this theorem will be briefly indicated below; a simple proof has been given in university lectures by the present author, using only elementary properties of projections in H.

We must first make a brief digression on real-valued functions $F(t)$ of a real variable t. Let S be the set of all $F(t)$ which are never decreasing and everywhere continuous to the left. Introduce in S a partial ordering by writing $F_1 >> F_2$ whenever F_2 is absolutely continuous with respect to F_1. If $F_1 >> F_2$, and at the same time $F_2 >> F_1$, we say that F_1 and F_2 are equivalent. The set of all F equivalent to a given F_1 is called the equivalence class C_1 of F_1. For two equivalence classes C_1 and C_2 we define order and equivalence in the obvious way. When two classes are equivalent, they are identical.

Now, let H be a real and separable Hilbert space (analogous arguments also apply, *mutatis mutandis,* in the complex case). Let H_t be a one-parameter family of subspaces of H such that

$$\begin{aligned} &H_t \subset H_u && \text{for } t < u, \\ &H_{t\text{-}0} = H_t && \text{for all } t, \\ &H_{-\infty} = \{0\}, && H_{+\infty} = H. \end{aligned}$$

Denoting by P_t the projection of H onto H_t, we further write for any element $z \in H$

$$z(t) = P_t z, \qquad ||\, z(t)\, ||^2 = F_z(t).$$

Then $F_z(t)$ as a function of t will belong to the set S defined above. We finally denote by $H(z)$ the cyclic subspace of H spanned by the elements $z(t)$ for all real t.

With the given family H_t of subspaces in H we may then associate a decomposition of H in a vector sum of mutually orthogonal cyclic subspaces, according to the following theorem.

It is possible to find a finite or enumerable sequence of elements $z_1, \ldots,$ $z_N \in H$ such that:

(1) $H(z_m) \perp H(z_n)$ *for* $m \neq n$;

(2) *if* $F_n(t) = ||z_n(t)||^2$, *then* $F_1 >> \cdots >> F_N$;

(3) $H = H(z_1) \oplus \cdots \oplus H(z_N)$,

where the expression on the right-hand side of (3) denotes the vector sum of the mutually orthogonal subspaces involved.

The number N and the equivalence classes of the functions $F_1, \ldots, F_N$ are uniquely determined by the given H_t family of subspaces of H. N is called the multiplicity of the H_t family.

Consider any stochastic process—a one-dimensional or a vector process of any order—the random variables of which depend on the single time parameter t. Under appropriate conditions, which it is not necessary to specify in this introduction, such a process will define a separable Hilbert space H, spanned by all the random variables of the process. Further, if H_t is the subspace of H spanned by all those variables that correspond to values $\leqq t$ of the time parameter, H_t will define a nondecreasing family of subspaces of H of the kind considered above.

The application of the Hilbert-space theorem stated above to the H_t family of subspaces defined by the given stochastic process will then directly yield the innovations representation of the process, in the form given in the relevant papers reproduced below. The innovations representation obtained in this way is a canonical representation in the precise sense indicated by the basic Hilbert-space theorem.

In the further study of the properties of this type of representation, and of its applications, many new problems arise, most of which are so far only partially solved. These are investigated in the papers reproduced below.

Comments on the Theory of Multiplicity for Gaussian Processes

TAKEYUKI HIDA

In this short note I would like to emphasize the importance of the concept of multiplicity in investigating Gaussian processes which are not necessarily assumed to be stationary.

Let $X(t)$, $t \subset T$, be a separable Gaussian process with mean zero and assume that it is purely nondeterministic in the sense that

$$\bigcap_{t \subset T} \mathbf{M}_t(X) = \{0\},$$

where $\mathbf{M}_t(X)$ is the closed linear manifold spanned by the $X(s)$, $s \leqslant t$. Associated with $\mathbf{M}_t(X)$ is a projection $E(t)$, and $\{E(t);\ t \subset T\}$ is a resolution of the identity in the Hilbert space $\mathbf{M}(X)$ spanned by all the $X(t)$, $t \subset T$. We can therefore speak of the multiplicity of $E(t)$, i.e., the multiplicity of the process $X(t)$, by appealing to the Hellinger–Hahn theorem. Generally, $X(t)$ can be expressed in the form (generalized canonical representation)

$$X(t) = \sum_{i=1}^{N} \int_0^t F_i(t, u) dB_i(u) \tag{1}$$

where N ($\leqslant +\infty$) is the multiplicity of $X(t)$ and $\{dB_i(u);\ 1 \leqslant i \leqslant N\}$ is a system of mutually independent Gaussian random measures.

An especially significant case arises when N is exactly unity. To make the situation simple we shall discuss the cases where the random measure is homogeneous, i.e., Brownian, and where $T = [0, \infty)$. Then, together with the assumption $N = 1$, the representation (1) becomes

$$X(t) = \int_0^t F(t, u)\, dB(u), \qquad E[dB(u)]^2 = du. \tag{2}$$

Now we may illustrate why the case $N = 1$ is of special importance and why the expression (2) is meaningful. Suppose that $X(t)$ has the canonical representation (2). Then

(1) the process $X(t)$ can be viewed as a linear functional of a single white-noise input;

(2) by the canonical property, the infinitesimal random variable $dB(t)$ represents the new information or new randomness that $X(t)$ gains during $(t, t+dt)$;

(3) the kernel $F(t, u)$ is of Volterra type and is a causal impulse response function, and, again, by the canonical property, F determines the causally invertible map;

(4) the canonical representation is essentially unique: once the explicit form of F is known, it tells us many important probabilistic properties of the process $X(t)$, e.g., the Markov property, sample function properties, etc.

Now the following two problems are proposed from our viewpoint: (a) to find sufficient conditions which guarantee that the multiplicity of $X(t)$ is one, and (b) to obtain the exact form of the canonical kernel F in (2). These two problems, however, have usually been discussed simultaneously.

So far as this writer knows, the following cases have been discussed by many authors.

(i) The most beautifully established results have been obtained in the case where $X(t)$ is mean-continuous and stationary. Karhunen's well-known theory is applied and his representation using the optimal kernel is simply our canonical representation. It is noted that the method of obtaining the canonical kernel is also indicated in his theory using the Fourier transform. Perhaps there is no need to review his theory in detail.

(ii) H. Cramér developed the multiplicity theory at the Fourth Berkeley Symposium held in 1960. Since then he has obtained many interesting basic results. For example, as for the problem (a), he has introduced a certain class of harmonizable processes such that their multiplicity is proved to be unity. Part of the proof depends on the technique used in obtaining the representation of stationary processes. Close to this but somewhat different has been the approach made by Kallianpur and Mandrekar, who have developed and also generalized the theory.

(iii) Suppose that $X(t)$ is $N - 1$ times differentiable (in the mean-square sense) and satisfies

$$LX(t) = \sum_{k=0}^{N} a_k(t) \frac{d^{N-k}}{dt^{N-k}} X(t) = B'(t) \tag{3}$$

$[B'(t)$ is the derivative of the Brownian motion $B(t)]$. Then $X(t)$ is called an N-tuple Markov Gaussian process in the restricted sense. The multiplicity of such a process is unity, and the solution of (3) gives the canonical representation of $X(t)$, where the canonical kernel is the Riemann function associated with the differential operator L. Roughly speaking, L gives the inverse map of (2). Observing the kernel function, we can immediately recognize the multiple Markov property for the process $X(t)$. Such a consideration leads us to a generalization of the notion of the multiple Markov property in the restricted sense: i.e., $X(t)$ is said to be N-tuple Markov (in the weak sense), if, for any t_0,

$$\text{the family } \{X(t_0\,;t_j)\,;\,t_j \geqslant t_0,\ 1 \leqslant j \leqslant N'\} \text{ contains exactly } N \text{ linearly independent elements for any different } t_j\text{'s with } N' \geqslant N, \tag{4}$$

where $X(t_0\,;\,t)$ denotes the projection of $X(t)$ onto $\mathbf{M}_{t_0}(X)$. If we further assume the existence of the canonical representation, then we prove that the canonical kernel is expressed in the form (Goursat kernel)

$$F(t, u) = \sum_{i=1}^{N} f_i(t)g_i(u)$$

which is helpful in discussing not only the Markov property but also the prediction problem (Lévy, 1955; Hida, 1960). It is expected that (4) together with our general assumption of a purely nondeterministic process implies that the multiplicity is unity.

(iv) P. Lévy, just after he originated the theory of canonical representation, discussed how to get the canonical kernel $F(t, u)$ directly from the covariance function $\Gamma(t_1, t_2)$ of the Gaussian process (as a consequence we have $N = 1$). To solve the problem, Lévy started with the nonlinear integral equation of the form

$$\Gamma(t_1, t_2) = \int_0^t F(t_1,u)F(t_2,u)du, \qquad t = \min(t_1,t_2). \tag{5}$$

Under some restricted assumptions on $\Gamma(t_1, t_2)$ he proceeded as follows: Set

$$\gamma(t_1, t_2) = \frac{\partial^2}{\partial t_1 \partial t_2}\Gamma(t_1, t_2)$$

and

$$\sigma^2(t) = \lim_{\tau \to 0} \frac{1}{\tau}\left[\Gamma(t, t) + \Gamma(t + \tau, t + \tau) - 2\Gamma(t, t + \tau)\right].$$

With these given functions one can determine $G(t, u)$ first, then $R(t, u)$, by solving the integral equations

$$\gamma(t, u) = \sigma^2(u)G(t, u) + \int_0^t G(t, v)\gamma(v, u)dv$$

and

$$G(t, u) + R(t, u) = \int_u^t G(t, v)R(v, u)dv.$$

Now the desired kernel $F(t, u)$ is given by

$$F(t, u) = \sigma(u)\left[1 - \int_u^t R(v, u)dv\right].$$

Although Lévy himself has never mentioned it specifically, these calculations have an intimate connection with the absolute continuity or singularity of two Gaussian measures derived from two Gaussian processes. I believe that further investigations from such a viewpoint are quite worthwhile.

Recently T. Kailath and his co-authors, as well as M. Hitsuda, have obtained several important results on this problem. Their work seems to have been done independently of Lévy.

Finally, I should like to add two remarks on multiplicity theory. As we observed in (iii), the Markov property is reflected in the canonical kernel. This relation is expected to hold even for Gaussian processes with a multidimensional time parameter. The derivation of the canonical representation for those processes would be the first step to this end. Another problem is the following: being inspired by the argument in (iv), we are interested in a transformation of a Gaussian process such that the multiplicity is kept invariant; at the least, we are looking for probabilistic characterizations of such processes which guarantee the minimum value of their multiplicity.

Editors' Comments on Paper 1

The authors of the first paper, H. Bode and C. Shannon, are well known for their outstanding contributions to statistical communication theory. The present paper appeared in 1950; unquestionably it represents a major landmark in the theory of optimal linear filtering. The idea of prewhitening the observation process, although implicit in the earlier work of Wiener, is clearly presented for the first time, and the advantages of the technique are described. The paper is a brilliant example of the expression of fundamental mathematical concepts in simple engineering terms.

Since it allows the whitening filter to be anticipatory (noncausal), their technique is not directly connected to the canonical decomposition theory. However, it represents one of the first major efforts to apply to a communications problem some of the mathematical ideas that preceded the derivation of the Cramér–Hida representation.

This paper is reprinted with the permission of the authors and the Director of Editorial Services of the Institute of Electrical and Electronics Engineers.

Reprinted from *Proceedings of the I.R.E.*, **38**, 417–425 (1950)

1

A Simplified Derivation of Linear Least Square Smoothing and Prediction Theory*

H. W. BODE†, SENIOR MEMBER, IRE, AND C. E. SHANNON†, FELLOW, IRE

***Summary*—The central results of the Wiener-Kolmogoroff smoothing and prediction theory for stationary time series are developed by a new method. The approach is motivated by physical considerations based on electric circuit theory and does not involve integral equations or the autocorrelation function. The cases treated are the "infinite lag" smoothing problem, the case of pure prediction (without noise), and the general smoothing prediction problem. Finally, the basic assumptions of the theory are discussed in order to clarify the question of when the theory will be appropriate, and to avoid possible misapplication.**

I. INTRODUCTION

IN A CLASSIC REPORT written for the National Defense Research Council,[1] Wiener has developed a mathematical theory of smoothing and prediction of considerable importance in communication theory. A similar theory was independently developed by Kolmogoroff[2] at about the same time. Unfortunately the work of Kolmogoroff and Wiener involves some rather formidable mathematics—Wiener's yellow-bound report soon came to be known among bewildered engineers as "The Yellow Peril"—and this has prevented the wide circulation and use that the theory deserves. In this paper the chief results of smoothing theory will be developed by a new method which, while not as rigorous or general as the methods of Wiener and Kolmogoroff, has the advantage of greater simplicity, particularly for readers with a background of electric circuit theory. The mathematical steps in the present derivation have, for the most part, a direct physical interpretation, which enables one to see intuitively what the mathematics is doing.

II. THE PROBLEM AND BASIC ASSUMPTIONS

The main problem to be considered may be formulated as follows. We are given a perturbed signal $f(t)$ which is the sum of a true signal $s(t)$, and a perturbing noise $n(t)$

$$f(t) = s(t) + n(t).$$

It is desired to operate on $f(t)$ in such a way as to obtain, as well as possible, the true signal $s(t)$. More generally, one may wish to combine this smoothing operation with prediction, i.e., to operate on $f(t)$ in such a way as to obtain a good approximation to what $s(t)$ will be in the future, say α seconds from now, or to what it was in the past, α seconds ago. In these cases we wish to approximate $s(t+\alpha)$ with α positive or negative, respectively. The situation is indicated schematically in Fig. 1; the problem is that of filling the box marked "?."

* Decimal classification: 510. Original manuscript received by the Institute, July 13, 1949; revised manuscript received, January 17, 1950.

† Bell Telephone Laboratories, Inc., Murray Hill, N. J.

[1] N. Wiener, "The Interpolation, Extrapolation, and Smoothing of Stationary Time Series," National Defense Research Committee; reprinted as a book, together with two expository papers by N. Levinson, published by John Wiley and Sons, Inc., New York, N. Y., 1949.

[2] A. Kolmogoroff, "Interpolation und Extrapolation von Stationären Zufälligen Folgen," *Bull. Acad. Sci.* (URSS) Sér. Math. 5, pp. 3–14; 1941.

It will be seen that this problem and its generalizations are of wide application, not only in communication theory, but also in such diverse fields as economic prediction, weather forecasting, gunnery, statistics, and the like.

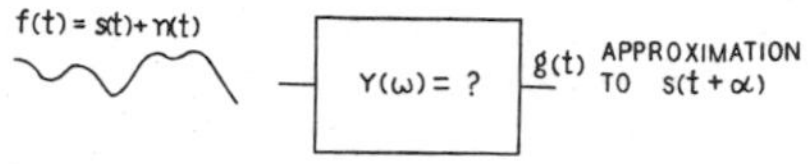

Fig. 1—The smoothing and prediction problem.

The Wiener-Kolmogoroff theory rests on three main assumptions which determine the range of application of the results. These assumptions are:

1. The time series represented by the signal $s(t)$ and the noise $n(t)$ are *stationary*. This means essentially that the statistical properties of the signal and of the noise do not change with time. The theory cannot properly be applied, for example, to *long-term* economic effects, since the statistics of, say, the stock market were not the same in 1850 as they are today.

2. The criterion of error of approximation is taken to be the *mean-square discrepancy* between the actual output and the desired output. In Fig. 1 this means that we fill the box "?" in such a way as to minimize the mean-square error $\overline{[g(t)-s(t+\alpha)]^2}$, the average being taken over all possible signal and noise functions with each weighted according to its probability of occurrence. This is called the *ensemble* average.

3. The operation to be used for prediction and smoothing is assumed to be a *linear* operation on the available information, or, in communication terms, the box is to be filled with a linear, physically realizable, filter. The available information consists of the past history of the perturbed signal, i.e., the function $f(t)$ with $t \leqq t_1$, where t_1 is the present time. A linear, physically realizable filter performs a linear operation on $f(t)$ over just this range, as we will see later in connection with equations (3) and (4).

The theory may therefore be described as *linear least square prediction and smoothing of stationary time series.* It should be clearly realized that the theory applies only when these three assumptions are satisfied, or at least are approximately satisfied. If any one of the conditions is changed or eliminated, the prediction and smoothing problem becomes very difficult mathematically, and little is known about usable explicit solutions. Some of the limitations imposed by these assumptions will be discussed later.

How is it possible to predict at all the future behavior of a function when all that is known is a perturbed version of its past history? This question is closely associated with the problems of causality and induction in philosophy and with the significance of physical laws. In general, physical prediction depends basically on an *assumption* that regularities which have been observed in the past will obtain in the future. This assumption can never be proved deductively, i.e., by purely mathematical argument, since we can easily conceive mathematical universes in which the assumption fails. Neither can it be established inductively, i.e., by a generalization from experiments, for this very generalization would assume the proposition we were attempting to establish. The assumption can be regarded only as a central postulate of physics.

Classical physics attempted to reduce the physical world to a set of strict causal laws. The future behavior of a physical system is then exactly predictable from a knowledge of its past history, and in fact all that is required is a knowledge of the present state of the system. Modern quantum physics has forced us to abandon this view as untenable. The laws of physics are now believed to be only statistical laws, and the only predictions are statistical predictions. The "exact" laws of classical physics are subject to uncertainties which are small when the objects involved are large, but are relatively large for objects on the atomic scale.

Linear least square smoothing and prediction theory is based on statistical prediction. The basic assumption that statistical regularities of the past will hold in the future appears in the mathematics as the assumption that the signal and noise are *stationary* time series. This implies, for example, that a statistical parameter of the signal averaged over the past will give the same value as this parameter averaged over the future.

The prediction depends essentially on the existence of correlations between the future value of the signal $s(t_1+\alpha)$ where t_1 is the present time, and the known data $f(t)=s(t)+n(t)$ for $t \leqq t_1$. The assumption that the prediction is to be done by a *linear* operation implies that the only type of correlation that can be used is *linear* correlation, i.e., $\overline{s(t_1+\alpha)\, f(t)}$. If this correlation were zero for all $t \leqq t_1$, no significant linear prediction would be possible, as will appear later. The best mean-square estimate of $s(t_1+\alpha)$ would then be zero.

III. Properties of Linear Filters

In this section, a number of well-known results concerning filters will be summarized for easy reference. A linear filter can be characterized in two different but equivalent ways. The first and most common description is in terms of the complex transfer function $Y(\omega)$. If a pure sine wave of angular frequency ω_1 and amplitude E is used as input to the filter, the output is also a pure sine wave of frequency ω_1 and amplitude $|Y(\omega_1)|E$. The phase of the output is advanced by the angle of $Y(\omega_1)$, the phase of the filter at this frequency. It is frequently convenient to write the complex transfer function $Y(\omega)$ in the form $Y(\omega)=e^{A(\omega)}e^{iB(\omega)}$ where $A(\omega)=\log|Y(\omega)|$ is the gain, and $B(\omega)=\text{angle}\,[Y(\omega)]$ is the phase. Since we will assume that the filter can contain an ideal amplifier as well as passive elements, we can add any constant to A to make the absolute level of the gain as high as we please.

The second characterization of a filter is in terms of time functions. Let $K(t)$ be the inverse Fourier transform of $Y(\omega)$

$$K(t) = \frac{1}{2\pi}\int_{-\infty}^{\infty} Y(\omega)e^{i\omega t}d\omega. \tag{1}$$

Then $Y(\omega)$ is the direct Fourier transform of $K(t)$

$$Y(\omega) = \int_{-\infty}^{\infty} K(t)e^{-i\omega t}dt. \tag{2}$$

Knowledge of $K(t)$ is completely equivalent to knowledge of $Y(\omega)$; either of these may be calculated if the other is known.

The time function $K(t)$ is equal to the output obtained from the filter in response to a unit impulse impressed upon its input at time $t=0$, as illustrated by Fig. 2. From this relation we can readily obtain the response

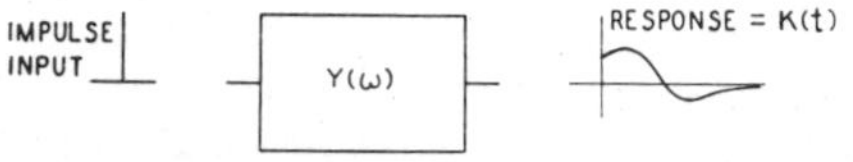

Fig. 2—Impulsive response of a network.

of the filter to any artibrary input $f(t)$. It is merely necessary to divide the input wave into a large number of thin vertical slices, as shown by Fig. 3. Each slice can be regarded as an impulse of strength $f(t)dt$, which will pro-

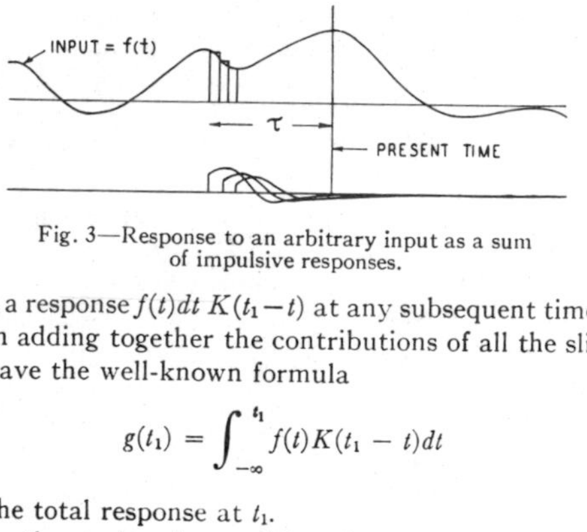

Fig. 3—Response to an arbitrary input as a sum of impulsive responses.

duce a response $f(t)dt\,K(t_1-t)$ at any subsequent time t_1. Upon adding together the contributions of all the slices we have the well-known formula

$$g(t_1) = \int_{-\infty}^{t_1} f(t)K(t_1 - t)dt \tag{3}$$

for the total response at t_1.

For the study of smoothing theory, (3) can conveniently be replaced by a slightly different expression. Setting $\tau = t_1 - t$, we have

$$g(t_1) = \int_0^{\infty} f(t_1 - \tau)K(\tau)d\tau. \tag{4}$$

In this formulation, τ stands for the age of the data, so that $f(t_1-\tau)$ represents the value of the input wave τ seconds ago. $K(\tau)$ is a function like the impulsive admittance, but projecting into the past rather than the future, as shown by Fig. 4. It is evidently a weighting function by which the voltage inputs in the past must be multiplied to determine their contributions to the present output.

Criteria for physical realizability can be given in terms of either the K function or Y. In terms of the impulsive response $K(t)$, it is necessary that $K(t)$ be zero for $t<0$; that is, the network cannot respond to an impulse before the impulse arrives. Furthermore, $K(t)$ must approach zero (with reasonable rapidity) as $t \to +\infty$. Thus the effect of an impulse at the present time should eventually die out.

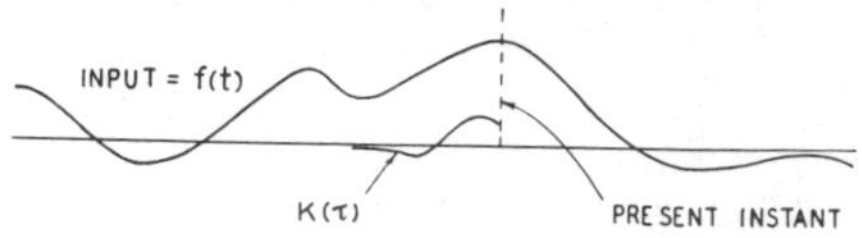

Fig. 4—Response as a weighted average of the past input.

These requirements are also meaningful in terms of the interpretation of K as a weighting function. Thus, the filter cannot apply a weighting to parts of the input that have yet to occur; hence, $K(\tau)=0$ for $\tau<0$. Also the effect of the very remote past should gradually die out, so that $K(\tau)$ should approach zero as $\tau \to \infty$. It may also be noted that these conditions are also sufficient for physical realizability in the sense that any impulsive response $K(t)$ satisfying them can be approximated as closely as desired with a passive lumped element network, together with a single amplifier.

In terms of frequency response, the principal condition for physical realizability is that $Y(\omega)$, considered as a function of the complex variable ω, must be an analytic function in the half plane defined by $Im(\omega)<0$. In addition, the function must behave on the real frequency axis[3] in such a way that

$$\int_0^{\infty} \frac{\log |Y(\omega)|}{1+\omega^2}d\omega \tag{5}$$

is a finite number.

The requirements of physical realizability lead to the well-known loss-phase relations. For a given gain $A = \log |Y(\omega)|$ satisfying (5), there is a minimum possible phase characteristic. This phase is given by

$$B(\omega_0) = \frac{2\omega_0}{\pi}\int_0^{\infty} \frac{A(\omega) - A(\omega_0)}{\omega^2 - \omega_0^2}d\omega. \tag{6}$$

If the square of the prescribed gain $|Y(\omega)|^2 = Y(\omega)\overline{Y}(\omega)$ is a rational function of ω, say, $P_1(\omega)/P_2(\omega)$ where $P_1(\omega)$ and $P_2(\omega)$ are polynomials, the minimum phase characteristic can be found as follows: Calculate the roots of $P_1(\omega)$ and $P_2(\omega)$ and write $|Y(\omega)|^2$ as

$$|Y(\omega)|^2 = k^2\frac{(\omega-\alpha_1)(\omega-\bar{\alpha}_1)(\omega-\alpha_2)(\omega-\bar{\alpha}_2)\cdots}{(\omega-\beta_1)(\omega-\bar{\beta}_1)(\omega-\beta_2)(\omega-\bar{\beta}_2)\cdots} \tag{7}$$

[3] Including the point at infinity. Actual physical networks will, of course, always have zero gain at infinite frequency, and the above requirement shows that the approach to zero cannot be too rapid. An approach of the type ω^{-n} ($6n$ db per octave) is possible but $e^{-|\omega|}$ or $e^{-\omega^2}$ causes the integral in (5) to diverge and is physically unrealizable.

where $\alpha_1, \alpha_2 \cdots \beta_1, \beta_2 \cdots$ all have imaginary parts >0. That is, these are the roots and poles of $|Y(\omega)|^2$ in the upper-half plane and the conjugate terms are the corresponding roots and poles in the lower-half plane. The minimum phase network then has the transfer function

$$Y(\omega) = k \frac{(\omega - \alpha_1)(\omega - \alpha_2) \cdots}{(\omega - \beta_1)(\omega - \beta_2) \cdots}. \tag{8}$$

A minimum phase network has the important property that its inverse, with the transfer function $Y^{-1}(\omega)$, is also physically realizable.[4] If we pass a signal $f(t)$ through the filter $Y(\omega)$, we can recover it in its original form by passing it through the inverse filter. Moreover, the recovery takes place *without loss of time*. On the other hand, there is no physically realizable exact inverse for a nonminimum phase network. The best we can do is to provide a structure which has all the properties of the theoretical inverse, except for an extra phase lag. The extra phase lag can be equalized to give a constant delay by the addition of a suitable phase equalizer, but it cannot be eliminated. Thus, if we transmit a signal through a nonminimum network, we can recover it only after a delay; that is, we obtain $f(t-\alpha)$ for some positive α.

IV. General Expression for the Mean-Square Error

Suppose we use for the predicting-smoothing filter in Fig. 1 a filter with transfer characteristic $Y(\omega)$. What is the mean-square error in the prediction? Since different frequencies are incoherent, we can calculate the average power in the error function

$$e(t) = s(t + \alpha) - g(t) \tag{9}$$

by adding the contributions due to different frequencies. Consider the components of the signal and noise of a particular frequency ω_1. It will be assumed that the signal and noise are incoherent at all frequencies. Then, at frequency ω_1 there will be a contribution to the error due to noise equal to $N(\omega_1)|Y(\omega_1)|^2$, where $N(\omega_1)$ is the average noise power at that frequency ω_1.

There is also a contribution to the error due to the failure of components of the signal, after passing through the filter, to be correct. A component of frequency ω_1 should be advanced in phase by $\alpha\omega_1$, and the amplitude of the output should be that of the input. Hence there will be a power error

$$|Y(\omega_1) - e^{i\alpha\omega_1}|^2 P(\omega_1) \tag{10}$$

where $P(\omega_1)$ is the power in the signal at frequency ω_1.

The total mean-square error due to components of frequency ω_1 is the sum of these two errors, or

$$E_{\omega_1} = |Y(\omega_1)|^2 N(\omega_1) + |Y(\omega_1) - e^{i\alpha\omega_1}|^2 P(\omega_1), \tag{11}$$

[4] If the original function has a zero at infinity, so that the required inverse has a pole there, there are complications, but an adequate approximation can be obtained in physical cases.

and the total mean-square error for all frequencies is

$$E = \int_{-\infty}^{\infty} [|Y(\omega)|^2 N(\omega) + |Y(\omega) - e^{i\alpha\omega}|^2 P(\omega)] d\omega. \tag{12}$$

The problem is to minimize E by proper choice of $Y(\omega)$, remembering that $Y(\omega)$ must be physically realizable.

Several important conclusions can be drawn merely from an inspection of (12). The only way in which the signal and noise enter this equation is through their power spectra. Hence, the only statistics of the signal and noise that are needed to solve the problem are these spectra. Two different types of signal with the same spectrum will lead to the same optimal prediction filter and to the same mean-square error. For example, if the signal is speech it will be predicted by the same filter as would be used for prediction of a thermal noise which has been passed through a filter to give it the same power spectrum as speech.

Speaking somewhat loosely, this means that a linear filter can make use only of statistical data pertaining to the *amplitudes* of the different frequency components; the statistics of the relative phase angles of these components cannot be used. Only by going to nonlinear prediction can such statistical effects be used to improve the prediction.

It is also clear that in the linear least square problem we can, if we choose, replace the signal and noise by any desired time series which have the same power spectra. This will not change the optimal filter or the mean square error in any way.

V. The Pure Smoothing Problem

The chief difficulty in minimizing (12) for the mean-square error lies in properly introducing the condition that $Y(\omega)$ must be a physically realizable transfer function. We will first solve the problem with this constraint waived and then from this solution construct the best physically realizable filter.

Waiving the condition of physical realizability is equivalent to admitting any $Y(\omega)$, or, equivalently, any impulsive response $K(t)$. Thus, $K(t)$ is not necessarily zero for $t<0$, and we are allowing a weighting function to be applied to both the past and future of $f(t)$. In other words, we assume that the entire function $f(t) = s(t)+n(t)$ from $t=-\infty$ to $t=+\infty$ is available for use in prediction.

In (12), suppose

$$Y(\omega) = C(\omega)e^{iB(\omega)} \tag{13}$$

with $C(\omega)$ and $B(\omega)$ real. Then (12) becomes

$$E = \int_{-\infty}^{\infty} [C^2N + P(C^2 + 1 - 2C\cos(\alpha\omega - B))] d\omega \tag{14}$$

where $C(\omega)$, $N(\omega)$, and the like are written as C, N, and so forth, for short. Clearly, the best choice of $B(\omega)$ is $B(\omega) = \alpha\omega$ since this maximizes $\cos(\alpha\omega - B(\omega))$. Then (14) becomes

$$E = \int_{-\infty}^{\infty} [C^2(P + N) - 2PC + P] d\omega. \tag{15}$$

Completing the square in C by adding and subtracting $P^2/(P+N)$ we obtain

$$E = \int_{-\infty}^{\infty} \left[C^2(P + N) - 2PC + \frac{P^2}{P + N} - \frac{P^2}{P + N} + P \right] d\omega \tag{16}$$

or

$$E = \int_{-\infty}^{\infty} \left(\left[\sqrt{P+N}\, C - \frac{P}{\sqrt{P+N}} \right]^2 + \frac{PN}{P+N} \right) d\omega. \tag{17}$$

The bracketed term is the square of a real number, and therefore positive or zero. Clearly, to minimize E we choose C to make this term everywhere zero, thus

$$C(\omega) = \frac{P(\omega)}{P(\omega) + N(\omega)}$$

and

$$Y(\omega) = \frac{P(\omega)}{P(\omega) + N(\omega)} e^{i\alpha\omega}. \tag{18}$$

With this choice of $Y(\omega)$ the mean square error will be, from (17),

$$E = \int_{-\infty}^{\infty} \frac{P(\omega)N(\omega)}{P(\omega) + N(\omega)} d\omega. \tag{19}$$

The best weighting function is given by the inverse Fourier transform of (18)

$$K(t) = \frac{1}{2\pi} \int_{-\infty}^{\infty} \frac{P(\omega)}{P(\omega) + N(\omega)} e^{i\omega(t+\alpha)} d\omega. \tag{20}$$

This $K(t)$ will, in general, extend from $t = -\infty$ to $t = +\infty$. It does not represent the impulsive response of a physical filter. However, it is a perfectly good weighting function. If we could wait until all the function $s(t) + n(t)$ is available, it would be the proper one to apply in estimating $s(t+\alpha)$.

To put the question in another way, the weighting $K(\tau)$ can be obtained in a physical filter if sufficient delay is allowed so that $K(\tau)$ is substantially zero for the future. Thus we have solved here the "infinite lag" smoothing problem. Although $Y(\omega)$ in (18) is nonphysical, $Y(\omega)e^{-i\beta\omega}$ will be physical, or nearly so, if β is taken sufficiently large.

VI. The Pure Prediction Problem

We will now consider another special case, that in which there is no perturbing noise. The problem is then one of pure prediction. What is the best estimate of $s(t+\alpha)$ when we know $s(t)$ from $t = -\infty$ up to $t = 0$?

We have seen that the solution will depend only on the power spectra of the signal and noise, and since we are now assuming the noise to be identically zero, the solution depends only on the power spectrum $P(\omega)$ of the signal. This being the case, we may replace the actual signal by any other having the same spectrum. The solution of the best predicting filter will be the same for the altered problem as for the original problem.

Any desired spectrum $P(\omega)$ can be obtained by passing wide-band resistance noise or "white" noise through a shaping filter whose gain characteristic is $\sqrt{P(\omega)}$. The spectrum of resistance noise is flat (at least out to frequencies higher than any of importance in communication work), and the filter merely multiplies this constant spectrum by the square of the filter gain $P(\omega)$. The phase characteristic of the filter can be chosen in any way consistent with the conditions of physical realizability. Let us choose the phase characteristic so that the filter is minimum phase for the gain $\sqrt{P(\omega)}$. Then the filter has a phase characteristic given by

$$B(\omega_0) = \frac{-\omega_0}{\pi} \int_0^{\infty} \frac{\log P(\omega) - \log P(\omega_0)}{\omega^2 - \omega_0^2} d\omega. \tag{21}$$

Furthermore, this minimum phase network has a physically realizable inverse.

We have now reduced the problem to the form shown in Fig. 5. What is actually available is the function $s(t)$ up to $t = 0$. However, this is equivalent to a knowledge of

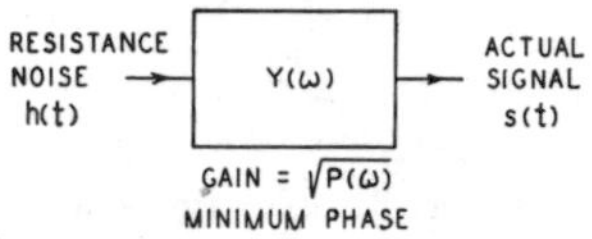

Fig. 5—Construction of actual signal spectrum from resistance noise.

the resistance noise $h(t)$ up to $t = 0$, since the filter Y has a physically realizable inverse and we can pass the available function $s(t)$ through the inverse Y^{-1} to obtain $h(t)$.

The problem, therefore, is equivalent to asking what is the best operation to apply to $h(t)$ in order to approximate $s(t+\alpha)$ in the least square sense? The question is easily answered. A resistance noise can be thought of as made up of a large number of closely spaced and very short impulses, as indicated in Fig. 6. The impulses have a Gaussian distribution of amplitudes and are statistically independent of each other. Each of these impulses entering the filter Y produces an output corresponding to the impulsive response of the filter, as shown at the right of Fig. 6, and the signal $s(t)$ is the sum of these elementary responses.

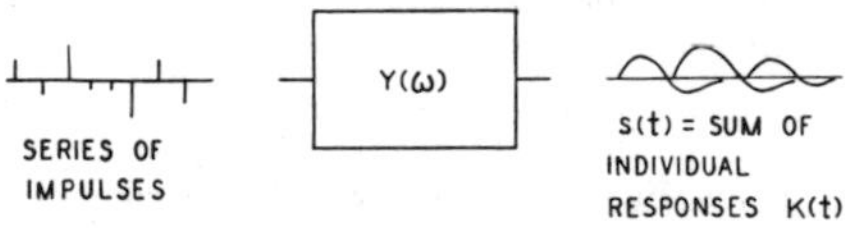

Fig. 6—Result of resistance noise input.

What is known is $h(t)$ up to the present; that is, we know effectively the impulses up to $t = 0$ and nothing about those after $t = 0$; these have not yet occurred. The future signal $s(t+\alpha)$ is thus made up of two parts; the tails of responses due to impulses that have already occurred, and a part due to impulses which will occur between the present time and time $t=\alpha$. The first part is completely predictable, while the second part is entirely unpredictable, being statistically independent of our available information at the present time.

The total result of the first part can be obtained by constructing a filter whose impulsive response is the tail of the impulsive response of filter Y moved ahead α seconds. This is shown in Fig. 7 where $K_1(t)$ is the new impulsive response and $K(t)$ the old one. The new filter responds to an impulse entering *now* as the filter Y will respond in α seconds. It responds to an impulse that entered one second ago as Y will respond in α seconds to one that entered it one second ago. In short, if $h(t)$ is used as input to this new filter Y_1, the output now will be the predictable part of the future response of Y to the same input α seconds from now.

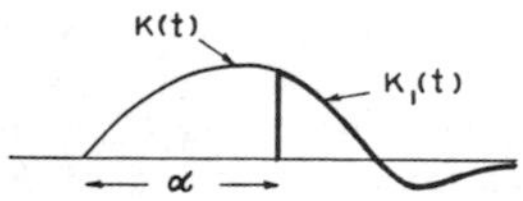

Fig. 7—Construction of the physical response $K_1(t)$ from $K(t)$.

The second, or unpredictable part of the future response, corresponding to impulses yet to occur, cannot, of course, be constructed. We know, however, that the mean value of this part must be zero, since future impulses are as likely to be of one sign as the other. Thus the arithmetic average, or center of gravity, of the possible future responses is the predictable part given by the output of Y_1. But it is well known that the arithmetic mean of any distribution is the point about which the mean-square error is the least. The output of Y_1 is thus the desired prediction of $s(t+\alpha)$.

In constructing Y_1 we assumed that we had available the white noise $h(t)$. Actually, however, our given data is the signal $s(t)$. Consequently, the best operation on the given data is $Y_1(\omega)\, Y^{-1}(\omega)$, the factor $Y^{-1}(\omega)$ reducing the function $s(t)$ to the white noise $h(t)$, and the second operation $Y_1(\omega)$ performing the best prediction based on $h(t)$.

The solution may be summarized as follows:

1. Determine the minimum phase network having the gain characteristic $\sqrt{P(\omega)}$. Let the complex transfer characteristic of this filter be $Y(\omega)$, and its impulsive response $K(t)$.

2. Construct a filter whose impulsive response is

$$\begin{aligned} K_1(t) &= K(t+\alpha) \quad \text{for} \quad t \geqq 0 \\ &= 0 \qquad\qquad \text{for} \quad < 0. \end{aligned} \tag{22}$$

Let the transfer characteristic of this network be $Y_1(\omega)$.

3. The optimal least square predicting filter then has a characteristic

$$Y_1(\omega)Y^{-1}(\omega). \tag{23}$$

The mean-square error E in the prediction is easily calculated. The error is due to impulses occurring from time $t=0$ to $t=\alpha$. Since these impulses are uncorrelated, the mean-square sum of the errors is the sum of the individual mean-square errors. The individual pulses are effective in causing mean-square error in proportion to the square of $K(\alpha-t)$. Hence, the total mean-square error will be given by

$$\begin{aligned} E^2 &= \rho \int_0^\alpha K^2(\alpha - t)dt \\ &= \rho \int_0^\alpha K^2(t)dt \end{aligned} \tag{24}$$

where $\rho = \int p(\omega)d\omega$ is the mean-square signal. By a similar argument the mean-square value of $s(t+\alpha)$ will be

$$U^2 = \rho \int_0^\infty K^2(t)dt, \tag{25}$$

and the relative error of the prediction may be measured by the ratio of the root-mean-square error to the root-mean-square value of $s(t+\alpha)$, i.e.,

$$\frac{E}{U} = \left[\frac{\int_0^\alpha K^2(t)dt}{\int_0^\infty K^2(t)dt}\right]^{1/2}. \tag{26}$$

The prediction will be relatively poor if the area under the curve $K(t)^2$ out to α is large compared to the total area, good if it is small compared to the total. It is evident from (26) that the relative error starts at zero for $\alpha=0$ and is a monotonic increasing function of α which approaches unity as $\alpha \to \infty$.

There is an important special case in which a great deal more can be shown by the argument just given. In our analysis, the actual problem was replaced by one in which the signal was a Gaussian type of time series, derived from a resistance noise by passing it through a filter with a gain $\sqrt{P(\omega)}$. Suppose the signal is already a time series of this type. Then the error in prediction, due to the tails of impulses occurring between $t=0$ and $t=\alpha$, will have a Gaussian distribution. This follows from the fact that each impulse has a Gaussian distribution of amplitudes and the sum of any number of effects, each Gaussian, will also be Gaussian. The standard deviation of this distribution of errors is just the root-mean-square error E obtained from (24).

Stated another way, on the basis of the available data, that is, $s(t)$ for $t<0$, the future value of the signal $s(t+\alpha)$ is distributed according to a Gaussian distribution. The best linear predictor selects the center of this distribution for the predicted value. The actual future value will differ from this as indicated in Fig. 8, where the future value is plotted horizontally, and the probability density for various values of $s(t+\alpha)$ is plotted vertically.

It is clear that in this special case the linear prediction method is in a sense the best possible. The center of the

Gaussian distribution remains the natural point to choose if we replace the least square criterion of the

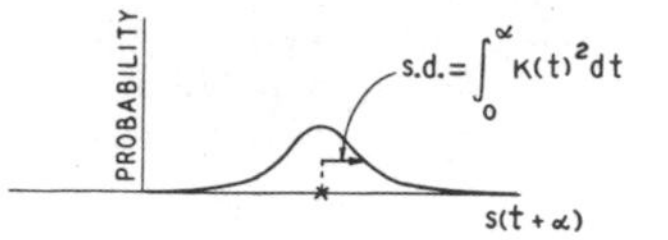

Fig. 8—Distribution of prediction errors in the Gaussian case.

best prediction by almost any other reasonable criterion, such as the median value or the most probable value. Thus in this case a nonlinear computation would offer nothing which the linear process does not already provide. In the general case, on the other hand, the distribution of future values will not be Gaussian, and the shape of the distribution curve may vary from point to point depending upon the particular past history of the curve. Under these circumstances, a nonlinear scheme may offer improvements upon the linear process and the exact characteristics of the optimal procedure will depend critically upon the criterion adopted for the best prediction.

VII. Prediction in the Presence of Noise

Now consider the general prediction and smoothing problem with noise present. The best estimate of $s(t+\alpha)$ is required when the function $s(t)+n(t)$ is known from $t=-\infty$ to the present. If $s(t)+n(t)$ is passed through a filter whose gain is $[P(\omega)+N(\omega)]^{-1/2}$, the result will be a flat spectrum which we can identify with white noise. Let $Y_1(\omega)$ be the transfer function of a filter having this gain characteristic and the associated minimum phase. Then both $Y_1(\omega)$ and the inverse $Y_1^{-1}(\omega)$ are physically realizable networks. Evidently, knowledge of the input of Y_1 and knowledge of its output are equivalent. The best linear operation on the output will give the same prediction as the corresponding best linear operation on the input.

If we knew the entire function $s(t)+n(t)$ from $t=-\infty$ to $t=+\infty$ the best operation to apply to the input of $Y_1(\omega)$ would be that specified by (18). If we let $B(\omega)$ be the phase component of Y_1, this corresponds to the equivalent operation

$$Y_2(\omega) = \frac{P(\omega)}{[P(\omega)+N(\omega)]^{1/2}} e^{i[\alpha\omega - B(\omega)]} \tag{27}$$

on the "white noise" output of Y_1.

Let the impulse response obtained from (27) be $K_2(t)$. As illustrated by Fig. 9, $K_2(t)$ will, in general, contain tails extending to both $t=+\infty$ and $t=-\infty$, the junction between the two halves of the curve being displaced from the origin by the prediction time α. The associated $K_2(\tau)$ of Fig. 10 is, of course, the ideal weighting function to be applied to the "white noise" output of Y_1. But the only data actually available at $\tau=0$ are the impulses which may be thought of as occurring during the past history of this output. What weights should be given these data to obtain the best

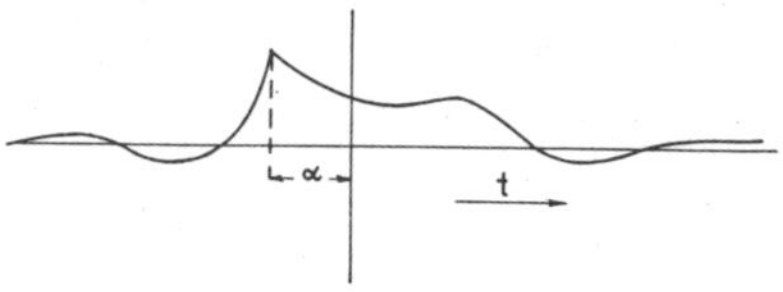

Fig. 9—Possible function $K_2(t)$.

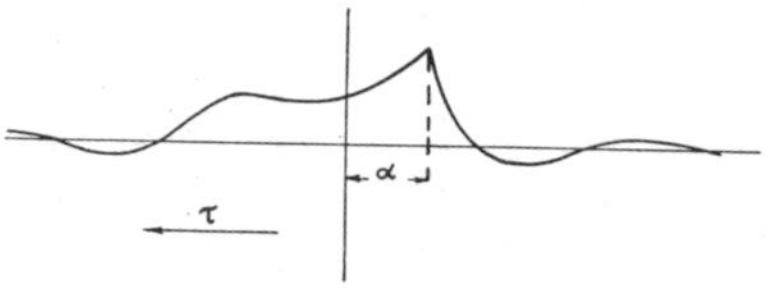

Fig. 10—Weighting function $K_2(\tau)$, corresponding to Fig. 9.

prediction? It seems natural to weight these as one would if all data were available, and to weight the future values zero (as we must to keep the filter physical). The fact that this is actually correct weighting when the various input impulses are statistically independent will now be shown as a consequence of a general statistical principle.

Suppose we have a number of chance variables, $x_1, x_2, \cdots, x_n$ which are statistically independent, or at least have the property that the mean product of any two, $\overline{x_m x_n}$, is equal to zero. These variables are to be interpreted as the amplitudes of the individual white noise impulses to which we are attempting to apply the weighting function of Fig. 10.

Let y be another chance variable, correlated with $x_1, \cdots, x_n$, which we wish to estimate in the least square sense by performing a linear operation on $x_1 \cdots x_n$. In the problem at hand y is the actual signal $s(t)$ at the time α seconds from now.

The predicted value will be

$$y_1 = \sum_{i=1}^{n} a_i x_i$$

and the mean-square error is

$$\begin{aligned} E &= \overline{(y-y_1)^2} = \overline{(y - \sum a_i x_i)^2} \\ &= \overline{y^2} - 2\sum_{i=1}^{n} a_i \overline{x_i y} + \sum_{i,j=1}^{n} a_i a_j \overline{x_i x_j} \\ &= \overline{y^2} - 2\sum a_i \overline{x_i y} + \sum_{i=1}^{n} a_i^2 \overline{x_i^2}, \end{aligned} \tag{28}$$

since all terms in the double sum vanish except those for which $i=j$. We seek to minimize E by proper choice of the a_i. Setting the partial derivatives with respect to a_i equal to zero, we have

$$\frac{\partial E}{\partial a_i} = -2\overline{x_i y} + 2a_i \overline{x_i^2} = 0$$

or

$$a_i = \frac{\overline{x_i y}}{\overline{x_i^2}}. \quad (29)$$

The important fact about this calculation is that each of the n minimizing equations involves only the a_1 in question; $\partial E/\partial a_1$ involves only a_1, etc. In other words, minimizing E on all the a_i is equivalent to minimizing separately on the individual a_i; a_1 should have the value $\overline{x_1 y}/\overline{x_1^2}$ whatever values are assigned to the other a's.

Returning now to the prediction and smoothing problem, the function $K_2(\tau)$ gives the proper weighting to be attached to the impulses if we could use them all. Requirements of physical realizability demand that future impulses corresponding to $\tau<0$ be given weight zero. From the above statistical principle those occurring in the past should still be given the weighting $K_2(\tau)$. In other words, the proper filter to apply to the input white noise has an impulse response zero for $t<0$ and $K_2(t)$ for $t>0$.

To summarize, the solution consists of the following steps:

1. Calculate the minimum phase transfer function for the gain $(P+N)^{-1/2}$. Let this be $Y_1(\omega)$.

2. Let

$$Y_2(\omega) = Y_1^{-1}(\omega)\frac{P}{P+N}.$$

This is a nonphysical transfer function. Let its Fourier transform be $K_2(t)$.

3. Set $K_3(t)=K_2(t+\alpha)$ for $t\geqq 0$ and $K_3(t)=0$ for $t<0$. That is, cut off the first α seconds of $K_2(t)$ and shift the remaining tail over to $t=0$. This is the impulse response of a physical network, and is the optimal operation on the past history of the white noise input. Let the corresponding transfer function be $Y_3(\omega)$.

4. Construct $Y_4(\omega)=Y_3(\omega)Y_1(\omega)$. This is the optimal smoothing and prediction filter, as applied to the actual given $s(t)+n(t)$.

As in the pure prediction problem, if the signal and noise happen to be Gaussian time series, the linear prediction is an absolute optimum among all prediction operations, linear or not. Furthermore, the distribution of values of $s(t+\alpha)$, when $f(t)$ is known for $t<0$, is a Gaussian distribution.

VIII. Generalizations

This theory is capable of generalization in several directions. These generalizations will be mentioned only briefly, but can all be obtained by methods similar to those used above.

In the first place, we assumed the true signal and the noise to be uncorrelated. A relatively simple extension of the argument used in Section IV allows one to account for correlation between these time series.

A second generalization is to the case where there are several correlated time series, say $f_1(t), f_2(t), \cdots, f_n(t)$. It is desired to predict, say, $s_1(t+\alpha)$ from a knowledge of $f_1, f_2, \cdots, f_n$.

Finally the desired quantity may not be $s(t+\alpha)$ but, for example, $s'(t+\alpha)$, the future derivative of the true signal. In such a case, one may effectively reduce the problem to that already solved by taking derivatives throughout. The function $f(t)$ is passed through a differentiator to produce $g(t)=f'(t)$. The best linear prediction for $g(t)$ is then determined.

IX. Discussion of the Basic Assumptions

A result in applied mathematics is only as reliable as the assumptions from which it is derived. The theory developed above is especially subject to misapplication because of the difficulty in deciding, in any particular instance, whether the basic assumptions are a reasonable description of the physical situation. Anyone using the theory should carefully consider each of the three main assumptions with regard to the particular smoothing or prediction problem involved.

The assumption that the signal and noise are stationary is perhaps the most innocuous of the three, for it is usually evident from the general nature of the problem when this assumption is violated. The determination of the required power spectra $P(\omega)$ and $N(\omega)$ will often disclose any time variation of the statistical structure of the time series. If the variation is slow compared to the other time constants involved, such nonstationary problems may still be solvable on a quasi-stationary basis. A linear predictor may be designed whose transfer function varies slowly in such a way as to be optimal for the "local" statistics.

The least square assumption is more troublesome, for it involves questions of values rather than questions of fact. When we minimize the mean-square error we are, in effect, paying principal attention to the very large errors. The prediction chosen is one which, on the whole, makes these errors as small as possible, without much regard to relatively minor errors. In many circumstances, however, it is more important to make as many very accurate predictions as possible, even if we make occasional gross errors as a consequence. When the distribution of future events is Gaussian, it does not matter which criterion is used since the most probable event is also the one with respect to which the mean-square error is the least. With lopsided or multimodal distributions, however, a real question is involved.

As a simple example, consider the problem of predicting whether tomorrow will be a clear day. Since clear days are in the majority, and there are no days with negative precipitation to balance days when it rains, we are concerned here with a very lopsided distribution. With such a curve, the average point, which is the one given by a prediction minimizing the mean-square error, might be represented by a day with a light drizzle. To a man planning a picnic, however, such a prediction would have no value. He is interested in the probability that the weather will really be clear. If the picnic must be called off because it in fact rains, the actual amount of precipitation is of comparatively little consequence.

As a second example, consider the problem of intercepting a bandit car attempting to flee down a network of roads. If the road on which the bandit car happens to be forks just ahead, it is clear that a would-be interceptor should station himself on one fork or the other, making the choice at random if necessary. The mean-square error in the interception would be least, however, if he placed himself in the fields beyond the fork. Problems similar to these may also arise in gunnery, where, in general, we are usually interested in the number of actual hits and "a miss is as good as a mile."

The third assumption, that of linearity, is neither a question of fact, nor of evaluation, but a self-imposed limitation on the types of operations or devices to be used in prediction. The mathematical reason for this assumption is clear; linear problems are always much simpler than their nonlinear generalizations. In certain applications the linear assumption may be justified for one or another of the following reasons:

1. The linear predictor may be an absolute optimal method, as in the Gaussian time series mentioned above.

2. Linear prediction may be dictated by the simplicity of mechanization. Linear filters are easy to synthesize and there is an extensive relevant theory, with no corresponding theory for nonlinear systems.

3. One may use the linear theory merely because of the lack of any better approach. An incomplete solution is better than none at all.

How much is lost by restricting ourselves to linear prediction? The fact that nonlinear effects may be important in a prediction can be illustrated by returning to the problem of forecasting tomorrow's weather. We are all familiar with the fact that the pattern of events over a period of time may be more important than the happenings taken individually in determining what will come. For example, the sequence of events in the passage of a cold or warm front is characteristic. Moreover, the significance of a given happening may depend largely upon the intensity with which it occurs. Thus, a sharp dip in the barometer may mean that moderately unpleasant weather is coming. Twice as great a drop in the same time, on the other hand, may not indicate that the weather will be merely twice as unpleasant; it may indicate a hurricane.

As a final point, we may notice that the requirement that the prediction be obtained from a linear device and the objective of minimizing the mean-square error are not, in all problems, quite compatible with one another. The absolute best mean-square prediction (ignoring the assumption of linearity) would, of course, always pick the mean of the future distribution, i.e., the "center of gravity," since in any case this minimizes the mean-square error. In general, however, the position of this center of gravity will be a nonlinear function of the past history. When we require that the prediction be a *linear* operation on the past history, the mathematics is forced to compromise among the conflicting demands of various possible past histories. The compromise amounts essentially to averaging over-all relative phases of the various components of the signal; any pertinent information contained in the relative phases cannot be used properly.

This can be illustrated by the familiar statistical problem of calculating a line or plane of regression to provide a linear least square estimation of one variable y from the knowledge of a set of variables correlated with y.[5] The simplest such problem occurs when there is just one known variable x, and one unknown variable y to be estimated from x. Fig. 11 shows three of the "scatter diagrams" used in statistics. The variable x may be, for example, a man's weight and y his height. A large population is sampled and plotted. It is then desired to

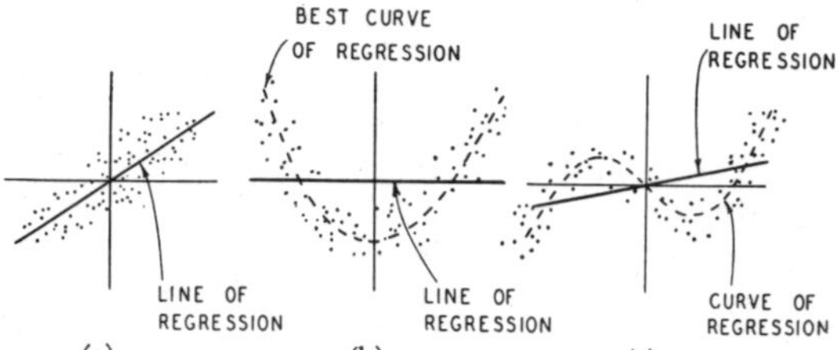

Fig. 11—Some scatter diagrams with lines and curves of regression.

estimate, or predict, a man's height, knowing only his weight. If we agree to use only linear operations y must be calculated in the form $y=ax$. The best choice of a for least square prediction is $\overline{xy}/\overline{x^2}$ and the corresponding straight line is known as the line of regression. The case of a normal distribution corresponds to the Gaussian type noise in which the linear prediction is an absolute optimum.

Figs. 11(b) and 11(c) are scatter diagrams for other distributions of two variables. The lines of regression are now not nearly as good in predicting y as they were in Fig. 11(a). The requirement that the predicted value be a *linear* function of the known data requires a compromise which may be very serious. It is obvious in Figs. 11(b) and 11(c) that a much better estimate of y could be formed if we allowed nonlinear operations on x. In particular, functions of the form ax^2+b and cx^3+dx would be more suitable.

In predicting y from two known variables x_1 and x_2 we can construct a scatter diagram in three dimensions. The linear prediction requires fitting the points with a plane of regression. If there are n known quantities $x_1, x_2, \cdots, x_n$ we need $(n+1)$ dimensional space and the linear theory corresponds to a hyperplane of n dimensions.

The problem of smoothing and prediction for time series is analogous. What we are now dealing with, however, is the function space defined by all the values of $f(t)$ for $t<0$. The optimal linear predictor corresponds to a hyperplane in this function space.

[5] P. G. Hoel, "Introduction to Mathematical Statistics," John Wiley and Sons, Inc., New York, N. Y.; 1947.

Editors' Comments on Paper 2

At approximately the same time that the Bode–Shannon paper was published, Olof Hanner, who is now with the Institute of Mathematics at the Chalmers Technical University in Göteborg, Sweden, produced this paper on the structure of wide-sense stationary processes.

Earlier work by Karhunen, Wold, and others had already established the basic unit multiplicity representation of a regular (pnd) wide-sense stationary process. Hanner's paper, though, by concentrating on the time-domain analysis and by noting the dichotomy between deterministic and nondeterministic components, places its emphasis on concepts that are closer to the central subject of the present volume.

The basic decomposition problem is still in early form. However, the ideas to be developed fully later are already taking shape here.

This paper is reprinted with the permission of the author and the Royal Swedish Academy of Science.

ARKIV FÖR MATEMATIK Band 1 nr 14

Communicated 12 January 1949 by MARCEL RIESZ and H. CRAMÉR

2

Deterministic and non-deterministic stationary random processes

By OLOF HANNER

1. Let $x(t)$ be a complex-valued random process depending on a real continuous parameter t, which may be regarded as representing time. We will assume that, for every t, the mean value $E\{x(t)\} = 0$ and the variance $E\{|x(t)|^2\}$ is finite. Then, in accordance with KHINTCHINE [3], we say that $x(t)$ is a stationary random process if the function

$$r(t) = E\{x(s+t)\,\overline{x(s)}\}$$

is independent of s. Then $r(t)$ has the properties

$$r(s-t) = E\{x(s)\,\overline{x(t)}\}$$

$$r(0) = E\{|x(t)|^2\} > 0.$$

We shall also assume that $r(t)$ is continuous for $t = 0$. Then

$$E\{|x(t+h) - x(t)|^2\} = 2\,r(0) - r(h) - r(-h) \to 0$$

when $h \to 0$, and

$$|r(t) - r(s)|^2 = |E\{(x(t) - x(s))\,\overline{x(0)}\}|^2 \leqq E\{|x(t) - x(s)|^2\}\,E\{|x(0)|^2\} \to 0$$

when $s \to t$, so that $r(t)$ is continuous for every t. The process will then be called a continuous stationary random process (KHINTCHINE [3]).

We shall in this paper study such processes and prove a decomposition theorem, which says that an arbitrary process of this type is the sum of two other processes of the same type, where one is deterministic and the other is completely non-deterministic (Theorem 1), and where the completely non-deterministic part can be expressed in terms of a random spectral function (Theorem 2).

The corresponding decomposition theorem for a stationary process depending on an integral parameter or, in other words, for a stationary sequence, has been stated by WOLD [5] in 1938. It has later been simplified and completed by KOLMOGOROFF [4] using the technique of Hilbert space. In a more systematic way KARHUNEN [2] introduced Hilbert space methods into the theory of probability. Using his results we shall be able to prove our theorems.

2. Consider a continuous stationary process $x(t)$. It may be interpreted as a curve in the Hilbert space $L_2(x)$ which consists of random variables of the type

$$\sum_{\nu=1}^{n} c_\nu x(t_\nu)$$

c_ν being constants, and of random variables which are limits of sequences of such sums in the sense of mean convergence. The scalar product of two random variables x and y is defined as $E\{x\bar{y}\}$ and the norm $\|x\|$ is defined by

$$\|x\|^2 = E\{|x|^2\}.$$

Hence, two random variables in $L_2(x)$ are orthogonal if. and only if they are uncorrelated.

That the process is continuous means that the curve is continuous, that is

$$\|x(t) - x(s)\| \to 0$$

when $s \to t$. As a consequence of the continuity we get that $L_2(x)$ is separable. For the set $\{x(r);\ r\ \mathit{rational}\}$ is a countable complete set in $L_2(x)$.

If A and B are closed linear subspaces of $L_2(x)$, orthogonal to each other, we denote by $A \oplus B$ the direct sum of A and B, and, if $A \supset B$, we denote by $A \ominus B$ the orthogonal complement of B with respect to A. Further let P_A be the projection of $L_2(x)$ onto A.

Let T_h be the unitary linear transformation defined by

$$T_h x(t) = x(t+h)$$

(KARHUNEN [2], p. 55). The transformations T_h constitute an abelian group: $T_h T_k = T_{h+k}$. The existence of such a unitary transformation is equivalent to the process being stationary.

We denote by $L_2(x; a)$ the closed linear manifold in $L_2(x)$ determined by the set $\{x(t);\ t \leqq a\}$, that is the least closed linear manifold containing all the $x(t)$ when $t \leqq a$. If $z \in L_2(x; a)$ then $T_h z \in L_2(x; a+h)$ and conversely, so that we write

$$T_h L_2(x; a) = L_2(x; a+h). \tag{2.1}$$

From the definition of $L_2(x; a)$ follows

$$L_2(x; a) \subset L_2(x; b) \qquad a < b. \tag{2.2}$$

Since $x(t)$ is continuous, $L_2(x; a)$ is continuous in a, that is

$$\lim_{b \to a+0} L_2(x; b) = \lim_{b \to a-0} L_2(x; b) = L_2(x; a).$$

Then also the projection $P_{L_2(x;\,a)}$ is continuous in a, that is for every element z, $P_{L_2(x;\,a)} z$ is continuous in a.

Now let us take a fixed t and an arbitrary a and consider

$$X_t(a) = P_{L_2(x;\,a)}\, x(t)$$

that is the projection of $x(t)$ in $L_2(x;\,a)$. $X_t(a)$ may be considered as that part of $x(t)$ that is determined by the process at the time a. We want to study $\|X_t(a)\|$. This is a continuous function of a, which in consequence of (2.2) is non decreasing and constant for $a \geqq t$. Since

$$T_h X_t(a) = X_{t+h}(a+h) \tag{2.3}$$

we have

$$\|X_t(a)\| = \|X_{t+h}(a+h)\|.$$

Thus it will be sufficient to take $t = 0$. Instead of $X_0(a)$ we then write $X(a)$.

There will be two extreme cases.

1) $\|X(a)\|$ is constant. Then $\|X(a)\| = \|X(0)\| = \|x(0)\|$ for every a so that $X(a) = x(0)$. This may be written

$$x(0) \in L_2(x;\,a) \tag{2.4}$$

for every a. Then also

$$x(h) = T_h\, x(0) \in T_h\, L_2(x;\,a-h) = L_2(x;\,a) \tag{2.5}$$

for every h so that

$$L_2(x) = L_2(x;\,a).$$

Definition. A stationary process for which $L_2(x;\,a) = L_2(x)$ will be called deterministic.

Thus we have proved

Proposition A. If $\|X(a)\|$ is constant, then $x(t)$ is deterministic.

The converse is obvious.

Remark. It would be sufficient to know that (2.4) holds for some negative number a. For then, in consequence of (2.2), (2.5) holds for every $h \leqq 0$, so that $L_2(x;\,0) = L_2(x;\,a)$. Hence

$$L_2(x;\,na) = L_2(x;\,(n-1)\,a) = \cdots = L_2(x;\,0)$$

for every integer n, positive or negative, so that the process is deterministic.

2) $\|X(a)\| \to 0$ when $a \to -\infty$. Put

$$M = \prod_a L_2(x;\,a) = \lim_{a\to-\infty} L_2(x;\,a).$$

Then

$$P_M\, x(0) = 0$$

and, since (2.1) implies $T_h M = M$, we have for every t

$$P_M x(t) = 0$$

so that $M \perp L_2(x)$. But $M \subset L_2(x)$, hence $M = 0$, that is M contains only the zero element.

DEFINITION. A stationary process for which $\prod_a L_2(x; a) = 0$ will be called completely non-deterministic.

Thus we have proved

PROPOSITION B. If $\|X(a)\| \to 0$ when $a \to -\infty$, then the process is completely non-deterministic.

Conversely if the process is completely non-deterministic, then $\|X(a)\| \to 0$, for if $X(a) \to z$, $\|z\| \neq 0$, then $z \in L_2(x; a)$ for every a, and thus $z \in M$, which is a contradiction.

We now shall prove that an arbitrary stationary process is the sum of two uncorrelated components, one deterministic and the other completely non-deterministic, or more precisely

THEOREM 1. $x(t)$ is a stationary process. Then there exist two other stationary processes $y(t)$ and $z(t)$ such that

a) $x(t) = y(t) + z(t)$

b) $y(t) \in L_2(x)$, $z(t) \in L_2(x)$

c) $y(s) \perp z(t)$ for every s and t

d) $y(t)$ is deterministic

e) $z(t)$ is completely non-deterministic.

PROOF. Take

$$M = \prod_a L_2(x; a) \quad \text{and} \quad N = L_2(x) \ominus M.$$

From (2.1), we have $T_h M = M$, and hence, since T_h is unitary, $T_h N = N$. Put

$$y(t) = P_M x(t) \quad \text{and} \quad z(t) = P_N x(t).$$

Then a), b) and c) are satisfied. From

$$y(t+h) + z(t+h) = x(t+h) = T_h x(t) = T_h [y(t) + z(t)] = T_h y(t) + T_h z(t)$$

we obtain

$$y(t+h) = T_h y(t) \quad \text{and} \quad z(t+h) = T_h z(t).$$

Thus $y(t)$ and $z(t)$ are stationary processes. It remains to prove d) and e). From a) and from

$$L_2(y) \subset M \quad \text{and} \quad L_2(z) \subset N$$

we have

$$L_2(x) \subset L_2(y) \oplus L_2(z) \subset M \oplus N = L_2(x)$$

so that

$$L_2(y) = M \quad \text{and} \quad L_2(z) = N. \tag{2.6}$$

When $t \leqq a$ we have

$$y(t) \in M \subset L_2(x; a) \quad \text{and} \quad z(t) = x(t) - y(t) \in L_2(x; a)$$

and hence

$$L_2(y; a) \subset L_2(x; a) \quad \text{and} \quad L_2(z; a) \subset L_2(x; a).$$

But from a)

$$L_2(x; a) \subset L_2(y; a) \oplus L_2(z; a)$$

so that

$$L_2(x; a) = L_2(y; a) \oplus L_2(z; a)$$

or, with the aid of (2.6),

$$L_2(y; a) = M\, L_2(x; a) \quad \text{and} \quad L_2(z; a) = N\, L_2(x; a).$$

Now we get

$$L_2(y; a) = M\, L_2(x; a) = M = L_2(y).$$

Hence $y(t)$ is deterministic. Further

$$M_z = \prod_a L_2(z; a) = \prod_a N\, L_2(x; a) = N \prod_a L_2(x; a) = N\, M = 0.$$

Hence $z(t)$ is completely non-deterministic. This completes the proof of the theorem.

3. Now we will study the completely non-deterministic processes. Therefore we assume that $y(t) = 0$, so that $x(t) = z(t)$ is a completely non-deterministic process.

For every pair (a, b) of real numbers, $a < b$, we construct

$$L_2(x; a, b) = L_2(x; b) \ominus L_2(x; a)$$

and

$$x(a, b) = P_{L_2(x; a, b)}\, x(b).$$

Then $x(a, b) \perp L_2(x; a)$, and $x(a, b)$ may be interpreted as that part of $x(b)$ that is not determined of the process at the time a.

From the Remark to Proposition A, we conclude

$$x(b) \notin L_2(x; a)$$

or

$$x(a, b) \neq 0. \tag{3.1}$$

Hence $L_2(x; a, b)$ always contains elements with positive norm. It is clear that

$$T_h x(a, b) = x(a + h, b + h).$$

The element $x(a, b)$ is continuous in (a, b), for take, for instance, $a_1 < a$ and $b_1 > b$, then

$$x(a_1, b_1) - x(a, b) = x(a_1, b_1) - x(a_1, b) + x(a_1, b) - x(a, b)$$
$$= P_{L_2(x; a_1, b_1)}[x(b_1) - x(b)] + P_{L_2(x; a_1, a)} x(b).$$

Here the last two terms tend to zero when $a_1 \to a$ and $b_1 \to b$ because

$$\| P_{L_2(x; a_1, b_1)}[x(b_1) - x(b)] \| \leqq \| x(b_1) - x(b) \| \to 0$$

independently of a_1, when $b_1 \to b$ and

$$P_{L_2(x; a_1, a)} x(b) \to 0$$

when $a_1 \to a$.

We are now going to define a random spectral function $Z(S)$ (KARHUNEN [2], p. 36). This means in this case, that for every measurable set S on the t-axis, with finite measure $m(S)$, there shall be a random variable $Z(S) \in L_2(x)$ satisfying

1) If S_1 and S_2 are disjoint sets

$$Z(S_1 + S_2) = Z(S_1) + Z(S_2).$$

2) If S_1 and S_2 are disjoint sets

$$Z(S_1) \perp Z(S_2).$$

3) $\| Z(S) \|^2 = m(S)$.

2) and 3) may be combined in

$$E\{Z(S_1) Z(S_2)\} = m(S_1 S_2)$$

for arbitrary S_1 and S_2.

If $Z(S)$ is defined and satisfies 1), 2) and 3), when S is an interval, then there is a unique extension to all measurable sets. Thus we have to define $Z(I_a^b)$ for every interval $I_a^b = (a, b)$.

Let u be a fixed positive number and take a $z \in L_2(x; 0, u)$. For every interval I_a^b we take

$$Z(I_a^b) = P_{L_2(x; a, b)} \int_A^B T_h z \, dh \tag{3.2}$$

where $A < a - u$ and $B > b$. (The integral is defined as a Riemann integral, CRAMÉR [1], p. 219.) $Z(I_a^b)$ is independent of A and B, since the variations of the integral, when A and B vary, are orthogonal to $L_2(x; a, b)$.

When $a < b < c$ we have

$$Z(I_a^b) + Z(I_b^c) = Z(I_a^c) \tag{3.3}$$

$$Z(I_a^b) \perp Z(I_b^c) \tag{3.4}$$

and for arbitrary h

$$T_h Z(I_a^b) = Z(I_{a+h}^{b+h}). \tag{3.5}$$

From these three properties we get

$$\|Z(I_a^b)\|^2 = \tau(b-a)$$

where $\tau \geqq 0$. τ depends on z, and we shall prove

PROPOSITION C. One can find a z, such that $\tau = 1$.

PROOF. It will be sufficient to find z so that $\tau > 0$, for then to $z/\sqrt{\tau}$ corresponds $\tau = 1$.

Suppose the contrary so that for every $z \in L_2(x; 0, u)$

$$\|Z(I_a^b)\|^2 = 0.$$

Then for every $z' \in L_2(x)$

$$E\{Z(I_a^b)\overline{z'}\} = 0. \tag{3.6}$$

We take

$$z = y(s_1, t_1) \quad \text{and} \quad z' = y(s_2, t_2)$$

where

$$0 \leqq s_1 < t_1 \leqq u \quad \text{and} \quad 0 \leqq s_2 < t_2 \leqq u \tag{3.7}$$

and where we have written

$$y(s, t) = x(s, u) - x(t, u) = P_{L_2(x; s, t)}\, x(u).$$

Thus z and z' are certain projections of $x(u)$, and they satisfy (3.6) for arbitrary (s_1, t_1) and (s_2, t_2) satisfying (3.7). But we know from (3.1) that

$$y(0, u) = x(0, u) = P_{L_2(x; 0, u)}\, x(u) \neq 0$$

and we shall show that this leads to a contradiction.

First we have $z \in L_2(x; 0, u)$ and $z' \in L_2(x; 0, u)$. We then obtain

$$0 = E\{Z(I_0^u)\overline{z'}\} = E\left\{P_{L_2(x; 0, u)} \int_{-u}^{u} T_h z\, dh\, \overline{z'}\right\}$$

$$= E\left\{\int_{-u}^{u} T_h z\, dh\, \overline{z'}\right\}$$

$$= \int_{-u}^{u} E\{T_h z\, \overline{z'}\}\, dh$$

that is

$$\int_{-u}^{u} E\{T_h y(s_1, t_1)\overline{y(s_2, t_2)}\} = 0. \tag{3.8}$$

Let δ be a positive number $< \frac{1}{2}u$, and consider

$$E\{T_h y(0, u)\, \overline{y(\delta, u-\delta)}\}.$$

This expression is a continuous function of h, which when $h=0$ takes the value $\|y(\delta, u-\delta)\|^2$, which is continuous in δ, and when $\delta \to 0$ we obtain

$$\|y(\delta, u-\delta)\|^2 \to \|y(0, u)\|^2 = \|x(0, u)\|^2 > 0.$$

Then it is possible to choose δ, such that $\|y(\delta, u-\delta)\|^2 > 0$ and $\gamma < \delta$ such that

$$L = \int_{-\gamma}^{\gamma} E\{T_h y(0, u)\, \overline{y(\delta, u-\delta)}\}\, dh \neq 0.$$

If we make a subdivision $t_0 = \delta, t_1, t_2, \ldots t_n = u - \delta$ of the interval $(\delta, u-\delta)$ by a finite number of points, then we have

$$L = \sum_{i=1}^{n} \int_{-\gamma}^{\gamma} E\{T_h y(0, u)\, \overline{y(t_{i-1}, t_i)}\}\, dh$$
$$= \sum_{i=1}^{n} \int_{-\gamma}^{\gamma} E\{T_h y(t_{i-1} - \gamma, t_i + \gamma)\, \overline{y(t_{i-1}, t_i)}\}\, dh.$$

Let us compare this with

$$M = \sum_{i=1}^{n} \int_{-\gamma-(t_i-t_{i-1})}^{\gamma+(t_i-t_{i-1})} E\{T_h y(t_{i-1} - \gamma, t_i + \gamma)\, \overline{y(t_{i-1}, t_i)}\}\, dh$$
$$= \sum_{i=1}^{n} \int_{-u}^{u} E\{T_h y(t_{i-1} - \gamma, t_i + \gamma)\, \overline{y(t_{i-1}, t_i)}\}\, dh$$
$$= 0$$

as a consequence of (3.8).

M and L are independent of the division of $(\delta, u-\delta)$. Since $L \neq 0$ and $M = 0$, if we can show $M = L$, we have the contradiction.

$$M - L = \sum_{i=1}^{n} \left[\int_{-\gamma-(t_i-t_{i-1})}^{-\gamma} + \int_{\gamma}^{\gamma+(t_i-t_{i-1})}\right] E\{T_h y(t_{i-1} - \gamma, t_i + \gamma)\, \overline{y(t_{i-1}, t_i)}\}\, dh.$$

$$|M - L| \leqq 2 \sum_{i=1}^{n} (t_i - t_{i-1}) \|y(0, u)\| \cdot \|y(l_{i-1}, t_i)\| \tag{3.9}$$
$$\leqq 2u \|y(0, u)\| \cdot \sup \|y(t_{i-1}, t_i)\|.$$

But

$$y(t_{i-1}, t_i) + x(t_i, u) = x(t_{i-1}, u)$$

and

$$y(t_{i-1}, t_i) \perp x(t_i, u).$$

Hence

$$\|y(t_{i-1}, t_i)\|^2 = \|x(t_{i-1}, u)\|^2 - \|x(t_i, u)\|^2.$$

Since $\|x(t, u)\|^2$ is a continuous function of t, we can make $\sup \|y(t_{i-1}, t_i)\|$ as small as we please, by making the division fine enough. Hence (3.9) yields

$$|M - L| = 0$$

and we have got the contradiction.

Then we can take some z for which $\tau = 1$. The corresponding $Z(I_a^b)$ defined by (3.2) will be the sought-for random spectral function.

4. Let $L_2(Z)$ be the closed linear manifold determined by the set $\{Z(I_a^b)\}$. That $L_2(Z) = L_2(x)$ will be shown in Proposition D.

Let us write

$$\begin{aligned} Z(a) &= -Z(I_a^0) \qquad a < 0 \\ Z(0) &= 0 \\ Z(a) &= Z(I_0^a) \qquad a > 0. \end{aligned}$$

Then

$$Z(I_a^b) = Z(b) - Z(a).$$

This may be written

$$Z(I_a^b) = \int_a^b d\,Z(u).$$

Thus we have defined

$$\zeta = \int_{-\infty}^{\infty} g(u)\,d\,Z(u) \tag{4.1}$$

when $g(u) = 1$ in a finite interval and $= 0$ elsewhere. We will define it for every complex-valued function $g(u)$ such that

$$\int_{-\infty}^{\infty} |g(u)|^2\,d\,u$$

is finite. If

$$\begin{aligned} g(u) &= c_\nu \quad \text{in} \quad (a_\nu, b_\nu) \\ &= 0 \ \text{elsewhere} \end{aligned}$$

where (a_ν, b_ν) are a finite number of finite intervals, we define

$$\int_{-\infty}^{\infty} g(u)\,d\,Z(u) = \Sigma\, c_\nu\,[Z(b_\nu) - Z(a_\nu)].$$

In the general case the integral is defined as the limit of a sequence of integrals of functions $g_n(u)$, such that

$$\int_{-\infty}^{\infty} |g(u) - g_n(u)|^2 du \to 0$$

that is

$$g(u) = \underset{n\to\infty}{\text{l. i. m.}}\, g_n(u)$$

where $g_n(u)$ are functions taking a constant value in each of a finite number of finite intervals. For a complete discussion of this kind of integral see KARHUNEN [2], p. 37.

We have

$$E\left\{\int_{-\infty}^{\infty} g_1(u)\, dZ(u) \overline{\int_{-\infty}^{\infty} g_2(u)\, dZ(u)}\right\} = \int_{-\infty}^{\infty} g_1(u)\, \overline{g_2(u)}\, du \tag{4.2}$$

and hence

$$\left\| \int_{-\infty}^{\infty} g(u)\, dZ(u) \right\|^2 = \int_{-\infty}^{\infty} |g(u)|^2 du.$$

It is clear that ζ in (4.1) is in $L_2(Z)$. Conversely, for every $\zeta \in L_2(Z)$, there is a function $g(u)$ such that (4.1) holds. For this is obvious if $\zeta = Z(I_a^b)$ and then also if $\zeta = \Sigma c_\nu Z(I_{a_\nu}^{b_\nu})$. To prove the general case, we only have to take a sequence ζ_n, $\zeta_n \to \zeta$, such that the ζ_n are sums of this type. Then the corresponding functions $g_n(u)$ converge in the mean to a function $g(u)$. And such a sequence always exists since $\zeta \in L_2(Z)$. The function $g(u)$ is uniquely determined almost everywhere, for suppose

$$\int_{-\infty}^{\infty} g(u)\, dZ(u) = \int_{-\infty}^{\infty} g_1(u)\, dZ(u).$$

Then

$$0 = \left\| \int_{-\infty}^{\infty} [g(u) - g_1(u)]\, dZ(u) \right\|^2 = \int_{-\infty}^{\infty} |g(u) - g_1(u)|^2 du \tag{4.3}$$

so that $g(u) = g_1(u)$ almost everywhere.

Now put

$$x_1(t) = P_{L_2(Z)}\, x(t).$$

Then $T_h x_1(t) = x_1(t+h)$, and hence $x_1(t)$ is a stationary process. For every t

$$L_2(Z) = L_2[Z(I_a^b);\quad a < b \leqq t] \oplus L_2[Z(I_a^b);\quad t \leqq a < b].$$

From the definition of $Z(I_a^b)$ we conclude

$$x(t) \perp L_2[Z(I_a^b);\quad t \leqq a < b]$$
$$Z(I_a^b) \in L_2(x;\, a, b) \subset L_2(x;\, b).$$

Hence

$$x_1(t) \in L_2[Z(I_a^b);\ a < b \leqq t] \subset L_2(x;\ t) \tag{4.4}$$

and hence $L_2(x_1;\ a) \subset L_2(x;\ a)$, so that

$$\prod_a L_2(x_1;\ a) \subset \prod_a L_2(x;\ a) = 0$$

that is, $x_1(t)$ is completely non-deterministic.

$\|x_1(t)\|$ is independent of t. We must have

$$\|x_1(t)\| \neq 0. \tag{4.5}$$

For if $x_1(t) = 0$ then $x(t) \perp L_2(Z)$, and hence $L_2(x) \perp L_2(Z)$. But as $L_2(Z) \subset L_2(x)$ and contains elements with positive norm, this is a contradiction.

PROPOSITION D. $L_2(Z) = L_2(x)$, or what is equivalent, $x_1(t) = x(t)$.

PROOF. Let

$$y(t) = x(t) - x_1(t).$$

Then

$$T_h y(t) = y(t + h)$$

so that $y(t)$ is a stationary process. From (4.4) we have

$$y(t) \in L_2(x;\ t).$$

Hence

$$L_2(y;\ a) \subset L_2(x;\ a) \tag{4.6}$$

so that $y(t)$ is also completely non-deterministic. From

$$y(t) = x(t) - x_1(t) = P_{L_2(x) \ominus L_2(Z)}\, x(t)$$

we conclude

$$y(t) \perp L_2(Z). \tag{4.7}$$

We now have proved that

$$x(t) = x_1(t) + y(t)$$

is a decomposition of $x(t)$ into two orthogonal completely non-deterministic processes, where $x_1(t) \neq 0$ (4.5). We shall prove that then we have $y(t) = 0$.

Suppose the contrary. Then we construct $y_1(t) \neq 0$ from $y(t)$ in the same manner as $x_1(t)$ from $x(t)$:

$$y(t) = y_1(t) + z(t).$$

Here $z(t) \perp y_1(t)$ may be $= 0$ or $\neq 0$.

$$x(t) = x_1(t) + y_1(t) + z(t).$$

Now $x_1(0) \in L_2(Z)$ so that

$$x_1(0) = \int_{-\infty}^{\infty} g(u)\, d\, Z(u)$$

for some function $g(u)$, for which

(4.8) $$\int_{-\infty}^{\infty} |g(u)|^2\, d\, u \text{ is finite.}$$

From (4.4) we conclude that

(4.9) $$g(u) = 0 \text{ for almost every } u > 0,$$

that is

$$x_1(0) = \int_{-\infty}^{0} g(u)\, d\, Z(u).$$

Then we get from (3.5)

(4.10) $$x_1(t) = \int_{-\infty}^{t} g(u - t)\, d\, Z(u).$$

Similarly there is a random spectral function $Z'(I_a^b) \in L_2(y)$ and a complex-valued function $g'(u)$, such that

$$y_1(t) = \int_{-\infty}^{t} g'(u - t)\, d\, Z'(u).$$

From (4.7) and from

$$L_2(Z') \subset L_2(y)$$

it follows that

$$L_2(Z') \perp L_2(Z).$$

We also have

$$z(t) \in L_2(y) \perp L_2(Z)$$

and analogously to (4.7)

$$z(t) \perp L_2(Z').$$

From (4.6) we conclude

$$Z'(I_a^b) \subset L_2(y; b) \subset L_2(x; b)$$

so that we have

(4.11) $$L_2[Z'(I_a^b);\ a < b \leqq t] \subset L_2(x; t).$$

Now take the element

$$\eta = \int_{s}^{0} \overline{g'(s - u)}\, d\, Z(u) - \int_{s}^{0} \overline{g(s - u)}\, d\, Z'(u)$$

where s is some negative number. We assert that for a convenient s

a) $\|\eta\| \neq 0$

b) $\eta \in L_2(x; 0)$

c) $\eta \perp x(t), \quad t \leqq 0.$

In fact

$$\|\eta\|^2 = \int_s^0 |g'(s-u)|^2 du + \int_s^0 |g(s-u)|^2 du > 0$$

for some s since

$$\lim_{s \to -\infty} \int_s^0 |g(s-u)|^2 du = \int_{-\infty}^0 |g(u)|^2 du = \|x_1(0)\|^2 > 0.$$

Thus a) holds. b) is an immediate consequence of (4.4) and (4.11), and to prove c) we write

$$x(t) = \int_{-\infty}^t g(u-t)\, dZ(u) + \int_{-\infty}^t g'(u-t)\, dZ'(u) + z(t).$$

Then c) is trivial when $t \leqq s$, and when $s < t \leqq 0$, it follows from

$$E\{x(t)\,\bar{\eta}\} = \int_s^t g(u-t)\, g'(s-u)\, du - \int_s^t g'(u-t)\, g(s-u)\, du = 0.$$

But it is a contradiction that a), b), and c) all hold. Hence $y(t) = 0$. This completes the proof of Proposition D.

5. We denote by $L^2(a, b)$ the class of complex-valued functions $f(u)$ defined for $a \leqq u \leqq b$, for which

$$\int_a^b |f(u)|^2 du$$

is finite. Let us make the convention that, for every $f(u) \in L^2(a, b)$, we write $f(u) = 0$ when $u < a$ or $u > b$. Then

$$f(u) \in L^2(-\infty, \infty) \text{ and } L^2(a, b) \subset L^2(-\infty, \infty).$$

Take the function $g(u)$ defined in the previous section. (4.8) and (4.9) yield that

$$g(u) \in L^2(-\infty, 0). \tag{5.1}$$

Hence $g(u-t) \in L^2(-\infty, 0)$ for every $t \leqq 0$. Denote by $L_2\{g(u-t);\ t \leqq 0\}$ the closed linear manifold in $L^2(-\infty, 0)$ determined by these elements.

Now we may sum up the results in

THEOREM 2. For every completely non-deterministic stationary process $x(t)$ we have an integral representation

(5.2) $$x(t) = \int_{-\infty}^{t} g(u-t)\, dZ(u)$$

where

a) $Z(I_a^b) = Z(b) - Z(a)$ is a random spectral function satisfying

(5.3) $$T_h Z(I_a^b) = Z(I_{a+h}^{b+h})$$

b) $g(u) \in L^2(-\infty, 0)$

c) $L_2\{g(u-t);\, t \leqq 0\} = L^2(-\infty, 0)$.

If

(5.4) $$x(t) = \int_{-\infty}^{t} g_1(u-t)\, dZ_1(u)$$

is another representation, $g_1(u)$ and $Z_1(I_a^b)$ satisfying a), b), and c), then there is a complex number ω, $|\omega| = 1$, such that

$$g_1(u) = \bar{\omega}\, g(u) \cdot \text{ for almost every } a$$
$$Z_1(I_a^b) = \omega Z(I_a^b),$$

and conversely, for every ω, $|\omega| = 1$, $g_1(u)$ and $Z_1(I_a^b)$ determined by these formulas satisfy a), b), c), and (5.4).

PROOF. The existence of the integral representation is an immediate consequence of Proposition D and (4.10). a) is (3.5) and b) is (5.1). We obtain from (5.2)

$$L_2(x;\, 0) \subset L_2[Z(I_a^b);\; a < b \leqq 0].$$

But (4.4) yields

$$L_2[Z(I_a^b);\; a < b \leqq 0] \subset L_2(x;\, 0)$$

so that

(5.5) $$L_2(x;\, 0) = L_2[Z(I_a^b);\; a < b \leqq 0].$$

Now let us prove c). Suppose that c) is false. Then there is a function $f(u) \in L^2(-\infty, 0)$, such that

$$\int_{-\infty}^{0} |f(u)|^2\, du > 0$$

(5.6) $$\int_{-\infty}^{0} g(u-t)\, \overline{f(u)}\, du = 0 \quad t \leqq 0.$$

Take the corresponding element in $L_2[Z(I_a^b);\ a < b \leqq 0]$

$$\zeta = \int_{-\infty}^{0} f(u)\, dZ(u).$$

Then

$$\|\zeta\|^2 = \int_{-\infty}^{0} |f(u)|^2\, du > 0$$

and from (5.6)

$$\zeta \perp L_2(x;\ 0) \tag{5.7}$$

in contradiction to (5.5).

Conversely c) implies (5.5). For the falseness of (5.5) yields the existence of a $\zeta \neq 0$ satisfying (5.7), and then we obtain from (5.6) the falseness of c).

From (5.5) we get

$$L_2(x;\ a,\ b) = L_2[Z(I_c^d);\ a \leqq c < d \leqq b]. \tag{5.8}$$

Now let $g_1(u)$ and $Z_1(I_a^b)$ satisfy the conditions in the theorem. Then $Z_1(I_a^b)$ satisfies (5.5) and then (5.8). Hence we obtain

$$L_2[Z(I_c^d);\ a \leqq c < d \leqq b] = L_2[Z_1(I_c^d);\ a \leqq c < d \leqq b].$$

In particular for every finite interval (a, b) there is a function $f(u) \in L^2(a, b)$, such that

$$Z_1(I_a^b) = \int_a^b f(u)\, dZ(u). \tag{5.9}$$

By virtue of (4.3) $f(u)$ is uniquely determined almost everywhere. This function $f(u)$, defined by (5.9) for the interval (a, b), will satisfy the same relation for every sub-interval. For since $Z(I_a^b)$ and $Z_1(I_a^b)$ satisfy (3.3), we have $(a < c < b)$

$$Z_1(I_a^c) + Z_1(I_c^b) = \int_a^c f(u)\, dZ(u) + \int_c^b f(u)\, dZ(u)$$

and from (3.4) we then get

$$Z_1(I_a^c) = \int_a^c f(u)\, dZ(u) \quad \text{and} \quad Z_1(I_c^b) = \int_c^b f(u)\, dZ(u). \tag{5.10}$$

Conversely, if $f(u)$ satisfies (5.10) then also (5.9) holds for this function. Therefore, if we define a function $f(u)$ for every real u, such that (5.9) holds for every interval of the type $(n, n+1)$ where n is an arbitrary integer, then this function satisfies (5.9) for every interval (a, b).

Now we shall prove that this function $f(u)$ is constant almost everywhere. Since $Z(I_a^b)$ and $Z_1(I_a^b)$ satisfy (5.3), we have from (5.9)

$$Z_1(I_a^b) = T_h Z_1(I_{a-h}^{b-h}) = T_h \int_{a-h}^{b-h} f(u)\, dZ(u) = \int_a^b f(u-h)\, dZ(u).$$

Hence for every h

$$f(u) = f(u-h)$$

for almost every u. Then it is a simple consequence that $f(u)$ is constant almost everywhere, that is there exists a number ω such that

$$f(u) = \omega$$

almost everywhere. Then (5.9) yields

$$Z_1(I_a^b) = \omega Z(I_a^b)$$

and from $\| Z_1(I_a^b) \|^2 = \| Z(I_a^b) \|^2$ we conclude that $|\omega| = 1$. Hence

$$x(t) = \int_{-\infty}^{t} g(u-t)\, dZ(u) = \int_{-\infty}^{t} \bar{\omega}\, g(u-t)\, dZ_1(u)$$

and from (5.4) we obtain

$$g_1(u-t) = \bar{\omega}\, g(u-t)$$

for almost every u.

Finally, if $|\omega| = 1$, then it is evident that $g_1(u)$ and $Z_1(I_a^b)$ satisfy a), b), c), and (5.4). This completes the proof of Theorem 2.

6. We may express Theorem 2 in another way. We obtain from (4.2)

$$r(s-t) = E\{x(s)\,\overline{x(t)}\} = \int_{-\infty}^{\infty} g(u-s)\,\overline{g(u-t)}\, du. \tag{6.1}$$

The integral formula in Theorem 2 then establishes an isomorphism between $L_2(x)$ and $L^2(-\infty, \infty)$, where the element $x(t) \in L_2(x)$ corresponds to $g(u-t) \in L^2(-\infty, \infty)$. $\{x(t)\}$ is a complete set in $L_2(x)$ on account of the definition of $L_2(x)$, and $\{g(u-t);\, t \textit{ arbitrary}\}$ is a complete set in $L_2(-\infty, \infty)$, which follows from c) in Theorem 2. We also conclude from c) that $L_2(x; a)$ corresponds to $L^2(-\infty, a)$ and hence that $L_2(x; a, b)$ corresponds to $L^2(a, b)$.

To prove Theorem 2 we have constructed $Z(I_a^b)$ and then obtained $g(u)$ with the properties b) and c). Now we see, that if we have found $g(u)$ satisfying b), c) and (6.1), then we are able to establish the isomorphism and denote by $Z(I_a^b)$ the element in $L_2(x)$ that corresponds to the function

$$f(u) = 1 \quad a < u < b$$
$$= 0 \quad \text{elsewhere}$$

in $L^2(-\infty, \infty)$.

REMARK. This construction of $Z(I_a^b)$ is possible even if $g(u)$ does not satisfy c). Then the following example shows that the uniqueness pronounced at the end of Theorem 2, is false if we omit c). For take

$$g(u) = 0 \quad u \geqq 0 \qquad \text{and} \qquad g_1(u) = 0 \quad u \geqq 0$$
$$= e^u \quad u < 0 \qquad\qquad\qquad = e^{2[u]-u+1} \quad u < 0.$$

Then

$$\int_{-\infty}^{\infty} g(u-s)\,\overline{g(u-t)}\,d\,u = \int_{-\infty}^{\infty} g_1(u-s)\,\overline{g_1(u-t)}\,d\,u.$$

Here c) holds for $g(u)$ but is false for $g_1(u)$.

7. A simple consequence of Theorem 2 is

THEOREM 3. Let $y(t) \neq 0$ and $z(t) \neq 0$ be two uncorrelated completely non-deterministic processes and $x(t) = y(t) + z(t)$. Then

$$L_2(x) \neq L_2(y) \oplus L_2(z).$$

PROOF. From Theorem 2 we have

$$y(t) = \int_{-\infty}^{t} g(u-t)\,d\,Z(u) \qquad \text{and} \qquad z(t) = \int_{-\infty}^{t} g'(u-t)\,d\,Z'(u).$$

Put

$$\eta = \int_{0}^{\infty} \overline{g'(-u)}\,d\,Z(u) - \int_{0}^{\infty} \overline{g(-u)}\,d\,Z'(u).$$

Obviously $E\{x(t)\,\bar\eta\} = 0$ for every t, so that $\eta \perp L_2(x)$. Since $\|\eta\| > 0$ this proves the theorem.

REFERENCES. [1] H. CRAMÉR, On the theory of stationary random processes, Annals of Mathematics, 41 (1940). — [2] K. KARHUNEN, Über lineare Methoden in der Wahrscheinlichkeitsrechnung, Annales Academiae Scientiarum Fennicae, Series A I, 37 (1947). — [3] A. KHINTCHINE, Korrelationstheorie der stationären stochastischen Prozesse, Mathematische Annalen, 109 (1934). — [4] A. N. KOLMOGOROFF, Stationary sequences in Hilbert's space, Bolletin Moskovskogo Gosudarstvennogo Universiteta, Matematika 2 (1941). — [5] H. WOLD, A study in the analysis of stationary time series, Inaugural Dissertation, Uppsala (1938).

Tryckt den 28 juni 1949

Uppsala 1949. Almqvist & Wiksells Boktryckeri AB

Editors' Comments on Paper 3

Paul Lévy, the well-known mathematician of the University of Paris, following earlier ideas of his own, studies here the problem of determining conditions for a random process to have unit multiplicity. Although not operating explicitly in the multiplicity context, he poses most of the questions that were later incorporated in the work of Cramér and Hida.

The approach is functional-analytic and the objective is the solution to an integral equation of the form

$$R(t,s) = \int_{-\infty}^{\min(t,s)} h(t,\tau)h(s,\tau)d\tau$$

where $R(t,s) = E[x(t)x(s)]$, which is another way of requiring the process $x(t)$ to have multiplicity equal to unity.

Lévy clearly points out the geometric second-order-moment nature of the problem, and raises subtle points on the separability of the Hilbert spaces spanned by the random processes in question and the role of their continuity. Furthermore, he obtains several conditions on the invertibility of linear systems in terms of their impulse-response functions $h(t,\tau)$. The connection of Lévy's work to the theory of multiplicity, developed later by Cramér and Hida, has been demonstrated in a later paper by Hitsuda (1968).

This paper is reprinted with the permission of the author and the Dunod/Gauthier-Villars Publishing Company.

Reference

Hitsuda, M., (1968). Representations of Gaussian processes equivalent to Wiener process. *Osaka J. Math.*, **5**, 299–312.

Reprinted from *Annales de Science de l'École Normale Superieure*, **73**, 121–156 (1956)

3

SUR UNE CLASSE

DE COURBES DE L'ESPACE DE HILBERT

ET SUR UNE

ÉQUATION INTÉGRALE NON LINÉAIRE

PAR M. PAUL LÉVY.

Introduction.

Le présent travail a son origine dans des recherches sur les fonctions aléatoires laplaciennes exposées en 1950 au second Symposium de Berkeley (1). Dans une Note récente (2), j'ai attiré l'attention sur le fait que le rapprochement de deux résultats de cette Communication de 1950 permet, dans des cas étendus, d'obtenir des solutions de l'équation intégrale non linéaire

$$\text{(E)} \qquad \Gamma(t_1, t_2) = \int_0^t F(t_1, u)\, F(t_2, u)\, du \qquad [t = \operatorname{Min}(t_1, t_2) \geqq 0].$$

Le but initial du présent travail était de développer cette Note, où j'avais surtout indiqué le moyen d'obtenir ce que j'appelle la *solution canonique*. Il y a toujours d'autres solutions qu'il restait à étudier. D'autre part il m'a paru désirable, pour exposer des résultats qui sont essentiellement analytiques de ne pas employer le langage des probabilités (3). Mais une simple vérification analytique des principaux résultats n'aurait pas mis en évidence ce qui fait l'intérêt de l'équation (E) : elle n'est pas formée artificiellement, mais s'introduit tout naturellement dans l'étude de la *covariance* d'une fonction aléatoire laplacienne.

---

(1) *Proc. of the Second Berkeley Symposium on math. Statistics and Probability* (University of California Press, 1951), p. 171-187. *Voir* aussi *Univ. of California Public. in Statistics*, t. 1, 1953, p. 331-390.

(2) *C. R. Acad. Sc.*, t. 242, 1956, p. 1252.

(3) Des notes en bas de page rappelleront brièvement l'interprétation probabiliste des résultats les plus importants.

Or il est possible de mettre ce fait en évidence en traduisant en termes géométriques la théorie des fonctions aléatoires laplaciennes exposée en 1955 au troisième Symposium de Berkeley ([4]), où, à l'occasion d'une application, j'avais développé ma théorie de 1950 et défini la *représentation canonique* de ces fonctions. On sait qu'une variable aléatoire laplacienne peut être représentée par un vecteur d'un espace de Hilbert Ω. Une fonction aléatoire laplacienne est alors représentée par une fonction vectorielle $\Phi(t)$ définie dans cet espace, et une équation différentielle stochastique vérifiée par la fonction aléatoire devient une équation différentielle vérifiée par la fonction vectorielle $\Phi(t)$.

C'est ce qui m'a conduit à faire précéder le chapitre consacré à l'intégration de l'équation (E) par deux chapitres où sont étudiées des courbes et des fonctions vectorielles dans l'espace de Hilbert. Le premier est consacré à la fonction particulière $X(t)$ qui correspond à la fonction aléatoire du mouvement brownien ; elle est définie par la propriété que le vecteur $X(t_1)-X(t_2)$ a pour longueur $|t_1-t_2|$. Dans le second sont étudiées les fonctions de la forme

$$\Phi(t)=\int_0^t F(t, u)\, dX(u), \tag{1}$$

et d'une forme un peu plus générale. Nous dirons que $F(t, u)$ est le *noyau* de $\Phi(t)$. Nous distinguerons des noyaux *propres* ou *impropres*, *canoniques* ou *non canoniques*. Une fonction $\Phi(t)$ a toujours une infinité de noyaux différents, groupés en *classes de noyaux équivalents*. A une transformation triviale près, chaque classe contient au plus un noyau propre. Il peut exister plusieurs classes distinctes; mais il n'y a qu'un noyau canonique.

Certains des résultats de ces deux premiers chapitres correspondent à des théorèmes déjà contenus dans ma Communication de 1955 à Berkeley. Mais il y a aussi plusieurs résultats nouveaux, notamment le théorème I.3 qui caractérise les fonctions que j'appelle *admissibles;* j'entends par là qu'il existe dans l'espace Ω une droite sur laquelle la projection de $X(t)$ coïncide avec la fonction considérée. Il en résulte une caractérisation des noyaux canoniques beaucoup plus précise que celle donnée antérieurement (théorème II.7.1). Les n°s II.7.4° et II.8.2° sont aussi nouveaux.

Si je n'utilise aucune notion de calcul des probabilités, j'en conserverai certaines notations. Ainsi les notations ξ_p, ξ_q (que j'ai l'habitude d'utiliser pour désigner deux variables laplaciennes réduites, indépendantes l'une de l'autre) représenteront deux vecteurs unitaires orthogonaux. La notation XY [remplaçant E(XY)] représentera le produit scalaire des vecteurs X et Y. La *covariance* de $\Phi(t)$ [en calcul des probabilités $E\{\Phi(t_1)\Phi(t_2)\}$] sera le produit scalaire $\Phi(t_1)\Phi(t_2)$.

([4]) Vol. II des *Proc.* en cours d'impression.

Quant à l'intégration de l'équation (E), disons seulement ici qu'elle repose sur une décomposition du problème, la première étape consistant à déduire de la covariance $\Gamma(t_1, t_2) = \Phi(t_1)\,\Phi(t_2)$ une équation de la forme

$$\delta\Phi(t) = dt \int_0^t G(t, u)\, d\Phi(u) + \sigma(t)\,\xi_t \sqrt{dt} \qquad (dt > 0), \tag{2}$$

dont l'intégration donnera ensuite $\Phi(t)$ sous la forme (1). En outre, après l'étude du cas régulier, qui ne fait que développer et préciser des résultats indiqués antérieurement, nous étudierons quelques exemples de cas singuliers.

I. — La fonction X(*t*) et la courbe C.

I.1. Remarques préliminaires. — Nous nous placerons dans des espaces de Hilbert réels, qui seront désignés par la lettre Ω, avec ou sans indice. Indiquons seulement que tous les résultats exposés s'étendent aisément au cas complexe; il n'y a qu'à remplacer les produits scalaires, et même le produit de deux facteurs scalaires qui figure dans l'équation (E), par des produits hermitiens.

Nous appellerons *plan* d'une figure l'intersection de toutes les variétés linéaires qui la contiennent. Même si un tel plan P coïncide avec l'espace Ω d'abord considéré, cela ne nous empêchera pas de considérer une ou plusieurs directions perpendiculaires à Ω; cela implique seulement l'introduction d'un espace plus vaste, dont Ω ne sera qu'une section. L'espace de Hilbert sera ainsi une construction créée et étendue en fonction de nos besoins. Comme l'ensemble des variables laplaciennes dont il est l'image, ou comme l'ensemble des nombres transfinis, c'est une construction dont on ne peut pas concevoir l'achèvement définitif.

Nous désignerons indifféremment par les lettres majuscules X, Y et Φ, avec ou sans indices, des points de Ω ou les vecteurs issus de l'origine O et aboutissant à ces points. Les notations $|X|$ et XY représenteront la longueur du vecteur X et le produit scalaire des vecteurs X et Y. Toutefois, il nous arrivera aussi de désigner par XY le segment qui joint les points X et Y, mais en le précisant chaque fois de manière à éviter toute ambiguïté.

La *covariance* d'un système de vecteurs X_n, c'est-à-dire l'ensemble des carrés X_n^2 et des produits $X_p X_q$, définit parfaitement la figure formée par ces vecteurs, indépendamment de son orientation dans l'espace. De même la covariance $\Gamma(t_1, t_2)$ d'une fonction vectorielle $\Phi(t)$ définit toutes les propriétés intrinsèques de cette fonction, c'est-à-dire la forme de la figure formée par l'origine O et la ligne L décrite par le point $\Phi(t)$, et la loi du mouvement sur cette ligne; mais elle ne donne aucune indication sur l'orientation de cette figure dans l'espace Ω. Cela n'a d'ailleurs aucun sens de parler de cette orientation si la ligne L n'est pas placée dans un espace où des axes sont préalablement définis. Aussi sera-t-il commode de considérer la fonction $\Phi(t)$ comme bien définie si l'on connaît sa covariance; nous prendrons en principe $\Phi(0)$ comme origine; la covariance $\Gamma(t_1, t_2)$ s'annulera alors avec $t_1 t_2$.

On sait que la covariance n'est pas une fonction quelconque. Il nous suffit de rappeler qu'elle est symétrique, que $\Gamma(t, t) \geqq 0$, et que, si $\Phi(t)$ est continu, elle est continue.

I.2. Définition de $X(t)$ et de la ligne C. Rappel de propriétés connues. — Nous désignerons par $X(t)$ la fonction vectorielle définie par la condition

$$|X(t') - X(t)| = \sqrt{|t' - t|}, \tag{I.2.1}$$

et par C la ligne lieu du point $X(t)$; le point $X(0)$ sera pris pour origine.

On déduit de cette définition que, si t_n croît avec n, tous les triangles $X(t_0)$ $X(t_1)$ $X(t_n)$ $(n > 1)$ sont rectangles, le sommet de l'angle droit étant $X(t_1)$. L'arc (t_1, ∞) de la ligne C est dans un plan perpendiculaire à la corde $X(t_0)$ $X(t_1)$, et à deux intervalles disjoints tels que (t_0, t_1) et (t_2, t_3) correspondent toujours deux cordes rectangulaires.

Deux points $X(t)$ et $X(t')$ de C divisent cette courbe en trois arcs situés, l'un sur la sphère de diamètre $X(t)$ $X(t')$, les deux autres dans les plans tangents perpendiculaires à ce diamètre. Le plan de n points ne recoupe jamais la courbe en d'autres points.

On peut définir comme suit la construction de la ligne C. On choisira une suite de nombres t_n variant de $-\infty$ à $+\infty$, et l'on construira la ligne polygonale de sommets $X(t_n)$, inscrite dans C; ses côtés sont deux à deux perpendiculaires, et leurs longueurs bien définies par la formule (I.2.1). Ensuite, dans chaque intervalle (t_n, t_{n+1}), on se donnera une suite partout dense de nombres $t_{n,\nu}$, et l'on déterminera successivement tous les $X(t_{n,\nu})$ par la règle suivante : pour déterminer le point $X(t)$ d'un arc $X(t')$ $X(t'')$ dont les extrémités sont connues et qui ne contient aucun autre point déjà connu, on se donnera une direction orientée perpendiculaire au plan défini par tous les points connus, et dans le demi-plan (à deux dimensions) contenant $X(t')$, $X(t'')$ et cette direction, on construira le triangle rectangle $X(t')$ $X(t)$ $X(t'')$, dont l'hypoténuse est placée, les deux autres côtés ayant des longueurs connues. On peut remarquer que, t variant de t' à t'', la projection de $X(t)$ sur la corde $X(t')$ $X(t'')$ de C se déplace avec une vitesse constante; $X(t)$ est donc, dans le demi-plan considéré, un point du demi-cercle de diamètre $X(t')$ $X(t'')$ bien défini par sa projection sur ce diamètre.

On démontre aisément que les $X(t_{n,\nu})$ ainsi obtenus définissent une fonction continue, et que la condition (I.2.1) est bien vérifiée quels que soient t et t' [5].

---

[5] Il est bien évident qu'à toute courbe continue définie dans Ω correspond une fonction aléatoire continue en moyenne quadratique. Le résultat obtenu ici suffit donc à établir l'existence et la continuité en moyenne quadratique de la fonction aléatoire du mouvement brownien. Les démonstrations

La covariance $X(t)\,X(t')$ est nulle si $tt' \leqq 0$, et dans le cas contraire égale au plus petit des modules $|t|$ et $|t'|$.

La définition de la courbe étant indépendante de l'instant pris comme origine, elle peut glisser sur elle-même sans se déformer, comme une hélice dans l'espace ordinaire. Naturellement, ce glissement implique une rotation extraordinairement irrégulière. Une corde donnée a un déplacement continu, mais cela n'empêche pas que pendant un temps τ arbitrairement petit, toute corde de longueur $\leqq \sqrt{\tau}$ a tourné d'un angle droit.

Pendant ce mouvement, le plan de la courbe C ne change pas. Mais si l'on ne considère que la moitié de la courbe (correspondant aux $t \geqq 0$), et qu'on la fasse glisser sur C dans le sens des t croissants, son plan perdra constamment une infinité de dimensions.

I.3. La projection de $X(t)$ sur une droite D et les fonctions admissibles. — Nous nous placerons sur un intervalle (t', t'') qui peut être fini ou infini; nous désignerons par Ω le plan de l'arc de C qui correspond à cet intervalle, et par Ω' un espace de Hilbert contenant Ω et au moins un axe perpendiculaire à Ω. Nous dirons qu'une fonction scalaire $x(t)$ est *admissible relativement à* Ω (ou Ω') s'il existe dans Ω (ou Ω') une droite D orientée sur laquelle la projection du point $X(t)$ soit $x(t)$. On peut naturellement supposer $x(0) = 0$; cela n'est pas essentiel.

Théorème I.3. — *La condition nécessaire et suffisante pour qu'une fonction donnée $x(t)$ soit admissible relativement à Ω et à l'intervalle (t', t'') $(t'' > t')$ est*

$$\text{(I.3.1)} \qquad I = \int_{t'}^{t''} \frac{[d\,x(t)]^2}{dt} = 1.$$

Relativement à Ω', elle est $I \leqq 1$.

Rappelons d'abord une propriété connue de l'intégrale de Hellinger I, qu'on peut définir comme limite de la somme riemannienne

$$\text{(I.3.2)} \qquad S = \sum \frac{[\Delta\,x(t)]^2}{\Delta t}.$$

Il est évident que l'addition d'un nouveau point de division τ dans un intervalle (τ', τ'') ne peut qu'augmenter cette somme, et l'augmente effectivement, à moins que $x(\tau) - x(\tau')$ et $x(\tau'') - x(\tau)$ ne soient proportionnels à $\tau - \tau'$ et $\tau'' - \tau$. On en déduit aisément, d'une part l'existence dans tous les cas d'une

---

plus compliquées ne sont utiles que si l'on veut établir la continuité presque sûre de cette fonction. D'une manière beaucoup plus générale, il suffit que la définition d'une fonction aléatoire soit correcte au point de vue de Bernoulli pour qu'on soit assuré de son existence, indépendamment de toute étude de ses propriétés probables ou presque sûres.

limite, finie ou infinie, indépendante du choix des points de division, d'autre part que le minimum de l'intégrale pour un intervalle de longueur l où la variation de $x(t)$ est h est réalisé quand $x(t)$ est linéaire, et égal à $\frac{h^2}{l}\cdot$ Ce résultat s'étend sans difficulté à n'importe quel ensemble de mesure l. L'intégrale I ne peut donc être finie que si $x(t)$ est absolument continu; en désignant alors par $x'(t)$ la dérivée de $x(t)$, elle s'écrit

$$\text{(I.3.3)} \qquad I = \int_{t'}^{t''} x'^2(t)\,dt.$$

Observons maintenant que, si $x(t)$ est la composante de $X(t)$ suivant une droite D orientée, la somme S peut s'écrire

$$\text{(I.3.4)} \qquad S_n = \sum_1^n \left[\frac{\Delta x(t_\nu)}{\Delta X(t_\nu)}\right]^2 = \sum_1^n \alpha_\nu^2,$$

les α_ν étant les cosinus des angles de D et des côtés successifs d'une ligne polygonale Π_n inscrite dans l'arc (t', t'') de C. Ces côtés étant deux à deux orthogonaux, on a $S_n = 1$ si D est dans le plan P_n de Π_n, et plus généralement $S_n = \cos^2\theta_n \leqq 1$ si D fait avec ce plan l'angle θ_n.

Si maintenant nous ajoutons de nouveaux points de division, le plan $\Omega_{t',t''}$ de l'arc considéré de la ligne C se trouve défini par une infinité d'axes deux à deux rectangulaires dont les n premiers définissent P_n. Si α'_ν est l'angle du $\nu^{\text{ième}}$ axe avec D, on a donc

$$S_n = \sum_1^n \alpha_{n,\nu}^2 = \sum_1^n \alpha_\nu'^2 = \cos^2\theta_n.$$

Pour n infini, S_n tend vers I. D'autre part l'angle θ de D avec $\Omega_{t',t''}$ est défini par

$$\cos^2\theta = \sum_1^\infty \alpha_\nu'^2,$$

de sorte que θ_n tend vers θ. On a donc à la limite $I = \cos^2\theta \leqq 1$, et, si D est dans $\Omega_{t',t''}$ (et dans ce seul cas), $I = 1$. Les conditions indiquées sont bien nécessaires.

Pour démontrer la réciproque, donnons-nous une fonction $x(t)$ vérifiant la condition $I \leqq 1$, et commençons par construire la figure constituée par Π_n et D. On peut construire Π_n dont les côtés ont des longueurs connues $\sqrt{\Delta t_\nu}$ et sont deux à deux orthogonaux. On définit ensuite $\alpha_{n,\nu}$ par la condition

$$\alpha_{n,\nu} = \frac{\Delta x_\nu(t)}{\sqrt{\Delta t_\nu}}.$$

(le radical étant positif). On a alors $S_n = \sum_1^n \alpha_{n,\nu}^2 \leq 1$, et l'on peut prendre pour D la direction dont les cosinus directeurs sont $\alpha_{n,1}, \alpha_{n,2}, \ldots, \alpha_{n,n}$, et, perpendiculairement au plan P_n de Π_n, $\sqrt{1 - S_n}$.

Si l'on supprime un des sommets de Π_n, on retrouve la figure analogue constituée par Π_{n-1} et D. Il en résulte qu'inversement, si l'on a d'abord construit cette dernière figure, on peut la conserver, et retrouver la première par l'adjonction d'un nouveau point convenablement placé. On peut répéter cette opération, pour des valeurs indéfiniment croissantes de n. On obtient à la limite la ligne C, et chaque $\Delta X(t)$ a bien pour projection $\Delta x(t)$, puisqu'il en est ainsi pour les côtés des lignes Π_n. On a naturellement toujours $I = \cos^2\theta$, donc $I = 1$ si D est dans $\Omega_{t',t''}$ et $I < 1$ dans le cas contraire. Le théorème est ainsi démontré.

I.4. Corollaires. — Nous allons déduire du théorème I.3 quelques propriétés simples des fonctions admissibles.

Corollaire I.4.1. — 1° *Toute fonction admissible $x(t)$ est continue et vérifie une condition Lipschitz de la forme*

$$|x(t+\tau) - x(t)| \leq \varphi(\tau), \tag{I.4.1}$$

où $\varphi(\tau) = o(\sqrt{|\tau|})$ $(\tau \to 0)$; (on peut de plus supposer que cette fonction soit paire, continue, et croisse avec $|\tau|$). 2° *Cet énoncé est le meilleur possible, en ce sens qu'on ne peut pas trouver une fonction particulière $\varphi(\tau)$, qui soit $o(\sqrt{|\tau|})$, et telle que la condition* (I.4.1) *soit vérifiée pour n'importe quelle fonction admissible et pour τ assez petit,*

1° Il résulte de la condition (I.3.1) qu'on peut partager l'intervalle (t', t'') de variation de t en un nombre fini d'intervalles partiels i_n dans chacun desquels on a

$$\int_i \frac{[dx(t)]^2}{dt} \leq \frac{\varepsilon^2}{2}.$$

Pour τ positif assez petit, $x(t)$ et $x(t+\tau)$ seront toujours dans le même intervalle ou dans deux intervalles consécutifs. On aura donc

$$\operatorname*{Sup}_{t \in (t', t''-\tau)} \frac{1}{\tau}[x(t+\tau) - x(t)]^2 \leq \varepsilon^2,$$

c'est-à-dire, ε étant arbitrairement petit, que le premier membre tend vers zéro avec τ, C. Q. F. D.

2° Supposons maintenant donnée une fonction $\varphi(|\tau|)$, qui soit continue, croissante, et $o(\sqrt{|\tau|})$ pour τ tendant vers zéro. Si une fonction $x(t)$ est égale à $2\varphi(t)$ pour des valeurs t_n de t qui tendent vers zéro en décroissant, et si elle varie linéairement dans chaque intervalle (t_n, t_{n-1}), on a

$$\int_0^{t_0} \frac{[dx(t)]^2}{dt} = 4 \sum \frac{[\varphi(t_{n-1}) - \varphi(t_n)]^2}{t_{n-1} - t_n},$$

et il suffit que cette série soit convergente pour que $x(t)$, quoique ne vérifiant pas la condition (I.4.1) pour $t = 0$ et $\tau = t_n$, soit admissible dans un intervalle $(0, T)$ assez petit, et que la deuxième partie de l'énoncé soit démontrée.

Or il en est bien ainsi si les t_n sont déterminés successivement de manière à vérifier par exemple les conditions

$$\frac{\varphi^2(t_n)}{t_n} \leq \frac{1}{2n^2}, \qquad \frac{[\varphi(t_n) - \varphi(t_{n+1})]^2}{t_n - t_{n+1}} \leq \frac{1}{n^2}, \qquad \frac{\varphi^2(t_{n+1})}{t_{n+1}} \leq \frac{1}{2(n+1)^2}, \qquad \ldots.$$

La première, puisque $\varphi^2(t_n) = o(t_n)$, est vérifiée si n est assez petit. La seconde et la troisième le sont aussi si, une fois t_n choisi, t_{n+1} est assez petit; et ainsi de suite. Le corollaire est donc démontré.

Corollaire I.4.2. — *On a toujours la condition de Lipschitz*

$$[x(t_1) - x(t_0)]^2 \leq |t_1 - t_0|,$$

et, si le maximum est atteint pour des valeurs particulières t_0 et $t_1 > 0$, $x(t)$ varie linéairement dans (t_0, t_1) et est constant dans (t', t_0) et dans (t_1, t'').

C'était évident, puisque l'égalité considérée implique que D soit parallèle à $X(t_0)\,X(t_1)$. Nous le rappelons pour faire remarquer que c'est aussi une conséquence du théorème I.3.

Corollaire I.4.3. — *Si* $\alpha \leq \frac{1}{2}$, t^α et $\mathcal{R}(ct^{\alpha+i\beta})$ *ne sont jamais des fonctions admissibles dans* $(0, T)$. *Si* $\alpha > \frac{1}{2}$, $\mathcal{R}(ct^{\alpha+i\beta})$ (et en particulier ct^α si $\beta = 0$), *sont admissibles dans* $(0, T)$ *si* $|c|$ *est assez petit* (le maximum possible tendant vers zéro pour T infini).

Naturellement, pour l'intervalle (T, ∞), les inégalités seraient retournées.

II. — Les fonctions $\Phi(t)$ et les courbes L.

II.1. Le cas particulier des fonctions additives. — Ce sont les fonctions de la forme

$$\Phi_0(t) = \int_0^t \sigma(u)\, dX(u), \qquad \text{(II.1.1)}$$

où $\sigma(u)$ est une fonction mesurable et de carré sommable dans tout intervalle fini (o, T). La ligne ainsi décrite par le point $\Phi_0(t)$ sera désignée par L_0.

La covariance de $\Phi_0(t)$ est

$$\Phi_0(t_1)\,\Phi_0(t_2)=\int_0^t \sigma^2(u)\,du=\omega(t) \qquad [t=\operatorname{Min}(t_1,\ t_2)], \tag{II.1.2}$$

tandis que la covariance mixte de $\Phi_0(t)$ et $X(t)$ est

$$\Phi_0(t_1)\,X(t_2)=\int_0^t \sigma(u)\,du. \tag{II.1.3}$$

Le signe de $\sigma(u)$ est donc sans influence sur les propriétés intrinsèques de $\Phi_0(t)$ et sur la forme de la courbe L_0. Il intervient au contraire si l'on veut la situer par rapport à C.

Désignons par Ω_t et Ω les plans des arcs (o, t) et (o, ∞) de C (au sens du n° I.1). $\Phi_0(t)$ est un point quelconque de Ω_t. En convenant de ne pas considérer comme distinctes deux fonctions qui ne diffèrent que sur un ensemble de mesure nulle, il y a correspondance biunivoque entre les points de Ω_t et les fonctions $\sigma(t)$ mesurables et de carrés sommables dans (o, t) (si l'on refuse la convention précédente, il faut introduire des classes de fonctions, deux fonctions d'une même classe ne différant que sur un ensemble de mesure nulle).

Pour un petit intervalle $(u,\ u+du)$ $(du>0)$, la variation $\delta X(u)$ est de la forme $\xi_u\sqrt{du}$, ξ_u désignant un vecteur unitaire dans Ω. En convenant qu'à deux valeurs distinctes de u, ou plus correctement à deux intervalles disjoints, correspondent des vecteurs orthogonaux, il est naturel de représenter $\Phi_0(t)$ par la notation

$$\Phi_0(t)=\int_0^t \sigma(u)\,\xi_u\sqrt{du}. \tag{II.1.4}$$

Il y a toutefois une différence importante entre les formules (II.1.1) et (II.1.4). La première définit les relations entre $X(t)$ et $\Phi(t)$; la seconde ne définit que la courbe C et les propriétés intrinsèques de $\Phi(t)$. Cette différence se manifeste lorsqu'on veut généraliser $\Phi_0(t)$ par l'introduction des *σ-fonctions*. Rappelons que ces fonctions généralisent les fonctions mesurables et de carrés sommables comme les distributions de L. Schwartz généralisent les fonctions sommables. Une σ-fonction est donc la racine carrée d'une distribution non négative, ce qui revient à dire que, dans (II.1.2), on peut prendre pour $\omega(t)$ n'importe quelle fonction non décroissante, et ayant la valeur initiale $\omega(-0)=0$. Alors que, si elle est absolument continue, les ξ_u n'interviennent en quelque sorte que par des moyennes, et que $\Phi_0(t)$ ne dépend finalement que de l'ensemble dénombrable des vecteurs qui servent à définir $X(t)$, dans le cas général, $\Phi_0(t)$ peut dépendre des valeurs de ξ_u sur un ensemble de mesure nulle, ou même en un point isolé, si $\omega(t)$ est discontinu en ce point; nous dirons dans ce dernier cas qu'il s'agit d'un ξ_u *instantané*. La formule (II.1.4) conserve dans ces cas un sens très clair; elle définit une fonction dont la covariance est $\omega(t)$. Au contraire, la formule (II.1.1), qui ne peut définir que des points de Ω, ne convient pas dans le cas général; nous ne l'utiliserons donc que dans le cas où $\sigma(u)$ est une vraie fonction [6].

II.2. Définitions et remarques préliminaires. — Les fonctions $\Phi(t)$ sont les fonctions définies par la formule

$$\Phi(t) = \int_0^t F(t, u)\, \xi_u \sqrt{du} \qquad (t > 0), \tag{II.2.1}$$

où, pour chaque t, $F(t, u)$ est une σ-fonction de u définie dans $(0, t)$. Précisons une fois pour toutes que les formules où intervient $F(t, u)$ supposent $0 \leq u \leq t$. Dans le cas où $F(t, u)$ est une vraie fonction, nous pouvons écrire

$$\Phi(t) = \int_0^t F(t, u)\, dX(u). \tag{II.2.2}$$

Dans tous les cas, nous désignerons par $X(t)$ la fonction

$$X(t) = \int_0^t \xi_u \sqrt{du}. \tag{II.2.3}$$

La fonction $F(t, u)$ sera appelée le *noyau* de la représentation considérée de $\Phi(t)$; $X(t)$ est donc la fonction de noyau unité. *Nous ne considérons ici que les valeurs positives de t. Les définitions de la fonction* $X(\,.\,)$, *de la ligne* C, *et de l'espace* Ω *qui la contient, sont supposées modifiées en conséquence.*

Nous désignerons par L la ligne représentant $\Phi(t)$; elle part toujours du point $\Phi(-0)$ pris comme origine. Nous désignerons par K et K* les classes de fonctions représentables respectivement par les formules (II.2.1) et (II.2.2). La première est plus étendue que la seconde. Cela n'est pas évident; il peut arriver qu'un noyau qui n'est pas une vraie fonction définisse une fonction $\Phi(t)$ qui soit tout de même de la classe K*. Mais un exemple tel que

$$\Phi(t) = \int_0^{t+0} \sqrt{\delta(u-t)}\, \xi_u \sqrt{du} = \xi_t \tag{II.2.4}$$

---

(6) Pour plus de détails sur les σ-fonctions, voir *Bull. Sc. math.*, t. 80, 1956, p. 83. Rappelons seulement que la décomposition classique de $\omega(t)$ en trois termes, un terme absolument continu $\omega_0(t)$, un terme continu $\omega_1(t)$ qui ne varie que sur un ensemble de mesure nulle, et un terme $\omega_2(t)$ qui ne varie que par sauts, conduit à une décomposition de $\sigma(t)$ en trois termes, deux à deux orthogonaux dans n'importe quel intervalle, et dont le premier seul est une vraie fonction; nous dirons que les deux autres sont la *partie singulière* et la *partie liée aux discontinuités* de $\omega(t)$.

[$\delta(.)$ étant la fonction de Dirac] montre que les vecteurs $\Phi(t)$ peuvent être tous deux à deux rectangulaires. La ligne L (qui dans ce cas n'est pas une courbe continue) ne peut être placée que dans un espace de Hilbert à une infinité non dénombrable de dimensions. Au contraire la ligne L représentant une fonction de la classe K^* est dans l'espace Ω défini par C, qui n'a qu'une infinité dénombrable de dimensions. Donc la fonction définie par la formule (II.2.4) n'appartient pas à K^*.

Nous désignerons par Ω_Φ l'espace de Hilbert défini par la ligne L, et par $\Omega_{t,\Phi}$ le plan de l'arc (o, t) de cette courbe. Si $F_0(t, u)$ est la partie de $F(t, u)$ qui est une vraie fonction [*cf.* note (6) ci-dessus], et $\Phi_0(t)$ la partie correspondante de $\Phi(t)$, la ligne L_0 lieu du point $\Phi_0(t)$ est la projection de L sur Ω, et le plan de l'arc (o, t) de cette ligne est l'intersection de Ω_t et $\Omega_{t,\Phi}$. Si $F(t, u)$ n'est pas une vraie fonction de t, u dans l'intervalle (o, T) (c'est-à-dire pour $o \leqq u \leqq t \leqq T$), $\Omega_{t,\Phi}$ ne se réduit pas à cette intersection. Inversement, Ω_t peut aussi ne pas être contenu dans $\Omega_{t,\Phi}$; c'est le cas, en particulier, s'il existe sur l'axe des t un ensemble $\mathcal{E}$ de mesure positive tel que $F(t, u)$ soit identiquement nul quand $u \in \mathcal{E}$.

II.3. La covariance. — La covariance de $\Phi(t)$, dont la donnée suffit à définir les propriétés intrinsèques de $\Phi(t)$, est

$$\Gamma(t_1, t_2) = \Phi(t_1)\Phi(t_2) = \int_0^t F(t_1, u)\, F(t_2, u)\, du \qquad [t = \mathrm{Min}(t_1, t_2)], \tag{II.3.1}$$

tandis que la covariance mixte de $\Phi(t)$ et $X(t)$ est

$$\Phi(t_1) X(t_2) = \int_0^t F(t_1, u)\, du = \int_0^t F_0(t_1, u)\, du. \tag{II.3.2}$$

La formule (II.3.1) montre l'importance de l'équation (E). La résolution permet de passer de la définition de $\Phi(t)$ par sa covariance à son expression explicite (II.2.1) ou (II.2.2).

On remarque que, pour l'étude intrinsèque de $\Phi(t)$, on ne change rien en remplaçant $F(t, u)$ par $\varepsilon(u)\, F(t, u)$ où $\varepsilon(u) = \pm 1$. Ce facteur, qui n'intervient que par son carré, n'a même pas besoin d'être une fonction mesurable de u. Au contraire, pour l'étude jointe de $\Phi(t)$ et $X(t)$, il faut préciser le signe de $F(t, u)$, qui doit être, pour chaque t fixe, une fonction de u, mesurable dans (o, t), donc sommable (puisqu'elle est de carré sommable). La condition que le signe $\varepsilon(u)$ de $F(t, u)$ soit, pour chaque t fixe, une fonction mesurable de $\omega(u)$ suffit, si elle est

$$\omega(t, u) = \int_0^u F^2(t, v)\, dv \tag{II.3.3}$$

vérifiée pour plusieurs noyaux F_n, pour l'étude jointe des fonctions correspondantes $\Phi_n(t)$ [dont l'une peut être $X(t)$].

II.4. Représentations propres et représentations impropres. — Nous dirons que ξ_u est *sans effet* sur $\Phi(t)$ jusqu'à l'instant $T > u$ s'il existe un petit intervalle $[u, u')$ qu'on puisse négliger dans le calcul de l'intégrale (II.1.1), quel que soit $t \in [u, T)$. En d'autres termes, cela veut dire que, pour tout $t \in [u, T)$,

$$\int_{u-0}^{t'} F^2(t, v)\,dv = \omega(t, t') - \omega(t, u, 0) = 0 \qquad [t' = \mathrm{Min}(t, u')]. \tag{II.4.1}$$

La borne supérieure $U = f(u)$ des T pour lesquels il en est ainsi définit l'*instant à partir duquel l'effet de ξ_u se fait sentir* (7).

Désignons par E_0 l'ensemble des $u \geqq 0$ pour lesquels $f(u) = u$, c'est-à-dire que l'effet de ξ_u est *immédiat;* par E_1 l'ensemble de ceux pour lesquels $u < f(u) < \infty$, c'est-à-dire que l'effet de ξ_u est *retardé;* enfin par E_2 l'ensemble des u pour lesquels $f(u) = \infty$, c'est-à-dire que ξ_u est sans aucun effet sur la fonction $\Phi(t)$ (8).

On remarque que la condition (II.4.1) reste nécessairement vérifiée si l'on remplace u par un autre nombre de l'intervalle (u, u'). Il en résulte évidemment que E_2, ainsi que la réunion de E_1 et E_2, sont des réunions d'intervalles ouverts à droite.

Nous dirons que *la représentation* (II.2.1) *de $\Phi(t)$ est propre* [ou que *le noyau* $F(t, u)$ *est propre*] dans un intervalle $(0, T)$ [ou sur toute la demi-droite $(0, \infty)$] si cet intervalle (ou cette demi-droite) est entièrement contenu dans E_0. Dans le cas contraire, elle sera *impropre*.

II.5. Classes de noyaux équivalents (9). — 1° Nous dirons qu'un noyau $F(t, u)$ est *équivalent à zéro* si la fonction $\omega(t, u)$ [formule (II.3.3)] est identiquement nulle.

Nous dirons que deux noyaux sont *équivalents (au sens strict)*, ou appartiennent à une même *classe*, si l'on peut passer de l'un à l'autre par une combinaison des trois opérations suivantes, qui sont sans effet sur la covariance, donc aussi sur la définition intrinsèque de $\Phi(t)$.

Opération a. — C'est l'addition à $F(t, u)$ d'un noyau équivalent à zéro.

---

(7) Précisons que nous ne distinguons pas ici le cas où il s'agit d'un ξ_u instantané et celui où il du ξ_u s'agit moyen relatif à un petit intervalle $(u, u + du)$ $(du > 0)$. Quand nous voudrons préciser qu'il s'agit du second cas, nous introduirons $\xi_u \sqrt{du}$ plutôt que ξ_u. Nous ne distinguerons pas non plus le cas où cet effet se fait sentir brusquement à l'instant $U = f(u)$ (ou $U + 0$) de celui où il se fait sentir progressivement à partir de cet instant.

(8) Le langage du calcul des probabilités est ici très intuitif. Chaque ξ_u définit une information, qui peut être, soit immédiatement utile pour l'étude de la fonction aléatoire $\Phi(t)$, soit utilisable avec retard, soit définitivement inutile.

(9) Les définitions introduites ici ne sont pas les mêmes que dans ma Note du 19 mars 1956 (*C. R. Acad. Sc.*, **242**, p. 1575).

Opération b. — C'est la multiplication de $F(t, u)$ par un facteur $\varepsilon(u)$ toujours égal à ± 1.

Ces deux opérations sont triviales, et il est commode de ne même pas considérer comme distincts deux noyaux déduits l'un de l'autre par ces opérations. Ils sont tous les deux propres, ou tous les deux impropres.

Opération c. — C'est un changement de variable $u = \varphi(v)$. Supposons d'abord la fonction $\varphi(v)$ absolument continue, croissant avec v de zéro à l'infini, toujours $\geqq v$, et non identique à v. En posant $t = \varphi(\tau)$, il vient

$$\Phi(t) = \int_0^{\tau} F[t, \varphi(v)]\, \eta_v \sqrt{\varphi'(v)\, dv} \qquad (\eta_v = \xi_u),$$

d'où, en désignant par $G(t, v)$ la fonction égale à $F[t, \varphi(v)]\sqrt{\varphi'(v)}$ si $v < \tau$ et nulle si $v > \tau$,

(II.5.1)
$$\varphi(t) = \int_0^{t} G(t, v)\, \eta_v \sqrt{dv}.$$

Par cette équation, on déduit de n'importe quel noyau donné, propre ou impropre, une infinité de noyaux équivalents, tous impropres. L'opération inverse permettant de retrouver le noyau initial, nous sommes conduits à supprimer la condition $u \geqq v$, qui n'est pas vérifiée par cette opération inverse, et la remplacer par la condition moins restrictive, mais dépendant du noyau initial : $v \leqq f(u)$ (notation du n° II.4). En d'autres termes, si l'on se donne d'abord la relation $u = \varphi(v)$ entre u et v, on définira une opération applicable seulement aux noyaux $F(t, u)$ qui sont nuls si $t < \text{Max}(u, v)$.

D'autre part, ni la continuité, ni même le caractère biunivoque de la relation entre u et v, ne sont essentiels. On peut se donner sur l'axe des u deux ensembles $\mathcal{E}_u$ et $\mathcal{E}_v$ de même puissance, et établir une correspondance biunivoque seulement entre les u qui appartiennent à $\mathcal{E}_u$ et les v qui appartiennent à $\mathcal{E}_v$; on aura alors une transformation applicable seulement aux noyaux qui sont nuls si u n'appartient pas à $\mathcal{E}_u$. Pour la correspondance entre $\mathcal{E}_u$ et $\mathcal{E}_v$, la continuité peut être remplacée par une condition de mesurabilité : il faut que $F^2(t, u)\, du$ puisse être remplacé par $G^2(t, v)\, dv$, autrement dit que, sur l'ensemble transformé de $(v_1, v_2) \cap \mathcal{E}_v$, l'intégrale de $F^2(t, u)\, du$ soit bien définie, quels que soient v_1, v_2, et $t > \text{Max}[v, \varphi(v)]$ $[v \in (v_1, v_2)]$. Cette condition est réalisée par exemple si $\mathcal{E}_u$ comprend une infinité d'intervalles disjoints e_ν, transformés d'intervalles e'_ν, la correspondance entre chaque e_ν et e'_ν étant continue et biunivoque; il n'est pas nécessaire que l'ordre des e_ν sur l'axe des u soit celui des e'_ν, et, pour chacun de ces intervalles, la fonction $\varphi(v)$ peut être indifféremment croissante ou décroissante.

On peut aussi établir une correspondance biunivoque entre un ensemble de mesure nulle et un ensemble de mesure positive, et transformer un noyau qui est une vraie fonction en un noyau singulier ou inversement.

2° Si un noyau $G(t, v)$ est équivalent à un noyau propre $F(t, u)$, on pourra toujours aisément retrouver ce noyau propre : le nombre $u = \varphi(v)$ qui doit être pris comme nouvelle variable est en effet l'instant à partir duquel l'effet de η_v se fait sentir. Cette remarque conduit alors à se poser le problème suivant : que donne cette opération appliquée à un noyau impropre quelconque $G(t, v)$? Peut-on toujours, sinon obtenir un noyau propre, du moins faire disparaître l'ensemble E_1?

La fonction $u = \varphi(v)$ étant donnée, il n'y a aucune difficulté si toutes ses valeurs sont distinctes : ou bien toutes les valeurs positives sont réalisées, et l'on a une représentation propre; ou bien il y a des lacunes, qui appartiennent à E_2, et l'axe des u se partage entre E_0 et E_2; E_1 est nécessairement vide.

Supposons maintenant qu'une même valeur de u corresponde à deux valeurs v' et v'' de v. On serait alors conduit à introduire au même instant u deux termes différents $\eta_{v'} G(t, v')\sqrt{dv'}$ et $\eta_{v''} G(t, v'')\sqrt{dv''}$, et l'on peut être tenté de penser que cette circonstance exclut toute représentation propre.

Or il y a des cas d'exception, dont le plus simple est celui où, après transformation, $\eta_{v'}$ et $\eta_{v''}$ ont des coefficients $c' F(t, u)\sqrt{du}$ et $c'' F(t, u)\sqrt{du}$ qui sont proportionnels, au moins pendant un intervalle de temps (u, u'). On posera alors

$$c'\eta_{v'} + c''\eta_{v''} = c\,\xi_u \qquad (c^2 = c'^2 + c''^2),$$

de manière à réunir les deux termes en un terme unique $c\xi_u F(t, u)\sqrt{du}$; puis, à l'instant t', on introduira une autre combinaison de $\eta_{v'}$ et $\eta_{v''}$.

Il peut arriver qu'on obtienne ainsi des représentations propres. On peut en effet définir une fonction $u = \varphi(v)$ de manière qu'à chaque u donné correspondent deux valeurs v' et v'' de v, distinctes, et $\in (0, u)$, puis, en partant d'une fonction définie par une représentation propre, décomposer ξ_u en une somme de deux termes orthogonaux $\eta_{v'} \cos\varphi$ et $\eta_{v''} \sin\varphi$. De la représentation impropre obtenue, on remontera à la représentation propre par le regroupement de ces termes.

Ces remarques peuvent être généralisées. Rien ne nous empêche en effet de décomposer chaque ξ_u en un nombre quelconque ou même une infinité dénombrable de vecteurs orthogonaux $c_\nu \eta_{v_\nu}$, ou, inversement, si un nombre quelconque de valeurs u_ν de u conduisent à une même valeur v de $f(u)$, d'effectuer sur les ξ_u qui leur correspondent une substitution orthogonale, qui permettra dans certains cas de les remplacer par des vecteurs η_v dont les effets ne se feraient sentir que successivement. Mais cette méthode de recherche d'une représentation propre peut échouer dans des cas où pourtant cette représentation existe. C'est ce que prouvent les exemples suivants.

Premier exemple. — Posons

$$\text{(II.5.2)} \qquad \Phi(t) = \int_0^t (\alpha t \xi'_u + \beta u \xi''_u)\sqrt{du} \qquad (\xi'_u \xi''_u = 0).$$

L'intervention simultanée, à chaque instant u, de deux vecteurs orthogonaux ξ'_u et ξ''_u, avec des coefficients qui ne sont proportionnels dans aucun intervalle (u, u'), pourrait faire croire qu'il n'existe pour cette fonction aucune représentation propre de la forme (II.2.1). Or le calcul de la covariance montre que cette fonction admet deux représentations propres, non équivalentes, de la forme

$$\text{(II.5.3)} \qquad \int_0^t (\lambda t + \mu u)\, \xi_u \sqrt{du},$$

λ et μ étant définis au signe près, pour chacune de ces représentations, par les formules

$$\text{(II.5.4)} \qquad \alpha^2 = \lambda^2 + \frac{\lambda\mu}{2}, \qquad \beta^2 = \frac{\lambda\mu}{2} + \frac{\mu^2}{3} \qquad [\text{d'où } (\lambda+\mu)^2 = \alpha^2 + 3\beta^2] \quad (^{10}).$$

Deuxième exemple. — Partons du fait connu que $X(t)$ peut être défini dans (0, 1) par la série de Fourier-Wiener ([11])

$$X(t) = \xi_0 t + \sum_1^{\infty} \frac{1}{\pi\nu\sqrt{2}} [\xi_\nu(\cos 2\pi\nu t - 1) + \xi_{-\nu} \sin 2\pi\nu t],$$

les ξ_ν étant deux à deux orthogonaux. Donnons-nous une suite infinie dans les deux sens de nombres t_n croissant de zéro à l'infini, et, dans chaque intervalle (t_n, t_{n+1}), définissons $\frac{X(t) - X(t_n)}{\sqrt{t_{n+1} - t_n}}$ par une telle série, t étant seulement remplacé par $\frac{t - t_n}{t_{n+1} - t_n}$, et ξ_ν par $\xi_{n,\nu}$. On peut d'ailleurs supposer que les $\xi_{n,\nu}$ sont des ξ_u instantanés relatifs à des instants $u_{n,\nu}$ qui, pour chaque n fixe, croissant avec ν de t_{n-1} à t_n. On obtient ainsi pour $X(t)$ une représentation impropre de la forme (II.2.1).

Essayons alors d'en déduire une représentation propre pour une substitution sur u, qui devrait remplacer chaque $u_{n,\nu}$ par t_n, puisque chaque $\xi_{n,\nu}$ a un effet qui commence à se manifester à l'instant t_n. Il y a ainsi une infinité de ξ_u qui devraient avoir le même indice, et dans n'importe quel intervalle $(t_n, t_n + \tau)$, quel que petit que soit τ, il est indispensable, pour y définir $X(t)$, d'introduire une infinité dénombrable de nouvelles dimensions. On ne peut donc pas définir une substitution effectuée sur les $\xi_{n,\nu}$, qui permettrait de les remplacer par des variables η_v n'intervenant qu'à des instants $t_{n,\nu}$ qui forment une suite discrète,

---

([10]) Pour une étude plus complète de cet exemple, *voir* P. Lévy, Communication au Symposium de Brooklyn (avril 1955, *Proc.* en cours d'impression). Il n'est d'ailleurs qu'un cas particulier d'un exemple plus général qui sera étudié plus loin (nos II.7.4° et II.8.2°).

([11]) *Voir* R. E. A. C. Paley et N. Wiener, *Fourier transforms in the complex domain* (New-York, 1934), formule (37.06) relative au cas complexe. *Voir* aussi notre Mémoire déjà cité (*Second Berkeley Symposium*, p. 171), ou encore P. Lévy, *Le mouvement brownien* [*Mémorial des Sciences math.*, fasc. 126, p. 22, formule (36)].

et le procédé indiqué tout à l'heure pour essayer d'obtenir une représentation sans ξ_u à effet différé échoue complètement. Pourtant cette représentation existe : $X(t)$ est la fonction $\Phi(t)$ de noyau unité.

Ces exemples font penser qu'il y a une difficulté réelle, et ce n'est qu'à titre d'hypothèse que nous suggérons l'idée d'une théorie qui aurait l'aspect suivant : on définirait des *classes généralisées*, contenant des noyaux qui seraient *équivalents au sens large*. Si une fonction $\Phi(t)$ peut être représentée par une formule du type (II.2.1) dans laquelle chaque ξ_u serait sans effet ou à effet immédiat (c'est-à-dire que l'ensemble E_1 serait vide), chaque classe généralisée de représentations de cette fonction en contiendrait une et une seule (aux opérations *a* et *b* près), qui vérifierait cette condition, et on aurait une méthode pour l'obtenir.

Pour le moment, la seule chose sûre, dans cet ordre d'idées, est que la classe des noyaux équivalents à un noyau donné, au sens strict défini d'abord, est une notion trop restrictive, et que, de n'importe quelle représentation d'une fonction $\Phi(t)$ par la formule (II.2.1), on peut en déduire d'autres formant une classe plus générale.

Le premier des exemples ci-dessus montre qu'en tout cas il peut y avoir pour une même fonction $\Phi(t)$ plusieurs représentations propres, donc plusieurs classes distinctes. Même la fonction $X(t)$ admet les deux représentations propres distinctes

$$\pm\int_0^t \xi_u\sqrt{du}, \qquad \pm\int_0^t\left(\frac{3u}{t}-2\right)\xi_u\sqrt{du},$$

et il résultera du n° II.8.2° que ce ne sont pas les seules.

II.6. Noyaux canoniques. — 1° Nous dirons que le noyau $F(t, u)$ est *canonique*, ou qu'il donne une *représentation canonique* de $\Phi(t)$, si le premier terme de la somme

$$\text{(II.6.1)} \qquad \Phi(t')=\int_0^t F(t', u)\,\xi_u\sqrt{du}+\int_t^{t'} F(t', u)\,\xi_u\sqrt{du} \qquad (0<t<t')$$

représente un vecteur contenu dans $\Omega_{t,\Phi}$. Le second terme étant toujours orthogonal à ce plan, cela revient à dire que le premier est la projection de $\Phi(t')$ sur $\Omega_{t,\Phi}$.

Théorème II.6.1. — *Si le noyau* $F(t, u)$ *est canonique, l'ensemble* E_1 (défini au n° II.4) *est vide.*

Si en effet il n'est pas vide, on peut, pour tout $u\in E_1$, déterminer un nombre $t'>f(u)>u$ auquel corresponde dans $\Phi(t')$ un élément non nul $F(t', u)\,\xi_u\sqrt{du}$.

Quel que soit t entre u et $f(u)$, cet élément est une partie du premier terme de $\Phi(t')$, et n'est pas contenu dans $\Omega_{t,\Phi}$; le noyau n'est donc pas canonique,

C. Q. F. D.

Par contre les intervalles $(u', u'') \in E_2$, auxquels correspondent des ξ_u qui n'interviennent ni dans $\Omega_{t,\Phi}$ ni dans $\Phi(t')$, n'empêchent pas un noyau d'être canonique. Donc :

Corollaire II.6.1. — *Il y a deux et seulement deux types de noyaux canoniques : les noyaux canoniques propres, pour lesquels chaque* ξ_u *a un effet immédiat sur* $\Phi(t)$, *et les noyaux canoniques impropres, pour lesquels l'axe des u est la réunion des deux ensembles non vides* E_0 *et* E_2 (en exceptant le seul cas d'un noyau équivalent à zéro, cas où $\Phi = 0$ et où E_0 est vide).

L'ensemble E_2 est dans ce dernier cas une réunion d'intervalles. L'ensemble E_0 est toujours fermé à gauche ; il peut d'ailleurs ne comprendre que des points isolés ([12]).

2° Théorème II.6.2. — *Pour que* Ω_t *et* $\Omega_{t,\Phi}$ *coïncident pour tout* $t < T$, *il faut et il suffit que, pour* $0 \leqq u \leqq t < T$, *le noyau* $F(t, u)$ *soit une vraie fonction, et soit un noyau canonique propre.*

Nous avons déjà vu (n° II.2) que la première condition est nécessaire et suffisante que pour Ω_t contienne $\Omega_{t,\Phi}$. Nous pouvons donc la supposer remplie, et représenter $\Phi(t)$ par la formule (II.2.2). Il s'agit de montrer que, dans ces conditions, les deux autres conditions sont nécessaires et suffisantes pour que $\Omega_{t,\Phi}$ contienne Ω_t.

Montrons d'abord qu'elles sont nécessaires. Si Ω_t et $\Omega_{t,\Phi}$ coïncident, le premier terme de $\Phi(t')$, contenu dans Ω_t (puisque le noyau est une vraie fonction), l'est aussi dans $\Omega_{t,\Phi}$, c'est-à-dire que le noyau est canonique. D'autre part, Ω_t étant constamment croissant, $\Omega_{t,\Phi}$ l'est aussi, ce qui n'aurait pas lieu si, E_1 étant vide (d'après le théorème II.6.1), E_2 ne l'était pas ; il y aurait alors des intervalles de temps pendant lesquels aucun nouveau ξ_u ne ferait croître $\Omega_{t,\Phi}$. Il faut donc que E_2 soit vide aussi, donc que le noyau soit propre.

Inversement, supposons le noyau canonique et propre. Puisqu'il est propre, à tout $u < T$ on peut associer un $t' < T$ tel que $F(t', u)\sqrt{du}$ ne soit pas nul. Puisqu'il est canonique, $F(t', u)\xi_u\sqrt{du}$, qui est l'accroissement du premier

([12]) En calcul des probabilités, si le noyau est canonique, le processus est déterministe dans tous les intervalles dont E_2 est la réunion. Au contraire le hasard intervient constamment dans E_0, que cet ensemble soit fini, dénombrable, ou ait la puissance du continu. Les informations ainsi recueillies dans tout intervalle où E_0 n'est pas vide correspondent à l'accroissement de $\Omega_{t,\Phi}$. Si l'on s'intéresse surtout aux nouvelles informations, on peut n'étudier d'abord le processus qu'aux instants $t \in E_0$. Dans E_2, les valeurs de $\Phi(t)$ sont des fonctions certaines de ses valeurs dans E_0.

terme de $\Phi(t')$ quand t varie de u à $u+du$, est contenu dans $\Omega_{t,\Phi}$ si $t>u$. Donc tous les $\xi_u\sqrt{du}$ d'indices $u<t$, et par suite Ω_t, sont contenus dans $\Omega_{t,\Phi}$,

C. Q. F. D.

Il est important d'observer que la dernière partie du raisonnement ne subsiste pas si le noyau n'est pas une vraie fonction. Donnons-nous par exemple une suite de nombres u_n partout dense dans $(0, \infty)$, et une série convergente $\sum a_n^2$. Le noyau $\sum\limits_{u_n<t} a_n \xi_n \sqrt{\delta(u-u_n)}$, où δ est la fonction de Dirac, est canonique et propre, et pourtant Ω_t et $\Omega_{t,\Phi}$ sont orthogonaux; aucun de ces deux plans ne contient l'autre.

3° Théorème II.6.3 (*théorème d'unicité*). — *En ne considérant pas comme distincts deux noyaux déduits l'un de l'autre par les opérations a et b du n°* II.5, *chaque fonction $\Phi(t)$ a au plus une représentation canonique.*

Cela résulte de ce que, pour un noyau canonique, le premier terme $\Phi(t', t)$ de l'expression (II.6.1) de $\Phi(t')$, étant la projection de $\Phi(t')$ sur $\Omega_{t,\Phi}$, ne dépend que des propriétés intrinsèques de la fonction Φ. Il en est donc de même de la covariance

$$\Phi(t'_1, t)\,\Phi(t'_2, t) = \int_0^t F(t'_1, u)\,F(t'_2, u)\,du. \tag{II.6.2}$$

Le produit $F(t'_1, u)F(t'_2, u)$ est donc connu, et par suite $F(t, u)$ est connu, à cela près qu'on peut effectuer les opérations *a* et *b* du n° II.5.

L'importance de ce théorème résulte du fait que deux noyaux non équivalents, tous les deux propres, et qui sont tous les deux de vraies fonctions continues, peuvent représenter une même fonction $\Phi(t)$. Ainsi, d'après le n° II.5 (premier exemple), les noyaux u et $2u-t$ représentent la même fonction; de même t et $3u-2t$.

Dans les cas de ce genre, $\Omega_{t,\Phi}$ est le même pour les deux noyaux; Ω_t coïncide avec $\Omega_{t,\Phi}$ pour le noyau canonique, tandis que, pour l'autre noyau, $\Omega_{t,\Phi}$ n'est qu'une partie de Ω_t. La distance de $\Phi(t')$ à Ω_t, dont le carré est (si le noyau est une vraie fonction)

$$\int_t^{t'} F^2(t', u)\,du \tag{II.6.3}$$

est donc au moins aussi grande, et en général plus grande, pour le noyau canonique que pour l'autre. On peut ainsi vérifier que les noyaux $2u-t$ et $3u-2t$ ne sont pas canoniques; les noyaux u et t peuvent l'être, et, en fait, le sont évidemment.

4° *Conditions d'existence de la représentation canonique.* — D'après le théorème II.6.1, une condition nécessaire est qu'on puisse obtenir une représentation pour laquelle l'ensemble E_1 soit vide. On pourrait chercher d'abord si cette condition est réalisée. Mais il ne semble pas plus compliqué d'étudier directement le problème posé, en utilisant le fait que, dans une représentation canonique, l'accroissement de $\Omega_{t,\Phi}$ est à chaque instant t, ou pour chaque intervalle dt, exactement défini par les nouveaux ξ_u relatifs à cet instant ou à cet intervalle. Au contraire, pour une représentation propre non canonique, l'introduction constante de ces ξ_u conduit à un ensemble (qui est Ω_t si le noyau est une vraie fonction) dont $\Omega_{t,\Phi}$ n'est qu'une partie.

Il s'agit donc de savoir si l'accroissement de $\Omega_{t,\Phi}$ peut être défini par une famille unique de vecteurs ξ_u, ou s'il est nécessaire d'introduire, au moins à certains instants, au moins un autre vecteur ξ'_u.

Il n'y a aucune difficulté en ce qui concerne les ξ_u instantanés, qui correspondent chacun à un accroissement instantané de $\Omega_{t,\Phi}$. Si nous posons

$$\Omega_{u-0,\Phi} \bigcup_{t<u} \Omega_{t,\Phi}, \qquad \Omega_{u+0,\Phi} = \bigcap_{t>u} \Omega_{t,\Phi},$$

la condition que $\Omega_{u+0,\Phi}$ contienne au plus une dimension de plus que $\Omega_{u-0,\Phi}$ (c'est-à-dire ne contienne qu'une direction orthogonale à $\Omega_{u-0,\Phi}$) est évidemment nécessaire pour l'existence d'une représentation canonique (puisque autrement il faudrait introduire instantanément au moins deux vecteurs indépendants ξ_u et ξ'_u). Elle est suffisante en ce qui concerne les ξ_u instantanés.

Il est ainsi évident qu'une fonction de la forme

$$\Phi(t) = \sum_{u_n<t} [F_n(t)\,\xi_n + G_n(t)\,\xi'_n]$$

(où $\{u_n\}$ est une suite quelconque de nombres positifs) n'a pas de représentation canonique si, au moins pour un n, ξ_n et ξ'_n étant indépendants, $F_n(t)$ et $G_n(t)$ ne varient pas proportionnellement au moins dans un petit intervalle $(u_n, u_n + \varepsilon)$. Au contraire chacun des deux termes est mis sous la forme canonique.

La question est beaucoup moins simple dans le cas des ξ_u non instantanés. On ne peut que considérer un petit intervalle $(u, u + du)$, pendant lequel $\Omega_{t,\Phi}$ s'accroît d'une infinité de dimensions, et la relation entre ξ_u et l'indice n'a pas le caractère précis qu'elle a dans le cas précédent. Il peut être alors difficile de savoir si les ξ_u qu'il faut introduire peuvent être rangés en une suite unique. Il peut être aussi difficile de déduire $\Omega_{t,\Phi}$ d'une expression analytique donnée de $\Phi(t)$. Aussi nous bornerons-nous au cas des fonctions

$$X(t) = \int_0^t \xi_u \sqrt{du}, \qquad Y(u) = \int_0^t \eta_u \sqrt{du},$$

$$\Phi(t) = \int_0^t [F(t, u)\, dX(u) + G(t, u)\, dY(u)],$$

où les ξ_u et η_u sont tous deux à deux orthogonaux, et où F et G sont de vraies fonctions. On pourrait penser que, si ces deux noyaux sont séparément canoniques, $\Omega_{t,\Phi}$ est l'espace produit de $\Omega_{t,X}$ et $\Omega_{t,Y}$, de sorte que son accroissement ne peut être défini qu'en introduisant à chaque instant deux vecteurs indépendants. Il n'en est pas toujours ainsi, puisque les intégrales (II.5.2) et (II.5.3) peuvent définir une même fonction $\Phi(t)$. Il en est au contraire ainsi, par exemple, dans le cas où $F(t, u)$ est égal à zéro ou un suivant que t est rationnel ou irrationnel, et $G(t, u) = 1 - F(t, u)$. La ligne L comprend alors, sur chacune des lignes C et C′ représentant $X(t)$ et $X(t)$, un ensemble de points partout dense sur cette ligne, de sorte que $\Omega_{t,\Phi}$ contient $\Omega_{t,X}$ et $\Omega_{t,Y}$, et, dans ce cas, $\Phi(t)$ n'admet pas de représentation canonique à un terme.

II.7. Caractérisation des noyaux canoniques. — 1° Nous avons déjà vu que le noyau canonique rend maxima l'intégrale (II.6.3), donc inversement minimalise l'intégrale

$$\int_0^t F^2(t', u)\,du \qquad (0 < t < t').$$

Mais cela ne résout pas le problème de savoir si un noyau donné individuellement est canonique ou non. Dans le cas, que nous considérerons d'abord, de noyaux propres qui soient de vraies fonctions, ce problème est résolu par le théorème suivant :

Théorème II.7.1. — *Si une vraie fonction* $F(t, u)$ *est un noyau propre, la condition nécessaire et suffisante pour que la formule* (II.2.2) *donne dans l'intervalle* (0, T) *la représentation canonique de la fonction* $\Phi(t)$ *qu'elle définit est qu'il n'existe aucune fonction* $x(t)$ *définie dans* (0, T), *qui n'y soit pas constante, et y vérifie les conditions*

$$\int_0^t F(t, u)\,dx(u) = 0, \tag{II.7.1}$$

$$\int_0^t \frac{[dx(u)]^2}{du} < \infty. \tag{II.7.2}$$

Cette dernière condition, compte tenu du théorème I.3, revient à dire qu'il existe une fonction $c\,x(u)$ admissible dans $(0, t)$. Il est essentiel de remarquer qu'elle peut cesser d'être vérifiée pour $t = T$; nous nous plaçons dans l'intervalle (0, T) ouvert à droite.

D'après le théorème II.6.2, il suffit de démontrer que la condition indiquée est nécessaire et suffisante pour que Ω_t et $\Omega_{t,\Phi}$ coïncident. Comme en tout cas $\Omega_{t,\Phi} \subset \Omega_t$, dire qu'ils ne coïncident pas, c'est dire qu'on peut trouver dans Ω_t une direction Ox perpendiculaire à $\Omega_{t,\Phi}$. Dans ce cas $x(u)$, composante de $X(u)$ suivant cette direction, est une fonction admissible dans $(0, t)$, n'y est pas constante, et n'intervient pas dans le calcul de la fonction Φ dans cet intervalle, puisque $\Phi(u)$ y est toujours contenu dans $\Omega_{t,\Phi}$. Elle vérifie donc l'équation (II.7.1) pour tout $t < T$. Les conditions indiquées dans l'énoncé pour la fonction $x(t)$ sont ainsi vérifiées. L'existence d'une telle fonction est bien nécessaire pour que le noyau $F(t, u)$ *ne soit pas* canonique.

Inversement, supposons qu'une telle fonction $x(u)$ existe; alors, pour chaque t fixe, la condition (II.7.2) exprime qu'on peut placer l'axe des x de manière que la composante de $X(u)$ suivant cet axe soit de la forme $c\,x(u)$. La condition (II.7.1) exprime que $\Phi(t)$ est toujours perpendiculaire à cet axe, qui est ainsi contenu dans Ω_t, mais non dans $\Omega_{t,\Phi}$, et, d'après le théorème II.6.2, le noyau n'est pas canonique, C. Q. F. D.

2° L'extension au cas du noyau impropre ne présente aucune difficulté. Il faut en tout cas que l'ensemble E_1 soit vide, et, les ξ_u d'indices $u \in E_2$ n'intervenant pas dans $\Phi(t)$, il n'y a qu'à en faire abstraction, en remplaçant $X(t)$ par

$$X_0(t) = \int_{E_0 \cap (0,\, t)} dX(u),$$

et Ω_t par Ω_{t, X_0}. On peut exprimer le même résultat en disant que, $\Omega_{t,\Phi}$ ne croissant pas dans E_2, on peut prendre comme nouvelle variable la mesure t_0 de $E_0 \cap (0, t)$ et appliquer le théorème précédent avec cette variable. Si le noyau est une vraie fonction, E_0 a une mesure positive, et il n'y a aucune difficulté

Il n'est pas aussi simple de supprimer la restriction que le noyau soit une vraie fonction. Remarquons simplement que, si E_1 est vide (condition en tout cas nécessaire), l'existence d'un ξ_u instantané et à effet immédiat ne peut créer aucune difficulté. Il appartient à $\Omega_{t,\Phi}$ dès que $t > u$, et $F(t', u)\,\xi_u$ appartiendra toujours à $\Omega_{t,\Phi}$. Un tel terme ne peut pas empêcher le noyau d'être canonique. L'extension au cas où la σ-fonction $F(t, u)$ a une partie singulière semble plus difficile.

3° Corollaire II.7.1. — *Si l'équation* (II.7.1) *n'admet aucune solution non constante, le noyau* $F(t, u)$ *est canonique.*

Ce corollaire, évident si le noyau est une vraie fonction, subsiste en tout cas. Son hypothèse implique en effet que $\Phi(u)$ ne soit identiquement nul en projection sur aucune droite de Ω_t. Donc Ω_t ne contient aucune droite perpendiculaire à $\Omega_{t,\Phi}$, C. Q. F. D.

Remarquons maintenant que, si $\dfrac{\partial F(t, u)}{\partial u}$ existe et est borné dans (0, T), et si l'on suppose $x(0) = 0$, l'équation (II.7.1) se réduit à l'équation de Volterra.

$$F(t, t)\,x(t) = \int_0^t \frac{\partial F(t, u)}{\partial u}\,x(u)\,du. \tag{II.7.3}$$

Des propriétés connues de cette équation résulte alors que :

COROLLAIRE II.7.2. — *Si* $F(t, t)$ *est continu et ne s'annule pas dans* $[0, T)$, *et si la dérivée* $\frac{\partial F(t, u)}{\partial u}$ *existe et est bornée dans cet intervalle, le noyau* $F(t, u)$ *y est canonique.*

En particulier, les conditions de continuité étant supposées remplies, si $F(0, 0) \neq 0$, le noyau est canonique au moins dans un petit intervalle $(0, T)$. En dehors du cas où la fonction $F(t, u)$ admet des singularités, le cas singulier est celui où $F(0, 0) = 0$. Nous allons étudier un exemple de ce cas.

4° *Exemple. Le cas d'un polynome homogène.* — Soit le noyau

$$F(t, u) = \sum_{0}^{p} a_h t^{p-h} u^h. \tag{II.7.4}$$

L'équation (II.7.1) est alors d'un type bien connu, réductible par des dérivations successives à une équation différentielle linéaire ; le caractère d'homogénéité se conservant, cette équation est du type d'Euler. On sait qu'on l'intègre en cherchant des solutions de la forme $x(u) = u^s$. Portant donc directement cette expression de $x(u)$ dans l'équation

$$\int_0^t \sum_{0}^{p} a_h t^{p-h} u^h x'(u)\, du = 0, \tag{II.7.5}$$

nous obtenons l'équation caractéristique

$$\sum_{0}^{p} \frac{a_h}{s+h} = 0. \tag{II.7.6}$$

La solution générale de l'équation (II.7.4) est alors une combinaison linéaire d'expressions de la forme u^s, ou éventuellement $u^s(\log^k u)(k > 0)$, s étant racine de l'équation (II.7.6). Si $s = \sigma + i\tau$, on obtient les solutions réelles en séparant $u^\sigma \cos(\tau \log u)$ et $u^\sigma \sin(\tau \log u)$. De toute façon, la condition (II.7.2) est vérifiée seulement si $\mathcal{R}(s) > \frac{1}{2}$ (*cf.* corollaire I.4.3). Si cette condition n'est vérifiée pour aucune des racines de l'équation (II.7.6), aucune solution non nulle de l'équation (II.7.5) ne vérifie la condition II.7.2. Donc :

THÉORÈME II.7.2. — *La condition nécessaire et suffisante pour que le noyau* (II.7.4) *soit canonique est que toutes les racines de l'équation* (II.7.6) *aient leurs parties réelles* $\leq \frac{1}{2}$.

Ainsi, pour $p = 1$, $a_0 + a_1 = 1$ [condition que, sauf pour le noyau canonique $c(t-u)$, on réalise en divisant par $a_0 + a_1$], on trouve $s = -a_0$, et la condition cherchée est $a_0 \geq -\frac{1}{2}$ [cf. *loc. cit.* ([4]) et ([10])].

Remarquons d'autre part que la covariance $\Gamma(t, t')$ liée au noyau (II.7.4) est, si $t' > t$,

$$\int_0^t \sum_0^p a_h t'^{p-h} u^h F(t, u)\, du = \sum_0^p c_h t'^{p-h} t^{p+h+1}, \tag{II.7.7}$$

avec

$$c_h = a_h \sum_0^p \frac{a_k}{h+k+1}. \tag{II.7.8}$$

La détermination des coefficients a_h quand on connaît la covariance dépend ainsi d'un système d'équations algébriques d'ordre 2^{p+1}. Les noyaux $F(t, u)$ et $-F(t, u)$ n'étant pas distincts, il y a au plus 2^p solutions distinctes. La réduction du système (II.7.8) à un système d'ordre 2^p est d'ailleurs facile. On a en effet évidemment

$$\delta\Phi(t) = dt \int_0^t \frac{\partial F(t, u)}{\partial t}\, dX(u) + F(t, t)\, \delta X(t)$$

$$= S t^p \xi_t \sqrt{dt} + O(dt) \qquad \left(S = \sum_0^p a_h\right),$$

$$S^2 = \lim_{dt \searrow 0} \frac{[\delta\Phi(t)]^2}{t^{2p}\, dt}.$$

Si l'on choisit pour S la valeur positive ou nulle, on est ramené à un système d'ordre 2^p.

Pour $p = 1$, $S = 1$, on obtient ainsi deux solutions, liées par la relation involutive $a_0' + a_0'' + 1 = 0$. Si elles sont réelles, le noyau canonique est celui qui correspond à la plus grande valeur de a_0. Dans le cas contraire, c'est que la fonction donnée n'est pas une covariance [cf. *loc. cit.* ([10])].

Si $p > 1$, et qu'il s'agisse d'une covariance effectivement obtenue en partant d'un noyau réel, il y a au moins deux solutions réelles, exceptionnellement confondues. Nous ne savons pas s'il peut y en avoir plus. Mais, comme nous le verrons au n° II.8, il y a une infinité d'autres noyaux propres, non canoniques, qui sans être de la forme (II.7.4) correspondent à la covariance (II.7.7).

Les solutions imaginaires n'ont ici aucune signification, puisque nous nous sommes placés dans un espace de Hilbert réel. D'ailleurs, dans le cas complexe, il faudrait, dans les équations (II.7.8), remplacer $a_h a_k$ par le produit hermitien $a_h \bar{a}_k$.

II.8. Les noyaux de Goursat. — 1° Ce sont les noyaux de la forme

$$F(t, u) = \sum_1^p f_h(t)\, \varphi_h(u). \tag{II.8.1}$$

L'ordre de ce noyau est égal au nombre p des termes, à condition qu'il ne soit pas possible de le représenter par une somme analogue avec moins de p termes.

La covariance liée à un tel noyau est

$$\text{(II.8.2)} \quad \begin{cases} \Gamma(t_1, t_2) = \sum_1^p f_h(t') \int_0^t \varphi_h(u)\, F(t, u)\, du = \sum_1^p f_h(t')\, g_h(t) \\ [t = \text{Min}(t_1, t_2),\ t' = \text{Max}(t_1, t_2)], \end{cases}$$

l'expression de $g_h(t)$ pouvant s'écrire

$$\text{(II.8.3)} \quad g_h(t) = \sum_1^p f_k(t)\, A_{h,k}(t) \qquad \left[A_{h,k}(t) = \int_0^t \varphi_h(u)\, \varphi_k(u)\, du\right].$$

Certains des $g_h(t)$ pouvant être identiquement nuls, cette formule démontre seulement la première partie du théorème suivant :

Théorème II.8.1. — 1° *Si* F(t, u) *est un noyau de Goursat d'ordre* p, *la covariance qui lui correspond, considérée comme fonction de* t *et* t', *est un noyau de Goursat d'ordre* $q \leq p$. 2° *Si de plus le noyau* F(t, u) *est canonique*, $q = p$; *s'il n'est pas canonique, les deux cas* $q = p$ *et* $q < p$ *sont possibles.*

Pour la démonstration de la deuxième partie, je renverrai au théorème 4.7 de mon Mémoire de Berkeley [*loc. cit.* ([4])], et aux exemples considérés dans le chapitre 3 dudit Mémoire. Indiquons aussi sans démonstration le théorème suivant, qui est la traduction géométrique du théorème 4.8 du même Mémoire.

Théorème II.8.2. — *Pour que l'intersection de* $\Omega_{t,\Phi}$ *et du plan* $\Omega_{(t,\infty)\Phi}$ *de l'arc* (t, ∞) *de la ligne* L *se réduise quel que soit* $t > 0$ *à un plan euclidien à* p *dimensions, il faut et il suffit que la covariance, considérée comme fonction de* t *et* t', *soit un noyau de Goursat d'ordre* p.

Ce théorème s'applique en particulier (et devient trivial) dans le cas d'une fonction $\Phi(t)$ solution d'une équation différentielle linéaire et homogène d'ordre $p - 1$ dont les coefficients sont des fonctions scalaires de t. En dehors de ce cas, $\Phi(t)$ est au plus $p - 1$ fois dérivable, et, s'il l'est, l'intersection de $\Omega_{t,\Phi}$ et $\Omega_{(t,\infty),\Phi}$ est le plan défini par le vecteur $\Phi(t)$ et ses $p - 1$ dérivées.

2° Le théorème précédent nous conduit à une méthode naturelle de recherche du noyau canonique lié à une covariance donnée de la forme (II.8.2). Le noyau cherché est de la forme (II.8.1), les fonctions $f_h(t)$ étant connues. On peut alors obtenir toutes les solutions de cette forme et d'ordre p, canoniques ou non, en intégrant le système d'équations intégrales (II.8.3), système qu'on peut, par des dérivations suivies de l'élimination des fonctions $A_{h,k}(t)$, de manière à ne conserver que les dérivées

$$A'_{h,k}(t) = \varphi_h(t)\, \varphi_k(t)$$

et des dérivées d'ordres plus élevés, ramener à un système d'équations différentielles, complété par des conditions initiales déduites des équations contenant les $A_{h,k}(t)$.

On n'est naturellement pas sûr d'obtenir ainsi tous les noyaux propres liés à la covariance donnée. S'il en existe qui soient d'ordre $p'=p+r>p$, on peut chercher à les obtenir en se donnant r autres fonctions $f_h(t)$, et les associant à des fonctions $g_h(t)$ identiquement nulles. On déduira toujours les $\varphi_h(t)$ des équations (II.8.3) où les indices h et k varieront maintenant de 1 à p'.

Reprenons à ce point de vue le problème traité au n° II.7.4°, et ajoutons au noyau (II.7.4) les termes $ct^{p-\alpha}u^{\alpha}$ et $c't^{p-\beta}u^{\beta}$. Le système (II.7.8) devient

$$\text{(II.8.4)} \qquad c_h = a_h\left[\sum_0^p \frac{a_k}{h+k+1} + \frac{c}{h+\alpha+1} + \frac{c'}{h+\beta+1}\right] \qquad (h=0,1,\ldots,p),$$

$$\text{(II.8.5)} \qquad \sum_0^p \frac{a_k}{k+\alpha+1} + \frac{c}{2\alpha+1} + \frac{c'}{\alpha+\beta+1} = \sum_0^p \frac{a_k}{k+\beta+1} + \frac{c}{\alpha+\beta+1} + \frac{c'}{2\beta+1} = 0.$$

Il est bien évident que, si $p>0$, on peut déterminer les a_k de manière à vérifier les équations (II.8.5). En prenant ensuite les c_h définis par les formules (II.8.4), on obtient une covariance de la forme (II.8.2), à laquelle correspond, en dehors des noyaux de la forme (II.8.1), au moins un noyau qui contient des termes à exposants non entiers.

Nous avons ajouté deux termes. Nous aurions pu en ajouter un nombre quelconque $\leq p+1$. Comme p n'est pas borné : *la différence entre l'ordre d'un noyau de Goursat et celui de la covariance correspondante n'est pas bornée supérieurement.*

Remarquons qu'en ajoutant ainsi, à un noyau de la forme (II.7.4), $p+1$ termes dont chacun contient deux paramètres c et α, et déterminant les a_h par les équations (II.8.5) convenablement généralisées, on obtient des noyaux dépendant effectivement de $2p+2$ paramètres, tandis que les covariances correspondantes sont de la forme (II.7.7) et ne dépendent que de $p+1$ paramètres. Donc : *il existe des covariances de la forme* (II.7.7) *à chacune desquelles correspondent des noyaux dépendant de* $p+1$ *paramètres.* Il semble même que ce soit le cas général, pour celles des fonctions de cette forme qui sont effectivement des covariances ([13]).

([13]) Un calcul analogue au précédent, qui sera développé ailleurs, montre que, quels que soient l'entier p et les exposants $\alpha_h > \frac{1}{2}$, la fonction $X(t)$ admet des représentations de la forme

$$\int_0^t \left[c_0 + \sum_1^p c_h\left(\frac{u}{t}\right)^{\alpha_h}\right]\xi_u\sqrt{du}.$$

Ainsi une même fonction peut avoir des représentations propres dépendant d'un nombre arbitrairement grand de paramètres (note rajoutée à la correction des épreuves).

III. — Intégration de l'équation (E).

III.1. Remarques préliminaires. — 1° Nous avons déjà vu (corollaire II.7.2) le rôle de la fonction $\sigma(t) = |F(t, t)|$: si le noyau est continu et qu'elle ne s'annule pas, le noyau est canonique.

Il est important d'observer qu'elle ne dépend que des propriétés intrinsèques de $\Phi(t)$, du moins si nous admettons la continuité du noyau et l'existence et la continuité de la dérivée première $\frac{\partial F(t, u)}{\partial t} = f(t, u)$. On a en effet dans ces conditions.

$$\delta \Phi(t) = dt \int_0^t f(t, u) \xi_u \sqrt{du} + F(t, t) \xi_t \sqrt{dt} \qquad (dt > 0), \tag{III.1.1}$$

et par suite

$$\sigma^2(t) = F^2(t, t) = \frac{[\delta \Phi(t)]^2}{dt}. \tag{III.1.2}$$

Il est d'ailleurs facile de calculer $\sigma^2(t)$ en fonction de la covariance. On a en effet, pour $t' \geqq t$,

$$\frac{\partial \Gamma(t, t')}{\partial t} = F(t, t) F(t', t) + \int_0^t f(t, u) F(t', u)\, du,$$

$$\frac{\partial \Gamma(t, t')}{\partial t'} = \int_0^t F(t, u) f(t', u)\, du,$$

et par suite

$$\left[\frac{\partial \Gamma(t, t')}{\partial t} - \frac{\partial \Gamma(t', t)}{\partial t'}\right]_{t'=t} = F^2(t, t) = \sigma^2(t). \tag{III.1.3}$$

En raison de la propriété de symétrie de $\Gamma(t_1, t_2)$, cela revient à dire que $\frac{\partial \Gamma(t_1, t_2)}{\partial t_1}$ augmente brusquement de $\sigma^2(t)$ quand le point (t_1, t_2) traverse la droite $t_1 = t_2$ au point t, $t_2 - t_1$ devenant positif à ce moment.

Dans un travail antérieur, nous avions indiqué la formule équivalente

$$\sigma^2(t) = \lim_{\tau \to 0} \frac{1}{|\tau|} [\Gamma(t, t) - 2\Gamma(t, t+\tau) + \Gamma(t+\tau, t+\tau)]. \tag{III.1.4}$$

2° Hypothèses relatives a la covariance donnée $\Gamma(t, t_2)$. — Il s'agit bien entendu d'une fonction réelle et symétrique. Puisque nous supposons le noyau continu et $\Phi(0) = 0$, nous devons supposer la covariance continue et s'annulant avec $t_1 t_2$. Nous supposerons d'autre part l'existence et la continuité, sauf à la traversée de la droite $t_1 = t_2$, des dérivées premières et de la dérivée seconde

$$\gamma(t_1, t_2) = \frac{\partial^2 \Gamma(t_1, t_2)}{\partial t_1 \partial t_2}. \tag{III.1.5}$$

A la traversée de cette droite, la seule discontinuité est alors le changement de signe de la dérivée normale.

Ces conditions suffisent pour que l'une ou l'autre des formules (III.1.3) et (III.1.4) donnent pour $\sigma^2(t)$ une valeur bien définie, mais peut-être négative. Elles sont nécessaires pour la continuité de $F(t, u)$ et de sa dérivée $f(t, u)$.

Introduisons maintenant la forme quadratique en $\varphi(t)$

$$Q = \int_0^\infty \sigma^2(t)\,\varphi^2(t)\,dt + \int_0^\infty \int_0^\infty \gamma(t, u)\,\varphi(t)\,\varphi(u)\,dt\,du, \tag{III.1.6}$$

où $\varphi(t)$ est une σ-fonction, nulle en dehors d'un intervalle que nous appellerons son support, et de carré sommable dans cet intervalle. La *condition de Loève*, nécessaire et suffisante pour que $\Gamma(t_1, t_2)$ soit une covariance, revient à dire que cette forme est non négative. Nous introduirons une condition un peu plus restrictive en la supposant strictement positive.

Lorsque toutes ces conditions sont réalisées, nous dirons qu'on est dans le *cas régulier*. Dans ce cas $\sigma^2(t)$ est toujours positif. En prenant en effet pour $\varphi^2(t)$ la fonction de Dirac $\delta(t - t_0)$, on trouve

$$Q = \sigma^2(t_0) > 0,$$

et, si les intégrales qui définissent Q sont prises de -0 à $+\infty$, cette conclusion subsiste pour $t_0 = 0$.

3° Conséquences de l'hypothèse $\sigma(t) > 0$. — Nous avons vu qu'un noyau continu vérifiant cette condition est nécessairement canonique. Il ne peut alors exister que deux noyaux continus, égaux et de signes contraires, qui soient solutions de l'équation (E). Nous dirons que les noyaux continus ainsi obtenus dans le cas régulier sont des *noyaux réguliers*.

Il y a naturellement une infinité d'autres solutions, déduites des noyaux réguliers par les opérations définies au n° II.5; mais elles sont toutes discontinues.

III.2. La première étape de l'intégration. — Cette première étape est la recherche d'une équation de la forme

$$\delta\Phi(t) = dt \int_0^t G(t, u)\,d\Phi(u) + \xi_t\sigma(t)\sqrt{dt} \qquad (dt > 0) \tag{III.2.1}$$

pour définir la variation de $\Phi(t)$. Nous supposons toujours ξ_t normal au plan $\Omega_{t,\Phi}$, de sorte que le premier terme, comme celui de la formule (III.1.1) dans le cas d'un noyau canonique, est la projection de $\delta\Phi(t)$ sur ce plan. De toute façon $\sigma^2(t)$ s'identifie avec la fonction définie par la formule (III.1.4) (puisque $\sigma^2(t)\,dt$ est la partie principale de $[\delta\Phi(t)]^2$), et que nous supposons toujours positive, et, en choisissant convenablement le sens positif sur l'axe des ξ_t, on peut encore supposer $\sigma(t) > 0$.

Il reste à déterminer $G(t, u)$. Nous allons dans ce but former une équation intégrale qui lie $\gamma(t, u)$ et $G(t, u)$. On l'obtient en multipliant les deux membres de l'équation (III.2.1) par $\delta\Phi(x)(0 < x < t)$. Dans l'intégrale, il faut distinguer des autres l'élément $G(t, x)\delta\Phi(x)$ dont le produit par $\delta\Phi(x)$ est

$$G(t, x)[\delta\Phi(x)]^2 = G(t, x)\sigma^2(x)\,dx + o(dx).$$

Comme d'autre part

$$\delta\Phi(t)\,\delta\Phi(x) = \gamma(t, x)\,dt\,dx,$$

et que $\delta\Phi(x)$, contenu dans $\Omega_{t,\Phi}$, est orthogonal à ξ_t, en divisant par $dt\,dx$, on obtient l'équation

$$\text{(III.2.2)} \qquad \gamma(t, x) = G(t, x)\sigma^2(x) + \int_0^t G(t, u)\gamma(u, x)\,du \qquad (0 < x < t).$$

La fonction γ étant définie par la formule (III.I.5), il s'agit de déduire $G(t, x)$ de cette équation. Pour chaque t fixe, c'est une équation de Fredholm, et son déterminant n'est pas nul. Si en effet l'équation sans second membre

$$\sigma^2(x)\varphi(x) + \int_0^t \gamma(u, x)\varphi(u)\,du = 0$$

avait une solution non identiquement nulle dans $(0, t)$, en multipliant par $\varphi(x)$ et intégrant dans $(0, t)$, on aurait un résultat en contradiction avec l'hypothèse $Q > 0$. La détermination de $G(t, x)$ dépend donc de la résolution d'une équation de Fredholm, de noyau $\frac{\gamma(u, x)}{\sigma^2(x)}$, dont le déterminant n'est pas nul. Le problème se simplifie dans le cas déjà considéré où $G(t, u)$, donc aussi $\frac{\gamma(u, x)}{\sigma^2(x)}$, est un noyau de Goursat.

Pour le problème inverse, qui consiste à déduire $\gamma(t, x)$ et $\Gamma(t, x)$ de l'équation (III.2.2), complétée par la propriété de symétrie de ces fonctions, la méthode indiquée dans notre Mémoire de Berkeley de 1950 [*loc. cit.* (1)], consiste à intégrer l'équation (III.2.1) comme nous allons le faire, et en déduire la covariance $\Gamma(t, x)$.

III.3. Intégration de l'équation (III.2.1). — Si l'on pose

$$\text{(III.3.1)} \qquad Y(t) = \int_0^t \xi_u\sigma(u)\sqrt{du} = \Phi(t) - V(t),$$

$$dV(t) = V'(t)\,dt \quad \text{et} \quad \delta Y(t) = \xi_t\sigma(t)\sqrt{dt}$$

sont respectivement égaux aux deux termes de l'expression (III.2.1) de $\Phi(t)$. Donc

$$\text{(III.3.2)} \qquad V'(t) = \int_0^t G(t, u)\,d\Phi(u),$$

et $V'(t) \in \Omega_{t,\Phi}$, d'où, pour $0 < u \leqq t$, $V'(u) \in \Omega_{u,\Phi} \subset \Omega_{t,\Phi}$ et $V(t) \in \Omega_{t,\Phi}$. Cette condition étant réalisée, dans le cas du noyau canonique, par la différence

$$\delta \Phi(t) - \xi_t F(t, t) \sqrt{dt} = dt \int_0^t F(t, u)\, dX(u),$$

nous sommes assurés que c'est bien au noyau canonique que nous conduit l'équation (III.2.1). Or, en y remplaçant $\Phi(u)$ par $Y(u) + V(u)$ et posant

$$Z(t) = \int_0^t G(t, u)\, dY(u), \tag{III.3.3}$$

on obtient l'équation de Volterra

$$V'(t) - \int_0^t G(t, u)\, V'(u)\, du = Z(t). \tag{III.3.4}$$

On sait qu'elle se résout par la formule

$$V'(t) = Z(t) - \int_0^t R(t, u)\, Z(u)\, du, \tag{III.3.4}$$

où $R(t, u)$ est le noyau résolvant, défini lui-même par l'équation

$$G(t, u) + R(t, u) = \int_u^t G(t, v)\, R(v, u)\, dv = \int_u^t R(t, v)\, G(v, u)\, dv. \tag{III.3.5}$$

En substituant alors à $Z(t)$ son expression (III.3.3), et tenant compte de l'équation (III.3.5) pour simplifier l'expression obtenue, il vient

$$V'(t) = -\int_0^t R(t, u)\, dY(u) = -\int_0^t R(t, u)\, \xi_u \sigma(u) \sqrt{du}, \tag{III.3.6}$$

d'où enfin, en intégrant, et tenant compte de la formule (III.3.1)

$$\Phi(t) = \int_0^t \xi_u \sigma(u) \sqrt{du} \left[1 - \int_u^t R(v, u)\, dv \right]. \tag{III.3.7}$$

Telle est la solution de l'équation (III.2.1), et : *le noyau canonique qui correspond à la covariance* $\Gamma(t_1, t_2)$ *donnée est*

$$F(t, u) = \sigma(u) \left[1 - \int_u^t R(v, u)\, dv \right]. \tag{III.3.8}$$

Il a été obtenu, en partant de cette covariance par la formule (III.1.4), *suivie de la résolution de l'équation de Fredholm* (III.2.2), *du calcul du noyau résolvant* $R(t, u)$ *de l'équation de Volterra* (III.3.3), *et enfin par la formule* (III.3.8).

Compte tenu des hypothèses indiquées au n° III.1.2°, on obtient ainsi un noyau régulier qui, au signe près, est la seule solution continue de l'équation (E) ([14]).

III.4. CALCUL DE VÉRIFICATION. — La fonction $\Gamma(t_1, t_2)$ s'annulant sur les axes, et ses dérivées premières ayant à la traversée de la droite $t_1 = t_2$ une discontinuité définie par $\sigma^2(t)$, elle est bien définie (pour t_1 et t_2 positifs) par sa dérivée seconde $\gamma(t_1, t_2)$. Il suffit donc de vérifier que la fonction $F(t, u)$ que nous venons d'obtenir, qui vérifie évidemment la condition $F(t, t) = \sigma(t)$, vérifie aussi l'équation

$$\gamma(t, u) = \sigma(u) f(t, u) + \int_0^u f(t, v) f(u, v)\, dv \qquad \left[0 < u < t,\ f(t, u) = \frac{\partial F(t, u)}{\partial t} \right],$$

déduite de l'équation (E) par deux dérivations. De l'expression (III.3.8) de $F(t, u)$ on déduit d'ailleurs

$$f(t, u) = -\sigma(u) R(t, u),$$

de sorte que l'équation à vérifier prend la forme

$$\text{(III.4.1)} \qquad \gamma(t, u) = -\sigma^2(u) R(t, u) + \int_0^u R(t, v) R(u, v) \sigma^2(v)\, dv \qquad (0 < u < t).$$

Il suffit à cet effet, en admettant provisoirement que l'équation (III.2.2), considérée comme équation en $\gamma(t, x)$, n'admet qu'une solution symétrique en t et x, de vérifier que les expressions de $\gamma(t, x)$ et $\gamma(u, x)$ déduites de la formule (III.4.1), vérifient bien l'équation (III.2.2). Il faut bien entendu, dans le dernier terme de cette équation, séparer les intervalles $(0, x)$ et (x, t). L'équation à vérifier s'écrit ainsi

$$\begin{aligned} -\sigma^2(x) R(t, x) + \int_0^x R(t, v) R(x, v) \sigma^2(v)\, dv - \sigma^2(x) G(t, x) \\ = -\int_0^x G(t, v) R(x, v) \sigma^2(v)\, dv - \sigma^2(x) \int_0^t G(t, u) R(u, x)\, du \\ + \int_0^x G(t, u)\, du \int_0^u R(x, v) R(u, v) \sigma^2(v)\, dv \\ + \int_x^t G(t, u)\, du \int_0^x R(x, v) R(u, v) \sigma^2(v)\, dv. \end{aligned}$$

Cette équation est de la forme

$$A \sigma^2(x) + \int_0^x B R(x, v) \sigma^2(v)\, dv = 0,$$

A et B étant bien nuls, d'après l'équation (III.3.5).

([14]) Les remarques finales du n° II.5.2° prouvent que le théorème d'unicité ne s'applique qu'aux noyaux continus. Non seulement il y a une classe très générale de noyaux impropres équivalents à ce noyau, mais, dans un cas aussi simple que celui de $X(t)$, il y a d'autres noyaux propres. Ma Note du 5 mars 1956 [*loc. cit.* ([2])] contenait à ce sujet une phrase inexacte (fin du n° 1).

Il reste à montrer que la fonction $\gamma(t, x)$ est bien définie, quand $G(t, x)$ et $\sigma^2(x)$ sont connues, par l'équation (III.2.2) et sa propriété de symétrie. Or cette équation et cette propriété expriment que γ est la dérivée seconde de la covariance d'une fonction $\Phi(t)$ solution de l'équation (III.2.1) qui s'annule avec t. Il suffit donc de montrer que ces dernières conditions déterminent parfaitement $\Phi(t)$. Cela résulte du calcul très simple qui a réduit l'équation (III.2.1) à l'équation de Volterra (III.3.4), qui n'a qu'une solution.

III.4. Exemples de cas singuliers. — 1° Nous désignons ainsi d'une manière générale les cas où les hypothèses du n° III.1.2° ne sont pas vérifiées, même s'il ne s'agit pas d'une discontinuité du noyau ou de la covariance. Il s'agira surtout de cas où $\sigma(t)$ est nul, en certains points ou même identiquement, ce qui rend au contraire possible la continuité, à la traversée de la droite $t_1 = t_2$, des dérivées premières de la covariance, et aussi celle du noyau à la traversée de la droite $t = u$, si on le suppose défini et nul pour $u > t$. Cette dernière remarque montre que l'hypothèse $\sigma(t) = 0$ permet l'existence de noyaux impropres et cependant continus, déduits du noyau propre par un changement de variable sur u (opération c du n° II.5).

L'hypothèse $\sigma(t) = 0$ implique l'existence d'une σ-fonction $\varphi(t)$ non identiquement nulle qui annule la forme quadratique Q du n° III.1.2°; cela est vrai même si $\sigma(t)$ ne s'annule qu'en un point t_0; en effet, dans ce cas, Q est nul pour $\varphi(t) = \sqrt{\delta(t - t_0)}$. Montrons par un exemple que la réciproque n'est pas vraie.

C'est l'exemple connu de la fonction

$$\Phi(t) = X(t) - X(1)\,\mathrm{Min}\,(1, t), \tag{III.4.1}$$

dont la covariance $\Gamma(t, u)$ pour $0 < u < t < 1$, est $u(1-t)$; alors $\gamma(u, t) = -1$, cette dérivée étant nulle pour $t > 1$. La variation de $\Phi(t)$ est, dans $(0, 1)$

$$\delta\Phi(t) = -\Phi(t)\frac{dt}{1-t} + \xi_t\sqrt{dt} = -\frac{dt}{1-t}\int_0^t d\Phi(u) + \xi_t\sqrt{dt}, \tag{III.4.2}$$

tandis que, pour $t \geq 1$, $\delta\Phi(t) = \xi_t\sqrt{dt}$. On a donc toujours $\sigma(t) = 1$. On a d'autre part, quel que soit le coefficient c,

$$c\,dt\,\Phi(1) = c\,dt\int_0^1 d\Phi(u) = 0, \tag{III.4.3}$$

de sorte que, pour $t \geq 1$, en ajoutant cette expression à $\delta\Phi(t)$, on obtient pour cette variation une infinité d'expressions de la forme (III.2.1); il n'y a plus de théorème d'unicité, et c'est lié au fait que les fonctions $\varphi(t)$ de support $(0, 1)$ et constantes dans cet intervalle annulent Q.

On remarque que, si $T' > T$, une fonction de support $(0, T)$ peut être considérée comme ayant pour support $(0, T')$. Il en résulte qu'il existe un nombre T_0 bien défini tel que la forme Q apparaisse comme strictement positive dans $(0, T)$ si $T < T_0$ et non si $T > T_0$. Si $T = T_0$, on ne peut rien affirmer *a priori*. Dans le cas de la fonction définie par la formule (III.4.1), Q n'est pas strictement positif dans $(0, T_0)$. Il l'est au contraire dans le cas d'une fonction définie jusqu'à l'instant T_0 par un noyau canonique propre, le noyau étant ensuite nul dans un intervalle (T_0, T_1) (c'est-à-dire si $T_0 < u < t < T_1$), où par suite $\Omega_{t,\Phi}$ cesse de croître.

2° Nous allons maintenant supposer le noyau de la forme $g(t-u)\,H(t, u)$, $H(t, u)$ étant un noyau régulier, et $g(\tau)$ une fonction qui s'annule avec τ. Nous écartons ainsi les cas où la singularité du noyau ne serait pas la même en tous les points de la ligne $t = u$ [15]. Nous étudierons surtout le cas où $g(\tau) = \tau^{\alpha}$, et mentionnerons brièvement celui où $g(\tau)$ contient en outre des facteurs logarithmiques. En désignant par p la partie entière de α, nous supposerons de plus que $H(t, u)$ est $p+1$ fois dérivable.

Supposons d'abord $g(t-u) = (t-u)^p$, p étant un entier positif. Alors

$$\text{(III.4.4)} \qquad \frac{\partial^p F(t, u)}{\partial t^p} = F_p(t, u)$$

est un noyau régulier qui, pour $t = u$, prend la valeur $p!\,H(t, t) \neq 0$. La covariance correspondante est

$$\text{(III.4.5)} \qquad \Gamma_p(t_1, t_2) = \frac{\partial^{2p} \Gamma(t_1, t_2)}{\partial t_1 \partial t_2}.$$

Si donc $\Gamma(t_1, t_2)$ est connu, cette covariance est connue; on en déduit $F_p(t, u)$ par la méthode indiquée pour le cas régulier, puis $F(t, u)$ par la formule

$$\text{(III.4.6)} \qquad F(t, u) = \int_u^t \frac{(t-v)^{p-1}}{(p-1)!} F_p(v, u)\, dv.$$

On saura d'ailleurs sans difficulté si une covariance donnée se rattache à ce cas. En effet la formule (III.1.3) appliquée à $F_p(t, u)$ donne

$$\text{(III.4.7)} \qquad \frac{\partial^{2p}}{\partial t^p \partial t'^p}\left[\frac{\partial \Gamma(t, t')}{\partial t} - \frac{\partial \Gamma(t, t')}{\partial t'}\right] \to F_p^2(t, t) > 0 \qquad (t' \searrow t),$$

tandis que pour un indice $q < p$ la limite serait nulle.

---

(15) L'exemple du polynome homogène, étudié à la fin du chapitre II, se rattache au cas écarté ici, à moins qu'il ne soit de la forme $c(t-u)^p$.

3° Supposons maintenant $g(\tau)=\tau^\alpha$ (α positif non entier; nous désignerons sa partie entière par p). La méthode précédente subsiste, sans autre changement que le remplacement de la dérivée d'ordre p par la dérivée d'ordre α au sens de Riemann

$$\text{(III.4.8)} \qquad D^\alpha F(t, u)=\frac{\partial}{\partial t}\int_0^t \frac{(t-v)^{p-\alpha}}{(p-\alpha)!}\frac{\partial^p F(v, u)}{\partial v^p}\,du,$$

la notation $(p-\alpha)!$ représentant bien entendu la fonction eulérienne $\Gamma(p-\alpha+1)$; de même, dans la formule (III.4.6), le remplacement de p par α implique l'introduction de la fonction eulérienne.

On traiterait de même le cas où $g(\tau)\sim\tau^\alpha\left(\log\frac{1}{\tau}\right)^\beta$ $(\tau\searrow 0)$, et les cas analogues obtenus en introduisant des logarithmes itérés. Dans tous ces cas, une opération généralisant la dérivation d'ordre fractionnaire de Riemann, que j'ai indiquée autrefois (*Bull. Sc. math.*, 1926), permet de ramener le noyau considéré à un noyau régulier, qui n'est ni infini ni nul pour $t=u$.

Pour reconnaître qu'on est dans ces cas, et déterminer α (et éventuellement β, ...), le plus simple semble être de partir de la formule

$$\text{(III.4.9)} \qquad \Gamma(t+\tau, t+\tau)-2\Gamma(t, t+\tau)+\Gamma(t, t)$$
$$=\int_t^{t+\tau} F^2(t+\tau, u)\,du+\int_0^t [F(t+\tau, u)-F(t, u)]^2 du \qquad (\tau>0).$$

Nous supposerons $p=0$ [si $p>0$, on est ramené à ce cas par des dérivations ordinaires, qu'on continuera tant qu'on trouvera par la formule (III.4.7) des valeurs nulles de $F_p(t, t)$].

Le premier terme au second membre de la formule (III.4.9) est, pour τ infiniment petit, équivalent à $\frac{\tau g^2(\tau)}{2\alpha+1}H^2(t, t)$, donc, en nous bornant au cas où $g(\tau)=\tau^\alpha$, à $\frac{\tau^{2\alpha+1}}{2\alpha+1}H(t, t)$. Pour évaluer la seconde intégrale, en supposant que $\frac{\partial H(t, u)}{\partial t}$ soit borné, nous poserons $u=t-v$. Pour $v\in(0, \tau)$, la fonction intégrée est $O(\tau^{2\alpha})$, et l'intégrale relative à cet intervalle est $O(\tau^{2\alpha+1})$. Comme elle est positive, elle s'ajoute au premier terme sans changer l'ordre de grandeur.

D'autre part, pour $v>\tau$, $g'(v)=\alpha v^{\alpha-1}$ et la valeur moyenne de $g'(.)$ dans l'intervalle $(v, v+\tau)$ sont du même ordre de grandeur. L'ordre de grandeur de l'intégrale relative à l'intervalle (τ, t) est donc, si $2\alpha-1\neq 0$, celui de

$$\tau^2\int_\tau^t v^{2\alpha-2}\,dv=\frac{\tau^2}{2\alpha-1}(t^{2\alpha-1}-\tau^{2\alpha-1}),$$

donc celui de $\tau^{2\alpha+1}$ si $\alpha<\frac{1}{2}$, et celui de τ^2 si $\frac{1}{2}<\alpha<1$. Pour $\alpha=\frac{1}{2}$, c'est celui de $\tau^2\log\frac{1}{\tau}$.

Tout compte fait, le premier membre de l'expression (III.4.9) est de l'ordre de grandeur de $\tau^{2\alpha+1}$ si $0 < \alpha < \frac{1}{2}$, de celui de $\tau^2 \log \frac{1}{\tau}$ si $\alpha = \frac{1}{2}$, et, si $\frac{1}{2} < \alpha < 1$, il est de l'ordre de grandeur de τ^2, mais par soustraction d'un terme de la forme $\tau^2 f(t)$, est réduit à l'ordre de grandeur de $\tau^{2\alpha+1}$. On peut donc ainsi dans tous les cas déduire α de l'expression donnée de $\Gamma(t)$, et appliquer la méthode indiquée.

Ces remarques s'étendent sans peine au cas où il y a des facteurs logarithmiques.

III.5. Remarques finales. — Comme on le voit, nous n'avons pu donner que des exemples de cas singuliers, et nous sommes loin de pouvoir indiquer une méthode générale qui permettrait d'obtenir le noyau canonique toutes les fois qu'il existe. Même pour les noyaux de la forme $g(t-u)\,H(t, u)$, $H(t-u)$ étant un noyau régulier, le cas où $g(\tau)$ tend irrégulièrement vers zéro avec τ, et celui où $g(\tau) = o(\tau^p)$ quel que soit p, échappent à notre méthode.

Nous allons maintenant terminer par quelques remarques sur le cas où, dans un espace Ω donné *a priori*, $\Phi(t)$ est défini par la formule

$$\Phi(t) = \sum_1^\infty \xi_n \varphi_n(t), \tag{III.5.1}$$

les fonctions $\varphi_n(t)$ devant bien entendu vérifier la condition

$$\sum_1^\infty \varphi_n^2(t) < \infty. \tag{III.5.2}$$

En d'autres termes, le point $\Phi(t)$ est défini par ses coordonnées $\varphi_n(t)$. C'est un mode de définition qui s'applique évidemment aux fonctions $\Phi(t)$ les plus générales ([16]).

La covariance d'une telle fonction $\Phi(t)$ est évidemment

$$\Phi(t_1)\Phi(t_2) = \sum_1^\infty \varphi_n(t_1)\,\varphi_n(t_2). \tag{III.5.3}$$

Nous supposerons que Ω est le plan de la ligne L lieu du point $\Phi(t)$, c'est-

---

([16]) Il n'est pas nécessaire de se limiter à une infinité dénombrable de vecteurs orthogonaux ξ_n. La formule (III.5.2) exige seulement que, pour chaque t, il y ait au plus une infinité dénombrable de $\varphi_n(t)$ qui soient différents de zéro.

Compte tenu de cette remarque, on voit que n'importe quelle fonction de la forme (II.2.1) peut être mise sous la forme (III.5.1).

à-dire qu'il n'existe aucune relation de la forme

$$\sum_{1}^{\infty} a_n \varphi_n(t) = c \tag{III.5.4}$$

à coefficients non tous nuls et rendant la série $\sum a_n^2$ convergente [17]. Nous ferons même une hypothèse plus restrictive : la fonction $\Phi(t)$ pouvant n'être définie que dans un certain intervalle (T, T′), aucune relation de la forme (III.5.4) n'est vérifiée dans aucun intervalle (t, t') intérieur à (T, T′). Alors le plan de n'importe quel arc de la ligne L coïncide avec le plan Ω de l'arc (T, T′).

Cette condition est vérifiée en particulier pour les fonctions $\Phi(t)$ qui sont *analytiques*. Nous entendons par là que, non seulement toutes les fonctions $\varphi_n(t)$ sont analytiques, mais qu'il en est de même de toutes les combinaisons linéaires $\sum a_n \varphi_n(t) \left(\sum a_n^2 < \infty\right)$; ainsi le caractère analytique est indépendant des axes choisis dans Ω.

Que la fonction $\Phi(t)$ soit analytique ou non, il résulte de l'absence de relation de la forme (III.5.4) dans n'importe quel intervalle intérieur à (T, T′) que $\Omega_{t,\Phi}$ est invariant dans cet intervalle. Si T est l'origine du processus, tous les ξ_n qui interviennent dans l'expression (III.5.1) de $\Phi(t)$ apparaissent instantanément à l'instant T, ce qui exclut évidemment toute représentation de la forme (II.2.1).

Il n'en est pas de même si T n'est pas l'instant origine. Tout ce qu'on peut dire, c'est que, si le plan de l'arc (T, T′) a plus d'une direction orthogonale à $\Omega_{T,\Phi}$, il n'y a pas de représentation canonique de $\Phi(t)$. Mais il peut arriver que ce plan soit continu dans $\Omega_{T,\Phi}$ ou ne contienne qu'une direction orthogonale à $\Omega_{T,\Phi}$; dans ce cas, si $\Phi(t)$ n'est pas analytique dans (o, T) et a une représentation canonique jusqu'à l'instant T, il y a au plus un nouveau ξ_T à introduire à cet instant, et l'on obtient une représentation canonique impropre valable jusqu'à l'instant T′.

Si $T > -\infty$, et qu'on considère une fonction $\Phi(t)$ définie dans (T, T′) par une formule du type (III.5.1), on peut toujours la prolonger à gauche de manière à réaliser le cas précédent, et obtenir une représentation canonique valable dans (T_0, T') $(T_0 < T)$. Cela est d'ailleurs artificiel. Une formule de la forme (III.5.1) représente dans (T, T′) des fonctions, analytiques ou non, qu'il n'y a en général pas intérêt à représenter par la formule (II.2.1). Ces deux formules, sans être incompatibles, ne sont pas utiles dans les mêmes cas,

---

(17) Si cette série est divergente, la condition (III.5.4) n'a de sens que dans une partie ω de Ω. Avec une topologie plus fine que celle de l'espace de Hilbert, on peut considérer qu'elle définit une variété linéaire fermée dans ω. Mais elle est partout dense dans Ω, et l'on ne peut pas la considérer comme une variété linéaire de cet espace.

et cela explique le fait que l'expression (III.5.1) n'implique en général aucune discontinuité de la covariance à la traversée de la ligne $t_1 = t_2$.

Au point de vue de l'équation (E), il résulte de ces remarques que, si $\Gamma(t, t')$ est donné dans $(0, \infty)$ et est une covariance, il peut n'exister aucune solution. Mais, si cette covariance est donnée seulement dans (t_0, ∞) $(t_0 > 0)$, c'est-à-dire si elle est donnée pour $t_0 < t < t'$, on peut toujours d'une infinité de manières la prolonger dans $(0, t_0)$ de manière à obtenir des solutions de l'équation (E).

Editors' Comments on Paper 4

This paper is an interesting contribution to the study of deterministic random processes. The subject of determinism has not been studied extensively, partly because of the analytical difficulties in establishing conditions on the autocovariance function for a process to be deterministic and partly because of the degenerate nature of such processes.

By placing the problem in a mathematical context, the author follows Wiener's approach and studies the class of analytic processes. He also provides interesting examples of such processes; e.g., the class of what might be called ''conjugate'' wide-sense stationary processes; that is, processes whose autocovariance function depends only on the sum of the two arguments.

This paper is reprinted with the permission of the author and the Society for Industrial and Applied Mathematics.

Reprinted from *Theory of Probability and Its Applications*, 402–409 (1959)

4

ANALYTIC RANDOM PROCESSES

YU. K. BELYAEV

(*Translated by R. A. Silverman*)

1. Below we shall study a rather narrow class of random processes, which it is natural to call analytic processes. Similar processes were studied by Loève [1], who gave some sufficient conditions for the analyticity of the sample functions of a random process, conditions which are not natural. The basic result of this paper is to obtain sufficient conditions for the analyticity of the sample functions of a random process in terms of its second moments. In the case of Gaussian processes these conditions are also necessary. In addition, we examine some problems connected with analytic random processes.

By $\xi(t)$, $t \in T$, we shall denote a real random process, given for the values $t \in T$, which is separable together with all its derivatives, if such exist. As shown by Doob [2], one can always construct a process which has this property and has given n-dimensional distributions. Moreover, we shall assume that the second moments $B(t, s) = \mathbf{M}\xi(t)\xi(s)$, called the covariance of the process, exist. For simplicity we shall henceforth assume that $\mathbf{M}\xi(t) = 0$.

We can also consider n-dimensional random processes $\vec{\xi}(t) = \{\xi_1(t), \cdots, \xi_n(t)\}$, $t \in T$, where we impose the same restrictions on the components $\xi_i(t)$ and assume the existence of the second moments $\mathbf{M}\xi_i(t)\xi_j(t) = B_{ij}(t, s)$. For T one can take either a segment or the whole real line.

Definition. The random process $\vec{\xi}(t)$ is called *analytic* in the region D if almost all sample functions of its components can be analytically continued in the region D. Since a random process $\eta(t)$ which takes complex values can be represented in the form $\eta(t) = \xi_1(t) + i\xi_2(t)$, where $\xi_1(t)$ and $\xi_2(t)$ are real random processes, and since for the analyticity of $\eta(t)$ in the region D it is necessary and sufficient that $\xi_1(t)$ and $\xi_2(t)$ be analytic in the same region, a complex analytic process can be regarded as a two-dimensional analytic random process.

Analytic random processes occur in many cases where certain random processes are approximated by using other simpler random processes. In particular, this is the case for the so-called canonical representation of a process [**3**], when the functions in terms of which the process is expanded are analytic and we consider as an approximation to the process a finite number of terms of the canonical expansion. We now give some examples of analytic random processes.

1°. A polynomial of degree N with random coefficients

$$\xi_N(t) = \sum_{n=0}^{N} \xi_n t^n.$$

2°. A trigonometric polynomial of degree N with random coefficients

$$\xi_N(t) = \sum_{n=0}^{N} (\xi_n \cos nt + \eta_n \sin nt).$$

3°. A process $\xi(t)$, $-\infty < t < +\infty$, with

$$B(t, s) = \sigma^2 e^{-\delta^2(t-s)^2}$$

as its covariance.

4°. A process $\xi(t)$, $-\infty < t < +\infty$, with a covariance of the form

$$B(t, s) = \int_{-w}^{w} e^{i(t-s)\lambda} dF(\lambda).$$

Such processes are called random processes with bounded spectrum.

5°. A process $\xi(t)$, $-\infty < t < +\infty$, whose covariance $B(t, s)$ is a continuous function with the property

$$B(t, s) = B(t+s).$$

While it is obvious in the first two examples that almost all the sample functions are analytic, in the other cases the assertion follows from Theorem 1 proved below.

2. We note that it follows from the analyticity of the random process $\xi(t)$ in the neighborhood of the point t_0 that the sample functions $\xi(t, \omega)$ of the process can be represented as a Taylor series with random coefficients

$$\xi(t, \omega) = \sum_{k=0}^{\infty} a_k(t_0, \omega) \frac{(t-t_0)^k}{k!},$$

where

$$a_k(t_0, \omega) = \lim_{h \downarrow 0} \frac{\Delta_h^{(k)} \xi(t_0, \omega)}{h^k}.$$

We also note that the analyticity of a random process does not in general imply the analyticity of its moments. It is easy to construct examples which confirm this fact. However, the analyticity of a process can sometimes be determined from its moments. The following theorem gives sufficient conditions for the analyticity of a process.

Theorem 1. *If the covariance $B(t, s)$ is an analytic function of two variables in a neighborhood of the point (t_0, t_0), then the roandm process is analytic in a neighborhood of this point.*

For the proof we need the assertion [**1**] to the effect that if the covariance $B(t, s)$ of the process has partial derivatives up to order $2(n+1)$ inclusive in a square $[a, b] \times [a, b]$, then almost all sample functions of the process $\xi(t)$ are differentiable n times on the segment $[a, b]$. In our case, $B(t, s)$ is infinitely differentiable for $|t-t_0| < r$ and $|s-t_0| < r$; consequently, almost all sample functions are infinitely differentiable for $|t-t_0| < r$. Thus

$$\xi^{(n)}(t, \omega) = \frac{d^n \xi(t, \omega)}{dt^n}$$

exists. From the condition that $B(t, s)$ is analytic it follows that

$$B(t, s) = \sum_{k, l=0}^{\infty} \left. \frac{\partial^{k+l} B(t, s)}{\partial t^k \partial s^l} \right|_{t=s=t_0} \cdot \frac{(t-t_0)^k (s-t_0)^l}{k! l!}$$

for

$$|t-t_0| < r, \qquad |s-t_0| < r.$$

Consider the analytic random processes

$$\xi_n(t, \omega) = \sum_{k=0}^{n} \xi^{(k)}(t_0, \omega) \frac{(t-t_0)^k}{k!}.$$

For them we have

$$\mathbf{M}|\xi(t)-\xi_n(t)|^2 = \sum_{k,\,l=n+1}^{\infty} \frac{\partial^{k+l} B(t, s)}{\partial t^k \partial s^l}\bigg|_{t=s=t_0} \cdot \frac{(t-t_0)^{k+l}}{k!\,l!}. \tag{1}$$

Consequently

$$\lim_{n\to\infty} \mathbf{M}|\xi(t)-\xi_n(t)|^2 = 0,$$

since in the right-hand side of (1) there occurs the remainder term of a convergent series. It follows from this that

$$\xi(t) = \underset{n\to\infty}{\text{l.i.m.}}\ \xi_n(t), \tag{2}$$

where l.i.m. denotes the limit in the mean square.

We now show that $\lim_{n\to\infty} \xi_n(t)$ exists with probability one. For this it is sufficient to show that

$$r_n(t) = \sum_{k=n}^{\infty} \xi^{(k)}(t_0) \frac{(t-t_0)^k}{k!} \to 0, \qquad n \to \infty,$$

with probability one. This will be the case if

$$\sum_{k=0}^{\infty} \mathbf{M} r_n^2(t) < \infty.$$

We recall that the convergence of a power series implies its absolute convergence in the circle of radius r, that a power series can be differentiated term by term, and that in doing so its radius of convergence is not changed. We have

$$\mathbf{M}[r_n(t)]^2 = \sum_{k,\,l=n}^{\infty} \frac{\partial^{k+l} B(t, s)}{\partial t^k \partial s^l}\bigg|_{t=s=t_0} \frac{(t-t_0)^{k+l}}{k!\,l!}$$

and

$$\sum_{n=0}^{\infty} \mathbf{M}[r_n(t)]^2 = \sum_{k,\,l=0}^{\infty} [\min(k, l)+1] \frac{\partial^{k+l} B(t, s)}{\partial t^k \partial s^l}\bigg|_{t=s=t_0} \frac{(t-t_0)^{k+l}}{k!\,l!}$$

$$\le \sum_{k,\,l=0}^{\infty} \left|\frac{\partial^{k+l} B(t, s)}{\partial t^k \partial s^l}\right|_{t=s=t_0} \frac{|t-t_0|^{k+l}}{(k-1)!\,(l-1)!} < \infty,$$

where the convergence of the last term follows from the absolute convergence of the series for $\partial^2 B(t, s)/\partial t \partial s$. Thus, $\tilde{\xi}(t) = \lim_{n\to\infty} \xi_n(t)$ exists with probability one, and consequently is an analytic random process. From (2) we find that

$$\xi(t) = \tilde{\xi}(t) \tag{3}$$

with probability one, for any fixed t, $|t-t_0| < r$, and then (3) is also satisfied with probability one for any countable everywhere dense set in the neighborhood $\{t:|t-t_0| < r\}$. Since the sample functions of $\xi(t)$ and $\tilde{\xi}(t)$ are continuous, the equality (3) is also valid with probability one for all t in this neighborhood. This proves the theorem.

The condition stated in Theorem 1 is necessary for Gaussian processes.

Theorem 2. *For a Gaussian process to be analytic in a neighborhood of the point t_0, it is necessary and sufficient that its covariance $B(t, s)$ be an analytic function in a neighborhood of the point (t_0, t_0).*

The sufficiency follows from the preceding theorem. We now prove the necessity. Because of the analyticity

$$\xi(t, \omega) = \sum_{k=0}^{\infty} \xi^{(k)}(t_0, \omega) \frac{(t-t_0)^k}{k!} \tag{4}$$

with probability one, and consequently (4) also converges in probability for fixed t. Since the process is Gaussian, it follows from this that (4) converges in the mean square. Thus we have

$$\sum_{k,l=0}^{\infty} \mathbf{M}\xi^{(k)}(t_0)\xi^{(l)}(t_0)\,\frac{(t-t_0)^{k+l}}{k!\,l!} < \infty.$$

It follows from this that

$$\sum_{k,l=0}^{\infty} \mathbf{M}\xi^{(k)}(t_0)\xi^{(l)}(t_0)\,\frac{(t-t_0)^{k}\cdot(s-t_0)^{l}}{k!\,l!} < \infty,$$

bearing in mind (4) and $\mathbf{M}[r_n(t)]^2 \to 0$ as $n \to \infty$, we obtain

$$B(t, s) = \mathbf{M}\xi(t)\xi(s) = \sum_{k,l=0}^{\infty} \mathbf{M}\xi^{(k)}(t_0)\xi^{(l)}(t_0)\cdot\frac{(t-t_0)^{k}(s-t_0)^{l}}{k!\,l!},$$

i.e., $B(t, s)$ is analytic in a neighborhood of (t_0, t_0). This proves the theorem.

We now consider stationary random processes. In this case the following theorem gives the criterion for the covariance to be analytic.

Theorem 3. *For the covariance*

$$B(\tau) = \int_{-\infty}^{+\infty} e^{i\tau\lambda}dF(\lambda)$$

of a stationary process to be analytic for $|\tau| \leqq r$ *it is necessary and sufficient that*

$$\int_{-\infty}^{+\infty} e^{r\lambda}dF(\lambda) < \infty.$$

The proof is obvious.

For random processes which are stationary in the wide sense the following corollary of Theorem 1 is valid.

Corollary. *If the covariance* $B(\tau)$ *of a stationary random process is analytic for* $|\tau| < r$, *then almost all sample functions are analytic in the strip* $|\mathrm{Im}\, t| < r$, $t = \tau + i\sigma$.

We also note that since in the case of an n-dimensional stationary process we have

$$\sum_{l=1}^{n} B_{ll}(\tau) = \sum_{l=1}^{n} \int_{-\infty}^{+\infty} e^{i\tau\lambda}dF_{ll}(\lambda) = \int_{-\infty}^{+\infty} e^{i\tau\lambda}dF(\lambda) = B(\tau),$$

where $F(\Delta) = \sum_{l=1}^{n} F_{ll}(\Delta)$, it follows from the analyticity of $B(\tau)$ that $B_{ll}(\tau)$ is analytic. Therefore a sufficient condition for the analyticity of an n-dimensional process is the analyticity of the trace of the covariance matrix $\|B_{ij}(\tau)\|$. In particular, it follows from this that a sufficient condition for the analyticity of a complex stationary process $\eta(t) = \xi_1(t) + \xi_2(t)$ is the analyticity of its covariance $B(t-s) = \mathbf{M}\eta(t)\eta(s)$.

We now return to the examples of analytic processes. Because of Theorem 1, for a random process to be analytic it is sufficient that its covariance be analytic. This condition is satisfied in examples 3° and 4°. In example 5°, it follows from the positive definiteness and continuity of $B(\tau)$ and also from the property

$$\Delta_h^{(2n)} B(\tau) = \mathbf{M}\left|\Delta_h^{(n)}\xi\left(\frac{\tau}{2}\right)\right|^2 \geqq 0$$

that $B(\tau)$ is analytic in the strip $z = \tau + i\sigma$, $\mathrm{Re}\, z \in (a, b)$, [4], where (a, b) is the interval in which $B(\tau)$ is continuous. If we assume that the process $\xi(t)$, $-\infty < t < +\infty$, is mean square continuous, then $B(t, s) = B(t+s)$ is everywhere continuous, and consequently is an entire function. Thus, in the last example we have a non-stationary analytic process.

3. We now turn to some problems related to analytic random processes. It follows from the foregoing that if a random process has a covariance which is an entire function, then almost all its sample functions are also entire. It is interesting to estimate the order of growth of the sample functions in terms of the covariance.

We recall that an entire function $f(z)$ is called a function of exponential type with exponent no greater than σ if for any $\varepsilon > 0$

$$|f(z)| \leqq e^{(\sigma+\varepsilon)|z|}$$

for any z with large enough modulus. A necessary and sufficient condition for this is that the inequality

$$\overline{\lim_{n\to\infty}} \sqrt[n]{f^{(n)}(0)} \leqq \sigma$$

hold [5]. This condition holds if and only if

$$|f^{(n)}(0)| \leqq \sigma^n \varphi(n)$$

for all sufficiently large n, where $\varphi(n)$ is such that

$$\overline{\lim_{n\to\infty}} \sqrt[n]{\varphi(n)} \leqq 1.$$

Theorem 4. *If the covariance $B(\tau)$ of a stationary process is an entire function of exponential type with exponent not exceeding σ, then almost all the sample functions are entire functions of exponential type with exponent not exceeding σ.*

PROOF. From the Corollary to Theorem 1 we obtain the fact that almost all sample functions are entire. We now prove the other properties. It is enough for us to show that beginning with some $m_0 = m_0(\omega)$

$$|\xi^{(m)}(0, \omega)| \leqq \sigma^m \psi(m) \tag{5}$$

for almost all ω, where $\psi(m)$ is such that

$$\overline{\lim_{m\to\infty}} \sqrt[m]{\psi(m)} \leqq 1.$$

Since $\mathbf{M}|\xi^{(n)}(0)|^2 = B^{(2n)}(0)$, then because of the hypothesis of the theorem,

$$\mathbf{M}|\xi^{(n)}(0)| \leqq \sigma^{(2n)}\varphi(n)$$

for $n \geqq n_0$, where

$$\overline{\lim_{n\to\infty}} \sqrt[n]{\varphi(n)} \leqq 1.$$

Applying the Chebyshev inequality, we obtain

$$\mathbf{P}\{|\xi^{(n)}(0, \omega)| > \sigma^n \sqrt{\varphi(n)} \cdot n \text{ for at least one } n \geqq n_0\}$$

$$\leqq \sum_{n=n_0}^{\infty} \mathbf{P}\{|\xi^{(n)}(0, \omega)| > \sigma^n \sqrt{\varphi(n)} \cdot n\} \leqq \sum_{n=n_0}^{\infty} \frac{\mathbf{M}|\xi^{(n)}(0)|^2}{\sigma^{2n}\varphi(n) n^2} < \infty.$$

From this we find that for almost all sample functions there exists an $m_0(\omega)$ such that for all $m > m_0(\omega)$ the inequality (5) holds, where $\psi(m) = \sqrt{\varphi(m)} \cdot m$, and consequently

$$\overline{\lim_{m\to\infty}} \sqrt[m]{\psi(m)} \leqq 1.$$

This proves the theorem.

This theorem is applicable to random processes with bounded spectra. We note that questions of this type are intimately connected with a rigorous proof of de la Vallée-Poussin's formula for the sample functions of stationary random processes with bounded spectra.

Theorem 5. *Let $\xi(t)$, $-\infty < t < \infty$, be a process with bounded spectrum whose covariance has the form*

$$B(\tau) = \int_{-\tilde{w}}^{\tilde{w}} e^{i\tau\lambda} dF(\lambda). \tag{6}$$

Then for almost all sample functions the formula

$$\xi(t, \omega) = \sum_{k=-\infty}^{+\infty} \xi\left(\frac{k\pi}{w}, \omega\right) \frac{\sin w\left(t - \frac{k\pi}{w}\right)}{w\left(t - \frac{k\pi}{w}\right)} \tag{7}$$

is valid, where $w > \tilde{w}$ is any fixed number.

This formula is often encountered in the literature (e.g. [6, 7]). However, a rigorous proof of the formula for the sample functions of random processes with bounded spectra has not been given.

It is well known (see e.g. [8]) that

$$e^{it\lambda} = \sum_{k=-\infty}^{+\infty} e^{ik\pi(\lambda/w)} \cdot \frac{\sin w\left(t-\frac{k\pi}{w}\right)}{w\left(t-\frac{k\pi}{w}\right)}$$

for $\lambda < w$. It can be shown that for the remainder of this series

$$R_n(t) = \sum_{|k|>n} e^{ik\pi(\lambda/w)} \frac{\sin w\left(t-\frac{k\pi}{w}\right)}{w\left(t-\frac{k\pi}{w}\right)},$$

the estimate

$$|R_n(t)| < \frac{C(t)}{n(w-\lambda)}, \qquad C(t) = \frac{4w[2\pi+|t|\cdot w]}{\pi^2}$$

is valid. In fact, since

$$e^{ik\pi(\lambda/w)} \frac{\sin w\left(t-\frac{k\pi}{w}\right)}{w\left(t-\frac{k\pi}{w}\right)} + e^{-ik\pi(\lambda/w)} \frac{\sin w\left(t+\frac{k\pi}{w}\right)}{w\left(t+\frac{k\pi}{w}\right)}$$

$$= \frac{\sin wt}{(wt)^2-(k\pi)^2}\left[2wt\cos k\pi\left(1-\frac{\lambda}{w}\right) - i2\pi k \sin k\pi\left(1-\frac{\lambda}{w}\right)\right],$$

we obtain, for any fixed t for large enough n,

$$|\mathrm{Re}\, R_n(t)| < \frac{1}{n}\,\frac{4w|t|}{\pi^2}$$

for the real part of $R_n(t)$. For the imaginary part of $R_n(t)$, using the Abel transformation and the boundedness of

$$\sum_{|k|<n} \sin k\pi\left(1-\frac{\lambda}{w}\right),$$

we obtain

$$|\mathrm{Im}\, R_n(t)| < \frac{1}{n}\cdot\frac{8}{\pi\left(1-\frac{\lambda}{w}\right)}.$$

From this we have

$$s_n(t) = \sup_{-\tilde{w}<\lambda<\tilde{w}} \left| e^{it\lambda} - \sum_{-n}^{n} e^{ik\pi(\lambda/w)} \frac{\sin w\left(t-\frac{k\pi}{w}\right)}{w\left(t-\frac{k\pi}{w}\right)} \right| < \frac{C(t)}{n\left(1-\frac{\tilde{w}}{w}\right)},$$

We now apply this to the proof of (7). It follows from (6) that the process can be represented in the form

$$\xi(t) = \int_{-\tilde{w}}^{\tilde{w}} e^{it\lambda}\Phi(d\lambda),$$

where $\Phi(\Delta)$ is a random measure on the line such that

$$\mathbf{M}\Phi(\Delta_1)\overline{\Phi(\Delta_2)} = F(\Delta_1 \cap \Delta_2).$$

Consider now the process

$$\xi_n(t) = \sum_{k=-n}^{n} \xi\left(\frac{k\pi}{w}\right) \frac{\sin w\left(t-\frac{k\pi}{w}\right)}{w\left(t-\frac{k\pi}{w}\right)} = \int_{-w}^{w}\left[\sum_{-n}^{n} e^{ik\pi(\lambda/w)} \frac{\sin w\cdot\left(t-\frac{k\pi}{w}\right)}{w\left(t-\frac{k\pi}{w}\right)}\right]\Phi(d\lambda).$$

For it we obtain

$$\mathbf{M}|\xi(t)-\xi_n(t)|^2 = \int_{-w}^{w} \left| e^{it\lambda} - \sum_{-n}^{n} e^{ik\pi(\lambda/w)} \frac{\sin w\left(t-\frac{k\pi}{w}\right)}{w\left(t-\frac{k\pi}{w}\right)} \right|^2 dF(\lambda) \tag{8}$$

$$\leqq |s_n(t)|^2 b^2 \leqq \frac{C(t)^2\cdot b^2}{n^2\left(1-\frac{\tilde w}{w}\right)^2}, \qquad (b^2 = \mathbf{M}|\xi(t)|^2).$$

From this it follows at once that (7) is valid in the mean square. We note that it is easy to show that (7) is valid in the mean square for $\tilde w = w$ also. From (8) for

$$\eta_n(t) = \sum_{|k|>n} \xi\left(\frac{k\pi}{w}\right) \frac{\sin w\left(t-\frac{k\pi}{w}\right)}{w\left(t-\frac{k\pi}{w}\right)}$$

we obtain

$$\sum_n \mathbf{M}|\eta_n(t)|^2 \leqq \frac{C(t)^2}{\left(1-\frac{\tilde w}{w}\right)^2} \cdot \sum_n \frac{1}{n^2} < \infty.$$

Thus, the series occurring in the right-hand side of (7) converges with probability one for any fixed t. From this it is easy to show that it converges uniformly if t lies in any bounded interval. If we write $\tilde\xi(t) = \lim_{n\to\infty} \xi_n(t)$, then, since $\xi(t) = \tilde\xi(t)$ with probability one for any fixed t and since the sample functions of $\xi(t)$ and $\tilde\xi(t)$ are continuous, we find that (7) is valid for any t, QED.

We now estimate the rate of convergence of the series in (7).

$$P\{|\eta_n(t)| > f(n) \text{ for at least one } n \geqq n_0\} \leqq \sum_{n\geqq n_0} \frac{\mathbf{M}|\eta_n(t)|^2}{f(n)^2} \leqq \frac{C(t)^2}{\left(1-\frac{\tilde w}{w}\right)^2} \sum_{n\geqq n_0} \frac{1}{n^2 f(n)^2}.$$

Thus, in order that with probability one there exist an $n_0 = n_0(\omega)$ such that $|\eta_n(t,\omega)| < f(n)$ for all $n > n_0(\omega)$, it is sufficient that

$$\sum_{k=1}^{\infty} \frac{1}{n^2 f(n)^2} < \infty.$$

From this we find, for example, that

$$|\eta_n(t,\omega)| < \frac{(\log n)^{(1+\varepsilon)/2}}{\sqrt{n}}$$

for all $n \geqq n_0(\omega)$.

Similarly, we can show that (7) is valid for almost all sample functions and for $\tilde w = w$ in the case of stationary Gaussian processes with bounded spectra, whose spectral functions $F(\lambda)$ have continuous densities.

4. We consider still another special problem. As is easily proved, a necessary and sufficient condition that an analytic process can be represented in the form

$$\xi(t) = \sum_{k=0}^{\infty} \xi_k \cdot t^k,$$

where

$$M\xi_k\xi_l = \sigma_k^2\delta_{kl}, \qquad \delta_{kl} = 1,\ k = l, \quad \delta_{kl} = 0,\ k \neq l,$$

is the requirement that its covariance is a function of the product of its arguments, i.e.

$$B(t,s) = B(t\cdot s).$$

Thus, processes which can be expanded in Taylor series with uncorrelated random coefficients are non-stationary. We shall say that the process $\xi(t)$ with covariance $B(t,s)$ can be normalized to stationary, if the process

$$\eta(t) = \frac{\xi(t)}{\sqrt{B(t,t)}}$$

is stationary in the wide sense.

Theorem 6. *Among the class of analytic random processes which can be expanded in Taylor series with uncorrelated random coefficients, the only process which can be normalized to stationary is the process which has as its covariance*

$$B(t \cdot s) = \sigma^2 e^{-\delta^2 \cdot t \cdot s}. \tag{9}$$

To show this, we study the functional equation

$$\frac{B(t \cdot s)}{\sqrt{B(t^2) B(s^2)}} = f(t-s)$$

and show that the analytic solutions of this equation have the form (9) and that there are no other analytic solutions.

5. Among the other characteristic properties of analytic random functions we note the following. Since we can reconstruct all the values of a sample function of a stationary analytic process from an arbitrarily small part of the sample function, by using the methods of analytic continuation, then for stationary ergodic analytic random processes the spectral function of the process can be determined from an arbitrarily small part of the sample function. This is the case, in particular, for stationary Gaussian processes with bounded spectra whose spectral functions are absolutely continuous.

In conclusion, the author expresses deep gratitude to V. Ya. Kozlov, under whose direction a doctoral dissertation was written concerned with the problems discussed above, and also to A. N. Kolmogorov for suggestions.

REFERENCES

[1] M. Loève, *Fonctions aléatoires du second ordre* (in P. Lévy's book *Processus Stochastiques et Mouvement Brownien*, Gauthier-Villars, Paris, 1948).

[2] J. L. Doob, *Stochastic Processes*, John Wiley, New York, 1953.

[3] V. S. Pugachev, *General correlation theory of random functions*, Izv. Akad. Nauk SSSR, Ser. Matem., 17, 6, 1953.

[4] S. Bernstein, *Sur les fonctions absolument monotones*, Acta Math., 52, pp. 1–66, 1928.

[5] N. I. Akhiezer, *Lectures on Approximation Theory*, Gostekhizdat, Moscow, 1947. T-1

[6] A. A. Kharkevich, *Outline of a General Theory of Communication*, Gostekhteoretizdat, Moscow, 1955.

[7] A. M. Yaglom, *The correlation theory of continuous processes and fields with application to the problem of statistical extrapolation of time series and to the theory of turbulence*, Doctoral Dissertation, 1955.

[8] B. Ya. Levin, *Distribution of the Zeros of Entire Functions*, Gostekhteoretizdat, Moscow, 1956.

ANALYTIC RANDOM PROCESSES

YU. K. BELYAEV *(MOSCOW)*

(Summary)

This paper is devoted to investigating the so-called analytic random processes. Random process $\xi(t)$ is called analytic in a region D if almost all its sample functions are analytic and possess an analytic continuation in the region D. Analyticity of the covariance function $B(t, s) = \mathbf{M}\xi(t)\xi(s)$ in the neighborhood of (t_0, t_0) is a sufficient condition for analyticity of $\xi(t)$ in the neighborhood of t_0. For Gaussian processes, this condition is also necessary. Some other problems connected with analytic processes are also investigated.

Editors' Comments on Paper 5

At the beginning of the last decade, Professor Harald Cramér, a distinguished member of the Royal Swedish Academy of Sciences and widely known for his outstanding contributions in the field of mathematical statistics, obtained the multiplicity representation for second-order random processes with separable linear spans.

This landmark paper offers considerable insight into the linear structure of random processes through various interpretational comments and a concise and complete presentation. Being a communication to a symposium, the paper is brief and focuses on the statement of results. It also juxtaposes the discrete and the continuous case.

This paper was originally published by the University of California Press; it is reprinted with the permission of the author and the Regents of the University of California.

Originally published by the University of California Press; reprinted by permission of the Regents of the University of California
Reprinted from *Proceedings of the Fourth Berkeley Symposium on Statistics and Applied Probability*, **II**, 57–78 (1960)

5 ON SOME CLASSES OF NONSTATIONARY STOCHASTIC PROCESSES

HARALD CRAMÉR
STOCKHOLM

1. Introduction

This paper will be concerned with stochastic processes with finite second-order moments. We start from a given probability space $(\Omega, \mathfrak{F}, \mathfrak{P})$, where Ω is a space of points ω, while $\mathfrak{F}$ is a Borel field of sets in Ω, and $\mathfrak{P}$ is a probability measure defined on sets of $\mathfrak{F}$.

Any $\mathfrak{F}$-measurable complex-valued function $X = x(\omega)$ defined for all $\omega \in \Omega$ will be denoted as a *random variable*. We shall always assume that

$$(1) \qquad \begin{aligned} Ex &= \int_\Omega x(\omega)\, d\mathfrak{P} = 0, \\ E|x|^2 &= \int_\Omega |x(\omega)|^2\, d\mathfrak{P} < \infty. \end{aligned}$$

Two random variables which are equal except on a null set with respect to $\mathfrak{P}$ will be regarded as identical, and equations containing random variables are always to be understood in this sense.

A family of random variables $x(t) = x(t, \omega)$, defined for all t belonging to some given set T, will be called a *stochastic process* defined on T. With respect to T, we shall consider only two cases:

(i) T is the set of all integers $n = 0, \pm 1, \pm 2, \cdots$,

(ii) T is the set of all real numbers t.

With the usual terminology borrowed from the applications, we shall in these cases talk respectively of a stochastic process with *discrete time*, or with *continuous time*. In the first case, where we are concerned with a sequence of random variables, we shall usually write x_n in place of $x(n)$.

With due modifications, the majority of our considerations may be extended to cases where T is some other set of real numbers.

We shall also consider *finite-dimensional vector-valued stochastic processes*, writing

$$(2) \qquad \mathbf{x}(t) = \{x^{(1)}(t), x^{(2)}(t), \cdots, x^{(q)}(t)\},$$

where $\mathbf{x}(t)$ is a q-dimensional column vector, while the components $x^{(1)}(t), \cdots, x^{(q)}(t)$ are stochastic processes in the above sense.

The *covariance functions* of the $\mathbf{x}(t)$ process are

$$R_{jk}(t, u) = E\{x^{(j)}(t)\bar{x}^{(k)}(u)\} = \bar{R}_{kj}(u, t), \tag{3}$$

where $j, k = 1, 2, \cdots, q$. These are all finite, since we consider only variables with finite second-order moments. In the particular case $q = 1$, we are concerned with one single stochastic process $x(t)$, and there is only one covariance function

$$R(t, u) = E\{x(t)\bar{x}(u)\}. \tag{4}$$

2. Stationary stochastic processes

In the important particular case when all R_{jk} are functions of the difference $t - u$, so that we have

$$R_{jk}(t, u) = r_{jk}(t - u), \tag{5}$$

the $\mathbf{x}(t)$ process is known as a *stationary stochastic process*. We shall not here be concerned with the so-called *strictly stationary* processes, which satisfy more restrictive conditions.

The class of stationary processes possesses useful and interesting properties, which have been thoroughly studied. The present paper is the outcome of an attempt to generalize some of these properties to certain classes of nonstationary processes. In particular, various properties related to the problem of *linear least squares prediction* will be considered. For the sake of brevity, we shall in the sequel always use the word "prediction" in the sense of "linear least squares prediction."

In the first part of the paper, the general case of a vector-valued stochastic process with finite second-order moments will be studied. For such a process, there exists a uniquely defined decomposition into a *deterministic* and a *purely nondeterministic* component, which are mutually orthogonal.

In the case of a vector process with discrete time, the properties of this decomposition are a straightforward generalization of the well-known Wold decomposition [12] for stationary one-variable processes with discrete time. This case will be treated in detail in section 4 of the present paper.

For a process with continuous time, on the other hand, the properties of the nondeterministic component are somewhat more complicated than in the case of a stationary process. This case will be dealt with in section 5. We shall here give only the main lines of the argument, and state our main results. Complete proofs will be given in a forthcoming publication in the *Arkiv för Matematik*.

The decomposition into a deterministic and a purely nondeterministic component forms the basis of a *time-domain analysis* of a given stochastic process, generalizing the well-known properties of stationary processes.

For the class of stationary processes, "the chief advantage of turning from the time-domain analysis of stochastic processes to the *frequency-domain* or *spectral analysis* is the possibility of using the powerful methods of harmonic analysis" (Wiener and Masani [11], p. 140). This can be done for stationary processes, since there exist for these processes *spectral representations* in the form of Fourier-Stieltjes integrals, both for the process variables themselves and for the associated covariance functions.

In the second part of the present paper, we shall consider certain classes of nonstationary processes admitting spectral representations of a similar kind. Our results in this part of the paper are of a much less definite character than those given in the first part. In fact, only some highly preliminary results concerned with the spectral analysis of processes will be given here. Also we shall here consider only one-variable processes with discrete time, although most of our results may be generalized to the vector case, and also to processes with continuous time.

Two classes of stochastic processes admitting spectral representations will be considered, each including the stationary processes as a particular case.

For a stationary process with discrete time, it is well known that there exists a representation in the form of a stochastic Fourier-Stieltjes integral

$$x_n = \int_0^{2\pi} e^{inu}\, dz(u), \tag{6}$$

where $z(u)$ is a stochastic process with *orthogonal increments*. If we consider a process x_n representable in the same form, but without requiring that $z(u)$ should necessarily have orthogonal increments, we shall be led to a class of stochastic processes first introduced by Loève [7], [8], and called by him *harmonizable processes*. Obviously this class contains the class of stationary processes. The harmonizable processes will be considered in section 6 of the present paper.

Finally, in section 7 we shall consider a different kind of generalization of the concept of a stationary process. With respect to a stationary process with discrete time, it is well known that there exists a *unitary shift operator* U, which takes every x_n into the immediately following variable x_{n+1}. The properties of this operator are intimately connected with the properties of the stationary process. The more general class of processes obtained when it is only assumed that the shift operator is *normal* has been studied by Getoor [5]. In section 7 we shall give some very preliminary results concerning the spectral analysis of this class of processes, restricting ourselves to the case of processes with discrete time.

PART I. THE GENERAL VECTOR-VALUED PROCESS WITH FINITE SECOND-ORDER MOMENTS

3. Notation, deterministic and nondeterministic processes

All random variables $x, y, \cdots$ defined on the given probability space $(\Omega, \mathfrak{F}, \mathfrak{P})$, and satisfying (1), form a Hilbert space $\mathfrak{H}$, if the inner product and the norm are defined by the usual expressions

$$(x, y) = E(x\bar{y}), \qquad ||x||^2 = E|x|^2. \tag{7}$$

Whenever we use the term *convergence* with respect to a sequence of random variables, it will be understood that we refer to convergence in the topology induced by this norm, that is, convergence in quadratic mean.

We now consider a vector-valued stochastic process $\mathbf{x}(t) = \{x^{(1)}(t), \cdots, x^{(q)}(t)\}$, where the $x^{(j)}(t)$ are complex-valued stochastic processes, defined for all $t \in T$. Every random variable $x^{(j)}(t)$ is assumed to satisfy (1), and is thus an element of $\mathfrak{H}$. In order to avoid trivial difficulties we shall always suppose that, for every $j = 1, \cdots, q$,

$$E|x^{(j)}(t)|^2 > 0 \tag{8}$$

for at least one $t \in T$. We shall say that two processes $\mathbf{x}(t)$ and $\mathbf{y}(t)$ of this type are *orthogonal*, in symbols $\mathbf{x}(t) \perp \mathbf{y}(t)$, if

$$E\{x^{(j)}(t)\bar{y}^{(k)}(u)\} = 0, \tag{9}$$

for $j, k = 1, \cdots, q$ and all $t, u \in T$.

Let $\mathfrak{H}(\mathbf{x}, t)$ denote the subspace of $\mathfrak{H}$ spanned by the random variables $x^{(j)}(u)$ for $j = 1, \cdots, q$ and all u such that $u \in T$ and $u \leqq t$. We shall write this

$$\mathfrak{H}(\mathbf{x}, t) = \mathfrak{S}\{x^{(j)}(u), j = 1, \cdots, q, u \in T, u \leqq t\}, \tag{10}$$

where $\mathfrak{S}$ stands for "span." Instead of $\mathfrak{H}(\mathbf{x}, +\infty)$, we shall write simply $\mathfrak{H}(\mathbf{x})$.

As t decreases, the set $\mathfrak{H}(\mathbf{x}, t)$ can never increase. It follows that, when $t \to -\infty$, the set $\mathfrak{H}(\mathbf{x}, t)$ must tend to a limiting set, which we denote by $\mathfrak{H}(\mathbf{x}, -\infty)$. We thus have for any $t_1 < t_2$

$$\mathfrak{H}(\mathbf{x}, -\infty) \subset \mathfrak{H}(\mathbf{x}, t_1) \subset \mathfrak{H}(\mathbf{x}, t_2) \subset \mathfrak{H}(\mathbf{x}) \subset \mathfrak{H}. \tag{11}$$

It can be said that $\mathfrak{H}(\mathbf{x}, t)$ contains all the information available when we know the development of all the component processes $x^{(j)}(u)$ up to and including the point t. In the terminology used by Wiener and Masani ([11], p. 135), we can say that $\mathfrak{H}(\mathbf{x}, t)$ represents the *past and present* of $\mathbf{x}(t)$, while $\mathfrak{H}(\mathbf{x}, -\infty)$ corresponds to the *remote past* of the process.

If $\mathfrak{M}$ is any subspace of $\mathfrak{H}$, that is, any closed linear manifold in $\mathfrak{H}$, we denote by $P_{\mathfrak{M}}y$ the projection on $\mathfrak{M}$ of an arbitrary element y of $\mathfrak{H}$. When $\mathfrak{M} = \mathfrak{H}(\mathbf{x}, t)$, we write simply P_t instead of $P_{\mathfrak{M}}$.

Similarly, if $y^{(1)}, \cdots, y^{(q)}$ are any elements of $\mathfrak{H}$, and $\mathbf{y}$ denotes the column vector

$$\mathbf{y} = (y^{(1)}, \cdots, y^{(q)}), \tag{12}$$

we shall write

$$P_{\mathfrak{M}}\mathbf{y} = (P_{\mathfrak{M}}y^{(1)}, \cdots, P_{\mathfrak{M}}y^{(q)}), \tag{13}$$

replacing $P_{\mathfrak{M}}$ by P_t in the particular case when $\mathfrak{M} = \mathfrak{H}(\mathbf{x}, t)$.

The projection $P_{t-h}x^{(j)}(t)$ is, among all elements y of the subspace $\mathfrak{H}(\mathbf{x}, t - h)$, that which minimizes the norm $\|x^{(j)}(t) - y\|$. Accordingly $P_{t-h}x^{(j)}(t)$ is, from the point of view of linear least squares prediction, the best possible prediction of $x^{(j)}(t)$ in terms of all variables $x^{(1)}(u), \cdots, x^{(q)}(u)$ with $u \leqq t - h$. The norm

(14) $$\sigma_{th}^{(j)} = ||x^{(j)}(t) - P_{t-h}x^{(j)}(t)||$$

is the corresponding *error of prediction*. For $0 < h < k$, we obviously have

(15) $$0 \leqq \sigma_{th}^{(j)} \leqq \sigma_{tk}^{(j)}.$$

Going back to relation (11), we shall now consider two extreme possibilities with respect to the subspace $\mathfrak{H}(\mathbf{x}, -\infty)$, namely

(16) $$\begin{aligned} &(A) \qquad \mathfrak{H}(\mathbf{x}, -\infty) = \mathfrak{H}(\mathbf{x}), \\ &(B) \qquad \mathfrak{H}(\mathbf{x}, -\infty) = 0. \end{aligned}$$

In case (A), it follows from (11) that $\mathfrak{H}(\mathbf{x}, t) = \mathfrak{H}(\mathbf{x}, -\infty)$ for all t. Thus, in particular, $x^{(j)}(t) \in \mathfrak{H}(\mathbf{x}, -\infty)$ for all j and t, and it follows that the prediction error $\sigma_{th}^{(j)}$ reduces to zero for $j = 1, \cdots, q$, for all $t \in T$ and all $h > 0$. Hence for every $t \in T$ the components $x^{(j)}(t)$ of $\mathbf{x}(t)$ all can be exactly predicted by means of the information provided by the arbitrarily remote past of the process. In this case we shall say that the $\mathbf{x}(t)$ process is *deterministic*. Every $\mathbf{x}(t)$ process not satisfying condition (A) will be called *nondeterministic*.

On the other hand, in case (B) we can say that the information provided by the remote past of the $\mathbf{x}(t)$ process is, in the limit, of no value for the prediction of the component variables $x^{(j)}(t)$ at any given time t. Thus every piece of information contained in the process at the instant t must have entered the process as an *innovation* at some definite instant $u \leqq t$ in the past or present. Accordingly, a process satisfying condition (B) will be called a *purely nondeterministic* process. (Wiener and Masani [11] use the term *regular* process.) For such a process, the irrelevance of the remote past for prediction purposes may be expressed by the relation

(17) $$\lim_{h \to \infty} \sigma_{th}^{(j)} = ||x^{(j)}(t)||,$$

which holds for $j = 1, \cdots, q$ and every $t \in T$.

4. The discrete case

When T is the set of all integers $n = 0, \pm 1, \cdots$, we are concerned with a discrete vector process

(18) $$\mathbf{x}_n = (x_n^{(1)}, \cdots, x_n^{(q)}),$$

where it is assumed only that each component $x_n^{(j)}$ is a complex-valued stochastic process with discrete time parameter n, zero mean-values, and finite second-order moments. Using the notations introduced above, we have

(19) $$P_{n-1}\mathbf{x}_n = (P_{n-1}x_n^{(1)}, \cdots, P_{n-1}x_n^{(q)}).$$

Writing

(20) $$\xi_n = \mathbf{x}_n - P_{n-1}\mathbf{x}_n, \qquad \xi_n^{(j)} = x_n^{(j)} - P_{n-1}x_n^{(j)}$$

it follows that

(21) $$\xi_n = (\xi_n^{(1)}, \cdots, \xi_n^{(q)}).$$

The sequence of vector-valued random variables $\boldsymbol{\xi}_n$ defines a vector-valued stochastic process with discrete time parameter n. The $\boldsymbol{\xi}_n$ process will be called the *innovation process* corresponding to the given $\mathbf{x}_n$ process. This name may be justified by the following remarks.

If, for a certain value of n, we have

$$\boldsymbol{\xi}_n = (0, \cdots, 0), \tag{22}$$

this signifies that every component $\xi_n^{(j)}$ is zero, that is, every $x_n^{(j)}$ is contained in the subspace $\mathfrak{H}(\mathbf{x}, n - 1)$. From the prediction point of view this means that every component $x_n^{(j)}$ of $\mathbf{x}_n$ can be predicted exactly by means of the information available when we know the development of the $\mathbf{x}$ process up to and including the instant $n - 1$. Obviously this can be expressed by saying that no innovation enters the $\mathbf{x}$ process at the instant n or, equivalently, that the innovation received by the process at this instant reduces to zero.

Suppose, on the other hand, that for a certain n, the vector variable $\boldsymbol{\xi}_n$ has at least one component $\xi_n^{(j)}$ such that $E|\xi_n^{(j)}|^2 > 0$. Then $\xi_n^{(j)} = x_n^{(j)} - P_{n-1}x_n^{(j)}$ does not reduce to zero, so that $x_n^{(j)}$ cannot be predicted exactly in terms of the variables $x_m^{(1)}, \cdots, x_m^{(q)}$ with $m \leqq n - 1$. The variable $\xi_n^{(j)}$ then represents the innovation received by the component $x_n^{(j)}$ at the instant n, and consequently $\boldsymbol{\xi}_n = (\xi_n^{(1)}, \cdots, \xi_n^{(q)})$, which is not identically zero, is the *innovation* entering into the vector process $\mathbf{x}$ at the instant n.

The set of all those values of n, for which the innovation $\boldsymbol{\xi}_n$ does not reduce to zero, may be said to form the *innovation spectrum* of the $\mathbf{x}_n$ process. This set contains precisely all those time points where a new impulse, or an innovation, enters into the process.

The innovation spectrum may be empty, finite, or infinite; and it will be readily seen that, to any given set of integers n, we can construct an $\mathbf{x}_n$ process having this set for its innovation spectrum. For a deterministic process, the innovation spectrum is evidently empty, while for any nondeterministic process it must contain at least one value of n. For a nondeterministic stationary process, the innovation spectrum includes all integers n.

Let now n be a given integer, and consider the set of q random variables $\xi_n^{(1)}, \cdots, \xi_n^{(q)}$, with covariance matrix

$$\mathbf{R}_n = \{E(\xi_n^{(j)}\bar{\xi}_n^{(k)})\}, \qquad j, k = 1, \cdots, q. \tag{23}$$

The rank r_n of $\mathbf{R}_n$ is equal to the maximum number of linearly independent variables among $\xi_n^{(1)}, \cdots, \xi_n^{(q)}$. Thus $0 \leqq r_n \leqq q$, and $r_n > 0$ if, and only if, n belongs to the innovation spectrum of the $\mathbf{x}$ process.

Let $\mathfrak{k}(\mathbf{x}, n)$ denote the r_n-dimensional space spanned by the variables $\xi_n^{(1)}, \cdots, \xi_n^{(q)}$. Since $\xi_n^{(j)} \in \mathfrak{H}(\mathbf{x}, n)$, it is seen that $\mathfrak{k}(\mathbf{x}, n)$ is a subspace of $\mathfrak{H}(\mathbf{x}, n)$.

It follows from (20) that $\xi_n^{(j)}$ is always orthogonal to $\mathfrak{H}(\mathbf{x}, n - 1)$, and consequently $\mathfrak{k}(\mathbf{x}, n) \perp \mathfrak{H}(\mathbf{x}, n - 1)$. It also follows from (20) that any two variables $\xi_m^{(j)}$ and $\xi_n^{(k)}$ with $m \neq n$ are orthogonal, so that $\mathfrak{k}(\mathbf{x}, m) \perp \mathfrak{k}(\mathbf{x}, n)$ when $m \neq n$.

If we orthogonalize the set of variables $\xi_n^{(1)}, \cdots, \xi_n^{(q)}$, we shall obtain a set of

r_n variables, say $\eta_n^{(1)}, \cdots, \eta_n^{(r_n)}$, forming a complete orthonormal system in $\mathfrak{k}(\mathbf{x}, n)$, and in addition $q - r_n$ zero variables. If n does not belong to the innovation spectrum, $r_n = 0$, and the space $\mathfrak{k}(\mathbf{x}, n)$ reduces to zero.

The vector sum of the orthogonal family of subspaces $\mathfrak{k}(\mathbf{x}, m)$ with $m \leqq n$,

$$\mathfrak{K}(\mathbf{x}, n) = \mathfrak{k}(\mathbf{x}, n) + \mathfrak{k}(\mathbf{x}, n-1) + \cdots \tag{24}$$

is the space spanned by all $\xi_m^{(j)}$ with $j = 1, \cdots, q$ and $m \leqq n$. Obviously this is a subspace of $\mathfrak{H}(\mathbf{x}, n)$. It will be seen that the set of all variables $\eta_m^{(j)}$, where $j = 1, \cdots, r_m$ and $m \leqq n$, forms a complete orthonormal system in $\mathfrak{K}(\mathbf{x}, n)$. This remark will be used later.

We now proceed to the proof of the following lemma which, in the stationary case, corresponds to part (b) of lemma 6.10 of Wiener and Masani [11].

LEMMA 1. *The space $\mathfrak{K}(\mathbf{x}, n)$ is, within $\mathfrak{H}(\mathbf{x}, n)$, the orthogonal complement of $\mathfrak{H}(\mathbf{x}, -\infty)$. In symbols*

$$\begin{aligned} &\mathfrak{K}(\mathbf{x}, n) \perp \mathfrak{H}(\mathbf{x}, -\infty), \\ &\mathfrak{H}(\mathbf{x}, n) = \mathfrak{K}(\mathbf{x}, n) + \mathfrak{H}(\mathbf{x}, -\infty). \end{aligned} \tag{25}$$

We have already seen that $\mathfrak{k}(\mathbf{x}, m)$ is orthogonal to $\mathfrak{H}(\mathbf{x}, m-1)$, and a fortiori orthogonal to $\mathfrak{H}(\mathbf{x}, -\infty)$. Consequently the vector sum $\mathfrak{K}(\mathbf{x}, n)$ is orthogonal to $\mathfrak{H}(\mathbf{x}, -\infty)$.

Further, since $\mathfrak{K}(\mathbf{x}, n)$ and $\mathfrak{H}(\mathbf{x}, -\infty)$ are both subspaces of $\mathfrak{H}(\mathbf{x}, n)$, we have

$$\mathfrak{K}(\mathbf{x}, n) + \mathfrak{H}(\mathbf{x}, -\infty) \subset \mathfrak{H}(\mathbf{x}, n). \tag{26}$$

On the other hand we have $x_n^{(j)} = \xi_n^{(j)} + P_{n-1}x_n^{(j)}$, so that every element of $\mathfrak{H}(\mathbf{x}, n)$ is the limit of a convergent sequence of variables, each of which is the sum of a linear combination of $\xi_n^{(1)}, \cdots, \xi_n^{(q)}$ and an element of $\mathfrak{H}(\mathbf{x}, n-1)$. Since $\mathfrak{k}(\mathbf{x}, n)$ and $\mathfrak{H}(\mathbf{x}, n-1)$ are orthogonal, it follows that every element of $\mathfrak{H}(\mathbf{x}, n)$ is the sum of one element of $\mathfrak{k}(\mathbf{x}, n)$ and one of $\mathfrak{H}(\mathbf{x}, n-1)$, so that

$$\mathfrak{H}(\mathbf{x}, n) \subset \mathfrak{k}(\mathbf{x}, n) + \mathfrak{H}(\mathbf{x}, n-1) \subset \mathfrak{K}(\mathbf{x}, n) + \mathfrak{H}(\mathbf{x}, n-1). \tag{27}$$

By repeated application of this relation we obtain

$$\mathfrak{H}(\mathbf{x}, n) \subset \mathfrak{K}(\mathbf{x}, n) + \mathfrak{H}(\mathbf{x}, n-p) \tag{28}$$

for every $p > 0$, and finally, as $p \to \infty$,

$$\mathfrak{H}(\mathbf{x}, n) \subset \mathfrak{K}(\mathbf{x}, n) + \mathfrak{H}(\mathbf{x}, -\infty). \tag{29}$$

This, together with (26), completes the proof of the lemma.

We can now prove the analogue of the Wold decomposition for the $\mathbf{x}_n$ process, thus generalizing theorem 6.11 of Wiener and Masani [11].

THEOREM 1. *For any given $\mathbf{x}_n$ process, there is a uniquely determined decomposition*

$$\mathbf{x}_n = \mathbf{u}_n + \mathbf{v}_n \tag{30}$$

having properties (a) *and* (b).

(a) $\mathbf{u}_n = (u_n^{(1)}, \cdots, u_n^{(q)})$ and $\mathbf{v}_n = (v_n^{(1)}, \cdots, v_n^{(q)})$, *where all $u_n^{(j)}$ and $v_n^{(j)}$ belong to* $\mathfrak{H}(\mathbf{x}, n)$.

(b) *The* $\mathbf{u}_n$ *and* $\mathbf{v}_n$ *processes are orthogonal, and* $\mathbf{u}_n$ *is purely nondeterministic, while* $\mathbf{v}_n$ *is deterministic. The nondeterministic component* $\mathbf{u}_n$ *has, in addition, property* (c).

(c) $\mathbf{u}_n$ *can be expressed as a linear combination of those innovations* $\boldsymbol{\xi}_p$ *of the* $\mathbf{x}_n$ *process that have entered into the process before or at the instant* n,

$$\mathbf{u}_n = \sum_{p=-\infty}^{n} \mathbf{A}_{np}\boldsymbol{\xi}_p, \tag{31}$$

where the $\mathbf{A}_{np} = \{a_{np}^{(jk)}\}$ *are* $q \times q$ *matrices, such that the development formally obtained for any component* $u_n^{(j)}$,

$$u_n^{(j)} = \sum_{p=-\infty}^{n} \sum_{k=1}^{q} a_{np}^{(jk)} \xi_p^{(k)}, \tag{32}$$

is convergent in the topology of $\mathfrak{H}$. *Thus, writing*

$$c_{np}^{(j)} = \left\| \sum_{k=1}^{q} a_{np}^{(jk)} \xi_p^{(k)} \right\| = \left\{ E \left| \sum_{k=1}^{q} a_{np}^{(jk)} \xi_p^{(k)} \right|^2 \right\}^{1/2}, \tag{33}$$

we have

$$\sum_{p=-\infty}^{n} (c_{np}^{(j)})^2 < \infty \tag{34}$$

for all n *and for* $j = 1, \cdots, q$. *The coefficients* $a_{np}^{(jk)}$ *are uniquely determined if, and only if, the rank* r_p *has the maximum value* q, *while the* $c_{np}^{(j)}$ *are uniquely determined for all* n, p, *and* j.

PROOF. For all n and for $j = 1, \cdots, q$ we take for $u_n^{(j)}$ and $v_n^{(j)}$ the projections of $x_n^{(j)}$ on the subspaces $\mathfrak{R}(\mathbf{x}, n)$ and $\mathfrak{H}(\mathbf{x}, -\infty)$ respectively. It then follows from lemma 1 that $u_n^{(j)}$ and $v_n^{(j)}$ belong to $\mathfrak{H}(\mathbf{x}, n)$, and that we have

$$\begin{aligned} x_n^{(j)} &= u_n^{(j)} + v_n^{(j)}, \\ \mathbf{x}_n &= \mathbf{u}_n + \mathbf{v}_n, \end{aligned} \tag{35}$$

where

$$\mathbf{u}_n = (u_n^{(1)}, \cdots, u_n^{(q)}), \qquad \mathbf{v}_n = (v_n^{(1)}, \cdots, v_n^{(q)}). \tag{36}$$

Since $u_m^{(j)}$ belongs to $\mathfrak{R}(\mathbf{x}, m)$, while $v_n^{(k)}$ belongs to $\mathfrak{H}(\mathbf{x}, -\infty)$, it further follows from lemma 1 that $u_m^{(j)}$ and $v_n^{(k)}$ are always orthogonal. Thus the $\mathbf{u}_n$ and $\mathbf{v}_n$ processes are orthogonal, according to the definition given in section 3. If, in accordance with section 3, we define

$$\begin{aligned} \mathfrak{H}(\mathbf{u}, n) &= \mathfrak{S}(u_m^{(j)}, j = 1, \cdots, q, m \leq n), \\ \mathfrak{H}(\mathbf{v}, n) &= \mathfrak{S}(v_m^{(j)}, j = 1, \cdots, q, m \leq n), \end{aligned} \tag{37}$$

we thus find that $\mathfrak{H}(\mathbf{u}, m)$ and $\mathfrak{H}(\mathbf{v}, n)$ are orthogonal for all m and n.

Since all $u_m^{(j)}$ and $v_m^{(j)}$ with $m \leq n$ belong to $\mathfrak{H}(\mathbf{x}, n)$, we have

$$\mathfrak{H}(\mathbf{u}, n) + \mathfrak{H}(\mathbf{v}, n) \subset \mathfrak{H}(\mathbf{x}, n). \tag{38}$$

On the other hand, it follows from (35) and from the orthogonality of the $\mathbf{u}_n$ and $\mathbf{v}_n$ processes that

$$\mathfrak{H}(\mathbf{x}, n) \subset \mathfrak{H}(\mathbf{u}, n) + \mathfrak{H}(\mathbf{v}, n). \tag{39}$$

Hence by lemma 1, $\mathfrak{H}(\mathbf{u}, n) + \mathfrak{H}(\mathbf{v}, n) = \mathfrak{H}(\mathbf{x}, n) = \mathfrak{K}(\mathbf{x}, n) + \mathfrak{H}(\mathbf{x}, -\infty)$. By the definition of the $u_n^{(j)}$ and $v_n^{(j)}$ we have, however, $\mathfrak{H}(\mathbf{u}, n) \subset \mathfrak{K}(\mathbf{x}, n)$ and $\mathfrak{H}(\mathbf{v}, n) \subset \mathfrak{H}(\mathbf{x}, -\infty)$, and thus obtain

$$\begin{aligned} \mathfrak{H}(\mathbf{u}, n) &= \mathfrak{K}(\mathbf{x}, n), \\ \mathfrak{H}(\mathbf{v}, n) &= \mathfrak{H}(\mathbf{x}, -\infty). \end{aligned} \tag{40}$$

From lemma 1 we then obtain $\mathfrak{H}(\mathbf{u}, -\infty) = \mathfrak{K}(\mathbf{x}, -\infty) = 0$, so that the $\mathbf{u}_n$ process is purely nondeterministic. On the other hand $\mathfrak{H}(\mathbf{v}, -\infty) = \mathfrak{H}(\mathbf{x}, -\infty) = \mathfrak{H}(\mathbf{v}, n)$ for every n, and so the $\mathbf{v}_n$ process is deterministic.

The properties (a) and (b) of the decomposition considered here are thus established, and we shall now prove that this is the only decomposition of the given $\mathbf{x}_n$ process that has these properties. Suppose, in fact, that $\mathbf{u}_n$ and $\mathbf{v}_n$ are any processes satisfying (35), and having the properties (a) and (b) stated in the theorem. Then (38) and (39) will still hold, so that we obtain as before

$$\mathfrak{H}(\mathbf{x}, n) = \mathfrak{H}(\mathbf{u}, n) + \mathfrak{H}(\mathbf{v}, n) \tag{41}$$

for all n. It is readily seen that, owing to the orthogonality of $\mathfrak{H}(\mathbf{u}, n)$ and $\mathfrak{H}(\mathbf{v}, n)$, this holds even for $n = -\infty$, and we obtain

$$\mathfrak{H}(\mathbf{x}, -\infty) = \mathfrak{H}(\mathbf{u}, -\infty) + \mathfrak{H}(\mathbf{v}, -\infty). \tag{42}$$

However, on account of property (b), we have $\mathfrak{H}(\mathbf{u}, -\infty) = 0$, and thus

$$\mathfrak{H}(\mathbf{x}, -\infty) = \mathfrak{H}(\mathbf{v}, -\infty) = \mathfrak{H}(\mathbf{v}, n) \tag{43}$$

for all n. Hence by (41) and lemma 1 we find that relations (40) will still hold. In the decomposition $x_n^{(j)} = u_n^{(j)} + v_n^{(j)}$, the first component must then be the projection of $x_n^{(j)}$ on $\mathfrak{K}(\mathbf{x}, n)$, and the second the projection on $\mathfrak{H}(\mathbf{x}, -\infty)$, so that the decomposition is unique.

It finally remains to prove property (c). In order to do this, we use the remark made above about the completeness of the orthonormal system $\eta_m^{(j)}$ in the space $\mathfrak{K}(\mathbf{x}, n)$. The corresponding Fourier development of the element $u_n^{(j)} \in \mathfrak{K}(\mathbf{x}, n)$ will have the form

$$u_n^{(j)} = \sum_{p=-\infty}^{n} \sum_{k=1}^{r_p} b_{np}^{(jk)} \eta_p^{(k)} \tag{44}$$

with

$$\sum_{p=-\infty}^{n} \sum_{k=1}^{r_p} |b_{np}^{(jk)}|^2 < \infty. \tag{45}$$

For any fixed p, the orthogonal variables $\eta_p^{(1)}, \cdots, \eta_p^{(r_p)}$ are certain linear combinations of the innovation components $\xi_p^{(1)}, \cdots, \xi_p^{(q)}$. The coefficients appearing in these linear combinations will be uniquely determined if, and only if, the rank r_p has its maximum value q. Replacing the $\eta_p^{(k)}$ in the above development of $u_n^{(j)}$ by their expressions in terms of the $\xi_p^{(k)}$, we obtain the development given under (c), and we find that

$$(c_{np}^{(j)})^2 = \sum_{k=1}^{r_p} |b_{np}^{(jk)}|^2. \tag{46}$$

The proof of theorem 1 is thus completed.

We can now immediately state the following result, which gives the application of theorem 1 to the prediction problem for the $\mathbf{x}_n$ process.

THEOREM 2. *Let h be any positive integer. With the notation of theorem 1, the best prediction of the component* $x_n^{(j)}$ *in terms of all variables* $x_p^{(1)}, \cdots, x_p^{(q)}$ *with* $p \leqq n - h$ *will be*

$$P_{n-h}x_n^{(j)} = \sum_{p=-\infty}^{n-h} \sum_{k=1}^{q} a_{np}^{(jk)} \xi_p^{(k)} + v_n^{(j)}, \tag{47}$$

with the corresponding error of prediction

$$\begin{aligned} \sigma_{nh}^{(j)} &= \|x_n^{(j)} - P_{n-h}x_n^{(j)}\| \\ &= \left\{ \sum_{p=n-h+1}^{n} (c_{np}^{(j)})^2 \right\}^{1/2} \end{aligned} \tag{48}$$

This follows directly from theorem 1 and from the definition (14) of the error of prediction, if we observe that $v_n^{(j)}$, as well as all $\xi_p^{(k)}$ with $p \leqq n - h$, belong to $\mathfrak{H}(\mathbf{x}, n - h)$, while all $\xi_p^{(k)}$ with $p > n - h$ are orthogonal to this space.

It should be noted that the coefficients $a_{np}^{(jk)}$ in the expression for the best prediction of $x_n^{(j)}$ given in theorem 2 depend on certain covariances of the x process up to the time n. Accordingly, any statistical estimation of this prediction by means of theorem 2 must be based on information concerning the covariance structure of the process up to the time n, either from a priori knowledge (as in the case when the process is assumed to be stationary), or from previous statistical experience.

5. The continuous case

REMARK. I am indebted to Professor K. Itô for the observation that there are interesting points of contact of this section and a work by T. Hida on "Canonical representations of Gaussian processes," which will shortly appear in the *Memoirs of the College of Science, University of Kyoto.*

We now consider a q-dimensional stochastic vector process

$$\mathbf{x}(t) = \{x^{(1)}(t), \cdots, x^{(q)}(t)\}, \tag{49}$$

the parameter set T being the set of all real numbers t. Each component $x^{(j)}(t)$ is a complex-valued stochastic process with continuous time t, and the covariance functions $R_{jk}(t, u)$ defined by (3) are all assumed to be finite.

When t increases from $-\infty$ to $+\infty$, the point $x^{(j)}(t)$ describes a curve in the Hilbert space $\mathfrak{H}(\mathbf{x})$, and the $\mathbf{x}(t)$ process is made up by the set of q curves corresponding to the components $x^{(1)}(t), \cdots, x^{(q)}(t)$. The subspace $\mathfrak{H}(\mathbf{x}, t_0)$ is spanned by the arcs of these curves that belong to the domain $t \leqq t_0$. The properties of the family of all subspaces $\mathfrak{H}(\mathbf{x}, t)$, where t ranges from $-\infty$ to $+\infty$, will play an important part in the sequel.

The following theorem is directly analogous to the first part of theorem 1, and can be proved along similar lines, so that we may content ourselves here with stating the theorem.

THEOREM 3. *There is a unique decomposition of the* $\mathbf{x}(t)$ *process,*

$$\mathbf{x}(t) = \mathbf{u}(t) + \mathbf{v}(t), \tag{50}$$

having properties (a) *and* (b).

(a) $\mathbf{u}(t) = \{u^{(1)}(t), \cdots, u^{(q)}(t)\}$ *and* $\mathbf{v}(t) = \{v^{(1)}(t), \cdots, v^{(q)}(t)\}$, *where all* $u^{(j)}(t)$ *and* $v^{(j)}(t)$ *belong to* $\mathfrak{H}(\mathbf{x}, t)$.

(b) *The* $\mathbf{u}(t)$ *and* $\mathbf{v}(t)$ *processes are orthogonal, and* $\mathbf{u}(t)$ *is purely nondeterministic, while* $\mathbf{v}(t)$ *is deterministic.*

The second part of theorem 1 is concerned with the representation of the nondeterministic component of a given process with *discrete* time as a linear function of the innovations associated with the past and present of the process. For the nondeterministic component of a process with *continuous* time there exists, in fact, an analogous representation. However, the circumstances are somewhat more complicated than in the discrete case, and we shall here only give some preliminary discussion and state our main results, reserving complete proofs for a forthcoming publication.

For our present purpose, it will be sufficient to deal with the purely nondeterministic component $\mathbf{u}(t)$ of the given $\mathbf{x}(t)$ process, and we may then as well assume that $\mathbf{x}(t)$ itself is purely nondeterministic, that is, the deterministic component $\mathbf{v}(t)$ is identically zero. Further, we shall find it convenient to introduce a certain regularity condition relating to the behavior of the $\mathbf{x}(t)$ process in points of discontinuity. Thus it will be assumed throughout the rest of the present section that we are dealing with a vector process $\mathbf{x}(t)$ satisfying the following two conditions.

(C_1) $\mathbf{x}(t)$ is purely nondeterministic, that is, $\mathfrak{H}(\mathbf{x}, -\infty) = 0$.

(C_2) The limits $x^{(j)}(t-0)$ and $x^{(j)}(t+0)$ exist (as always in the $\mathfrak{H}$ topology) for $j = 1, \cdots, q$ and for every real t.

We shall then write

$$\mathbf{x}(t-0) = \{x^{(1)}(t-0), \cdots, x^{(q)}(t-0)\}, \tag{51}$$

and similarly for $\mathbf{x}(t+0)$.

It follows without difficulty from condition (C_2) that *the space* $\mathfrak{H}(\mathbf{x})$ *is separable,* and that the set of *points of discontinuity of* $\mathbf{x}(t)$, that is, the set of all t such that at least one of the relations

$$\mathbf{x}(t-0) = \mathbf{x}(t) = \mathbf{x}(t+0) \tag{52}$$

is not satisfied, is at most enumerable.

Let us now consider the family of subspaces $\mathfrak{H}(\mathbf{x}, t)$ of the space $\mathfrak{H}(\mathbf{x})$. As t increases from $-\infty$ to $+\infty$, the $\mathfrak{H}(\mathbf{x}, t)$ form a never decreasing set of subspaces, with $\mathfrak{H}(\mathbf{x}, -\infty) = 0$ and $\mathfrak{H}(\mathbf{x}, +\infty) = \mathfrak{H}(\mathbf{x})$. The limits $\mathfrak{H}(\mathbf{x}, t \pm 0)$ will exist for all t. If $(t, t + h)$ and $(u, u + k)$ are disjoint intervals, the orthogonal complements

$$\mathfrak{H}(\mathbf{x}, t + h) - \mathfrak{H}(\mathbf{x}, t) \quad \text{and} \quad \mathfrak{H}(\mathbf{x}, u + k) - \mathfrak{H}(\mathbf{x}, u) \tag{53}$$

are mutually orthogonal.

The set of all t such that for any $h > 0$ we have

$$\mathfrak{H}(\mathbf{x}, t + h) - \mathfrak{H}(\mathbf{x}, t - h) \neq 0 \tag{54}$$

will be called the *innovation spectrum* of the $\mathbf{x}(t)$ process. A point t such that at least one of the relations

$$\mathfrak{H}(\mathbf{x}, t - 0) = \mathfrak{H}(\mathbf{x}, t) = \mathfrak{H}(\mathbf{x}, t + 0) \tag{55}$$

is not satisfied, is a *point of discontinuity* of the innovation spectrum. The space $\mathfrak{H}(\mathbf{x})$ being separable, it follows immediately that the set of all discontinuity points is at most enumerable.

A discontinuity point of the innovation spectrum will not necessarily be a discontinuity point of the process, nor conversely. We shall make some remarks concerning the relations between these two kinds of discontinuities.

Let us first consider the case of a left discontinuity of the innovation spectrum, that is, a point t such that

$$\mathfrak{M}(t) = \mathfrak{H}(\mathbf{x}, t) - \mathfrak{H}(\mathbf{x}, t - 0) \neq 0. \tag{56}$$

Then it is easily shown that

$$\mathbf{x}(t) - \mathbf{x}(t - 0) \neq 0, \tag{57}$$

so that t is also a left discontinuity point of the process. Further, if $y^{(j)}$ denotes the projection of $x^{(j)}(t) - x^{(j)}(t - 0)$ on $\mathfrak{M}(t)$, we have $y^{(j)} \neq 0$ for at least one j, and the subspace $\mathfrak{M}(t)$ is spanned by the variables $y^{(1)}, \cdots, y^{(q)}$, and has thus at most q dimensions.

Thus in particular (56) implies (57). The converse statement is however not true: a left discontinuity of the process may, in fact, be a continuity point of the innovation spectrum.

Proceeding now to the case of a right discontinuity, it can be shown that neither of the two relations

$$\mathfrak{N}(t) = \mathfrak{H}(\mathbf{x}, t + 0) - \mathfrak{H}(\mathbf{x}, t) \neq 0 \tag{58}$$

and

$$\mathbf{x}(t + 0) - \mathbf{x}(t) \neq 0 \tag{59}$$

implies the other. In fact, it can be shown by examples that (58) may be satisfied even in a continuity point of the process, while on the other hand (59) may be satisfied even in a continuity point of the innovation spectrum.

The only implication that exists between the relations (56), (57), (58), and (59) is thus that (56) implies (57).

As in section 3, we now denote by $P_t z$ the projection of any point $z \in \mathfrak{H}(\mathbf{x})$ on the subspace $\mathfrak{H}(\mathbf{x}, t)$. When t increases from $-\infty$ to $+\infty$, the P_t form a never decreasing set of projections, with $P_{-\infty} = 0$ and $P_{+\infty} = I$. For $h > 0$, the difference $P_{t+h} - P_t$ is the projection on $\mathfrak{H}(\mathbf{x}, t + h) - \mathfrak{H}(\mathbf{x}, t)$. The limits $P_{t\pm 0}$ exist for every t, and are the projections on $\mathfrak{H}(\mathbf{x}, t \pm 0)$ respectively.

For an arbitrary random variable z in $\mathfrak{H}(\mathbf{x})$, we now define a stochastic process by writing for all real t

$$z(t) = P_t z. \tag{60}$$

It then follows from the above that $z(t)$ defines a complex-valued stochastic process with *orthogonal increments*, such that

$$\begin{aligned} z(-\infty) &= 0, \qquad z(+\infty) = z, \\ Ez(t) &= 0, \qquad E|z(t)|^2 = F(t, z), \end{aligned} \tag{61}$$

where $F(t, z)$ is, for any fixed z, a real, never decreasing and bounded function of t, such that

$$F(-\infty, z) = 0, \qquad F(+\infty, z) = E|z|^2. \tag{62}$$

The points of increase of $z(t)$, that is, the points t such that for any $h > 0$

$$E|z(t + h) - z(t - h)|^2 = F(t + h, z) - F(t - h, z) > 0, \tag{63}$$

form a subset of the innovation spectrum of $\mathbf{x}(t)$. Similarly, the left (right) discontinuities of $z(t)$ form a subset of the left (right) discontinuities of the innovation spectrum. Any increment $z(t + h) - z(t)$ belongs to the subspace $\mathfrak{H}(\mathbf{x}, t + h) - \mathfrak{H}(\mathbf{x}, t)$, and may thus be regarded as a part of the innovation received by the $\mathbf{x}(t)$ process during the interval $(t, t + h)$.

We now denote by $\mathfrak{L}(z)$ the subspace of $\mathfrak{H}(\mathbf{x})$ spanned by all the variables $z(u)$ for $-\infty < u < +\infty$, and by $\mathfrak{L}^*(z)$ the set of all random variables y representable in the form

$$y = \int_{-\infty}^{\infty} g(u)\, dz(u) \tag{64}$$

with

$$E|y|^2 = \int_{-\infty}^{\infty} |g(u)|^2\, dF(u, z) < \infty. \tag{65}$$

If no u is at the same time a left and a right discontinuity of $z(u)$, then $\mathfrak{L}(z)$ and $\mathfrak{L}^*(z)$ are identical (see Doob [4], pp. 425–429). The variable y given by (64 will belong to $\mathfrak{H}(\mathbf{x}, t)$ if, and only if, we have $g(u) = 0$ for almost all $u > t$, "almost all" referring to the $F(u, z)$-measure on the u-axis.

By means of the theory of spectral multiplicity in Hilbert space (see, for example, Stone [10], chapter VII, and Halmos [6]), we can now show that with any $\mathbf{x}(t)$ process satisfying (C_1) and (C_2) it is possible to associate a number N, which may be a finite positive integer or equal to $+\infty$, such that we can find N random variables $z_1, \cdots, z_N$ belonging to $\mathfrak{H}(\mathbf{x})$, with the properties

(a) $\qquad \mathfrak{L}(z_n) = \mathfrak{L}^*(z_n), \qquad n = 1, \cdots, N.$

(b) $\qquad \mathfrak{L}(z_m) \perp \mathfrak{L}(z_n), \qquad m \neq n.$

(c) $$\mathfrak{H}(\mathbf{x}) = \mathfrak{L}(z_1) + \cdots + \mathfrak{L}(z_N).$$

(d) N is the smallest number having the properties (a), (b), and (c).

In particular, since any component variable $x^{(j)}(t)$ of $\mathbf{x}(t)$ evidently belongs to $\mathfrak{H}(\mathbf{x}, t)$, we have the expression

$$x^{(j)}(t) = \sum_{k=1}^{N} \int_{-\infty}^{t} g_k^{(j)}(t, u)\, dz_k(u) \tag{66}$$

for $j = 1, \cdots, q$. If $N = \infty$, the series appearing here will converge in quadratic mean, so that

$$\sum_{k=1}^{N} \int_{-\infty}^{t} |g_k^{(j)}(t, u)|^2\, dF(u, z_k) < \infty. \tag{67}$$

If now we define a column vector

$$\mathbf{z}(u) = \{z_1(u), \cdots, z_N(u)\} \tag{68}$$

and a $q \times N$ matrix

$$\mathbf{G}(t, u) = \{g_k^{(j)}(t, u)\}, \qquad j = 1, \cdots, q;\ k = 1, \cdots, N, \tag{69}$$

we finally obtain the required expression for the vector variable $\mathbf{x}(t)$ in terms of past and present innovations of the process, as stated in the following theorem.

THEOREM 4. *The vector variable* $\mathbf{x}(t)$ *of any stochastic process satisfying conditions* (C_1) *and* (C_2) *can be expressed in the form*

$$\mathbf{x}(t) = \int_{-\infty}^{t} \mathbf{G}(t, u)\, d\mathbf{z}(u), \tag{70}$$

where $\mathbf{z}(u)$ *is an* N*-dimensional vector process with orthogonal increments, while* $\mathbf{G}(t, u)$ *is a* $q \times N$ *matrix, in accordance with* (68) *and* (69). *The development* (66) *formally obtained for the component* $x^{(j)}(t)$ *is then convergent as shown by* (67).

It will be seen that this is directly analogous to the last part of theorem 1, except that certain sums have been replaced by integrals, and that the q-dimensional random vector $\boldsymbol{\xi}_p$ has been replaced by the N-dimensional vector $\mathbf{z}(u)$. It is the fact that the multiplicity N may have any integral value from 1 to ∞ that introduces additional complication into the continuous case. It is possible to construct examples corresponding to any given value of N, even when it is required that $\mathbf{x}(t)$ be everywhere continuous (or even differentiable in quadratic mean). We finally state the following theorem, which is the continuous analogy of theorem 2.

THEOREM 5. *Let* $h > 0$ *be given. For any* $\mathbf{x}(t)$ *process satisfying conditions* (C_1) *and* (C_2), *the best prediction of the component* $x^{(j)}(t)$ *in terms of all variables* $x^{(1)}(u), \cdots, x^{(q)}(u)$ *with* $u \leqq t - h$ *will be*

$$P_{t-h}x^{(j)}(t) = \sum_{k=1}^{N} \int_{-\infty}^{t-h} g_k^{(j)}(t, u)\, dz_k(u), \tag{71}$$

with the corresponding error of prediction

(72)
$$\sigma_{th}^{(j)} = \|x^{(j)}(t) - P_{t-h}x^{(j)}(t)\|$$
$$= \left\{\sum_{k=1}^{N} \int_{t-h}^{t} |g_k^{(j)}(t, u)|^2 \, dF(u, z_k)\right\}^{1/2}$$

PART II. ON TWO CLASSES OF PROCESSES ADMITTING SPECTRAL REPRESENTATIONS

6. Harmonizable processes

We shall now consider a one-dimensional process with discrete time, such that x_n is given by a stochastic Fourier-Stieltjes integral

(73)
$$x_n = \int_0^{2\pi} e^{inu} \, dz(u),$$

where $z(u)$ denotes, for $0 \leqq u \leqq 2\pi$, a complex-valued random variable satisfying the conditions

(74)
$$Ez(u) = 0, \qquad E\{z(u)\bar{z}(v)\} = F(u, v).$$

It will be assumed that the complex-valued covariance function $F(u, v)$ is of bounded variation over the square C defined by $0 \leqq u, v \leqq 2\pi$, in the sense that, for every subdivision of C in a finite number of rectangles, we have

(75)
$$\sum |\Delta_2 F| < K,$$

the sum being extended over all the rectangles, and the constant K being independent of the subdivision.

The integral (73) will then exist as a limit in quadratic mean of certain Riemann sums. Processes of this type have been introduced by Loève [7], [8], and have been called by him *harmonizable processes*. The covariance function corresponding to the process defined by (73) is

(76)
$$R(m, n) = E(x_m\bar{x}_n) = \int_0^{2\pi} \int_0^{2\pi} e^{i(mu-nv)} \, dF(u, v).$$

Conversely, if the covariance function of a certain x_n process is given by (76), where $F(u, v)$ is a covariance function satisfying (75) it is known (Loève [8], Cramér [1]) that there exists a process $z(u)$ satisfying (74), and such that x_n is given by the integral (73).

Without changing the value of the integral (73), we can always suppose that $z(u)$ is everywhere continuous to the right in quadratic mean, so that $z(u + 0) = z(u)$. The function $F(u, v)$ then defines a *complex mass distribution* over C, such that the mass carried by any rectangle $h < u \leqq h + \Delta h$, $k < v \leqq k + \Delta k$ is equal to the second-order difference $\Delta_2 F(u, v)$ corresponding to this rectangle.

It follows from the Hermite-symmetric properties of covariances that the masses carried by two sets of points symmetrically situated with respect to the diagonal $u = v$ of the square C are always complex conjugates. If a point set belonging to the diagonal $u = v$ carries a mass different from zero, this mass will be real and positive.

The function $F(u, v)$ will be called the *spectral function* of the x_n process, while the distribution defined by F is the *spectral distribution* of the process.

In the particular case when the whole spectral mass is situated on the diagonal $u = v$, it follows from the general symbolic relation

$$E\{dz(u)\, d\bar{z}(v)\} = d_{u,v}F(u, v) \tag{77}$$

that the $z(u)$ process has orthogonal increments, and so in this case the x_n process is *stationary*.

In the general case, $F(u, v)$ may be represented as a sum of three components, each of which is a covariance of bounded variation over C,

$$F = F_1 + F_2 + F_3. \tag{78}$$

Here F_1 is absolutely continuous, with a *spectral density* $f_1(u, v)$ such that

$$F_1(u, v) = \int_0^u \int_0^v f_1(s, t)\, ds\, dt. \tag{79}$$

On the other hand, the F_2 and F_3 distributions both have their total masses concentrated in sets of two-dimensional Lebesgue measure zero. For F_2 this set is at most enumerable, each point carrying a mass different from zero, while the F_3 set is nonenumerable, and each single point carries the mass zero. In the stationary case, the F_1 component is absent, while the F_2 and F_3 components have their total masses situated on the diagonal $u = v$.

A sufficient condition that the harmonizable x_n process given by (73) will be deterministic can be obtained in the following way. The x_n process will be deterministic if, and only if, for every n and every $h > 0$ we can find a finite number of constants $c_0, c_1, \cdots, c_r$ such that the quantity

$$W = E|x_n - c_0x_{n-h} - c_1x_{n-h-1} - \cdots - c_rx_{n-h-r}|^2 \geqq 0 \tag{80}$$

will be arbitrarily small. Writing

$$g(u) = e^{inu} - c_0e^{i(n-h)u} - \cdots - c_re^{i(n-h-r)u}, \tag{81}$$

it follows from (76) that we have

$$W = \int_0^{2\pi} \int_0^{2\pi} g(u)\bar{g}(v)\, dF(u, v) \tag{82}$$

and hence by the Schwarz inequality

$$W^2 \leqq \int_0^{2\pi} \int_0^{2\pi} |g(u)|^2\, |dF(u, v)| \int_0^{2\pi} \int_0^{2\pi} |g(v)|^2\, |dF(u, v)|. \tag{83}$$

By the symmetry of the spectral distribution, the two factors in the last member are equal, so that we obtain

$$W \leqq \int_0^{2\pi} |g(u)|^2\, dG(u), \tag{84}$$

where

(85) $$G(u) = \int_0^u \int_0^{2\pi} |dF(s, t)|.$$

Now $G(u)$, being a never decreasing and bounded function of u, has almost everywhere in $(0, 2\pi)$ a nonnegative derivative $G'(u)$, and the integral

(86) $$\int_0^{2\pi} \log G'(u)\, du$$

will be finite or equal to $-\infty$. In particular, if $G'(u) = 0$ on a set of positive measure, the integral will certainly have the value $-\infty$.

If the integral (86) has the value $-\infty$, it follows from well-known theorems in the prediction theory for stationary processes that the coefficients c_j can be chosen so as to make the second member of (80) as small as we please. Thus we have the following result (Cramér [2]).

THEOREM 6. *If we have*

(87) $$\int_0^{2\pi} \log G'(u)\, du = -\infty$$

the x_n process is deterministic.

In particular, if the F_1 and F_3 components in (78) are absent, so that the whole mass of the F distribution is concentrated in isolated points, it will be seen that $G'(u) = 0$ almost everywhere, so that (87) will certainly hold, and the x_n process will be deterministic.

Consider now, on the other hand, a process x_n with a spectral function F having an absolutely continuous component F_1 not identically zero. Moreover, let us suppose that the spectral density $f_1(u, v)$ corresponding to F_1 belongs to L_2 over the square C. For such a process, we shall give a sufficient condition that it is nondeterministic. Let

(88) $$f_1(u, v) = \sum_{p=1}^{\infty} \mu_p \varphi_p(u) \bar{\varphi}_p(v)$$

be the expansion of $f_1(u, v)$ in terms of its eigenvalues μ_p and eigenfunctions $\varphi_p(u)$. The μ_p are real and positive, the $\varphi_p(u)$ are a set of orthonormal functions in $(0, 2\pi)$, and the series converges in quadratic mean over C. We then have

THEOREM 7. *Suppose that, in the expansion* (88), *there is a p such that the Fourier series of the eigenfunction $\varphi_p(u)$*

(89) $$\varphi_p(u) \sim \sum_{q=-\infty}^{\infty} b_{pq} e^{-iqu}, \qquad \sum_{q=-\infty}^{\infty} |b_{pq}|^2 < \infty$$

is "one-sided" in the sense that for a certain m it satisfies the conditions

(90) $$b_{pq} = 0 \text{ for } q < m, \qquad b_{pm} \neq 0.$$

Then the x_n process is nondeterministic, and the point $n = m$ belongs to its innovation spectrum.

Taking $n = m$ and $h = 1$ in the expressions (80) and (81) for W and $g(u)$ we have, in fact (see Riesz and Nagy [9], p. 240),

$$W = \int_0^{2\pi} \int_0^{2\pi} g(u)\bar{g}(v)\, dF(u, v) \geqq \int_0^{2\pi} \int_0^{2\pi} g(u)\bar{g}(v) f_1(u, v)\, du\, dv = \sum_{p=1}^{\infty} \mu_p \left| \int_0^{2\pi} g(u)\varphi_p(u)\, du \right|^2 \tag{91}$$

and thus by hypothesis

$$W \geqq \mu_p \left| \int_0^{2\pi} g(u)\varphi_p(u)\, du \right|^2 = 4\pi^2 \mu_p |b_{pm}|^2 \tag{92}$$

independently of the choice of the coefficients c_j. This obviously signifies that x_m cannot be predicted exactly in terms of the variables $x_{m-1}, x_{m-2}, \cdots$, so that the prediction error

$$||x_m - P_{m-1}x_m|| \tag{93}$$

is positive, and m belongs to the innovation spectrum of the x_n process, which is thus nondeterministic.

It follows from well-known theorems that, when the conditions of theorem 7 are satisfied, we have

$$\int_0^{2\pi} \log |\varphi_p(u)|\, du > -\infty. \tag{94}$$

The converse of this statement is, however, not true; (94) may be satisfied even in a case when $\varphi_p(u)$ does not have a one-sided Fourier expansion. A simple example is obtained by taking

$$2\pi f_1(u, v) = \varphi(u)\bar{\varphi}(v) \tag{95}$$

with

$$\varphi(u) = \begin{cases} e^{iu}, & 0 \leqq u \leqq \pi, \\ e^{-iu}, & \pi < u \leqq 2\pi. \end{cases} \tag{96}$$

It can also be shown by examples that there are nondeterministic processes with a spectral density $f_1(u, v)$ belonging to L_2 that do not satisfy the conditions of theorem 7 for any value of p.

By imposing a further restrictive condition on the behavior of the spectral density it is possible, however, to obtain a criterion which is both necessary and sufficient in order that a given harmonizable process be nondeterministic, and even have an a priori given set of integers as its innovation spectrum (Cramér [3]). Thus, in particular, it follows that, any set of integers being given, there always exists a harmonizable process having this set as its innovation spectrum.

7. Processes with normal shift operator

For a stationary process with discrete time, we have the integral representation

$$x_n = \int_0^{2\pi} e^{inu}\, dz(u), \tag{97}$$

where $z(u)$ has orthogonal increments. In the preceding section we considered the generalization obtained by dropping the assumption that $z(u)$ has orthogonal increments, and we saw that this leads to the class of harmonizable processes.

We now consider a different kind of generalization of (97), which leads to a different class of processes. To this effect, we now regard the integration variable u in (97) as a *complex* variable, and suppose that the integration is extended over a certain domain D in the plane of u, and that $z(u)$ is defined for all u belonging to D.

After an appropriate change of variables, the integral corresponding to (97) then takes the form

$$x_n = \int_D w^n \, dz(\rho, \lambda), \tag{98}$$

where $w = \rho \exp(i\lambda)$, while $z(\rho, \lambda)$ is a random variable satisfying the conditions

$$Ez(\rho, \lambda) = 0, \qquad E|z(\rho, \lambda)|^2 < K \tag{99}$$

for all ρ, λ such that w belongs to D. As in the stationary case, we still suppose that $z(\rho, \lambda)$ has orthogonal increments, so that we have in the usual symbolism

$$\begin{aligned} E\{dz(\rho_1, \lambda_1)\, \overline{dz}(\rho_2, \lambda_2)\} &= 0, \qquad w_1 \neq w_2, \\ E|dz(\rho, \lambda)|^2 &= dF(\rho, \lambda), \end{aligned} \tag{100}$$

where $F(\rho, \lambda)$ is a nonnegative and never decreasing function of ρ and λ, which is bounded throughout D. The integral (98) can then be defined in the same way as before, and we obtain

$$R(m, n) = E(x_m \bar{x}_n) = \int_D w^m \overline{w^n} \, dF(\rho, \lambda). \tag{101}$$

The function $F(\rho, \lambda)$ will be called the *spectral function* of the x_n process, and defines the *spectral distribution* of the process, which is a distribution of real and positive mass over the domain D.

We now introduce the further assumption that the domain D is entirely situated within the ring

$$\rho_1 \geqq \rho \geqq \rho_0 > 0. \tag{102}$$

We observe that this includes the particular case of a stationary process, when the domain D reduces to the unit circle.

In the present case we obtain from (101) for any complex constants c_j and any positive integer Q

$$0 \leqq E\left|\sum_{j=-Q}^{Q} c_j x_j\right|^2 = \sum_{j,k=-Q}^{Q} c_j \bar{c}_k R(j, k) = \int_D \left|\sum_{-Q}^{Q} c_j w^j\right|^2 dF, \tag{103}$$

and hence

$$\begin{aligned} E\left|\sum_{-Q}^{Q} c_j x_{j+1}\right|^2 &\leqq \rho_1^2 \, E\left|\sum_{-Q}^{Q} c_j x_j\right|^2, \\ E\left|\sum_{-Q}^{Q} c_j x_{j-1}\right|^2 &\leqq \frac{1}{\rho_0^2} E\left|\sum_{-Q}^{Q} c_j x_j\right|^2. \end{aligned} \tag{104}$$

According to Getoor [5], these inequalities imply that there is a *shift operator* N uniquely defined and bounded throughout the Hilbert space $\mathfrak{H}(x)$ of the x_n process, and such that

$$N^m x_n = x_{m+n} \tag{105}$$

for all $m, n = 0, \pm 1, \cdots$. It also follows from the work of Getoor that in our case N is a *normal* operator in $\mathfrak{H}(x)$. We have, in fact,

$$\begin{aligned}(Nx_m, x_n) &= E(x_{m+1}\bar{x}_n) = \int_D w^{m+1}\overline{w^n}\,dF \\ &= \int_D w^m(\overline{\bar{w}w^n})\,dF = E(x_m\bar{y}_n),\end{aligned} \tag{106}$$

where

$$y_n = N^* x_n = \int_D \bar{w}w^n\,dz. \tag{107}$$

It follows that

$$NN^*x_n = N^*Nx_n = \int_D |w|^2 w^n\,dz. \tag{108}$$

Thus N commutes with its adjoint N^*, and consequently N is normal. In the particular case of a stationary process, when the spectral mass is wholly situated on the unit circle, it is well known that N is even a unitary operator.

By an argument quite similar to that used for the deduction of the inequalities (104), we obtain for $n > 0$

$$\begin{aligned}\rho_0^{2n} E\left|x_0 - \sum_{j=1}^{Q} c_j x_{-j}\right|^2 &\leqq E\left|x_n - \sum_{j=1}^{Q} c_j x_{n-j}\right|^2 \\ &\leqq \rho_1^{2n} E\left|x_0 - \sum_{j=1}^{Q} c_j x_{-j}\right|^2.\end{aligned} \tag{109}$$

The coefficients c_j being arbitrary, this shows that we have for the prediction errors σ_{nh}, where h is any positive integer,

$$\rho_0^{2n}\sigma_{0h}^2 \leqq \sigma_{nh}^2 \leqq \rho_1^{2n}\sigma_{0h}^2. \tag{110}$$

For $n < 0$ we obtain in the same way

$$\rho_1^{2n}\sigma_{0h}^2 \leqq \sigma_{nh}^2 \leqq \rho_0^{2n}\sigma_{0h}^2. \tag{111}$$

From these inequalities, we obtain directly the following theorem.

THEOREM 8. *If the x_n process defined by* (98) *and* (102) *is nondeterministic, we have $\sigma_{nh} > 0$ for all n and all $h > 0$. In particular, the innovation spectrum of the process then contains all $n = 0, \pm 1, \cdots$.*

Suppose now that the spectral distribution defined by $F(\rho, \lambda)$ has a nonvanishing absolutely continuous component. We may then write

$$\begin{aligned}J = E\left|x_0 - \sum_1^Q c_j x_{-j}\right|^2 &= \int_D \left|1 - \sum_1^Q \frac{c_j}{w^j}\right|^2 dF \\ &\geqq \int_{\rho_0}^{\rho_1}\int_0^{2\pi} \left|1 - \sum_1^Q \frac{c_j}{w^j}\right|^2 f(\rho, \lambda)\,d\rho\,d\lambda,\end{aligned} \tag{112}$$

where $f(\rho, \lambda)$ is nonnegative and integrable. From the inequalities between arithmetic and geometric means we further obtain

(113)
$$\frac{1}{2\pi(\rho_1 - \rho_0)} J$$
$$\geqq \exp\left\{\frac{1}{2\pi(\rho_1 - \rho_0)} \int_{\rho_0}^{\rho_1} \int_0^{2\pi} \left[\log f(\rho, \lambda) + 2 \log\left|1 - \sum_1^Q \frac{c_j}{w^j}\right|\right] d\rho\, d\lambda\right\}$$
$$= G(f) \exp\left\{\frac{1}{\pi(\rho_1 - \rho_0)} \int_{\rho_0}^{\rho_1} \int_0^{2\pi} \log\left|1 - \sum_1^Q \frac{c_j}{w^j}\right| d\rho\, d\lambda\right\},$$

where

(114)
$$G(f) = \exp\left\{\frac{1}{2\pi(\rho_1 - \rho_0)} \int_{\rho_0}^{\rho_1} \int_0^{2\pi} \log f(\rho, \lambda)\, d\rho\, d\lambda\right\}.$$

From Jensen's theorem we obtain, however,

(115)
$$\exp\left\{\frac{1}{\pi(\rho_1 - \rho_0)} \int_{\rho_0}^{\rho_1} \int_0^{2\pi} \log\left|1 - \sum_1^Q \frac{c_j}{w^j}\right| d\rho\, d\lambda\right\}$$
$$= \exp\left\{\frac{2}{\rho_1 - \rho_0} \int_{\rho_0}^{\rho_1} \log \frac{|w_1 \cdots w_k|}{\rho^k}\, d\rho\right\} \geqq 1,$$

where $w_1, \cdots, w_k$ are the zeros of $1 - \sum_1^Q c_j w^{-j}$ outside the circle $|w| = \rho$. Consequently

(116)
$$\frac{1}{2\pi(\rho_1 - \rho_0)} J \geqq G(f),$$

and it follows that, if $G(f) > 0$, then the x_n process is nondeterministic, and we have

(117)
$$\sigma_{n1}^2 \geqq 2\pi(\rho_1 - \rho_0)G(f).$$

In the particular case when there is an expansion

(118)
$$\log f(\rho, \lambda) = \sum_{-\infty}^{\infty} \frac{b_j}{\rho^{|j|}} e^{-ij\lambda}, \qquad b_{-j} = \bar{b}_j,$$

absolutely convergent for $\rho \geqq \rho_0$, it can even be proved that the sign of equality holds in (117).

Finally, we may observe that it is also possible to give a sufficient condition for a deterministic process, corresponding at least partly to theorem 6. In fact, it can be shown that if the spectral distribution defined by $F(\rho, \lambda)$ is discrete, and if the set of points carrying a positive mass has at most a finite number of limiting points, then the x_n process is deterministic.

REFERENCES

[1] H. Cramér, "A contribution to the theory of stochastic processes," *Proceedings of the Second Berkeley Symposium on Mathematical Statistics and Probability*, Berkeley and Los Angeles, University of California Press, 1951, pp. 329–339.

[2] ———, "Remarques sur le problème de prédiction pour certaines classes de processus stochastiques," *Colloque sur le Calcul des Probabilités*, Paris, 1959, pp. 103–112.

[3] ———, "On the linear prediction problem for certain stochastic processes," *Ark. Mat.*, Vol. 4 (1959), pp. 45–53.

[4] J. L. DOOB, *Stochastic Processes*, New York, Wiley, 1953.

[5] R. K. GETOOR, "The shift operator for non-stationary stochastic processes," *Duke Math. J.*, Vol. 23 (1956), pp. 175–187.

[6] P. R. HALMOS, *Introduction to Hilbert Space and the Theory of Spectral Multiplicity*, New York, Chelsea, 1951.

[7] M. LOÈVE, "Fonctions aléatoires du second ordre," appendix to P. Lévy, *Processus Stochastiques et Mouvement Brownien*, Paris, Gauthier-Villars, 1948.

[8] ———, *Probability Theory*, New York, Van Nostrand, 1955.

[9] F. RIESZ and B. SZ.-NAGY, *Leçons d'Analyse Fonctionnelle*, Paris, 1955 (3rd ed.).

[10] M. H. STONE, *Linear Transformations in Hilbert Space and their Applications to Analysis*, New York, The American Mathematical Society, 1932.

[11] N. WIENER and P. MASANI, "The prediction theory of multivariate stochastic processes, I," *Acta Math.*, Vol. 98 (1957), pp. 111–150.

[12] H. WOLD, *A Study in the Analysis of Stationary Time Series*, Stockholm, Almqvist & Wiksell, 1954.

Editors' Comments on Paper 6

At about the same time as Cramér, Professor T. Hida of Nagoya University in Japan obtained independently the canonical decomposition of essentially the same class of random processes. Hida's interest was in Gaussian processes; thus the restriction of normality was carried along in his derivation. In this case the linear or canonical structure of a process fully describes the process itself. Furthermore, Hida treats extensively the class of multiple Markov processes, for which he derives canonical representations.

The development parallels that of Cramér in many respects; it shows once more that, when the solution to a problem reaches its maturity, more than one person may conceive of it. Whenever such independent contemporary derivations are obtained, each is distinguished by the author's individual scientific past, and, of course, by his own philosophy and style. There is, therefore, no risk of overlap when the contributions of both authors are included in this collection.

This paper is reprinted with the permission of the author and the Department of Mathematics of the University of Kyoto. Professor Hida has kindly provided an errata sheet, which is reproduced at the end of the article.

MEMOIRS OF THE COLLEGE OF SCIENCE, UNIVERSITY OF KYOTO, SERIES A
Vol. XXXIII, Mathematics No. 1, 1960.

Canonical representations of Gaussian processes and their applications

By

Takeyuki HIDA

6

(Received April 2, 1960)

Introduction

Let $B(t)$, $0 \leqq t < \infty$, be an additive real Gaussian process with $E(B(t))=0$ and $F(t, u)$ be a real-valued function of (t, u). The process $\tilde{X}(t)$ defined as

$$\tilde{X}(t) = \int_0^t F(t, u) dB(u) \tag{0. 1}$$

is a real Gaussian process with mean 0, and enjoys the property

$$\mathfrak{M}_t(\tilde{X}) \subset \mathfrak{M}_t(B), \qquad 0 \leqq t < \infty, \tag{0. 2}$$

where $\mathfrak{M}_t(\tilde{X})$ and $\mathfrak{M}_t(B)$ denote the closed linear manifolds generated by $\{\tilde{X}(\tau) ; \tau \leqq t\}$ and $\{B(\tau) ; \tau \leqq t\}$ respectively. Given a Gaussian process $X(t)$, P. Lévy called the expression (0. 1) a *representation* of $X(t)$, if $\tilde{X}(t)$ is version of $X(t)$ and he introduced the concept of *canonical representation*. Roughly speaking, a canonical representation is one for which the equality holds instead of the inclusion relation in (0. 2) (cf. Definition I. 2 and Theorem I. 2). In this case, $F(t, u)$ is called a canonical kernel.

P. Lévy has recently published several important papers concerning the canonical representation of Gaussian processes. However his pioneering works contain some points difficult for us to follow. The main aim of this paper is to establish his theory systematically and to prove some new facts.

We shall here give a brief account of the contents of this paper. In this paper we shall treat only real Gaussian processes and often omit the adjective "real Gaussian".

Section I. General theory.

As P. Lévy proved in [4],[1] a canonical representation of any process is uniquely determined if it exists. We shall prove this fact in detail in § I. 2. Further we shall give a necessary and sufficient condition for the existence of the canonical representation, using Hellinger-Hahn's Theorem in the theory of Hilbert Space. This fact is not found in Lévy's paper.

As to whether a given representation is canonical or not, Lévy gave a criterion using Hellinger integral (P. Lévy [6]). But we shall give another criterion which proves to be a generalization of Karhunen's kernel criterion for the moving average representation of stationary processes.

Section II. Multiple Markov process.

J. L. Doob (for stationary processes) and P. Lévy defined N-ple Markov processes[2] using the derivatives up to the $(N-1)$-th order of the processes. We generalize this notion to treat the processes which are not always differentiable.

The main results obtained here are as follows. The canonical kernel of the N-ple Markov process is a Goursat kernel of order N. This generalizes the fact obtained by Lévy. In § II. 3, we shall prove that the stationary N-ple Markov process is the sum of special simple Markov processes which are to be called general Ornstein-Uhlenbeck's Brownian motions. This also generalizes Doob's Theorem [1].

Section III. Lévy's $M(t)$ process.

Let $X(A)$, $A \in \boldsymbol{E}^N$ ($\boldsymbol{E}^N$ is N-dimensional Euclidean space), be an ordinary Brownian motion with N-dimensional parameter (cf. P. Lévy [4]) and $M_N(t)$ be the average of $X(A)$ over the sphere with center O (origin of $\boldsymbol{E}^N$) and radius $t(\geqq 0)$ in the parameter space $\boldsymbol{E}^N$. $M_N(t)$ is clearly a Gaussian process with time parameter t. Lévy discussed the canonical representation and the multiple Markov property for this process $M_N(t)$ only in the case N is odd. We shall simplify his proof by transforming $M_N(t)$ into a stationary

1) Numbers in square brackets refer to the list of references at the end of the paper.

2) In Lévy's terminology "Markov process of order N in the restricted sense".

process. Our present method is applicable to the case N is even. We shall prove that $M_{2p}(t)$ is not a multiple Markov process but the limiting process of multiple Markov processes. Furthermore we shall prove that there are 2^{p-1} different representation of $M_{2p-1}(t)$, which provides an affirmative answer to Lévy's problem (Lévy [4, p. 146]).

I would like to express my hearty thanks to Professor K. Itô for his encouragement and valuable suggestions and to Mr. N. Ikeda who helped me with valuable discussions in overcoming the difficulties in the course of this paper; in particular, the idea of using the reproducing kernel in the proof of Theorem I. 4 is due to Mr. Ikeda.

Section I. General theory of representation.

§I. 1. Definitions.

In order to define a representation of the given Gaussian process precisely, it is necessary for us to consider integrals with respect to certain random measures. The parameter space T of a process that will be treated here may be a closed interval, $(-\infty, \infty)$ or $[0, \infty)$. The symbol $\boldsymbol{B}_T$ denotes the Borel field of subsets of T.

Let $B(M)$, $M \in \boldsymbol{B}_T$, be a real Gaussian random measure such that

$$(\text{I. }1) \qquad E(B(M)) = 0 \quad \text{and} \quad E(B(M)^2) = v(M) \qquad \text{for every } M \in \boldsymbol{B}_T,$$

where v is a (non-negative) measure defined on $\boldsymbol{B}_T$. Then, $B(M)$ can be decomposed into two parts in the following way

$$B(M) = B_1(M) + \sum_{t_j \in M} X_{t_j},$$

where $B_1(\cdot)$ is a random measure associated with the continuous measure $v_1(\cdot) = E(B_1(\cdot)^2)$ and X_{t_j}'s are mutually independent Gaussian random variables with mean 0, each one of which corresponds to the jump point t_j of $v(u) = v((-\infty, u] \cap T)$. $B_1(\cdot)$ and $\{X_{t_j}\}$ will be called the continuous part and discontinuous part of $B(\cdot)$ respectively.

Let $f(u)$ be a Borel measurable function. Then, if $f \in \boldsymbol{L}^2(v)$, that is, $f \in \boldsymbol{L}^2(v_1)$ and $\sum_{t_j} f(t_j)^2 E(X_{t_j}^2) < \infty$, then

$$\int_M f(u)dB_1(u) \quad \text{and} \quad \sum_{t_j \in M} f(t_j)X_{t_j}$$

are well defined for any Borel set M. The integral of f over M may be written in the form

$$\int_M f(u)dB(u) = \int_M f(u)dB_1(u) + \sum_{t_j \in M} f(t_j)X_{t_j}, \tag{I. 2}$$

where the integrals which appear are interpreted in the usual way with respect to random measures dB, dB_1, Doob [2; pp. 426-433.]

Now we can give

Definition I. 1. Let $Y(t)$, $t \in T$, be a real Gaussian process with $E(Y(t))=0$ for every $t \in T$. Then the triple $(dB(t), \mathfrak{M}_t, F(t, u))$, or simply the pair $(dB(t), F(t, u))$, is called a *representation* of $Y(t)$, if

i) $B(\cdot)$ is a random measure satisfying (I. 1);

ii) $F(t, u)$ is a real Borel measurable function of u vanishing for $u > t$ and belonging to $\boldsymbol{L}^2(v)$ for every t;

iii)
$$X(t) = \int^t F(t, u)dB(u)^{3)}$$
is a version of $Y(t)$;

iv) $\mathfrak{M}_t$ is the closed linear manifold generated by $\{X(\tau); \tau \leq t\}$.

The function $F(t, u)$ is called a *kernel* of the representation.

There are many examples of processes which have no representation. Furthermore, even if a process has a representation, it may not be uniquely determined. The following examples will serve to illustrate such circumstances.

Example I. 1. Let $Y_1(t)$ be a Gaussian process with covariance function $\Gamma(s, t)=1$ for $s=t$, and $=0$ for $s \neq t$, and with $E(Y_1(t))=0$. Then $Y_1(t)$ has no representation.

Example I. 2. Let $B_1(t)$ and $B_2(t)$, $0 \leq t < \infty$, be standard Brownian motions which are independent of each other. Define

$$Y_2(t) = \begin{cases} B_1(t) & \text{if } t \text{ is rational,} \\ B_2(t) & \text{if } t \text{ is irrational.} \end{cases}$$

Then $Y_2(t)$ has no representation. Detailed discussions concerning this will be given later.

---

3) The notation $\int^t \cdots$ means the integral $\int_{(-\infty, t] \cap T} \cdots$ in the sense of (I. 2).

Example I. 3. Let $B(t)$, $0\leqq t<\infty$, be a standard Brownian motion. Then, for every positive integer n, we can determine constants $c_0, c_1, \cdots, c_n$ so that

$$\tilde{B}(t) = \int_0^t \left(c_0 + c_1 \frac{u}{t} + c_2\left(\frac{u}{t}\right)^2 + \cdots + c_n\left(\frac{u}{t}\right)^n\right) dB(u)$$

is again a standard Brownian motion (P. Lévy [6]). This proves that $B(t)$ has infinitely many representations.

We have now to determine the best class of representations for our purpose among all posible representations of a given process.

Definition. I. 2. The representation $(dB(t),\ \mathfrak{M}_t,\ F(t, u))$ is called *canonical*, if

$$E(X(t)/\boldsymbol{B}_s) = \int^s F(t, u)dB(u)$$

holds for every $s\leqq t$, where $\boldsymbol{B}_s$ is the smallest Borel field of measurable ω-sets with respect to which all the $X(\tau)$'s $(\tau\leqq s)$ are measurable. In this case, $F(t, u)$ is called a *canonical kernel.*

Definition. I. 3. Two representations $(dB^{(i)}(t),\ \mathfrak{M}_t^{(i)},\ F^{(i)}(t, u))$, $i=1, 2$, are called *equivalent*, if

$$\int_M F^{(1)}(t, u)^2 dv^{(1)}(u) = \int_M F^{(2)}(t, u)^2 dv^{(2)}(u) \qquad \text{for every } M\in \boldsymbol{B}_T,$$

considering them as measures, where

$$dv^{(i)}(u) = E(dB^{(i)}(u)^2), \qquad i = 1, 2.$$

This relation obviously satisfies the equivalence relations and therefore we can get the classes of representations.

Theorem I. 1. *(P. Lévy) For every $Y(t)$, there exists at most one class of canonical representations.*

Proof. Let $(dB^{(i)}(t),\ \mathfrak{M}_t^{(i)},\ F^{(i)}(t, u))$ $i=1, 2$, be canonical representations of $Y(t)$. Writing

$$X^{(i)}(t) = \int^t F^{(i)}(t, u)dB^{(i)}(u), \qquad i = 1, 2,$$

we have, for every t and every $s(\leqq t)$,

$$E((X^{(i)}(t)/\boldsymbol{B}_s^{(i)}) = \int^s F^{(i)}(t, u)dB^{(i)}(u),$$

where $\boldsymbol{B}_t^{(i)}$ dcnotes the Borel field corresponding to $X^{(i)}(t)$.

The equality

$$E(E(X^{(1)}(t)/\boldsymbol{B}_s^{(1)})^2) = E(E(X^{(2)}(t)/\boldsymbol{B}_s^{(2)})^2)$$

holds, since both sides are determined only by the probability law of $X(t)$. Hence we have

$$\int^s F^{(1)}(t, u)^2 dv^{(1)}(u) = \int^s F^{(2)}(t, u)^2 dv^{(2)}(u), \qquad \text{for every } s(\leqq t),$$

which proves the theorem.

§ I. 2. Canonical representations.

In this article we shall study important properties of a canonical representation.

Definition I. 4. A canonical representation $(dB(t), \mathfrak{M}_t, F(t, u))$ is called *proper* if

$$\text{(I. 3)} \qquad \mathfrak{M}_t = \mathfrak{M}_t(B) \qquad \text{for every } \ t \in T,$$

where $\mathfrak{M}_t(B)$ is the closed linear manifold generated by

$$\left\{ \int_{(-\infty, t] \cap M} dB(u) \, ; \ M \in \boldsymbol{B}_T \right\}.$$

Theorem I. 2. *For any given canonical representation, we can construct an equivalent proper canonical representation.*

Proof. It is sufficient to consider the case in which v is a continuous measure. Let $(dB(t), \mathfrak{M}_t, \widetilde{F}(t, u))$ be a given canonical representation of $Y(t)$. We shall show that we can construct a proper canonical representation of $Y(t)$ by deforming the given one.

1°) *Deformation.* First define $F(t, u) = \widetilde{F}(t, u)$. Put

$$\mu(M) = \bigvee_t \left(\int_M F(t, u)^2 dv(u) \right), \text{[4]}$$

$$v(M) \equiv \int_M dv(u) = E\left(\int_M dB(u) \right)^2, \qquad M \in \boldsymbol{B}_T.$$

Here we may suppose that v is a continuous measure. Then

$$\mu \ll v \, ; \text{[5]}$$

hence, by Radon-Nykodym's Theorem, there exists a Borel measurable function $f(u) \geqq 0$ such that

---

4) $\bigvee$ means the lattice sum.

5) $\mu \ll \tilde{v}$ means that the measure μ is absolutely continuous with respect to the measure $\tilde{v}$.

$$\mu(M) = \int_M f(u)dv(u) .$$

Since $N=N(f)=\{u\,;\,f(u)>0\}$ is Borel measurable, we can construct a random measure

$$B(M) \equiv \int_M dB(u) = \int_M \chi_N(u)d\tilde{B}(u), \qquad M\in \boldsymbol{B}_T, \tag{I. 4}$$

where $\chi_N(u)$ is the indicator function of N.

2°) The triple $(dB(t),\ \mathfrak{M}_t,\ F(t, u))$ is a representation of $Y(t)$. To prove this, it is sufficient to show that $X(t)=\int^t F(t, u)dB(u)$ is the same process as $Y(t)$, in the sense of equivalence in law. This can be proved as follows.

$$\begin{aligned} & E\Big(\int^t \tilde{F}(t, u)d\tilde{B}(u)-\int^t F(t, u)dB(u)\Big)^2 \\ &= E\Big(\int^t F(t, u)\,(1-\chi_N(u))d\tilde{B}(u)\Big)^2 \\ &= \int^t F(t, u)^2(1-\chi_N(u))^2 d\tilde{v}(u) \\ &= \int_{(-\infty,\, t]\cap N} (1-\chi_N(u))^2 F(t, u)^2 d\tilde{v}(u) = 0 , \end{aligned} \tag{I. 5}$$

which shows that $X(t)=\tilde{X}(t)$ $\Big(=\int^t \tilde{F}(t, u)d\tilde{B}(u)\Big)$ with probability one. Since $\tilde{X}(t)$ is the same process as $Y(t)$ by assumption, the relation above proves our assertion.

3°) Finally we shall prove that the representation $(dB(t),\ \mathfrak{M}_t,\ F(t, u))$ is proper canonical. Now we prove

$$\mathfrak{M}_t \supset \mathfrak{M}_t(B) . \tag{I. 6}$$

Suppose an element Z of $\mathfrak{M}_t(B)$ is orthogonal to (hence independent of) $\mathfrak{M}_t$, that is, $E(Z\cdot X(s))=0$ for every $s\leqq t$. Then Z is orthogonal to $E(X(s'')/\boldsymbol{B}_{s'})$ for every $s'\leqq t$ and every s'', since it is an element of $\mathfrak{M}_t$.

On the other hand Z can be written as

$$Z=\int^t h(u)dB(u) ,$$

with a Borel measurable function h. Noting that

$$E(X(s'')/\boldsymbol{B}_{s'}) = \text{Projection of } X(s'') \text{ on } \mathfrak{M}_{s'} \qquad (\text{in the } \boldsymbol{L}^2 \text{ sense})$$
$$= \text{Projection of } \tilde{X}(s'') \text{ on } \mathfrak{M}_{s'}$$
$$(\text{since } X(t)=\tilde{X}(t) \text{ with probability } 1)$$
$$= E(\tilde{X}(s'')/\boldsymbol{B}_{s'}) = \int^{s'} \tilde{F}(s'', u)d\tilde{B}(u)$$
$$(\text{from canonical property})$$
$$= \int^{s'} F(s'', u)dB(u)\,,$$

we have

$$E(Z\cdot E(X(s''))/\boldsymbol{B}_{s'})) = \int^{s'} h(u)F(s'', u)d\tilde{v}(u) = 0\,.$$

Therefore, for every s, $s'(<t)$

$$\int_s^{s'} h(u)F(s'', u)d\tilde{v}(u) = 0\,.$$

Since s'' is arbitrary, we can prove

$$\mu(N(h)) = 0\,,$$

where $N(h)=\{u\,;\ h(u)\neq 0\}$. Hence we have

$$E(Z\cdot W) = 0 \qquad \text{for every } W\in\mathfrak{M}_t(B)\,,$$

Thus we have proved (I. 6). Consequently we have (I. 3).

Now from (I. 3),

$$E(X(t)/B_s) = \text{projection of } X(t) \text{ on } \mathfrak{M}_s$$
$$= \text{Projection of } X(t) \text{ on } \mathfrak{M}_s(B) = \int^s F(t, u)dB(u)\,,$$

which proves that the representation is canonical and (I. 6) implies that it is proper. Thus we have proved the theorem.

By the argument used in the proof of the theorem, we hav

Corollary. If a representation $(dB(t), \mathfrak{M}_t, F(t, u))$ (not necessarily canonical) satisfies the condition

$$\mathfrak{M}_t(B) = \mathfrak{M}_t \qquad \text{for every } t,$$

then it is (proper) canonical.

As is well known, a stationary process $X(t)$ which is purely non-deterministic and M_2-continuous can be expressed as

$$X(t) = \int_{-\infty}^{t} F(t-u)dB(u)$$

and there exists one and only one representation having the property (I. 3). (Karhunen [1]). In our case, in which $X(t)$ is Gaussian, this means that it has a proper canonical representation.

§ I. 3. Existence of representation.

In order to study the existence of the canonical representation we shall summarize some known theorems of the theory of Hilbert space.

Let $\Gamma(s, t)$, $s, t \in T$, be a real non-negative definite function. Then there exists a Hilbert space $\mathfrak{H}$ satisfying

i) $\Gamma(s, t)$ belongs to $\mathfrak{H}$ as a function of s,

ii) $\langle f(s), \Gamma(s, t) \rangle$[6] $= f(t)$ for every $f \in \mathfrak{H}$,

iii) $\mathfrak{H}$ is the closed (in the topology $\| \ \|$) linear manifold generated by $\{\Gamma(\cdot, t)\,;\, t \in T\}$.

$\Gamma(s, t)$ is the *reproducing kernel* of that Hilbert space.

The construction and the important properties of $\mathfrak{H}$ may be seen in Aronszajn [1].

We can construct sub-spaces $\mathfrak{H}_t$ and $\mathfrak{H}_t^*$ of $\mathfrak{H}$:

$$\mathfrak{H}_t = \text{sub-space of } \mathfrak{H} \text{ generated by } \{\Gamma(\cdot, \tau)\,;\, \tau \leqq t\}\,;$$

$$\mathfrak{H}_t^* = \bigcap_n \mathfrak{H}_{t+\frac{1}{n}}\,.$$

Now let us assume that

(H. 1) $\qquad \mathfrak{H}$ is separable,

(H. 2) $\qquad \bigcap_{t \in T} \mathfrak{H}_t = \{0\} \qquad$ (hence $\bigcap_{t \in T} \mathfrak{H}_t^* = \{0\}$).

Noting that

$$\bigcup_{t \in T} \mathfrak{H}_t^* = \mathfrak{H} \quad \text{and} \quad \mathfrak{H}_s^* \subset \mathfrak{H}_t^*, \qquad s \leqq t,$$

we can see that there exists a resolution of the identity $\{E(t)\,;\, t \in T\}$ such that

$$\mathfrak{H}_t^* = E(t)\mathfrak{H}\,, \tag{I. 7}$$

by assumption. Then, by Hellinger-Hahn's Theorem,[7] there exist two denumerable sets $\{f^{(i)}\}$, $i = 1, 2, \cdots$ and $g^{(j)l}$, $j, l = 1, 2, \cdots$ in $\mathfrak{H}$ satisfying the following conditions (I. 8) to (I. 10).

6) The symbol $\langle\ ,\ \rangle$ donotes the inner product. We shall use $\|\ \|$ to denote the norm, i.e. $\|f\| = \sqrt{\langle f, f \rangle}$.

7) For proof see M. H. Stone [1] or S. Itô [1].

(I. 8)

i) For any intervals Δ_1, Δ_2

$$\langle \Delta_1 E f^{(i)}, \Delta_2 E f^{(j)} \rangle = 0, \qquad i \neq j,$$

with $\Delta E = E(b) - E(a)$ for $\Delta = (a, b]$;

ii) if $\Delta_1 \cap \Delta_2 = \emptyset$

$$\langle \Delta_1 E f^{(i)}, \Delta_2 E f^{(i)} \rangle = 0 ;$$

iii) for any i, $\rho_i(t) = \|E(t) f^{(i)}\|$ is continuous, non-decreasing and $\rho_{i+1} \ll \rho_i$ (considered as measures);

iv) $g^{(j)l}$ is the eigenvector of the self-adjoint operator $H = \int t dE(t)$ corresponding to the eigenvalue $t_j (l = 1, 2, \cdots)$.

$$\mathfrak{H} = \mathfrak{M} \oplus \mathfrak{N} \qquad \text{(direct sum)}, \tag{I. 9}$$

where $\mathfrak{M}$ and $\mathfrak{N}$ are defined by

$$\mathfrak{M} = \sum_i \oplus \mathfrak{M}(f^{(i)}),$$

$$\mathfrak{M}(f^{(i)}) = \left\{ f ; f = \int \varphi(t) dE(t) f^{(i)}, \qquad \varphi \in L^2(\rho_i) \right\} ;$$

$$\mathfrak{N} = \sum_{t_j} \sum_l \oplus \mathfrak{N}(g^{(i)l}),$$

$$\mathfrak{N}(g^{(j)l}) = \text{one dimensional sub-space generated by } g^{(j)l} ;$$

$$E(t) \mathfrak{M}(f^{(i)}) \subset \mathfrak{M}(f^{(i)}), \tag{I. 10}$$

which is equivalent to

$$E(t) P_i = P_i E(t), \qquad P_i = \text{Projection on } \mathfrak{M}(f^{(i)}). \tag{I. 10$'$}$$

Furthermore, though there may be many ways of choosing such $\{f^{(i)}\}$ and $\{g^{(j)l}\}$, their numbers are always the same.

By virtue of this theorem, we can define the multiplicity of $E(t)$.

Definition I. 5. The supremum of the number of $f^{(i)}$'s and the numbers of linearly independent eigenvectors corresponding to each t_j is called the *multiplicity* of $E(t)$.

Theorem I. 3. $\Gamma(\cdot, t)$ *is expressible as*

$$\Gamma(\cdot, t) = \sum_i \int^t F_i(t, u) dE(u) f^{(i)} + \sum_{t_j \leq t} \sum_l b_j^l(t) g^{(j)l}. \tag{I. 11}$$

Proof. From (I. 9), $\Gamma(\cdot, t)$ is written in the form

$$\Gamma(\cdot, t) = \sum_i \int F_i(t,u) dE(u) f^{(i)} + \sum_{t_j} \sum_l b_j^l(t) g^{(j)l}.$$

Applying (I. 10), we have

$$\Gamma(\cdot, t) = E(t)\Gamma(\cdot, t) = \sum_i E(t)\int F_i(t, u)dE(u)f^{(i)} + \sum_{t_j}\sum_l b_j^l(t)E(t)g^{(j)l}$$

$$= \sum_i \int^t F_i(t, u)dE(u)f^{(i)} + \sum_{t_j \leqq t}\sum_l b_j^l(t)E(t)g^{(j)l},$$

where the summation in the second term in the equation above extends over those $g^{(j)l}$'s the eigenvalue of which are not larger than t.

The case in which the function $F_i(t, u)$ in (I. 11) is degenerate, for example,

(I. 12) $$\sum_{k=1}^{N} f_k(t)g_k(u),$$

is of special interest, as we shall see in the next section.

After the preparation above, we can now discuss the existence of the canonical representation. Given a process $Y(t)$, let $\Gamma(s, t)$ be its covariance function. Let $\mathfrak{M}_t$ be the closed linear manifold generated by $\{Y(\tau); \tau \leqq t\}$ and

$$\mathfrak{M} = \bigvee_{t \in T} \mathfrak{M}_t, \quad \mathfrak{M}_t^* = \bigcap_n \mathfrak{M}_{t+1/n}.$$

We shall assume that

($\mathfrak{M}$. 1) $\mathfrak{M}$ is separable (as a sub-space of $L^2(\Omega)$),

($\mathfrak{M}$. 2) $\bigcap_{t \in T} \mathfrak{M}_t = \{0\}$.

We shall prove the following preliminary theorem leading to the fundamental Theorems I. 5 and I. 6.

Theorem I. 4. *There exists an isometric transformation from $\mathfrak{H}$ onto $\mathfrak{M}$ defined by*

(I. 13) $$\mathfrak{H} \ni \Gamma(\cdot, t) \leftrightarrow Y(t) \in \mathfrak{M}.$$

This isometry induces the following correspondence:

i) $\mathfrak{H}_t \leftrightarrow \mathfrak{M}_t$, $\mathfrak{H}_t^* \leftrightarrow \mathfrak{M}_t^*$,

ii) $E(t)f \leftrightarrow E(X/\boldsymbol{B}_t^*)$ provided that $f \leftrightarrow X$, with $\boldsymbol{B}_t^* = \boldsymbol{B}(\mathfrak{M}_t^*)$.

Proof. Define a mapping S from $\boldsymbol{L} = \{\Gamma(\cdot, t); t \in T\}$[8] into $\mathfrak{M}$ by

8) $\{\cdots\}$ denotes the linear space generated by the elements that are written in the bracket.

$$S: \quad \Gamma(\cdot, t) \to Y(t)$$
$$\sum a_i \Gamma(\cdot, t_i) \to \sum a_i Y(t_i) \qquad (a_i: \text{ real})$$

Then S is a linear transformation from a linear space $\boldsymbol{L}$ into the linear space $\mathfrak{L}=\{Y(t),\ t\in T\}\subset\mathfrak{M}$. Suppose $\sum a_i Y(t_i)=0$. Then $\sum a_i\Gamma(t, t_i)=0$ for every $t\in T$, which implies

$$f(t) = \langle f(\cdot), \Gamma(\cdot, t)\rangle = 0, \quad \text{if } f(\cdot) = \sum a_i \Gamma(\cdot, t)\,.$$

This shows that S is a one-to-one mapping from $\boldsymbol{L}$ onto $\mathfrak{L}$. And further

$$\langle \Gamma(\cdot, t), \Gamma(\cdot, s)\rangle = \Gamma(s, t) = E(Y(t)\cdot Y(s))\,,$$

which proves that S is isometric.

Since $\boldsymbol{L}$ and $\mathfrak{L}$ are dense in $\mathfrak{H}$ and $\mathfrak{M}$ respectively, S can be extended to an isometric one-to-one linear transformation $\bar{S}$ from $\mathfrak{H}$ onto $\mathfrak{M}$. Hence we have proved the existence of the isometry.

Next, $\bar{S}\mathfrak{H}_t=\mathfrak{M}_t$ is obvious. Therefore if $f\leftrightarrow X$,

$$E(t)f \leftrightarrow \text{Projection of } X \text{ on } \mathfrak{M}_t^* = E(X/\boldsymbol{B}_t^*).$$

Thus we have proved ii).

Theorem I. 4 along with the assumptions ($\mathfrak{M}$. 1), ($\mathfrak{M}$. 2) implies that (H. 1) and (H. 2) hold. We can therefore appeal to the Hellinger-Hahn's Theorem. Let $f^{(i)}$ be as defined in the statement of that theorem. Then there exists a continuous additive process $B^{(i)}(t)$ such that

$$dE(t)f^{(i)} \leftrightarrow dB^{(i)}(t)$$

under the correspondence mentioned in Theorem I. 4.

Theorem I. 5. *If $Y(t)$ satisfies ($\mathfrak{M}$. 1) and ($\mathfrak{M}$. 2), there exist Gaussian random measures $\{B^{(i)}(\cdot)\}$ and random variables $Y_{t_j}^l$ such that*

i) $B^{(i)}(\cdot)$, $Y_{t_j}^l$, $i, j, l=1, 2, \cdots$ *are all independent,*

ii) $E(B^{(i+1)}(\cdot)^2)\ll E(B^{(i)}(\cdot)^2)$, $i=1, 2, \cdots$,

iii) $Y(t)=\sum_i\int^t F_i(t, u)dB^{(i)}(u) + \sum_{t_j\leqq t}\sum_l b_j^l(t) Y_{t_j}^l$,

iv) $E(Y(t)/\boldsymbol{B}_s)=\sum_i\int^s F(t, u)dB^{(i)}(u) + \sum^*_{t_j\leqq s}\sum_l b_j^l(t) Y_{t_j}^l$,[9] $s\leqq t$.

9) $\sum^*_{t_j\leqq s}\sum_l = \sum_{t_j<s}\sum_l + \sum_{l(b_j^l(s)\neq 0)}$

Proof. If follows from (I. 8), (I. 9) and Theorem I. 4 that i), ii) and iii) are satisfied. Considering the isometry $\bar{S}$ in the proof of Theorem I. 4, we have

$$E(Y(t)/\boldsymbol{B}_s^*) = \bar{S}^{-1}(E(s)\Gamma(\cdot, t))$$
$$= \bar{S}^{-1}(E(s)\sum_i \int^t F_i(t, u)dE(u)f^{(i)} + E(s)\sum_{t_j \leqq t}\sum_l b_j^l(t)g^{(j)l}) .$$

By (I. 10), this is equal to

$$\bar{S}^{-1}(\sum_i \int^s F_i(t, u)dE(u)f^{(i)} + \sum_{t_j \leqq s}^* \sum_l b_j^l(t)g^{(j)l})$$
$$= \sum_i \int^s F_i(t, u)dB^{(i)}(u) + \sum_{t_j \leqq s}^* \sum_l b_j^l(t)Y_{t_j}^l$$
$$(E(X(t)/\boldsymbol{B}_s) = E(E(X(t)/\boldsymbol{B}_s^*)/\boldsymbol{B}_s) ,$$

which proves iv).

Definition I. 6. The system $(dB^{(i)}(t), \mathfrak{M}_t^{(i)}, F(t, u), b_j^l(t)\ Y_{t_j}^l; i, j, l = 1, 2, \cdots)$ obtained above is called a *generalized canonical representation* of $Y(t)$. The multiplicity of $E(t)$ is called the *multiplicity* of $Y(t)$.

The classification of generalized canonical representations is very complicated, because, for one thing, the choice of the system $\{f^{(i)}\}$ is not unique.

Theorem I. 6. *A necessary and sufficient condition that* $Y(t)$ *has a canonical representation is that* $Y(t)$ *satisfies the conditions* ($\mathfrak{M}$. 1), ($\mathfrak{M}$. 2) *and*

($\mathfrak{M}$. 3) *The multiplicity of* $Y(t)$ *is one.*

Proof. Necessity. Suppose that $(dB(t), \mathfrak{M}_t, F(t, u))$ is a canonical representation of $Y(t)$ and

$$\text{(I. 14)} \qquad X(t) = \int^t F(t, u)dB(u) \sim Y(t), \qquad (\sim: \text{equivalent in law})$$

The separability of $\mathfrak{M} = \bigvee_{t \in T} \mathfrak{M}_t$ is easily deduced from the definition of the integral.

Let $\mathfrak{M}_t(B)$ be the same as in Definition I. 4. Then

$$\mathfrak{M}_t \subset \mathfrak{M}_t(B) = \left\{\int^t f(u)dB(u)\ ;\ f \text{ is Borel measurable and } \in \boldsymbol{L}^2(v)\right\}$$

and

$$\bigcap_{t \in T} \mathfrak{M}_t(B) = \{0\}$$

imply the condition ($\mathfrak{M}$. 2). ($\mathfrak{M}$. 1) follows at once from the separability of $\bigcup_{t\in T}\mathfrak{M}_t(B)$.

Finally, if we decompose the random measure $B(\cdot)$ into two parts in the same way as in (I. 1), we can easily see that the multiplicity of $Y(t)$ is one by Definition I. 6.

Sufficiency. From Theorem I. 5 and the assumption that the multiplicity of $Y(t)$ is one, $Y(t)$ can be expressed in the form

$$Y(t) = \int^t F_1(t, u)dB^{(1)}(u) + \sum_{t_j \leqq t} b_j(t)X_{t_j}.$$

Now we can define a random measure $B(\cdot)$ by

$$B((a, b]) \equiv B(b) - B(a) = \int_a^b dB^{(1)}(u) + \sum_{a<t_j\leqq b} a_j X_{t_j}, \quad (\sum_j a_j^2 E(X_{t_j}^2) < \infty).$$

And define a function $F(t, u)$ of u for every fixed t, by

$$F(t, u) = \begin{cases} F_1(t, u) & \text{if } u \in T \bigwedge (-\infty, t] - \{t_j\}, \\ b_j(t)/a_j & \text{if } u = t_j. \end{cases}$$

Then we have

$$\int^t F(t, u)dB(u) \equiv \int^t F(t, u)dB^{(1)}(u) + \sum_{t_j\leqq t} a_j F(t, t_j)X_{t_j}.$$

$$= \int^t F_1(t, u)dB^{(1)}(u) + \int^t (F(t, u) - F_1(t, u))dB^{(1)}(u) + \sum_{t_j\leqq t} b_j(t)X_{t_j},$$

which is equal to $Y(t)$, since the second term of the last expression is zero with probability one.

The canonical property of $(dB(t), \mathfrak{M}_t, F(t, u))$ follows from iv) in Theorem I. 5.

The process $Y_2(t)$ which was given in Example I. 2 has no canonical representation, because the multiplicity of $Y_2(t)$ is 2, as is easily seen. But a generalized canonical representation exists. In fact it can be expressed in the form

$$Y_2(t) = \int_0^t F(t, u)dB_1(u) + \int_0^t (1 - F(t, u))dB_2(u),$$

where

$$F(t, u) = \begin{cases} 1, & \text{if } t \text{ is rational}, \\ 0, & \text{if } t \text{ is irrational}. \end{cases}$$

Corollary. *An additive process has a canonical representation.*

Proof. If $B(t)$ is an additive process, it can be expressed in the form

$$B(t)=\int^{t}dB_1(t)+\sum_{t_j\leqq t}^{*}X_{t_j}.$$

This shows that the multiplicity is one, and our assertion follows.

§I. 4. Kernel criterion for canonical representation.

It is important to give a criterion to determine whether a given representation is canonical or not. P. Lévy gave a criterion involving a Hellinger's integral (Lévy [6]), but we shall give another.

By Theorem I. 2, it is sufficient to give a criterion for a proper canonical representation.

Theorem I. 7. *A representation* $(dB(t),\ \mathfrak{M}_t,\ F(t,u))$ *is proper canonical if and only if, for any fixed* $t_0\in T$

(I. 15) $$\int^{t}F(t,u)f(u)dv(u)=0 \qquad \text{for every } t\leqq t_0.$$

implies

(I. 16) $$f(u)=0 \qquad \text{almost everywhere } (v) \text{ on } (-\infty,t_0]\cap T.$$

Proof. Suppose that the given representation is not proper canonical. Then by (I. 3) there exists an element $Z(\neq 0)$ of $\mathfrak{M}_{t_0}(B)$ which is independent of every $X(t)$, $t\leqq t_0$. Noting that Z can be expressed in the form

$$Z=\int^{t_0}f(u)dB(u),\qquad f\in \boldsymbol{L}^2(v),$$

we have

$$E(Z\cdot X(t))=0,\qquad \text{for every } t\leqq t_0,$$

which is identical with (I. 15). On the other hand

$$0\neq E(Z^2)=\int^{t_0}f(u)^2dv(u).$$

This shows that (I. 16) does not hold.

Conversely, if there is a function $f(u)$ satisfying (I. 15) but not satisfying (I. 16) for some t_0, then

$$Z=\int^{t_0}f(u)dB(u)$$

belongs to $\mathfrak{M}_t(B)$ but does not belong to $\mathfrak{M}_t$. Hence the representation is not proper canonical.

According to Karhunen [1], a stationary process which is purely non-deterministic and M_2-continuous always has its (moving average) representation, the kernel of which is a function of $t-u$. He also gave a kernel criterion for canonical (in our terminology) representation. One can easily see that our Theorem I. 7. is a generalization of Karhunen's theorem

Example 1. 4. Let $X_1(t)$ and $X_2(t)$ be defined by

$$X_1(t) = \int_0^t (2t-u)dB_1(u)$$

$$X_2(t) = \int_0^t (-3t+4u)dB_2(u)$$

where $B_i(t)$, $i=1, 2$, are ordinary Brownian motions. Then the two processes have the same probability distribution (cf. P. Lévy [4]), since they have the common covariance function $3ts-2s^2/3$ $(t>s)$. Using Theorem I. 7, we can prove that $(dB_1(t), 2t-u)$ is a proper canonical representation of $X_1(t)$. On the other hand

$$Z = \int_0^{t_0} u dB_2(u)$$

is independent of every $X_2(t)$ $(t \leqq t_0)$, which proves that the representation $(dB_2(t), -3t+4u)$ of $X_2(t)$ is not proper canonical—indeed it is not canonical.

Example 1. 5. (Particular case of Example I. 3). If we denote an ordinary Brownian motion by $B_0(t)$,

$$X(t) = \int_0^t (3-12u/t+10u^2/t^2)dB_0(u)$$

is again a Brownian motion. Here

$$Z_1 = \int_0^{t_0} u dB_0(u) \quad \text{and} \quad Z_2 = \int_0^{t_0} u^2 dB_0(u)$$

are independent of every $X(t)$ $(t \leqq t_0)$. Hence $(dB_0(t), 3-12u/t +10u^2/t^2)$ is not proper canonical. In fact, the canonical representation of $B_0(t)$ is $(dB_0(t), 1)$.

Section II. Multiple Markov Gaussian Processes.

§ II. 1. Simple Markov Gaussian Processes.

We intend to study multiple Markov Gaussian processes in this section, using the general theory of representation. All the processes to be discussed here are Gaussian processes with mean 0 satisfying the conditions ($\mathfrak{M}$. 1) and ($\mathfrak{M}$. 2). Furthermore we may assume that

($\mathfrak{M}$. 4) $\qquad \mathfrak{M}_t$ is continuous in t,

that is,

$$\lim_{t\to t_0} \mathfrak{M}_t \quad \text{exists and is equal to} \quad \mathfrak{M}_{t_0},$$

since we can easily remove the discontinuity of $\mathfrak{M}_t$.

First we shall treat a simple Markov Gaussian process. Though some of the results are well known, our presentation of the results will stress their specific probabilistic significance form our standpoint.

Let $Y(t)$ be a simple Markov process. As $Y(t)$ is Gaussian the simple Markov property is equivalent to the condition that, if $s \leqq t$,

$$E(Y(t)/\boldsymbol{B}_s) = \varphi(t, s)Y(s)\,, \tag{II. 1}$$

where $\varphi(t, s)$ is a real valued ordinary function of (t, s) (Doob [2]). This is also equivalent to

(II. 2) $\qquad Y(t)-\varphi(t, s)\, Y(s)$ is independent of every $Y(\tau)$, $\qquad \tau \leqq s$.

To avoid the case in which $Y(t)$ and $Y(s)$ are independent for $s \neq t$, let us assume that

(II. 3) $\qquad \Gamma(s, t) = E(Y(t)\cdot Y(s))$ never vanishes.

Then the equality

$$E(E(Y(t)/\boldsymbol{B}_{s'})/\boldsymbol{B}_s) = E(Y(t)/\boldsymbol{B}_s), \qquad \text{for every } s \leqq s' \leqq t$$

implies

$$\varphi(t, s')\varphi(s', s)Y(s) = \varphi(t, s)Y(s)\,.$$

Since $\Gamma(s, s) \neq 0$ by the assumption (II. 3), we have

$$\begin{cases} \varphi(t, s')\varphi(s', s) = \varphi(t, s), \\ \qquad \varphi(t, t) = 1, \end{cases} \tag{II. 4}$$

and we can prove that $\varphi(t, s)$ never vanishes. If we use the convention

$$\varphi(t, s) = \varphi(s, t)^{-1} \qquad \text{for } s > t ,$$

$\varphi(t, s)$ may be written as

$$\varphi(t, s) = f(t)/f(s) ,$$

where $f(t) = \varphi(t, s_0)$ with some fixed s_0.
Hence we have, from (II. 1),

$$E(f(t)^{-1}Y(t)/\boldsymbol{B}_s) = f(s)^{-1}Y(s) ,$$

which proves that $U(t) = f(t)^{-1}Y(t)$ is an additive process. Here we should note that the system of Borel fields relative to $U(t)$ is the same as that relative to $Y(t)$, since $f(t)$ never vanishes.

According to the Corollary to Theorem I. 6, $U(t)$ has a canonical representation, which has no discontinuous part as we assume ($\mathfrak{M}$. 4). Hence so does $Y(t)$:

$$Y(t) = f(t)U(t) = f(t)\int^t dU(u) .$$

Conversely, a process expressed in this form is obviously a simple Markov process provided that $f(t)$ never vanishes.

Summing up, we have

Theorem II. 1. *Under the assumption* ($\mathfrak{M}$. 1), ($\mathfrak{M}$. 2), ($\mathfrak{M}$. 4) *and* (II. 3), *a necessry and sufficient condition that $Y(t)$ is a simple Markov process is that it can be expressed in the form*

$$Y(t) = f(t)U(t) = f(t)\int^t dU(u) = \int^t f(t)g(u)dB(u) , \tag{II. 5}$$

where $U(t)$ is an additive process with the property ($\mathfrak{M}$. 4) (*$dB(t)$ is a continuous random measure*) *and $f(t)$ never vanishes.*

Making use of this theorem, we have (under the same assumptions)

Corollary 1. *Let $Y(t)$ be expressed in the form* (II. 5). *If $Y(t)$ is continuous in the mean, then $f(t)$ is continuous and $U(t)$ is continuous in the mean.*

Proof. If $Y(t)$ is continuous in the mean, then

$$\lim_{t \to t_0} E(Y(t)Y(s)) = E(Y(t_0)Y(s))$$

by the continuity of inner product in $\boldsymbol{L}^2(\Omega)$. By (II. 5), this can be written as,

$$\lim_{t\to t_0} f(t)f(s)E(U(t)U(s)) = f(t_0)f(s)E(U(s)^2), \qquad t>s,$$

since $U(t)$ is additive. Noting that $f(s)\neq 0$ and $E(U(s)^2)\neq 0$, we can see the continuity of $f(t)$. The continuity of $U(t)$ follows from $E(U(t)U(s))=\Gamma(t, s)/(f(t)f(s))$.

In particular, if $T=[0, \infty)$ and $E(U(t)^2)$ has a continuous derivative, $U(t)$ becomes an ordinary Brownian motion by the change of time scale.[1] In other words, $Y(t)$ has a canonical representation $(dB_0(t), f(t)\sigma(u))$ with Wiener's random measure $B_0(\cdot)$ and a proper canonical kernel $f(t)\sigma(u)$.

Corollary 2. *If $Y(t)$ is a stationary simple Markov process satisfying the conditions* ($\mathfrak{M}$. 1), ($\mathfrak{M}$. 2) *and* (II. 3), *then it has a version*

$$c\int_{-\infty}^{t} e^{-\lambda(t-u)}dB_0(u), \qquad \lambda>0. \tag{II. 6}$$

Proof. As is easily seen in the proof of Theorem II. 1, the covariance function γ of $Y(t)$ can be written in the form

$$\gamma(h) = f(t+h)f(t)\sigma(t)^2, \qquad h>0, \tag{II. 7}$$

even though we do not assume ($\mathfrak{M}$. 4). Putting $t=0$ in (II. 7), we have

$$f(h) = c_1\gamma(h),$$

and putting $h=0$ in (II. 7), we have

$$\sigma(t)^2 = c_2 f(t)^{-2} = c_3\gamma(h)^{-2}.$$

Hence it follows from (II. 7) that

$$c_3\gamma(h+t) = c_4\gamma(h)\cdot c_4\gamma(t).$$

Since γ is bounded above, we have

$$\gamma(h) = c^2e^{-\lambda|h|}, \qquad \lambda>0.$$

§ II. 2. Multiple Markov Gaussian processes.

In this article we shall define N-ple Markov process as a generalization of simple Markov process and study its properties. The property (II. 2) for a simple Markov process suggests that it is natural to give the following

Definition II. 1. If $\{E(Y(t_i)/\boldsymbol{B}_{t_0})\}$, $i=1, 2, \cdots, N$, are linearly

1) See Seguchi-Ikeda [1].

independent for any $\{t_i\}$ with $t_0 \leqq t_1 < t_2 < \cdots < t_N$, and if $\{E(Y(t_i)/\boldsymbol{B}_{t_0})\}$, $i=1, 2, \cdots, N, N+1$, are linearly dependent for any $\{t_i\}$ with $t_1 < t_2 < \cdots < t_{N+1}$, then $Y(t)$ is called *N-ple Markov process*.

A simple Markov process is a 1-ple Markov process in this sense only if it satisfies (II. 3).

Theorem II. 2. *If $Y(t)$ is an N-ple Markov process satisfying* ($\mathfrak{M}$. 1), ($\mathfrak{M}$. 2), ($\mathfrak{M}$. 3) *and* ($\mathfrak{M}$. 4), *it has a version $X(t)$ expressed in the following form*

$$X(t) = \int^t \sum_{i=1}^N f_i(t) g_i(u) dB(u) \tag{II. 8}$$

with a proper canonical kernel $\sum_{i=1}^N f_i(t) g_i(u)$, where $\{f_i(t)\}$, $i=1, 2, \cdots, N$, satisfy

$$\det(f_i(t_j)) \neq 0, \qquad \text{for any } N \text{ different } t_j, \tag{II. 9}$$

and $\{g_i(u)\}$, $i=1, 2, \cdots, N$, are linearly independent as the elements of $L^2(v\,;\,t)$[2] *for every t.*

Further the covariance function Γ of $Y(t)$ can be written in the form

$$\Gamma(s, t) = \sum_{i=1}^N f_i(t) h_i(s), \qquad s < t,$$

where $\{f_i(t)\}$, $i=1, 2, \cdots, N$, are the same as above and $\{h_i(s)\}$, $i=1, 2, \cdots, N$, are linearly independent.

Proof. By the assumptions there exists a proper canonical representation $(dB(t), F(t, u))$ of $Y(t)$:

$$Y(t) \sim X(t) = \int^t F(t, u) dB(u), \qquad (\sim : \text{equivalent in law})$$

It is sufficient to determine the form of $F(t, u)$ in the region $D = \{(u\ t,)\,;\, u < t\}$ instead of $\bar{D}$ on account of the assumption ($\mathfrak{M}$. 4).

If $Y(t)$ is an N-ple Markov process, then we can prove that, for any $\{t_j\}$ with $t_1 < t_2 < \cdots < t_N$ and for any $\tau > t_N$ there exist $\{a_j(\tau\,;\, t_1, t_2, \cdots, t_N)\}$ $j=1, 2, \cdots, N$, such that

$$Y(\tau) - \sum_{j=1}^N a_j(\tau\,;\, t_1, t_2, \cdots, t_N) Y(t_j) \tag{II. 10}$$

is independent of every $Y(\sigma)$, $\sigma \leqq t_1$. Therefore we have

2) $L^2(v\,;\,t) = \{\varphi\,;\, \varphi \in \boldsymbol{L}^2(v)$ and $\varphi(u) = 0$ for $u > t\}$.

$$\int^{\sigma} F(\sigma, u)\{F(\tau, u) - \sum_{j=1}^{N} a_j(\tau\,;\, t_1, t_2, \cdots, t_N)F(t_j, u)\}\, dv(u) = 0\,,$$

using the representation of $Y(t)$. Since $F(\sigma, u)$ is a proper canonical kernel, it is equivalent to

(II. 11) $$F(\tau, u) = \sum_{j=1}^{N} a_j(\tau\,;\, t_1, t_2, \cdots, t_N)F(t_j, u)$$

as an element of $\boldsymbol{L}^2(v\,;\, t_1)$ (see Theorem I. 7).

Take N different $\{s_j\}$ with $s_1 < s_2 < \cdots s_N$, arbitrarily in the interval $(-\infty, t_1) \cap T$. Expressing $F(\tau, u)$ and $\{F(t_j, u)\}$, $j=1, 2, \cdots$, N, in (II. 11) by $\{F(s_j, u)\}$, $j=1, 2, \cdots, N$, in the same way as in (II. 11), we get

$$\begin{aligned}&\sum_{j=1}^{N} a_j(\tau\,;\, s_1, s_2, \cdots, s_N)F(s_j, u)\\ &\quad = \sum_{k,j=1}^{N} a_k(\tau\,;\, t_1, t_2, \cdots, t_N)a_j(t_k\,;\, s_1, s_2, \cdots, s_N)F(s_j, u)\end{aligned}$$

as an element of $L^1(v\,;\, s_1)$.

Here $\{F(s_j, u)\}$, $j=1, 2, \cdots, N$, must be linearly independent functions in $\boldsymbol{L}^2(v\,;\, s_1)$; in fact, if this is not true, then $\{E(Y(s_j)/\boldsymbol{B}_{s_1})\}$, $j=1, 2, \cdots, N$, are linearly dependent, which contradicts our assumption. Hence we have

(II. 12) $$\begin{aligned}&\sum_{k=1}^{N} a_k(\tau\,;\, t_1, t_2, \cdots, t_N)a_j(t_k\,;\, s_1, s_2, \cdots, s_N)\\ &\quad = a_j(\tau\,;\, s_1, s_2, \cdots, s_N), \qquad \text{for every } j.\end{aligned}$$

Now we can prove

(II. 13) $$\det\,(a_j(t_k\,;\, s_1, s_2, \cdots, s_N)) \neq 0\,,$$

because

$$F(t_j, u) = \sum_{k=1}^{N} a_k(t_j\,;\, s_1, s_2, \cdots, s_N)F(s_k, u), \qquad j = 1, 2, \cdots, N,$$

are linearly independent functions in $\boldsymbol{L}^2(v\,;\, s_1)$. Therefore we have

(II. 14) $$\boldsymbol{a}(\tau, \boldsymbol{t}) = \boldsymbol{a}(\tau, \boldsymbol{s})B(\boldsymbol{s}, \boldsymbol{t})$$

by (II. 12) and (II. 13), where

$$\boldsymbol{a}(\tau, \boldsymbol{s}) = (a_1(\tau\,;\, s_1, s_2, \cdots, s_N), \cdots, a_N(\tau\,;\, s_1, s_2, \cdots, s_N))$$

and

$$B(\boldsymbol{t}, \boldsymbol{s}) = (b_{jk}(t_1, t_2, \cdots, t_N\,;\, s_1, s_2, \cdots, s_N)), \qquad j, k = 1, 2, \cdots, N,$$

with det $(B(\boldsymbol{t}, \boldsymbol{s})) \neq 0$. Taking N different $s_i'(<s_1)$, we have

$$\boldsymbol{a}(\tau, \boldsymbol{t}) = a(\tau, \boldsymbol{s})B(\boldsymbol{s}, \boldsymbol{t}) = \boldsymbol{a}(\tau, \boldsymbol{s}')B(\boldsymbol{s}', \boldsymbol{s})B(\boldsymbol{s}, \boldsymbol{t}),$$
$$\boldsymbol{a}(\tau, \boldsymbol{t}) = a(\tau, \boldsymbol{s}')B(\boldsymbol{s}', \boldsymbol{t}),$$

by (II. 14). Hence we have

(II. 15) $$B(\boldsymbol{s}', \boldsymbol{s})B(\boldsymbol{s}, \boldsymbol{t}) = B(\boldsymbol{s}', \boldsymbol{t}).$$

Fix all t_j's and define $\boldsymbol{f}_{\boldsymbol{s}}(\tau)$, $\boldsymbol{s}=(s_1, s_2, \cdots, s_N)$, by

$$\boldsymbol{f}_{\boldsymbol{s}}(\tau) = \boldsymbol{a}(\tau, \boldsymbol{s})B(\boldsymbol{s}, \boldsymbol{t}) \qquad \text{for} \quad \tau > s_N,$$

where $\boldsymbol{s}$ is any N-ple $(s_1, s_2, \cdots, s_N)$ such that $t_N > t_{N-1} \cdots > t_1 > s_N > s_{N-1} > \cdots > s_1$. Then we can use (II. 15) to see that $\boldsymbol{f}_{\boldsymbol{s}'}$ is an extension of $\boldsymbol{f}_{\boldsymbol{s}}(\tau)$ if $s_N > s_{N-1} > \cdots > s_1 > s_N' > s_{N-1}' > \cdots > s_1'$. Hence there exists a common extension for all $\boldsymbol{f}_{\boldsymbol{s}}(\tau)$'s. We denote this common extension with $\boldsymbol{f}(t) = (f_1(t), \cdots, f_N(t))$. Obviously these $f_i(t)$ satisfy (II. 9) on account of (II. 13) and the definition of $\boldsymbol{f}_{\boldsymbol{s}}(\tau)$.

Take $u \in T^0$ and fix it. If $\tau > t_N > \cdots > t_1 > s_N > \cdots > s_1 > u$, then we have

$$\begin{aligned} F(\tau, u) &\Big(\equiv \sum_{j=1}^{N} a_j(\tau\,;\, t_1, t_2, \cdots, t_N) F(t_j, u)\Big) \\ &= \boldsymbol{a}(\tau, \boldsymbol{t})\boldsymbol{F}(\boldsymbol{t}, u)^* \qquad (\boldsymbol{F}(\boldsymbol{t}, u) = (F(t_1, u), \cdots, F(t_N, u)) \\ &= \boldsymbol{f}(\tau)B(\boldsymbol{s}, \boldsymbol{t})^{-1}\boldsymbol{F}(\boldsymbol{t}, u)^* \\ &= \boldsymbol{f}(\tau)\boldsymbol{g}(u, \boldsymbol{s}, \boldsymbol{t})^*, \qquad (\boldsymbol{g}(u, \boldsymbol{s}, \boldsymbol{t}) = \boldsymbol{F}(\boldsymbol{t}, u)B(\boldsymbol{s}, \boldsymbol{t})^{*-1}). \end{aligned}$$

For $\tau > t_N' > \cdots > t_1' > s_N' \cdots > s_1'$, this is equal to

$$\boldsymbol{f}(\tau)\boldsymbol{g}(u, \boldsymbol{s}', \boldsymbol{t}')^*,$$

so that

$$\boldsymbol{f}(\tau)\boldsymbol{g}(u, \boldsymbol{s}, \boldsymbol{t})^* = \boldsymbol{f}(\tau)\boldsymbol{g}(u, \boldsymbol{s}', \boldsymbol{t}')^*$$

for $\tau > \max(t_N', t_N)$. Since $\boldsymbol{f}$ satisfies (II. 9), we have

$$\boldsymbol{g}(u, \boldsymbol{s}, \boldsymbol{t}) = \boldsymbol{g}(u, \boldsymbol{s}', \boldsymbol{t}').$$

Therefore $\boldsymbol{g}(u) = \boldsymbol{g}(u, \boldsymbol{s}, \boldsymbol{t})$ is well defined as a function of u, and

$$F(t, u) = \boldsymbol{f}(t)\boldsymbol{g}(u)^* = \sum_{i=1}^{N} f_i(t) g_i(u),$$

where $\{g_i(u)\}$, $i=1, 2, \cdots, N$, are linearly independent as elements of $\boldsymbol{L}^2(v\,;\, t)$, since $\{F(t_j, u)\}$, $j=1, 2, \cdots, N$, are linearly independent.

Further we have

$$\Gamma(t, s) = \sum_{i=1}^{N} f_i(t)(\sum_{j=1}^{N} f_j(s)\int^s g_i(u)g_j(u)dv(u))$$
$$\equiv \sum_{i=1}^{N} f_i(t)h_i(s) .$$

Then if

$$\sum_{i=1}^{N} a_i h_i(s) \equiv 0$$

for some constants $a_1, a_2, \cdots, a_N$,

$$\int^s (\sum_{j=1}^{N} f_j(s)g_j(u))(\sum_{i=1}^{N} a_i g_i(u))dv(u) \equiv 0 .$$

Noting that $\sum_{j=1}^{N} f_j(s)g_j(u)$ is a proper canonical kernel,- we have

$$\sum_{i=1}^{N} a_i g_i(u) = 0 .$$

Hence all the a_i must be 0. Thus we have proved the theorem completely.

A kernel $\sum_{i=1}^{N} f_i(t)g_i(u)$ satisfying the conditions stated in Theorem II. 2 is called a *Goursat kernel* of order N.

It should be noted that the expression (II. 13) is not uniquely determined, but the number of the summand is independent of the special way of expression as we have seen in the proof above.

As another remark, we should note that a process with a version of the form (II. 9) is not always an N-ple Markov process. In order that the converse of this theorem holds, it is sufficient to impose some regularity condition on the kernel as we shall see in § II. 4. 3°).

Example II. 1. If $f(t)$ is a function which is 1 for rational t and 0 for irrational t, then $X(t) = f(t)B_0(t)$, $0 \leq t < \infty$, is not a 1-ple Markov process, though it is expressed in the form

$$X(t) = \int_0^t f(t)dB_0(u) .$$

§ II. 3. Stationary multiple Markov Gaussian processes.

Let $Y(t)$, $t \in T = (-\infty, \infty)$, be a stationary Gaussian process with mean 0 satisfying the conditions ($\mathfrak{M}$. 1), ($\mathfrak{M}$. 2) and ($\mathfrak{M}$. 4′).

($\mathfrak{M}$. 4′) $Y(t)$ is continuous in the mean.

Then by Karhunen's theory, we can see that $Y(t)$ has a canonical representation and that it is expressed in the form

$$Y(t) = \int_{-\infty}^{t} F(t-u)dB_0(u), \qquad E(dB_0(u))^2 = du ,$$

using a canonical kernel $F(t-u)$. This canonical kernel is uniquely determined up to sign and is proper canonical.[3] Thus, in order to study the multiple Markov process for stationary case, it is sufficient for us to study the canonical kernel $F(t-u)$.

Lemma II. 1. *Let* $\{f_i(t)\}$, $i=1, 2, \cdots, N$, *satisfy* (II. 9) *and let* $\{g_i(u)\}$, $i=1, 2, \cdots, N$, *be linearly independent as elements of* $L^2((-\infty, c])$ *for every c. If* $\sum_{i=1}^{N} f_i(t)g_i(u)$ *is a function of* $t-u$ *in the domain* $\bar{D}=\{(u, t) ; u \leqq t\}$, *then* $\{f_i(t)\}$ *is a fundamental system of solutions of a certain linear differential equation of order N with constant coefficients, and* $\{g_i(u)\}$ *is also a fundamental system of solutions of its adjoint differential equation.*

Proof. First we consider $F(t-u)=\sum_{i=1}^{N} f_i(t)g_i(u)$ in the region $D_0=\{(u, t) ; u\leqq 0, t\geqq 0\}$. Let $\mathfrak{D}_0$ be the set of all C^∞-functions whose carriers are compact sets lying in the interval $(-\infty, 0]$. Then

$$(F*\varphi)(t) = \int_{-\infty}^{\infty} F(t-u)\varphi(u)du$$

is well defined by the assumption and belongs to $C^\infty((0, \infty))$ for every $\varphi \in \mathfrak{D}_0$ (Schwartz [1]).

Next we shall prove that there exist functions $\varphi_j(u)$, $j=1, 2, \cdots, N$ in $\mathfrak{D}_0$ such that

$$\text{(II. 16)} \qquad \det((g_i, \varphi_j)) \neq 0, \qquad i, j = 1, 2, \cdots, N,$$

where (g, φ) denotes the inner product of g and φ in $L^2(T)$. In fact there exists a function $\varphi_1(u) \in \mathfrak{D}_0$ such that $(g_1, \varphi_1) \neq 0$. (If there were no such function, $g_1(u)$ must vanish on $(-\infty, 0)$). Inductively, suppose that $\varphi_1, \varphi_2, \cdots, \varphi_n \in \mathfrak{D}_0$ are chosen so that

$$\det((g_i, \varphi_j)) \neq 0, \qquad i, j = 1, 2, \cdots, n .$$

And consider the determinant

3) For proof see Karhunen [1]. Also, M. Nisio gave another proof, which I knew by private communication.

$$\begin{vmatrix} (g_1, \varphi_1) & (g_1, \varphi_2) & \cdots & (g_1, \varphi_n) & (g_1, \varphi) \\ (g_2, \varphi_1) & (g_2, \varphi_2) & \cdots & (g_2, \varphi_n) & (g_2, \varphi) \\ \cdots & \cdots & \cdots & \cdots & \cdots \\ (g_{n+1}, \varphi_1) & (g_{n+1}, \varphi_2) & \cdots & (g_{n+1}, \varphi_n) & (g_{n+1}, \varphi) \end{vmatrix}.$$

If this determinant vanishes for every $\varphi \in \mathfrak{D}_0$, we get

$$\Delta_1(g_1, \varphi) + \Delta_2(g_2, \varphi) + \cdots + \Delta_{n+1}(g_{n+1}, \varphi) = 0, \qquad \varphi \in \mathfrak{D}_0,$$

by expanding this determinant with respect to the last column; since φ is arbitrary, we have

$$\Delta_1 g_1(u) + \Delta_2 g_2(u) + \cdots + \Delta_{n+1} g_{n+1}(u) = 0 \qquad \text{a.e. in } (-\infty, 0).$$

This contradicts the assumption that $\{g_i(u)\}$, $i=1, 2, \cdots, n+1$, are linearly independent, since $\Delta_{n+1} \neq 0$ by the assumption of induction. Thus we can take $\{\varphi_j(u)\}$, $j=1, 2, \cdots, N$, so that (II. 16) holds.

On the other hand, considering

$$(F*\varphi_j)(t) = \sum_{i=1}^{N} (g_i, \varphi_j) f_i(t)$$

and (II. 16), we can see that $f_i(t)$ is a linear combination of $(F*\varphi_j)(t)$ which belongs to $\mathbf{C}^\infty((0, \infty))$ Hence $f_i(t) \in \mathbf{C}^\infty((0, \infty))$ for every i, so that $F(t) \in \mathbf{C}^\infty((0, \infty))$. From these facts we can see that $g_i(u) \in \mathbf{C}^\infty((-\infty, 0))$ for every i.

Applying similar arguments to every region $D_a = \{(u, t); u \leqq a, t \geqq a\}$, $a \in T$, we can see that

$$f_i(\cdot),\ g(\cdot) \in \mathbf{C}^\infty(T^\circ), \qquad i = 1, 2, \cdots, N.$$

Thus we have

$$\text{(II. 17)} \qquad \sum_{i=1}^{N} f_i^{(k)}(t) g_i(u) = \frac{\partial^k}{\partial t^k} F(t-u) = (-1)^k \frac{\partial^k}{\partial u^k} F(t-u)$$
$$= (-1)^k \sum_{i=1}^{N} f_i(t) g_i^{(k)}(u), \qquad k = 0, 1, \cdots, N.$$

Putting $u=0$, we get

$$\frac{d^k}{dt^k} F(t) = (-1)^k \sum_{i=1}^{N} f_i(t) g_i^{(k)}(0), \qquad k = 0, 1, \cdots, N.$$

Therefore there exist $b_0, b_1, \cdots, b_N$ such the $\sum_{i=1}^{N} |b_i| > 0$ and that

$$b_0 F^{(N)}(t) + b_1 F^{(N-1)}(t) + \cdots + b_N F(t) = 0, \qquad t > 0,$$

so that

$$b_0F^{(N)}(t-u)+b_1F^{(N-1)}(t-u)+\cdots+b_NF(t-u)=0, \qquad t>u,$$

namely

$$\sum_{i=1}^{N}(b_0f_i^{(N)}(t)+b_1f_i^{(N-1)}(t)+\cdots+b_Nf_i(t))g_i(u)=0, \qquad t>u.$$

Since $\{g_i(u)\}$ are linearly independent in $\boldsymbol{L}^2((-\infty, t])$,

$$b_0f_i^{(N)}(t)+b_1f_i^{(N-1)}(t)+\cdots+b_Nf_i(t)=0, \qquad i=1, 2, \cdots, N.$$

If $b_0=0$, then $f_i(t)$ satisfies

$$c_1f_1(t)+c_2f_2(t)+\cdots+c_Nf_N(t)=0. \qquad \sum_{i=1}^{N}|c_i|\neq 0,$$

as a system of N solutions of linear differential equation of at most order $N-1$, which contradicts (II. 9). Hence $\{f_i(t)\}$ is a fundamental system of solutions of linear differential equation of order N with constant coefficients.

Exactly in the same way, we can prove the assertion for $\{g_i(u)\}$.

By the well-known fact in the theory of linear ordinary differential equations, $F(t-u)$ is a linear combination of the functions of the following types:

$$\begin{aligned} &e^{-\lambda(t-u)}\sin\mu(t-u), \quad t^ku^{n-k}e^{-\lambda(t-u)}\sin\mu(t-u), \qquad (\mu\neq 0)\\ \text{(II. 18)}\quad &e^{-\lambda(t-u)}\cos\mu(t-u), \quad t^ku^{n-k}e^{-\lambda(t-u)}\cos\mu(t-u), \qquad (\mu \text{ may be } 0)\\ &\qquad 0\leq k\leq n, \quad n\leq N. \end{aligned}$$

By Theorem II. 2 and Lemma II. 1 we have

Theorem II. 3. *If $Y(t)$ is a stationary N-ple Markov process satisfying the conditions* ($\mathfrak{M}$. 1), ($\mathfrak{M}$. 2) *and* ($\mathfrak{M}$. 4′), *then its canonical kernel is a linear combination of the functions described in* (II. 18) *with* $\lambda>0$.

The functions in (II. 18) (for $\mu\neq 0$) are split into two terms of the form $f(t)g(u)$. Therefore the number of the terms in the expression of the kernel is exactly N.

Corollary. *The spectral measure of a stationary N-ple Markov process is absolutely continuous with a density function of the following type:*

$$|Q(i\lambda)/P(i\lambda)|^2,$$

where P is a polynomial of degree N and Q is also a polynomial of degree at most $N-1$.

Proof. The spectral density function is obtained by Fourier transform of $F(\cdot)$. Hence our assertion is obvious.

This process is a component process of an N-dimensional stationary simple markov process in Doob's sense, Doob [1].

§ II. 4. Some special multiple Markov Gaussian processes.

1°) Let $Y(t)$ be a stationary N-ple Markov Gaussian process which is differentiable (with respect to $L^2(\Omega)$-norm) up to $N-1$ times. Such process plays an important role in the study of N-ple Markov processes as is seen in Doob's work [1].

Now let us assume that $Y(t)$ is expressed in the form

$$Y(t) \sim X(t) = \int_{-\infty}^{t} F(t-u)dB_0(u)$$

with a proper canonical kernel

$$F(t-u) = \sum_{i=1}^{N} f_i(t)g_i(u) .$$

Then we have the following

Theorem II. 4. *Let $X(t)$ be a stationary N-ple Markov process. Then*

i) a necessary and sufficient condition that $X(t)$ is differentiable is $F(0)=0$,

ii) in this case, there exists a complex number λ such that

$$e^{\lambda t}\frac{d}{dt}e^{-\lambda t}X(t) \tag{II. 19}$$

exists and it is a stationary $(N-1)$-ple Markov process.

Proof. i) If $h>0$,

$$\frac{1}{h}(X(t+h)-X(t)) = \frac{1}{h}\int_{t}^{t+h} F(t+h-u)dB_0(u) + \frac{1}{h}\int_{-\infty}^{t} \{F(t+h-u)-F(t-u)\}dB_0(u) .$$

Since $F(t-u)$ is analytic in D, the first term of the right hand side tends to 0 (in the mean) as h tends to 0 under the assumption $F(0)=0$. Hence $\lim_{h\to 0+} h^{-1}(X(t+h)-X(t))$ exists. Similarly $\lim_{h\to 0-} h^{-1}(X(t+h)-X(t))$ exists and

$$X'(t) = \int_{-\infty}^{t} \frac{\partial}{\partial t}F(t-u)dB_0(u) . \tag{II. 20}$$

Conversely, if $X(t)$ is differentiable, then

$$dX(t) = F(0)dB_0(t) + dt\int_{-\infty}^{t} \frac{\partial}{\partial t}F(t-u)dB_0(u)$$

will be of order dt, so that $F(0)$ should vanish.

ii) As we have seen in Theorem II. 3, $f_i(t)$ is a solution of a linear differential equation with constant coefficients. If we choose one of the characteristic roots of the differential equation, say λ,

$$e^{\lambda t}\frac{d}{dt}e^{-\lambda t}F(t-u)$$

is obviously a proper canonical Goursat kernel of order $N-1$. The existence of (II. 19) and the stationary property are obvious.

When λ is real (II. 19) is real valued process. When λ is complex, say $\lambda=\lambda_1+i\lambda_2$, (II. 19) is complex valued process, but

$$f_\lambda(t)\frac{d}{dt}f_\lambda(t)^{-1}X(t), \qquad f_\lambda(t) = e^{\lambda_1 t}\cos\lambda_2 t\,,$$

is a real valued stationary process.

If $F(t)$ satisfies the conditions

$$F(0) = F'(0) = \cdots = F^{(N-1)}(0) = 0\,,$$

$X(t)$ is differentiable $N-1$ times. Then we can take a sequence of complex numbers $\lambda_1, \lambda_2, \cdots, \lambda_{N-1}$ such that

$$\text{(II. 21)} \qquad e^{\lambda_i t}\frac{d}{dt}e^{(\lambda_{i-1}-\lambda_i)t}\cdots\frac{d}{dt}e^{(\lambda_1-\lambda_2)t}\frac{d}{dt}e^{-\lambda_1 t}X(t) \equiv X^{[i]}(t)$$

exists and it is a stationary $(N-i)$-ple Markov process.

Such a process was studied by Doob [1] and the formula (II. 21) suggests more general differential operator which will appear in 3°)

2°) We shall now discuss a multiple Markov process with a homogeneous canonical kernel; $F(t, u)$ is called to be homogeneous function of degree α if $F(ct, cu)=c^\alpha F(t, u)$. This process can be transformed into a stationary process by time change by virtue of the following

Lemma II. 2. *(P. Lévy) Let $X(t)$ be expressed as*

$$\text{(II. 22)} \qquad X(t) = \int_0^t F(t, u)dB_0(u)$$

with a proper canonical homogeneous (of degree α) kernel $F(t, u)$.

Then $t^{-\alpha-1/2}X(t)$ *is a stationary process of* $\log t$ *(P. Lévy [4 ; p. 141]).*

Applying this lemma to N-ple Markov process, we get

Theorem II. 5. *Let* $X(t)$ *be an* N*-ple Markov process and be expressed as in Lemma* II. 2. *Then* $e^{-(2\alpha+1)t}X(e^{2t})$ *is a stationary* N*-ple Markov process. Conversely any stationary* N*-ple Markov process* $\tilde{X}(t)$ *is an* N*-ple Markov process with homogeneous kernel of degree* 0 *changing the time parameter from* t *to* e^t *:*

$$\sqrt{t}\,\tilde{X}((\log t)/2) = X(t)\,.$$

Proof. The first part of the theorem is an immediate consequence of Lévy's lemma, if we notice that N-ple Markov property is invariant under such time change.

If $X(t)$ is a stationary N-ple Markov process, it is the sum of the processes of the following types:

$$\text{(II. 23)}\qquad \int_{-\infty}^{t} e^{-\lambda(t-u)}dB_0(u), \quad \int_{-\infty}^{t}(t-u)^k e^{-\lambda(t-u)}dB_0(u)\,.$$

Changing the time scale, they become

$$\text{(II. 23}'\text{)}\qquad \frac{1}{\sqrt{2}}\int_0^t (u/t)^{(\lambda-1)/2}d\tilde{B}_0(u), \quad \left(\frac{1}{\sqrt{2}}\right)^{k+1}\int_0^t (\log(u/t))^k(u/t)^{(\lambda-1)/2}d\tilde{B}_0(u)$$

respectively. Hence $X(t)$ is an N-ple Markov process with homogeneous kernel of degree 0. If λ is complex, these expressions are not real, but we can reduce them to real ones by the same procedure as used in the proof of Theorem II. 4. i).

3°) Let us generalize the results obtained in 1°) and 2°) to the case in which $X(t)$ is a general N-ple Markov process: our results include also those which were discussed by Dolph-Woodbury [1] and Lévy [4]. For the sake of brevity we shall assume $T=[0, \infty)$. We can discuss stationary N-ple Markov processes with parameter $\in(-\infty, \infty)$ in this scheme, if we apply to it the time change used in 2°).

We shall consider a process $X(t)$ which is expressed in the form

$$\text{(II. 24)}\qquad X(t) = \int_0^t \sum_{i=1}^{N} f_i(t)g_i(u)dB_0(u)\,,$$

with a proper canonical Goursat kernel $\sum_{i=1}^{N} f_i(t)g_i(u)$.

Hereafter (throughout this article), we shall always impose the following conditions on the kernel:

(A. 1) $f_i, g_i \in \boldsymbol{C}^{\infty}(T^{\circ})$, for every i,

(A. 2) f_i and $W(g_1, g_2, \cdots, g_i)$ never vanish for every i,

where $W(g_1, g_2, \cdots, g_i)$ is the Wronskian of $\{g_j\}$, $j=1, 2, \cdots, i$.

By these assumptions, we can find functions $v_0, v_1, \cdots, v_{N-1}$ such that

$$g_i(u) = (-1)^{N-i} v_0(u) \int_0^u v_1(u_1) \int_0^{u_1} v_2(u_2) \int_0^{u_2} \cdots \int_0^{u_{N-i-1}} v_{N-i}(u_{N-i})(du)^{N-i} \tag{II. 25}$$

and that

$$\begin{cases} v_i(u) \in \boldsymbol{C}^{\infty}(T^0), \\ v_i(u) \text{ never vanishes}, \qquad i = 0, 1, \cdots, N-1. \end{cases} \tag{II. 26}$$

Using these functions $\{v_i(u)\}$, $i=0, 1, \cdots, N-1$, and a function $v_N(u)$ satisfying (II. 26), we can define measures

$$m_i(M) = \int_M v_i(u)du, \quad M \in \boldsymbol{B}_T, \quad i = 0, 1, \cdots, N,$$

and the following differential operators:

$$L_t = \frac{d}{dm_0} \frac{d}{dm_1} \cdots \frac{d}{dm_{N-1}} \cdot \frac{1}{v_N(t)} \cdot,$$

$$L_t^{(j)} = \frac{d}{dm_j} \frac{d}{dm_{j+1}} \cdots \frac{d}{dm_{N-1}} \cdot \frac{1}{v_N(t)} \cdot,$$

$$L_u^* = \text{the adjoint operator of } L_t$$

$$\equiv \frac{d}{dm_N} \frac{d}{dm_{N-1}} \cdots \frac{d}{dm_1} \cdot \frac{1}{v_0(u)} \cdot,$$

$$L_u^{*(j)} = \frac{d}{dm_{N-j}} \frac{d}{dm_{N-j-1}} \cdots \frac{d}{dm_1} \cdot \frac{1}{v_0(u)} \cdot .$$

We shall often use the following notations:

$$F(t, u) = \sum_{i=1}^{N} f_i(t) g_i(u), \qquad u \leqq t,$$

$$F^{(i)}(t, u) = \frac{\partial^i}{\partial t^i} F(t, u),$$

$$F^{[j]}(t, u) = L_t^{(N-j)} F(t, u).$$

Theorem II. 6. *Let* $X(t)$ *be defined by* (II. 24) *and let the canonical kernel of its representation satisfies the conditions* (A.1) *and* (A.2). *If*

(II. 27) $F(t, t) \equiv F^{(1)}(t, t) \equiv \cdots F^{(N-2)}(t, t) \equiv 0$ *and* $F^{(N-1)}(t, t)$ *never vanishes,*

then we have

i) $X^{(i)}(t) = \frac{d^i}{dt^i} X(t)$ *exists for every* $i \leq N-1$,

ii) $X(t)$ *satisfies the equation*

$$L_t X(t) = B_0'(t)\,, \tag{II. 28}$$

where $B_0'(t)$ *is a derivative of* $B_0(t)$ *in the symbolic sense, so that* (II. 28) *means* $dL_t^{(1)}X(t) = v_0(t)dB_0(t)$, *and the measure* m_N *associated witn* L_t *should be taken appropriately.*

Proof. i) is proved in the same way as in the stationary case (Theorem II. 4, i)).

ii) Define $v_N(t) = f_1(t)$. Since $v_N(t)^{-1} \cdot F(t,t) \equiv 0$ by assumption, we can prove the existence of $\frac{d}{dt} v_N(t)^{-1} X(t)$; namely $L_t^{(N-1)} X(t)$ exists. Similarly, we can prove the existence of $L_t^{(i)} X(t)$, $i = N-2, N-3, \cdots, 1$, since $F^{[1]}(t, t) \equiv F^{[2]}(t, t) \equiv \cdots \equiv F^{[N-2]}(t, t) \equiv 0$.

Rewriting (II. 27) in the following forms

$$\sum_{i=1}^{N} f_i^{(k)}(t) g_i(t) = 0, \qquad k = 0, 1, \cdots, N-1,$$

$$\sum_{i=1}^{N} f_i^{(N-1)}(t) g_i(t) = a(t), \qquad \text{with } a(t) \equiv F^{(N-1)}(t, t)\,,$$

we can see that $F(t, u)/a(u)$ is a Riemann function for a certain linear differential equation

$$\widetilde{L}_t f = 0$$

of order N. The fundamental system of solutions of its adjoint differential equation

$$\widetilde{L}_u^* g = 0$$

is $\{g_i(u)/a(u)\}$, $i = 1, 2, \cdots, N$, as is well known. Hence $\widetilde{L}_u^* = L_u^* \cdot v(u)$ with a certain function $v(u)$. Thus we can prove that $\widetilde{L}_t = v(t) \cdot L_t$. By the property of Riemann function $v(t)$ must be 1, and therefore we have

$$f_i(t) = v_N(t) \int_0^t \mathrm{dm}_{N-1} \int \mathrm{dm}_{N-2} \int \cdots \int \mathrm{dm}_{N-i+1}, \tag{II. 29}$$

$$i = 1, 2, \cdots, N,$$

which proves that

$$L_t^{(1)}F(t, u) = g_N(u) \equiv v_0(u) .$$

Hence

$$L_t^{(1)}X(t) = \int_0^t L_t^{(1)}F(t, u)dB_0(u) = \int_0^t v_0(u)dB_0(u) .$$

Thus we can prove ii).

Combining this Theorem II. 6 with Theorem II. 9. in §II. 5 we can see that this $X(t)$ process is an N-ple *Markov process in the restricted sense* in Lévy's terminology

Corollary. *Under the same assumptions as in Theorem* II. 6, $L_t^{(i)}X(t)$ *is an* $(N-i)$*-ple Markov process in the restricted sense.*

Proof. As is easily seen in the proof of the theorem above,

$$L_t^{(i)}X(t) = \int_0^t L_t^{(i)}F(t, u)dB_0(u) .$$

Since $L_t^{(i)}F(t, u)$ is a Riemann function for the differential equation

$$\frac{d}{dm_1} \cdots \frac{d}{dm_{i-1}} f = 0 ,$$

our assertion is obvious.

Theorem II. 7. *Let* $\{v_i(u)\}$, $i=0, 1, \cdots, N$, *be functions satisfying the condition* (II. 26). *If we define* $f_i(t)$*'s and* $g_i(u)$*'s by* (II. 29) *and* (II. 25) *respectively, then*

i) $F(t, u) = \sum_{i=1}^{N} f_i(t)g_i(u)$ *is a proper canonical kernel,*

ii) *a process defined by*

$$X(t) = \int_0^t \sum_{i=1}^{N} f_i(t)g_i(u)dB_0(u)$$

is an N-ple Markov process,

iii) $F(t, t) \equiv F^{(1)}(t, t) \equiv \cdots \equiv F^{(N-2)}(t, t) \equiv 0$ *and* $F^{(N-1)}(t, t)$ *never vanishes; namely Theorem* II. 6 *holds for this process.*

Proof. i). We shall prove i) by using the kernel criterion which was given in §I. 4.

Suppose that

$$\int_0^t F(t, u)\varphi(u)du \equiv 0 \qquad \text{in } (0, t_0)$$

for some $t_0 \in T$ and some $\varphi \in \boldsymbol{L}^2([0, t_0])$. Writing it in the form

$$\sum_{i=1}^{N} f_i(t)\int_0^t g_i(u)\varphi(u)du \equiv 0 ,$$

and multiplying $v_N(t)^{-1}$, we can obtain its derivative (in Radon-Nikodym sense)

$$\sum_{i=2}^{N} f_i^{[1]}(t)\int_0^t g_i(u)\varphi(u)du + (\sum_{i=1}^{N} f_i(t)g_i(t))\varphi(t) = 0$$

a.e. in $(0, t_0)$.

Since the second term of the left hand side vanishes (because $\sum_{i=1}^{N} f_i(t)g_i(u)$ is a Riemann function corresponding to L_t), we have

$$\sum_{i=1}^{N} f_i^{(1)}(t)\int_0^t g_i(u)\varphi(u)du \equiv 0, \qquad \text{in } (0, t_0),$$

by taking an appropriate version.

Repeating such procedures, we can prove

$$v_1(t)\int_0^t v_0(u)\varphi(u)du \equiv 0, \qquad \text{in } (0, t_0).$$

Hence $\varphi(u)$ must be 0 as an element of $\boldsymbol{L}^2([0, t_0])$.

ii) First we shall prove that

(II. 30) $\quad \Delta(t_1, t_2, \cdots, t_N) \equiv \det(f_i(t_j)) \neq 0$ for any different t_j's.

If $\Delta(t_1, t_2, \cdots, t_N) = 0$ for some $t_1 < t_2 < \cdots < t_N$, we can prove

$$D(t_1, t_2, \cdots, t_N) \equiv \begin{vmatrix} \int_{t_2}^{t_1} v_{N-1}du, & \int_{t_3}^{t_2} v_{N-1}du, & \cdots, & \int_{t_N}^{t_{N-1}} v_{N-1}du \\ \int_{t_2}^{t_1} v_{N-1}\int v_{N-2}(du)^2, & \int_{t_3}^{t_2} v_{N-1}\int v_{N-2}(du)^2, & \cdots, & \int_{t_N}^{t_{N-1}} v_{N-1}\int v_{N-2}(du)^2 \\ \cdots\cdots & & & \\ \int_{t_2}^{t_1} v_{N-1}\int\cdots\int v_1(du)^{N-1}, & \int_{t_3}^{t_2} v_{N-1}\int\cdots\int v_1(du)^{N-1}, & \cdots, & \int_{t_N}^{t_{N-1}} v_{N-1}\int\cdots\int v_1(du)^{N-1} \end{vmatrix} = 0,$$

since v_N never vanishes. By the mean value theorem, we have

$$\begin{vmatrix} v_{N-1}(t_1'), & v_{N-1}(t_2'), & \cdots, & v_{N-1}(t_{N-1}') \\ v_{N-1}(t_1')\int_0^{t_1'} v_{N-2}du, & v_{N-1}(t_2')\int_0^{t_2'} v_{N-2}du, & \cdots, & v_{N-1}(t_{N-1}')\int_0^{t_{N-1}'} v_{N-2}du \\ \cdots\cdots & & & \\ v_{N-1}(t_1')\int_0^{t_1'}\cdots\int v_1(du)^{N-2}, & v_{N-1}(t_2')\int_0^{t_2'}\cdots\int v_1(du)^{N-2}, & \cdots, & v_{N-1}(t_{N-1}')\int_0^{t_{N-1}'}\cdots\int v_1(du)^{N-2} \end{vmatrix} = 0$$

for some $\{t'_i\}$, $i=1, 2, \cdots, N-1$, with $t'_i \in (t_i, t_{i+1})$. Since v_{N-1} never vaishes, it is proved that

$$D(t'_1, t'_2, \cdots, t'_{N-1}) = 0 .$$

Successively we have

$$D(t''_1, t''_2, \cdots, t''_{N-2}) = \cdots = D(t_1^{(N-2)}, t_2^{(N-2)}) = 0$$

for some $\{t_i^{(k)}\}$'s with $t_i^{(k)} \in (t_i^{(k-1)}, t_{i+1}^{(k-1)})$.

Finally we have

$$\int_{t_2^{(N-2)}}^{t_1^{(N-2)}} v_1(u)du = 0 ,$$

which contradicts the assumption (II. 26).

Therefore, by (II. 30), we can find functions $\{a_j(t\,;\, t_1, t_2, \cdots, t_N)\}$, $j=1, 2, \cdots, N$, such that

$$\sum_{j=1}^{N} a_j(t\,;\, t_1, t_2, \cdots, t_N) f_i(t_j) = f_i(t) \tag{II. 31}$$

holds for every i. Hence

$$X(t) - \sum_{j=1}^{N} a_j(t\,;\, t_1, t_2, \cdots, t_N) X(t_j)$$

is independent of every $X(\tau)$; $\tau \leqq t_1$.

On the other hand $\{E(X(t_j)/\boldsymbol{B}_{t_0})\}$, $j=1, 2, \cdots, N$, is linearly independent for any choice of t_j's with $t_0 \leqq t_1 < \cdots < t_N$, since $g_i(u)$'s are linearly independent as elements of $\boldsymbol{L}^2([0, t_0])$. Thus the assertion ii) is proved.

iii) is easily proved, noting that $F(t, u)$ is the Riemann function corresponding to L_t.

Theorem II. 8. *Let $X(t)$ be defined by* (II. 24) *with a canonical kernel $F(t, u) = \sum_{i=1}^{N} f_i(t) g_i(u)$, and let $f_i(t)$'s and $g_i(u)$'s satisfy the conditions* (A. 1) *and* (A. 2). *If*

$$F(t, t) \equiv F^{(1)}(t, t) \equiv \cdots \equiv F^{(k-1)}(t, t) \equiv 0 \text{ and } F^{(k)}(t, t) \not\equiv 0$$

for some $k (< N-1)$ independent of t, then there exist $Y(t)$ which is an N-ple Markov process in the restricted sense and $(N-1)$-th order differential operator M_t such that

$$X(t) = M_t Y(t) . \tag{II. 32}$$

Proof. By assumptions (A. 1) and (A. 2), $g_i(u)$'s can be expressed in the form (II. 25). Therefore there exist a differential operator L_t and a Riemann function $R(t, u)=\sum_{i=1}^{N}\tilde{f}_i(t)g_i(u)$ corresponding to L_t, where $\tilde{f}_i(t)$ is defined by

$$\tilde{f}_i(t) = v_N(t)\int_0^t dm_{N-1}\int dm_{N-2}\int\cdots\int dm_{N-i+1}, \qquad i = 1, 2, \cdots, N.$$

Define $Y(t)$ by

$$Y(t) = \int_0^t R(t, u)dB_0(u)\,.$$

Then, this $Y(t)$ will be the one to be obtained.

The assumption (A. 2) implies $W(\tilde{f}_1, \tilde{f}_2, \cdots, \tilde{f}_N)\neq 0$ for every t. Therefore we can find a differential operator M_t such that

$$M_t\tilde{f}_i(t) = \sum_{j=0}^{N-1} b_j(t)\tilde{f}_i^{(j)}(t) = f_i(t), \qquad i = 1, 2, \cdots, N.$$

Noting that $Y(t)$ has the j-th order derivative

$$Y^{(j)}(t) = \int_0^t R^{(j)}(t, u)dB_0(u)$$

for every $j\leq N-1$, by Theorem II. 6, i), M_t can be operated to $Y(t)$ and we have

$$\begin{aligned} M_tY(t) &\equiv \sum_{j=1}^{N-1} b_j(t)Y^{(i)}(t) \\ &= \int_0^t\sum_{j=1}^{N-1} b_j(t)R^{(j)}(t, u)dB_0(u) \\ &= \int_0^t\sum_{i=1}^{N}\Big(\sum_{j=1}^{N} b_j(t)\tilde{f}_i(t)\Big)g_i(u)dB_0(u) \\ &= \int_0^t\sum_{i=1}^{N} f_i(t)g_i(u)dB_0(u) \end{aligned}$$

This completes the proof.

Example II. 2. Lévy's example $X_1(t)$ which we discussed in Example 1. 5 satisfies all assumptions imposed on the canonical kernel in Theorem II. 8, where $N=2$, $v_0=v_1=1$ and $k=0$. In this case

$$Y(t) = \int_0^t (t-u)dB_1(u)$$

and $M_t=\dfrac{d}{dt}t\cdot$.

§ II. 5. Prediction of multiple Markov Gaussian processes.

For a Gaussian process $Y(t)$, the least square linear prediction on the basis of its values before $s(<t)$ is obtained by the conditional expectation $E(Y(t)/\boldsymbol{B}_s)$, as is well known. If there exists a canonical representation $(dB(t), F(t, u))$ of $Y(t)$, then, by definition,

$$E(X(t)/\boldsymbol{B}_s) = \int^s F(t, u)dB(u), \quad \left(X(t) = \int^t F(t, u)dB(u)\right). \tag{II. 33}$$

It is our aim to express it in terms of $X(\tau)$, $\tau \leq s$.

Theorem II. 9. *Let* $Y(t)$ *be a process defined in Theorem* II. 8. *Using the same notations, we have*

$$E(Y(t)/\boldsymbol{B}_s) = \sum_{j=1}^{N} b_j(t,s) Y^{[j-1]}(s), \qquad s<t, \tag{II. 34}$$

where

$$b_j(t, s) = \sum_{j=1}^{N} \tilde{f}_i(t)\Delta_{ji}/\Delta(s) \tag{II. 35}$$

with

$$\Delta(s) = \det\,(\tilde{f}_{(s)}^{[j-1]}(s)) \text{ and } \Delta_{ji} = (j, i)\text{-cofactor of } \Delta(s).$$

Proof. Putting $U_i(s) = \int_0^s g_i(u)dB_0(u)$, *we have*

$$E(Y(t)/\boldsymbol{B}_s) = \int_0^s \sum_{i=1}^{N} \tilde{f}_i(t) g_i(u) dB_0(u) = \sum_{i=1}^{N} \tilde{f}_i(t) U_i(s). \tag{II. 36}$$

On the other hand,

$$Y^{[k]}(s) = L_s^{(N-k)} Y(s) = \sum_{i=k+1}^{N} \tilde{f}_i^{[k]}(s) U_i(s), \qquad k = 0, 1, \cdots, N-1.$$

Since

$$\Delta(s) = (\tilde{f}_i^{[k]}(s))_{\substack{i=1,2,\cdots,N \\ k=0,1,\cdots,N-1}} = \prod_{i=1}^{N} v_i(s) \neq 0,$$

$U_i(s)$ can be written in terms of $Y^{[k]}(s)$'s, $k=0, 1, \cdots, N-1$, that is

$$U_i(s) = \frac{1}{\Delta(s)} \begin{vmatrix} \tilde{f}_1(s) & \tilde{f}_2(s) & \cdots & Y^i(s) & \cdots & \tilde{f}_N(s) \\ 0 & \tilde{f}_2^{[1]}(s) & \cdots & Y^{[1]}(s) & \cdots & \tilde{f}_N^{[1]}(s) \\ 0 & 0 & \cdots & Y^{[2]}(s) & \cdots & \tilde{f}_N^{[2]}(s) \\ \cdots & \cdots & \cdots & \cdots & \cdots & \cdots \\ 0 & 0 & \cdots & Y^{[N-1]}(s) & \cdots & \tilde{f}_N^{[N-1]}(s) \end{vmatrix}. \tag{II. 37}$$

Combinig this with (II. 36), we have

$$E(Y(t)/\boldsymbol{B}_s) = \sum_{i=1}^{N} \tilde{f}_i(t)\left(\sum_{j=1}^{N} \frac{\Delta_{ji}}{\Delta(s)} Y^{[j-1]}(s)\right)$$

$$= \sum_{j=1}^{N}\left(\sum_{i=1}^{N} \tilde{f}_i(t)\frac{\Delta_{ji}}{\Delta(s)}\right) Y^{[j-1]}(s),$$

which was to be proved.

Corollary. *Let* $X(t)$ *and* $Y(t)$ *be the same processes as in Theorem* II. 8. *Then we have*

$$E(X(t)/\boldsymbol{B}_s) = \sum_{j=1}^{N} c_j(t, s) Y^{[j-1]}(s), \tag{II. 38}$$

where

$$c_j(t, s) = \sum_{i=1}^{N} f_i(t) \frac{\Delta_{ji}}{\Delta(s)},$$

$\Delta(s)$ *and* Δ_{ji} *being the same as in* (II. 35).

Proof. Noting that

$$E(X(t)/\boldsymbol{B}_s) = \sum_{i=1}^{N} f_i(t) U_i(s) \qquad \text{for every } s < t,$$

we can easily prove (II. 38).

This corollary suggests the following symbolic calculous of determing the predictor. Using the differential operator M_t defined in (II. 33), (II. 38) becomes

$$E(M_t Y(t)/\boldsymbol{B}_s) = \sum_{j=1}^{N} c_j(t, s) Y^{[j-1]}(s)$$

$$= \sum_{j=1}^{N} M_t b_j(t,s) Y^{[j-1]}(s) = M_t E(Y(t)/\boldsymbol{B}_s).$$

This means that M_t and $E(\cdot/\boldsymbol{B}_s)$ are commutative.

On the other hand, (II. 38) may be denoted as

$$E(X(t)/\boldsymbol{B}_s) = \sum_{j=1}^{N} c_j(t, s) (M_t^{-1} X(t))_{t=s}^{[j-1]},$$

where M_t^{-1} is an integral operator such as $M_t(M_t^{-1}X(t)) = X(t)$. Hence formally speaking, the prediction operator for $X(t)$ is composed of differential and integral operators.

Theorem II. 10. *Under the same assumption as in Theorem* II. 9,

$$\lim_{s_N \uparrow s} \sum_{i=1}^{N} a_i(t, s_1, s_2, \cdots, s_N) Y(s_i)$$

exists and equals $E(Y(t)/\boldsymbol{B}_s)$, *where* $a_i(t, s_1, s_2, \cdots, s_N)$, $i=1, 2, \cdots, N$ *are the functions determined by Theorem* II. 7, (II. 31).

Proof. Refering to the proof of Theorem II. 7, we have

$$\sum_{i=1}^{N} a_i(t, s_1, s_2, \cdots, s_N) Y(s_i) = \sum_{i=1}^{N}\left(\frac{1}{\Delta_N(s_1, \cdots, s_N)} \sum_{j=1}^{N} \tilde{f}_j(t) \Delta_{ji}^{(N)}\right) Y(s_i)$$

$$= \sum_{j=1}^{N} \frac{1}{\Delta_N(s_1, \cdots, s_N)} \tilde{f}_j(t) \left(\sum_{i=1}^{N} \Delta_{ji}^{(N)} Y(s_i)\right),$$

where $\Delta_N(s_1, s_N, \cdots, s_N)$ has been defined in Lemma II. 5 and $\Delta_{ji}^{(N)}$ is its (j, i)-cofactor. Letting $s_1, s_2, \cdots, s_N$ tend to s successively, we can easily prove that $\sum_{i=1}^{N} a_i(t, s_1, s_2, \cdots, s_N) Y(s_i)$ tends to

$$\sum_{i=1}^{N} \frac{1}{\Delta(s)} \tilde{f}_i(t) \sum_{j=1}^{N} \Delta_{ji} Y^{[j-1]}(s) = \sum_{i=1}^{N} b_i(t, s) Y^{[i-1]}(s)$$

as was to be proved.

For stationry case, such prediction problem is well known (cf. J.L. Doob [1], [2])

§ II. 6. Sum of stationary multiple Markov Gaussian processes.

As we discussed in § II. 3, any stationary N-ple Markov process is considered as the sum of stationary N_i-ple Markov processes in the restricted sense with $\sum N_i = N$. The converse problem will be discussed here.

For the sake of simplicity we shall consider the sum of stationary simple Markov processes, which is 1-ple Markov process in the restricted sense. General cases are treated similarly.

Let $Y_j(t)$, $j=1, 2, \cdots, N$, be stationary simple Markov processes. Taking appropriate versions we can express $Y_j(t)$ with respect to the same random measure $B_0(\cdot)$ as follows

(II. 39) $$Y_j(t) = c_j \int_{-\infty}^{t} e^{-\lambda_j(t-u)} dB_0(u), \qquad \lambda_j (>0),\ c_j: \text{ constants}$$
$$j = 1, 2, \cdots, N.$$

Now let us consider

(II. 40) $$Y(t) = \sum_{j=1}^{N} Y_j(t) = \int_{-\infty}^{t} \sum_{j=1}^{N} c_j e^{-\lambda_j(t-u)} dB_0(u)$$

Obviously it is at most N-ple Markov stationary process. Even in the cast that all the λ_j's are distinct, the kernel $F(t-u) = \sum_{j=1}^{N} c_j e^{-\lambda_j(t-u)}$

is not always canonical (Example II. 3), and $Y(t)$ is not always N-ple Markov process (Example III. 3.).

Let $\hat{F}(\lambda)$ be the Fourier transform of $F(x)$; $\hat{F}(\lambda)=\frac{1}{\sqrt{2\pi}}\int_0^\infty e^{-i\lambda x}\cdot F(x)dx$. Then

$$\hat{F}(\lambda)=\frac{Q(i\lambda)}{\prod_{j=1}^{N}(i\lambda+\lambda_j)}\,,\quad (i=\sqrt{-1})\,, \tag{II. 41}$$

where $Q(i\lambda)$ is the polynomial of $i\lambda$ at most of degree $N-1$. Writing

$$Q(i\lambda)=\sum_{j=0}^{N-1}a_j(i\lambda)^{N-1-j}\,, \tag{II. 42}$$

we have

Theorem II. 11. *$F(t-u)$ is the proper canonical kernel if and only if $Q(x)$ has no zero point with positive real part.*

Proof. We use the kernel criterion proved in §I. 4. Define the numbers b_ν, $\nu=0, 1, \cdots, N$, by

$$\prod_{j=1}^{N}(x+\lambda_j)=\sum_{\nu=0}^{N}b_\nu x^{N-\nu}\,.$$

And define the differential operator L_t by

$$L_t=\sum_{\nu=0}^{N}b_\nu\left(\frac{d}{dt}\right)^{N-\nu}\,.$$

Then we can easily see that

$$L_te^{-\lambda_j(t-u)}=0,\quad\text{consequently}\quad L_tF(u-t)=0\,.$$

Now suppose

$$\int_{-\infty}^{t}F(t-u)\varphi(u)\equiv 0,\quad\text{for some}\quad \varphi\in \boldsymbol{L}^2((-\infty,a])\cap\mathfrak{D}\,, \tag{II. 43}$$

where $\mathfrak{D}=\{\varphi\,;\ \varphi\in\boldsymbol{C}^\infty$ and has compact carrier$\}$
The $(k+1)$-th derivative of it is

$$\begin{aligned}&F(0)\varphi^{(k)}(t)+F'(0)\varphi^{(k-1)}(t)+\cdots+F^{(k)}(0)\varphi(t)\\&\quad+\int_{-\infty}^{t}F^{(k+1)}(t-u)\varphi(u)du=0,\qquad k=0, 1, \cdots, N-1\,.\end{aligned} \tag{II. 44}$$

Then from (II. 43) and (II. 44), we have

$$\begin{aligned}&F(0)\Big(\sum_{k=0}^{N-1}b_{N-k-1}\varphi^{(k)}(t)\Big)+F'(0)\Big(\sum_{k=1}^{N-1}b_{N-k-1}\varphi^{(k-1)}(t)\Big)+\cdots\\&\quad+F^{(N-1)}(0)b_0\varphi(t)+\int_{-\infty}^{t}L_tF(t-u)\varphi(u)du=0\,,\end{aligned}$$

that is,

$$\sum_{\nu=0}^{N-1} F^{(\nu)}(0)\Big(\sum_{k=\nu}^{N-1} b_{N-k-1}\varphi^{(k-\nu)}(t)\Big) = 0 .$$

If we introduce a new differential operator

$$\widetilde{L}_t = \sum_{j=0}^{N-1} \tilde{a}_j\Big(\frac{d}{dt}\Big)^{N-1-j} ,$$

with $\tilde{a}_j = \sum_{\nu=0}^{j} F^{(\nu)}(0)b_{j-\nu}$, then the above equality can be written as

(II. 45) $$\widetilde{L}_t\varphi(t) = 0 .$$

Non trivial function φ satisfying (II. 45) exists and belongs to $L^2((-\infty, a])$ if and only if the characteristic equation

(II. 46) $$\sum_{j=0}^{N-1} \tilde{a}_j x^{N-1-j} = 0$$

of $\widetilde{L}_t$ has at least one root with positive real part.

On the other hand, noting that

$$\int_0^\infty e^{-i\lambda x}F^{(k)}(x)dx = F^{(k-1)}(0)+(i\lambda)F^{(k-2)}(0)+(i\lambda)^2F^{(k-3)}(0)+\cdots + (i\lambda)^{k-1}\hat{F}(\lambda)$$

and

$$\hat{F}(\lambda)\sum_{j=0}^{N} b_j(i\lambda)^{N-j} = Q(i\lambda) ,$$

we can prove

$$a_j = \tilde{a}_j \qquad \text{for every } j ,$$

Hence the desired condition is equivalent to the one that

(II. 46′) $$\sum_{j=0}^{N-1} a_j x^{N-1-j} = Q(x) = 0 .$$

has no root with positive real part.

Generally, not assuming that $\varphi \in \mathfrak{D}$ in (II. 43), the same assertion is true, since $(F*\varphi)(t)=0$ is equivalent to $F*(\varphi*\alpha)(t)=0$ for every $\alpha \in \mathfrak{D}$. (Note that $(\varphi*\alpha)(t) \in \mathbf{C}^\infty$). This completes the proof.

If we observe the proof of this theorem, we can see that its proposition may be improved to the case that all the $Y_j(t)$'s are stationary multiple Markov processes in the restricted sense.

As an obvious consequence of this theorem, we can say that $Y(t)$ defined by (II. 40) is an N-ple Markov stationary process, if the condition of the theorem is fullfilled. In particular, it is an

N-ple Markov process in the restricted sense if and only if all the a_j's are zero except a_0.

Example II. 3. Consider a process

$$X(t) = 3\int_{-\infty}^{t} e^{-(t-u)}dB_0(u) - 4\int_{-\infty}^{t} e^{-2(t-u)}dB_0(u) .$$

The kernel $3e^{-(t-u)} - 4e^{-2(t-u)}$ is not a proper canonical kernel.

Sction III. Lévy's M(t) process.

§ III. 1. Definition and known results.

Let $X(A, \omega)$, $A \in \boldsymbol{E}^N$ (N-dimensional Euclidean space), $\omega \in \Omega$, be a *Brownian motion with a parameter space* $\boldsymbol{E}^N$, that is

(III. 1)
- i) $X(A)$ is a Gaussian random variable with mean 0 for every A,
- ii) $X(O) = 0$, where O is the origin of $\boldsymbol{E}^N$,
- iii) $E(X(A) - X(B))^2 = r(A, B)$, where $r(A, B)$ denotes the distance between A and B.

Since $X(A, \omega)$ is continuous in A for almost all ω, (P. Lévy [1], T. Sirao [1]), the following integral is well defined and we have a Gaussian process $M_N(t)$ with a parameter space $T = [0, \infty)$,

$$M_N(t) = \int_{S(t)} X(A)d\sigma(A) , \tag{III. 2}$$

where $S_N(t)$ is the sphere in E^N with radius t and $d\sigma$ is the uniform measure on $S_N(t)$ with $\sigma(S_N(t)) = 1$.

P. Lévy studied the canonical representation and the Markov property of this process when N is odd (P. Lévy [3], [4]). Since $E(M_N(t)) = 0$, the covariance function of $M_N(t)$ is

$$\begin{aligned} \Gamma_N(t, s) &= E(M_N(t)M_N(s)) \\ &= \int_{A \in S_N(t)} \int_{B \in S_N(s)} E(X(A)X(B))d\sigma(A)d\sigma(B) \\ &= (t + s - \rho_N(t, s))/2 \end{aligned} \tag{III. 3}$$

where

$$\rho_N(t, s) = \int_{A \in S_N(t)} \int_{B \in S_N(s)} r(A, B)d\sigma(A)d\sigma(B) .$$

By the simple computations we have, for $t = s$,

$$\rho_N(t, t) = tJ_{N-2}/I_{N-2} \qquad (N \geqq 3) \tag{III. 4}$$

with $I_k=\int_0^{\pi/2}\sin^k\theta\, d\theta$ and $J_k=\int_0^{\pi}\sin^k\theta \sin\frac{\theta}{2}\, d\theta$, and for $s\neq t$

(III. 5) $$\rho_N(t, s) = \frac{1}{2I_{N-2}}\int_0^{\pi} r \sin^{N-2}\theta\, d\theta \qquad (N\geqq 3)$$

with $r=(t^2+s^2-2ts\cos\theta)^{1/2}$.

Using the analytic property of $\Gamma_N(t, s)$ and others, P. Lévy [4] obtained many important results concerning $M_N(t)$. First, if $N=2p+1$, $M_N(t)$ may be expressed as

(III. 6) $$M_N(t) = \int_0^t P_N(u/t)dB_0(u),$$

where $P_N\left(\frac{u}{t}\right)$ is a canonical kernal defined by

(III. 7) $$P_N(u) = \frac{2p}{\sqrt{\pi}}\sqrt{I_{2p}}\int_u^1(1-x^2)^{p-1}dx$$
$$= \text{polynomial of degree } 2p-1.$$

For example

Example III. 1.

$$M_5(t) = \int_0^t(2/3-u/t+u^3/3t^3)\sqrt{3}\,dB_0(u)$$

Example III. 2.

$$M_7(t) = \int_0^t(2/5-3u/4t+u^3/2t^3-3u^5/20t^5)\sqrt{10}\,dB_0(u).$$

Concerning the Markov property, it was proved that $M_{2p+1}(t)$ has continuous derivatives of orders $1, 2, \cdots, p$ and it is a $(p+1)$-ple Markov process in the restricted sense.

§ III. 2. Canonical representation of $M_N(t)$ process.

We are now interested in the canonical representation of $M_N(t)$ for the case that N is even particularly. First we shall consider some properties of $\Gamma_N(t, s)$ for odd and even N, and then we shall study $M_N(t)$ process.

As P. Lévy pointed out $e^{-t}M(e^{2t})$, which will be denoted by $X_N(t)$, becomes a stationary Gaussian process with parameter space $(-\infty, \infty)$. In fact

(III. 8) $$E(X_N(t)X_N(t+h)) = \frac{1}{2}\left(2\cosh h-\frac{1}{\sqrt{2}\,I_{N-2}}\int_0^{\pi}(\cosh(2h)-\cos\theta)^{1/2}\sin^{N-2}\theta\, d\theta\right) \qquad h\geqq 0.$$

It is a function of h and will be denoted by $\gamma_N(h)$.

Lemma. *If* $N \geqq 4$, $\gamma_N(h)$ *belongs to* $\mathbf{C}^2$ *and satisfies the following equation*

(III. 9) $$(2N-3)^2\gamma_N(h) - \gamma''_N(h) = 4(N-1)(N-2)\gamma_{N-2}(h).$$

Proof. If we note only the differentiability under the integral sign in the formula (III. 8), we can easily prove the existence of $\gamma'_N(h)$ and $\gamma''_N(h)$. Exact forms of them are

$$\gamma'_N(h) = \sinh h - \frac{\sinh 2h}{2I_{N-2}}\int_0^\pi \{2(\cosh 2h - \cos\theta)\}^{-1/2}\sin^{N-2}\theta\, d\theta$$

$$\gamma''_N(h) = \cosh h - \frac{1}{2\sqrt{2}\, I_{N-2}}\int_0^\pi \{2\cosh 2h(\cosh 2h - \cos\theta) - \cosh^2 2h + 1\}\{\cosh 2h - \cos\theta\}^{-3/2}\sin^{N-2}\theta\, d\theta$$

Thus we obtain (III. 9).

Theorem III. 1. *If* $N \geqq 4$, *we have*

(III. 10) $$c_N X_{N-2}(t) = e^{-(2N-3)t}\frac{d}{dt}e^{(2N-3)t}X_N(t), \quad c_N = 2\sqrt{(N-1)(N-2)}.$$

(Here a process and its version are identified)

Proof. From the above lemma, we can see that $e^{(2N-3)t}X_N(t)$ is differentiable. On the other hand, $X_N(t)$ is purely non-deterministic as is easily seen from the definition, and it is expressed as

$$X_N(t) = \int e^{it\lambda} dZ_N(\lambda)$$

with a Gaussian random measure $Z_N(\cdot)$. Hence

$$X^{[1]}(t) = e^{-(2N-3)t}\frac{d}{dt}\int e^{(i\lambda+2N-3)t} dZ_N(\lambda)$$

exists and is

$$\int (i\lambda + 2N-3)e^{it\lambda} dZ_N(\lambda).$$

The covariance function of $X^{[1]}(t)$ is

$$\int e^{ih\lambda}\{\lambda^2 + (2N-3)^2\}\,|\hat{F}^N(\lambda)|^2 d\lambda$$

$$= -\gamma''_N(h) + (2N-3)^2\gamma_N(h) = c_N^2\gamma_{N-2}(h),$$

where $|F_N(\lambda)|^2$ is the spectral density function of $X_N(t)$. Hence $X_N^{[1]}(t)$ can be regarded as a version of $c_N X_{N-2}(t)$.

Thus we can see by (III. 9) and (III, 10) that the study of $X_N(t)$ is reduced to the study of $X_2(t)$ or $X_3(t)$ so far as we consider. More exactly, if we denote the operator $c_N^{-1}e^{-(2N-3)t}\frac{d}{dt}e^{(2N-3)t}$ by D_N, it is easily proved that D_3 can be operated to $X_3(t)$ and

$$D_3D_5\cdots D_{2p+1}X_{2p+1}(t) = X_1(t)\,,$$

which is the Ornstein-Uhlenbeck's Brownian motion. Hence $X_{2p+1}(t)$ and therefore $M_{2p+1}(t)$ is a $(p+1)$-ple Markov process in the restricted sense. (This fact was proved by Lévy by another method). The spectral distribution function of $X_{2p+1}(t)$ has a density

$$\text{(III. 11)}\qquad |\hat{F}_{2p+1}(\lambda)|^2 = \frac{\prod_{k=1}^{p} c_{2k+1}}{\{(4p-1)^2+\lambda^2\}\{(4p-5)^2+\lambda^2\}\cdots\{3^2+\lambda^2\}}\cdot\frac{1}{1+\lambda^2}$$

Therefore we obtain the following

$$X_{2p+1}(t) = \int_{-\infty}^{t}\left(\sum_{k=1}^{p} a_k e^{-(4k-1)(t-u)} + a_0 e^{-(t-u)}\right)dB_0(u)$$

Here the kernel of the representation is to be determined so that the square of its Fourier transform is equal to $|\hat{F}_{2p+1}(\lambda)|^2$ and it satisfies the condition that $Q(\lambda)=$constant. This is posible.

Changing the time scale, we have the canonical representation of $M_N(t)$.

$$M_N(t) = \int_0^t \frac{1}{\sqrt{2}}\left(\sum_{k=1}^{p} a_k(u/t)^{2k-1} + a_0\right)dB_0(u)\,.$$

Obviously thus obtained representation coincides with the Lévy's result.

The problem to obtain all non canonical representations of $M_{2p+1}(t)$, where kernels are polynomials of (u/t) of degree $2p-1$ is easily solved, if we observe the spectral density function of $X_{2p+1}(t)$. The answer of the problem is that "*the number of different representations of above stated form of* $M_{2p+1}(t)$ *is just* 2^{p-1} *including canonical one*".

Proof. If a kernel is a polynomial of (u/t) of degree $2p-1$, it turns into the sum of exponential functions such as $e^{-(2k+1)(t-u)}$, $k\leq 2p-1$, the Fourier transform of which is a sum of the functions of the form $\frac{1}{i\lambda+(2k+1)}$. Hence the number of posible functions of $\hat{F}_{2p+1}(\lambda)$ is just 2^{p-1}. For example we obtain such a function multiplying $\frac{i\lambda-(4p-7)}{i\lambda+(4p-7)}$ to the function $\hat{F}_{2p+1}(\lambda)$ which corresponds to the canonical kernel.

Example III. 3. One of the non-canonical representation of $M_7(t)$ different from the Lévy's one (cf. P. Lévy [4] p. 146) is given as follows: let

$$\hat{F}_7(\lambda)=\left(\frac{c}{(i\lambda+1)(i\lambda+3)(i\lambda+7)(i\lambda+11)}\right)\cdot\frac{i\lambda-5}{i\lambda+5}$$

The rational function in the bracket corresponds to the canonical kernel. Then $M_7(t)$ is expressed as

$$\int_0^t(3/5-3u/t+5u^2/t^2-3u^3/t^3+2u^5/5t^5)\sqrt{10}dB_0(u)\,.$$

If N is even, we also have

$$D_4D_6\cdots D_{2p}X_{2p}(t)=X_2(t)\,.$$

Hence, if we know the canonical kernel of the representation of $X_2(t)$, then we can obtain that of $X_{2p}(t)$ easily and know the properties of it.

We have

$$\text{(III. 12)}\qquad |\hat{F}_{2p}(\lambda)|^2=\frac{\prod_{k=1}^{p}c_{2k}}{\{\lambda^2+(4p-3)^2\}\{\lambda^2+(4p-7)^2\}\cdots\{\lambda^2+5^2\}}|\hat{F}_2(\lambda)|^2$$

in the way similar to the case in which N is odd. Now it is our purpose to obtain the exact form of $|\hat{F}_2(\lambda)|^2$. To do so, let us consider $\gamma_2(h)$. If $h>0$,

$$\gamma_2(h)=\cosh h/2-\frac{1}{2\pi}\int_0^\pi(2\cosh(2h)-2\cos\theta)^{1/2}d\theta$$

Using the Legendre's polynomials the integral term of it may be expanded as follows

$$\frac{1}{\sqrt{2}\,\pi}\int_0^\pi\{\cosh(2h)-\cos\theta\}^{1/2}d\theta=\frac{1}{2}\{e^h+\sum_{k=0}^{\infty}(a_{k+1}^2+a_k^2-2a_ka_{k+1})e^{-(4k+3)h}\}\,,$$

where

$$a_k=\frac{1\cdot3\cdot5\cdots(2k-1)}{2\cdot4\cdot6\cdots2k.}$$

Taking the Fourier transform, we have

$$|\hat{F}_2(\lambda)|^2=\frac{1}{2\pi}\left(\frac{1}{1+\lambda^2}-\sum_{k=0}^{\infty}\frac{b_k}{\lambda^2+(4k+3)^2}\right),$$

where

$$b_k = (4k+3)(a_{k+1}-a_k)^2 \qquad (>0).$$

This proves that $X_2(t)$ is not a multiple Markov process, but, as it were, ∞-ple Markov process.

Let $\hat{F}^{(n)}(\lambda)$ be given by

$$\hat{F}^{(n)}(\lambda) = \left(\frac{1}{1+\lambda^2} - \sum_{k=0}^{n-1} \frac{b_k}{\lambda^2+(4k+3)^2}\right)/2\pi .$$

Then $\hat{F}^{(n)}(\lambda)>0$ (since $\hat{F}^{(n)}(\lambda)>|\hat{F}_2(\lambda)|^2\geqq 0$) and

$$\int \frac{\log \hat{F}^{(n)}(\lambda)}{1+\lambda^2} d\lambda > -\infty .$$

Therefore, there exists a stationary Gaussian process $X^{(n)}(t)$ which is expressed in the form

$$X^{(n)}(t) = \int_{-\infty}^{t} F_n(t-u)dB_0(u)$$

and has spectral density function $\hat{F}^{(n)}(\lambda)$.

Obviously $X^{(n)}(t)$ is a stationary $(n+1)$-ple Markov process and its covariance function $\gamma^{(n)}(h)$ converges to $\gamma_2(h)$ uniformly in any finite interval of h.

Summing up we have

Theorem III. 2. *$X_{2p}(t)$ is not a multiple Markov process, but it is a limiting process of n-ple Markov process* $(n\to\infty)$.

Proof. For $p=1$, we have already proved. Noting the formula (III. 12), we can easily prove our theorem.

From this theorem we can see that $M_{2p}(t)$ is not a multiple Markov process, but is a limiting process of n-ple Markov process with homogeneous canonical kernel of degree 0, which was to be obtained.

REFERENCES

N. Aronszajn [1]: *Theory of reproducing kernels.* Trans. Amer. Math. Soc. vol. 68, 337-404 (1950).

J. L. Doob [1]: *The elementary Gaussian processes.* Annals of Math. Stat. vol. 15, 229-282 (1944).

[2]: *Stochastic processes.* Willey, (1952).

C. L. Dolph and M. A. Woodbury

[1]: *On the relation between Green's functions and covariances of certain stochastic processes and its application to unbiased linear prediction.* Trans. Amer. Math. Soc. vol. 72, 519-550 (1952).

E. L. Ince [1]: *Ordinary differential equations.* Dover Pub. INC (1926).

K. Itô [1]: *Stochastic integral.* Proc. Imp. Acad. Tokyo, vol. 20, 519–524 (1944).

[2]: *Theory of Probability.* (In Japanese). Tokyo, (1953).

[3]: *Stochastic process.* Tata Institute Note, Bombay, (1959).

S. Itô [1]: *On Hellinger-Hahn's Theorem.* (In Japanese). Sūgaku, vol. 5 no. 2, 90–91 (1953).

K. Karhunen [1]: *Über die Struktur stationärer zufälliger Funktionen.* Arkiv för Matematik, 1. nr. 13, 144–160 (1950).

P. Lévy [1]: *Processus stochastiques et mouvement Brownien.* Gautier-Villars, Paris, (1948).

[2]: *Le mouvement brownien.* Memorial des Science Math. fasc. 129, (1954)

[3]: *Brownian motion depending on n parameters: The particular case* $n=5$. Proceedings of Symposia in Applied Math. vol. 7, 1–20 (1957).

[4]: *A special problem of Brownian motion, and a general theory of Gaussian random functions.* Proceedings of the Third Berkeley Symposium on Math. Stat. and Prob. vol. II. 133–175 (1956).

[5]: *Sur une classe de courbes de l'espace de Hilbert et sur une équation intégrale non linéaire.* Annales École Norm. Sup. tom. 73, 121–156 (1956).

[6]: *Fonction aléatoires à corrélation linéaire.* Illinois Journal of Math. vol. 1, 217–258 (1957).

[7]: *Fonctions linéairement Markoviennes d'ordre n.* Math. Japonicae, vol. 4, no. 3, 113–121 (1957).

[8]: *Sur quelques chasses de fonctions aléatoires.* Journal de Math. pures et appliquées. tom. 38, 1–23 (1959).

L. Schwartz [1]: *Théorie des distribution I.* Hermann & Cie, Paris, (1950).

T. Seguchi and N. Ikeda

[1]: *Note on the statistical inferences of certain continuous stochastic processes.* Memoires of the Faculty of Sci. Kyūsyū Univ. Ser. A, vol. 8, No. 2, 187–199 (1954).

T. Sirao [1]: *On the continuity of Brownian motion with a multidimensional parameter.* Nagoya Math. Journal, Vol. 16, 135 156 (1960).

M. H. Stone [1]: *Linear transformations in Hilbert space and its applications to analysis.* Amer. Math. Soc. Colloq. Pub. vol. 15. (1932).

Errata

page	line				
118	3	Change	X(t)	to	Y(t)
	↑10,5	Change	dB(t)	to	$d\tilde{B}(t)$
	↑3,4,5,6	Change	ν	to	$\tilde{\nu}$
119	1	Change	ν	to	$\tilde{\nu}$
	12,13	Change	$\tilde{\nu}$	to	ν

122	6	Change	$\|\|E(t)f^{(i)}\|\|$	to	$\|\|E(t)f^{(i)}\|\|^2$
123	↑5	Change	$\epsilon\Gamma(\cdot,t)$	to	$\ni\Gamma(\cdot,t)$
125	4,5,7	Change	$\bar{S}^{-1}$	to	$\bar{S}$
	6	Change	, this	to	, $E(Y(t)/\mathbf{B}_s)$
	11	Change	$F(t,u)$	to	$F_i(t,u)$
126	10	Change	Σ	to	Σ^*
134	2,3	Change	$=a(\tau,$	to	$=\mathbf{a}(\tau,$
	18,19	Change	$\mathbf{F}(\mathbf{t},u)^*$	to	$\mathbf{F}(\mathbf{s},u)^*$
	↑19	Change	$t'_N>\ldots>t'_1$	to	$t_N>\ldots>t_1$
	↑7,9,11	Change	$\mathbf{t}'$	to	$\mathbf{t}$
	↑8	Change	$(t'_N\ ,\ t_N)$	to	$(s'_N\ ,\ s_N)$
137	↑9	Change	$g(\cdot)$	to	$g_i(\cdot)$
147	15	Change	$Y^{(i)}(t)$	to	$Y^{(j)}(t)$
	17	Change	$\tilde{f}_i(t)$	to	$\tilde{f}_i^{(j)}(t)$
151	↑8	Insert after $\phi(u)$: du			
154	↑5	Change	$M(e^{2t})$	to	$M_N(e^{2t})$
155	↑2	Change	$\|F_N(\lambda)\|^2$	to	$\|\hat{F}_N(\lambda)\|^2$
157	↑8	Change	cosh h/2	to	cosh h

Editors' Comments on Paper 7

In this paper Professor Cramér presents his multiplicity representation in greater detail. Attention is focused on the purely nondeterministic (pnd) component of a random process and exclusively on the continuous case. The proofs are complete and rigorous and provide a sound foundation for the theory presented earlier by Cramér (in paper 5 of this volume).

This paper is reprinted with the permission of the author and the Royal Swedish Academy of Science.

ARKIV FÖR MATEMATIK Band 4 nr 19

Read 14 September 1960

7

On the structure of purely non-deterministic stochastic processes

By Harald Cramér

Introduction

1. The purpose of this paper is to give the proofs of some results recently communicated in a lecture at the Fourth Berkeley Symposium on Mathematical Statistics and Probability [1].

Although these results were expressed in the language of mathematical probability, they may equally well be regarded as concerned with the properties of certain curves in Hilbert space. In this first section of the Introduction we shall briefly state some results of the paper in Hilbert space language, and then in the following sections recur to the "mixed" language which seems convenient when probability questions are treated with the methods of Hilbert space geometry.

Let $\mathfrak{H}$ be a complex Hilbert space, and let, for every real τ, a set of q elements $x_1(\tau), x_2(\tau), \ldots, x_q(\tau)$ of $\mathfrak{H}$ be given. As τ runs through all real values, each element $x_j(\tau)$ describes a "curve" C_j in the space $\mathfrak{H}$. Let $C_j(t)$ denote the "arc" of C_j corresponding to values of $\tau \leqslant t$, and denote by $\mathfrak{H}(\mathbf{x}, t)$ the smallest subspace of $\mathfrak{H}$ containing the arcs $C_1(t), \ldots, C_q(t)$.

As t increases, the $\mathfrak{H}(\mathbf{x}, t)$ form a never decreasing family of subspaces, and the limiting spaces $\mathfrak{H}(\mathbf{x}, +\infty)$ and $\mathfrak{H}(\mathbf{x}, -\infty)$ will exist. It will be assumed that the following two conditions are satisfied:

(A) The strong limits $x_j(t \pm 0)$ exist for $j = 1, \ldots, q$ and for all real t.

(B) The space $\mathfrak{H}(\mathbf{x}, -\infty)$ contains only the zero element of $\mathfrak{H}$.

The projection of an arbitrary element z of $\mathfrak{H}(\mathbf{x}, +\infty)$ on the subspace $\mathfrak{H}(\mathbf{x}, t)$ will be denoted by $P_t z$.

We propose to show that the $x_j(t)$ can be simultaneously and linearly expressed in terms of certain mutually orthogonal elements. For this purpose we shall use considerations closely related to the theory of spectral multiplicity of self-adjoint transformations in a separable Hilbert space (cf. e.g. [7], Chapter VII).

It will be shown that it is possible to find a sequence $z_1, \ldots, z_N$ of elements of $\mathfrak{H}(\mathbf{x}, +\infty)$ such that we have for every $j = 1, \ldots, q$ and for all real t

$$x_j(t) = \sum_{n=1}^{N} \int_{-\infty}^{t} g_{jn}(t, \lambda)\, d z_n(\lambda), \tag{1}$$

where $z_n(\lambda) = P_\lambda z_n$. Here N may be a finite integer or equal to $\aleph_0$, the g_{jn} are complex-valued functions of the real variables t, λ, and the integrals are appropriately defined. Two increments $\Delta z_m(\lambda)$ and $\Delta z_n(\mu)$ are always orthogonal if $m \neq n$, while for $m = n$ they are orthogonal if they correspond to disjoint intervals.

We shall study the properties of the expansion (1), and in particular it will be shown that we have for any $u < t$

$$P_u x_j(t) = \sum_{n=1}^{N} \int_{-\infty}^{u} g_{jn}(t, \lambda)\, d z_n(\lambda). \tag{2}$$

When certain additional conditions are imposed, N is the smallest cardinal number such that a representation of the form (1) holds.

If the elements of $\mathfrak{H}$ are interpreted as random variables, the set of curves $C_1, \ldots, C_q$ will correspond to q simultaneously considered stochastic processes with a continuous time parameter. The above results then yield a representation of such a set of processes in terms of past and present "innovations", as well as an explicit expression for the linear least squares prediction, as will be shown in the sequel.

2. Consider a random variable x defined on a once for all given probability space, and satisfying the relations

$$E\,x = 0, \quad E\,|x|^2 < \infty. \tag{3}$$

The set of all random variables defined on the given probability space and satisfying (3) forms a Hilbert space $\mathfrak{H}$, if the inner product and the norm are defined in the usual way:

$$(x, y) = E\,(x\,\bar{y}), \quad \|x\|^2 = E\,|x|^2.$$

If two random variables x, y belonging to $\mathfrak{H}$ are such that

$$\|x - y\|^2 = E\,|x - y|^2 = 0,$$

they will be considered as identical, and we shall write

$$x = y.$$

In the sequel, any equation between random variables should be interpreted in this sense.

Whenever we are dealing with the *convergence* of a sequence of random variables, it will always be understood that we are concerned with convergence in the topology induced by the norm in $\mathfrak{H}$, which in probabilistic terminology corresponds to convergence in quadratic mean.

A family of complex-valued random variables $x(t)$, where t is a real parameter ranging from $-\infty$ to $+\infty$, and $x(t) \in \mathfrak{H}$ for every t, will be called a *one-dimensional stochastic process* with continuous time parameter t. Further, if

$x_1(t), \ldots, x_q(t)$ are q random variables, each of which is associated with a process of this type, the (column) vector

$$\mathbf{x}(t) = (x_1(t), \ldots, x_q(t)) \tag{4}$$

defines a *q-dimensional stochastic vector process* with continuous time t.

For every fixed t, each component $x_j(t)$ of the vector process (4) is a point in the Hilbert space $\mathfrak{H}$. As t increases from $-\infty$ to $+\infty$, this point describes a curve C_j in $\mathfrak{H}$, and so we are led to consider the set of curves $C_1 \ldots, C_q$ mentioned in the preceding section. The subspace $\mathfrak{H}(\mathbf{x}, t)$ is, from the present point of view, the subspace spanned by the random variables $x_1(\tau), \ldots, x_q(\tau)$ for all $\tau \leqslant t$, and we shall write this

$$\mathfrak{H}(\mathbf{x}, t) = \mathfrak{S}\{x_1(\tau), \ldots, x_q(\tau); \quad \tau \leqslant t\}.$$

$\mathfrak{H}(\mathbf{x}, t)$ may be regarded as the set of all random variables that can be obtained by means of linear operations acting on the components of $\mathbf{x}(\tau)$ for all $\tau \leqslant t$. We evidently have

$$\mathfrak{H}(\mathbf{x}, t_1) \subset \mathfrak{H}(\mathbf{x}, t_2)$$

whenever $t_1 < t_2$. It follows that the limiting spaces $\mathfrak{H}(\mathbf{x}, +\infty)$ and $\mathfrak{H}(\mathbf{x}, -\infty)$ exist, and also that the limiting spaces $\mathfrak{H}(\mathbf{x}, t \pm 0)$ exist for all t.

Following Wiener & Masani [8], we shall say that the space $\mathfrak{H}(\mathbf{x}, t)$ represents the *past and present* of the vector process (4), as seen from the point of view of the instant t. The limiting space $\mathfrak{H}(\mathbf{x}, -\infty)$ will be called the *remote past* of the process, while the space $\mathfrak{H}(\mathbf{x}, +\infty)$ will be briefly denoted by $\mathfrak{H}(\mathbf{x})$, and called the *space of the* $\mathbf{x}(t)$ *process*. We then have for any t

$$\mathfrak{H}(\mathbf{x}, -\infty) \subset \mathfrak{H}(\mathbf{x}, t) \subset \mathfrak{H}(\mathbf{x}, +\infty) = \mathfrak{H}(\mathbf{x}) \subset \mathfrak{H}.$$

In the particular case of a vector process $\mathbf{x}(t)$ satisfying

$$\mathfrak{H}(\mathbf{x}, -\infty) = \mathfrak{H}(\mathbf{x}), \tag{5}$$

it will be seen that complete information concerning the process is already contained in the remote past. Accordingly such a process will be called a *deterministic* process.

Any process not satisfying (5) will be called *non-deterministic*. In the extreme case when we have

$$\mathfrak{H}(\mathbf{x}, -\infty) = 0, \tag{6}$$

the remote past does not contain any information at all, and the process is said to be *purely non-deterministic*.

It is known [1] that any vector process (4) can be represented as the sum of a deterministic and a purely non-deterministic component, which are mutually orthogonal. In the present paper, we shall mainly be concerned with the structure of the latter component, and we may then as well assume that the given $\mathbf{x}(t)$ process itself is purely non-deterministic, i.e. that $\mathfrak{H}(\mathbf{x}, -\infty) = 0$.

Before introducing this assumption we shall, however, in sections 4–5 study the properties of the general $\mathbf{x}(t)$ process, without imposing more than the mildly restrictive condition (A), relating to the behaviour of the process in its points of discontinuity.

In section 6 we shall then, besides condition (A), introduce condition (B) which states that the $\mathbf{x}(t)$ process is purely non-deterministic. Throughout the rest of the paper, we shall then be concerned with processes satisfying both conditions (A) and (B).

In the particular case of a *stationary* one-dimensional process $x(t)$ satisfying (A) and (B), it is known that $x(t)$ can be linearly represented in terms of past innovations by an expression of the form [1]

$$x(t) = \int_{-\infty}^{t} g(t-\lambda)\, dz(\lambda),. \tag{7}$$

where $z(\lambda)$ is a process *with orthogonal increments*, which may be called an *innovation process* of $x(t)$.

In sections 8–9, we shall be concerned with representations of a similar kind, but generalized in two directions: the assumption of stationarity will be dropped, and a vector process $\mathbf{x}(t)$ will be considered instead of the one-dimensional $x(t)$. It will be shown that, for any $\mathbf{x}(t)$ process satisfying (A) and (B), we have a representation of the form (1) indicated in section 1 above. The $z_n(\lambda)$ occurring in (1) will now be one-dimensional stochastic processes with orthogonal increments. Accordingly we may say that, in the general case, we are concerned with an innovation process $(z_1(\lambda), \ldots, z_N(\lambda))$ which is multi-dimensional, and possibly even infinite-dimensional.

3. In a series of papers, P. Lévy (cf. [4–6], and further references there given) has investigated the properties of stochastic processes representable, in the notation of the present paper, in the form

$$x(t) = \int_{0}^{t} g(t, \lambda)\, dz(\lambda),$$

where $z(\lambda)$ is a *normal* (i.e Gaussian or Laplacian) process with independent increments. His investigations have been continued in a recent paper by T. Hida [3], who has also considered the more general representation

$$x(t) = \sum_{n=1}^{N} \int_{0}^{t} g_n(t, \lambda)\, dz_n(\lambda)$$

derived from considerations of spectral multiplicity, as well as the case when the lower limits of the integrals are $-\infty$ instead of zero.

[1] Cf. e.g. Doob [2], p. 588, and the references there given. The corresponding representation for a one-dimensional stationary process with *discrete* time parameter was first found by Wold [9], and was generalized to the vector case by Wiener & Masani [8].

In connection with my Berkeley lecture [1], Professor K. Itô kindly drew my attention to the work of Mr. Hida, which was then in the course of being printed. Evidently there are interesting points of contact between Mr. Hida's line of investigation and the one pursued in the present paper.

Discontinuities and innovations

4. Consider a stochastic vector process as defined in section 2,

$$\mathbf{x}(t) = (x_1(t), \ldots, x_q(t)),$$

and let us suppose that the following condition is satisfied:

(A) *The limits* $x_j(t-0)$ *and* $x_j(t+0)$ *exist for* $j=1, \ldots, q$ *and for all real* t.

We shall then write

$$\mathbf{x}(t-0) = (x_1(t-0), \ldots, x_q(t-0))$$

and similarly for $\mathbf{x}(t+0)$. Any point t such that at least one of the relations

$$\mathbf{x}(t-0) = \mathbf{x}(t) = \mathbf{x}(t+0)$$

is not satisfied, is a *discontinuity point* of the $\mathbf{x}$ process. The point t is a *left* or *right* discontinuity, or both, according as $\mathbf{x}(t-0) \neq \mathbf{x}(t)$, or $\mathbf{x}(t) \neq \mathbf{x}(t+0)$, or both. We shall now prove the following Lemma.

Lemma 1. *For any* $\mathbf{x}(t)$ *process satisfying* (A), *we have*

(*a*) *For* $j=1, \ldots, q$, *the functions* $E\,|x_j(t)|^2$ *are bounded throughout every finite* t*-interval.*
(*b*) *The set of all discontinuity points of the* $\mathbf{x}$ *process is at most enumerable.*
(c) *The Hilbert space* $\mathfrak{H}(\mathbf{x})$ *is separable.*

In order to prove (*a*), let us suppose that the non-negative function $E\,|x_j(t)|^2$ were not bounded in a certain finite interval I. Then it would be possible to find a sequence of points $\{t_n\}$ in I, and converging monotonely to a limit t^*, such that $E\,|x_j(t_n)|^2 \to \infty$. Clearly this is not compatible with condition (A), so that our hypothesis must be wrong, and point (*a*) is proved.

Point (*b*) will be proved if we can show that, for each j, the one-dimensional process $x_j(t)$ has at most an enumerable set of discontinuity points. A discontinuity point of $x_j(t)$ is then, of course, a point t such that at least one of the relations

$$x_j(t-0) = x_j(t) = x_j(t+0)$$

is not satisfied. Clearly it will be enough to show, e.g., that the inequality

$$x_j(t-0) \neq x_j(t) \tag{8}$$

cannot be satisfied in more than an enumerable set of points.

According to our conventions, (8) is equivalent to the relation

$$s(t) = E\,|\,x_j(t-0) - x_j(t)\,|^2 > 0.$$

We are going to show first that, given any positive number h and any finite interval I, there is at most a finite number of points t in I such that $s(t) > h$. This being shown, the desired result follows immediately by allowing first h to tend to zero, and then I to tend the whole real axis.

Suppose, in fact, that I contains an infinite set of points t with $s(t) > h$. We can then find an infinite sequence of points $\{t_n\}$ in I, converging to a limit t^*, and such that

$$s(t_n) = E\,|\,x_j(t_n - 0) - x_j(t_n)\,|^2 > h \tag{9}$$

for all n. Evidently we can even find a *monotone* sequence $\{t_n\}$ having these properties. Let us suppose, e.g., that t_n converges *decreasingly* to t^*. (The increasing case can, of course, be treated in the same way.) On account of condition (A) we can then find a sequence of positive numbers $\{\varepsilon_n\}$ tending to zero, such that

$$t_n - \varepsilon_n > t_{n+1},$$

$$E\,|\,x_j(t_n - \varepsilon_n) - x_j(t_n - 0)\,|^2 < \frac{c^2 h^2}{16\,K} \tag{10}$$

for all n, where c and K are constants such that $0 < c < 1$ and $E\,|\,x_j(t)\,|^2 < K$ throughout I. The existence of such a constant K follows from point (a) of the Lemma, which has already been established.

Now for any random variables u and v we have

$$E\,|\,u+v\,|^2 = E\,|\,u\,|^2 + E\,|\,v\,|^2 + E(u\,\bar{v}) + E(\bar{u}\,v)$$

$$\geqslant E\,|\,u\,|^2 + E\,|\,v\,|^2 - 2\sqrt{E\,|\,u\,|^2\,E\,|\,v\,|^2}.$$

Taking here

$$u = x_j(t_n - 0) - x_j(t_n),$$

$$v = x_j(t_n - \varepsilon_n) - x_j(t_n - 0),$$

we obtain from (9) and (10)

$$E\,|\,x_j(t_n - \varepsilon_n) - x_j(t_n)\,|^2 \geqslant h - 2\sqrt{4\,K \cdot \frac{c^2 h^2}{16\,K}} = (1-c)\,h \tag{11}$$

for all n. On the other hand, the sequences $\{t_n - \varepsilon_n\}$ and $\{t_n\}$ both converge decreasingly to t^*. Consequently by condition (A) we have (convergence, as usual, in the $\mathfrak{H}$ topology, i.e. in quadratic mean)

$$x_j(t_n - \varepsilon_n) \to x_j(t^* + 0),$$

$$x_j(t_n) \to x_j(t^* + 0),$$

and thus $$x_j(t_n - \varepsilon_n) - x_j(t_n) \to 0,$$

as n tends to infinity. However, this is incompatible with (11), so that our hypothesis must be wrong, and thus point (*b*) of the Lemma is proved.

The last part of the Lemma now follows immediately, if we consider an enumerable set $\{t_n\}$ including all discontinuity points of $\mathbf{x}(t)$ as well as an everywhere dense set of continuity points. The set of all finite linear combinations of the random variables $x_j(t_n)$ for $j=1, \ldots, q$ and $n=1, 2, \ldots$, with coefficients whose real and imaginary parts are both rational, is then an enumerable set dense in $\mathfrak{H}(\mathbf{x})$.

5. Consider now the family of subspaces $\mathfrak{H}(\mathbf{x}, t)$ associated with an $\mathbf{x}(t)$ process satisfying condition (A) of the preceding section.

As observed in section 2, $\mathfrak{H}(\mathbf{x}. t)$ never decreases as t increases. If $\mathfrak{H}(\mathbf{x}, t)$ effectively increases when t ranges over some interval $t_1 < t \leqslant t_2$, so that we have

$$\mathfrak{H}(\mathbf{x}, t_1) \neq \mathfrak{H}(\mathbf{x}, t_2),$$

this means that some new information has entered into the process during that interval. Accordingly we shall then say that the process has received an *innovation* during the interval $t_1 < t \leqslant t_2$, and we shall regard this innovation as being represented by the orthogonal complement

$$\mathfrak{H}(\mathbf{x}, t_2) - \mathfrak{H}(\mathbf{x}, t_1). \tag{12}$$

In fact, if this complement reduces to the zero element, the two spaces are identical, and no innovation has entered during the interval, while in the opposite case (12) is the set of all differences between an element of $\mathfrak{H}(\mathbf{x}, t_2)$ and its projection on $\mathfrak{H}(\mathbf{x}, t_1)$.

If, for a certain value of t, we have

$$\mathfrak{H}(\mathbf{x}, t-h) \neq \mathfrak{H}(\mathbf{x}, t+h)$$

for any $h>0$, this means that there is a non-vanishing innovation associated with any interval containing t as an interior point. The set of all points t having this property will be called the *innovation spectrum* of the $\mathbf{x}(t)$ process.

$\mathfrak{H}(\mathbf{x}, t)$ being never decreasing as t increases, it follows that the limiting spaces $\mathfrak{H}(\mathbf{x}, t \pm 0)$ will exist for every t. Any point such that at least one of the relations

$$\mathfrak{H}(\mathbf{x}, t-0) = \mathfrak{H}(\mathbf{x}, t) = \mathfrak{H}(\mathbf{x}, t+0)$$

is not satisfied, will certainly belong to the innovation spectrum, and will be called a *discontinuity point* of that spectrum. As in section 4, the terms *left* and *right* discontinuities will be used in the obvious sense.

The set of all discontinuity points constitutes the *discontinuous part* of the innovation spectrum. Let $t_1, t_2, \ldots$, be the points of this set (it will be shown below that the set is at most enumerable), and form the vector sum

$$\mathfrak{G}(\mathbf{x}, t) = \sum_{t_k \leqslant t} \mathfrak{H}(\mathbf{x}, t_k + 0) - \mathfrak{H}(\mathbf{x}, t_k - 0)$$

and the orthogonal complement $\mathfrak{F}(\mathbf{x}, t) = \mathfrak{H}(\mathbf{x}, t) - \mathfrak{G}(\mathbf{x}, t)$. The set of all points t such that $\mathfrak{F}(\mathbf{x}, t-h) \neq \mathfrak{F}(\mathbf{x}, t+h)$ for any $h > 0$ constitutes the *continuous part* of the innovation spectrum of $\mathbf{x}(t)$. If the discontinuous part is not a closed set, its limiting points will belong to the innovation spectrum, without necessarily belonging to any of the two parts here defined.

When t is a discontinuity point of the innovation spectrum, the number of dimensions of the subspace $\mathfrak{H}(\mathbf{x}, t) - \mathfrak{H}(\mathbf{x}, t-0)$ will be called the *left multiplicity* of the point t. Similarly the *right multiplicity* of t is the number of dimensions of $\mathfrak{H}(\mathbf{x}, t+0) - \mathfrak{H}(\mathbf{x}, t)$. If t is not a left (right) discontinuity, the left (right) multiplicity of t is of course equal to zero. We shall now prove the following Lemma.

Lemma 2. *For any* $\mathbf{x}(t)$ *process satisfying* (A), *we have*

(a) *The set of discontinuity points of the innovation spectrum is at most enumerable.*

(b) *In a left discontinuity point, the left multiplicity is at most equal to* q.

(c) *In a right discontinuity point, the right multiplicity may be any finite integer, or equal to* $\aleph_0$.

(d) *A left discontinuity of the innovation spectrum is always at the same time a left discontinuity of the process. On the other hand, a right discontinuity of the innovation spectrum is not necessarily a discontinuity of the process.*

We first observe that the two subspaces $\mathfrak{H}(\mathbf{x}, t_2) - \mathfrak{H}(\mathbf{x}, t_1)$ and $\mathfrak{H}(\mathbf{x}, u_2) - \mathfrak{H}(\mathbf{x}, u_1)$ corresponding to disjoint time intervals are always orthogonal. It follows that the subspaces

$$\mathfrak{H}(\mathbf{x}, t+0) - \mathfrak{H}(\mathbf{x}, t-0)$$

corresponding to different discontinuity points are orthogonal. By Lemma 1, the space $\mathfrak{H}(\mathbf{x})$ is separable, and cannot include more than an enumerable set of mutually orthogonal subspaces. Hence follows the truth of point (a) of the Lemma.

Point (b) of the Lemma asserts that the orthogonal complement $\mathfrak{H}(\mathbf{x}, t) - \mathfrak{H}(\mathbf{x}, t-0)$ has at most q dimensions. By definition, the space $\mathfrak{H}(\mathbf{x}, t)$ is spanned by the variables $x_j(\tau)$ with $j = 1, \ldots, q$ and $\tau \leqslant t$. Writing

$$x_j(t) = y_j + P_{t-0}\, x_j(t), \quad (j = 1, \ldots, q), \tag{13}$$

where P_{t-0} denotes the projection on $\mathfrak{H}(\mathbf{x}, t-0)$, it will be seen that every element of $\mathfrak{H}(\mathbf{x}, t)$ is the sum of an element of $\mathfrak{H}(\mathbf{x}, t-0)$ and a linear combination of the variables $y_1, \ldots, y_q$, which belong to $\mathfrak{H}(\mathbf{x}, t)$ and are orthogonal to $\mathfrak{H}(\mathbf{x}, t-0)$. Consequently the orthogonal complement $\mathfrak{H}(\mathbf{x}, t) - \mathfrak{H}(\mathbf{x}, t-0)$ is identical with the space spanned by $y_1, \ldots, y_q$, and thus has at most q dimensions. It will also be seen that, choosing the variables $y_1, \ldots, y_q$ in an appropriate way, we can construct examples of processes having, in a given point, a left discontinuity of the innovation spectrum with any given multiplicity not exceeding q. Thus (b) is proved.

We shall now prove (c) by constructing an example of an $\mathbf{x}(t)$ process, the innovation spectrum of which has, in a given point, a right discontinuity of infinite multiplicity. It will then be easily seen how the example may be modified in order to produce a discontinuity of any given finite multiplicity.

Let $z_1, z_2, \ldots$ be an infinite orthonormal sequence of random variables belonging to $\mathfrak{H}$. Denoting by $p_1, p_2, \ldots$ the successive prime numbers ($p_1 = 2$, $p_2 = 3, \ldots$), we define a process $\mathbf{x}(t) = (x_1(t), \ldots, x_q(t))$ by taking

$$x_2(t) = \cdots = x_q(t) = 0 \quad \text{for all } t,$$

$$x_1(t) = 0 \quad \text{for} \quad t \leqslant 0,$$

$$x_1(t) = t\, z_n \quad \text{for} \quad t = p_n^{-k}, \quad k = 1, 2, \ldots$$

$$x_1(t) = t\, z_1 \quad \text{for} \quad t > \tfrac{1}{2}.$$

For values of t in the interval $0 < t < \frac{1}{2}$, which are not of the form p_n^{-k}, we define $x_1(t)$ by linear interpolation:

$$x_1(t) = \frac{(t_2 - t)\, x_1(t_1) + (t - t_1)\, x_1(t_2)}{t_2 - t_1},$$

where $t_1 = p_{n_1}^{-k_1}$ and $t_2 = p_{n_2}^{-k_2}$ are the nearest values below and above t, for which $x_1(t)$ has been defined above. Some easy calculation shows that we have for all $t > 0$

$$E\,|x_1(t)|^2 \leqslant t^2.$$

This shows that the point $t = 0$ is a continuity point for the vector process $\mathbf{x}(t)$. Since every other real t is evidently also a continuity point, the $\mathbf{x}(t)$ process is everywhere continuous, and *a fortiori* satisfies condition (A).

On the other hand, we obviously have $\mathfrak{H}(\mathbf{x}, t) = 0$ for $t \leqslant 0$, while for every $t > 0$ the space $\mathfrak{H}(\mathbf{x}, t)$ will include the infinite orthonormal sequence $z_1, z_2, \ldots$. The point $t = 0$ will thus be a right discontinuity of the innovation spectrum of $\mathbf{x}(t)$, with an infinite right multiplicity. A case with any given finite multiplicity n is obtained if the variables $z_{n+1}, z_{n+2}, \ldots$ are replaced by zero. We have thus proved point (c) of the Lemma.

We observe that, by some further elaboration of the example given above, we may construct a process having the same multiplicity properties in each point of an everywhere dense set of values of t. We can even arrange the example so that the mean square derivative of $\mathbf{x}(t)$ exists for every t.

The last part of the Lemma follows simply from the above proofs of (b) and (c). If t is a left discontinuity of the innovation spectrum, at least one of the variables y_j occurring in (13) must be different from zero. Suppose, e.g., that $y_1 \neq 0$. Then

$$\| x_1(t) - x_1(t-0) \| \geqslant \| x_1(t) - P_{t-0}\, x_1(t) \| = \| y_1 \| > 0,$$

and so t is a left discontinuity of $x_1(t)$, and consequently also of $\mathbf{x}(t)$. On the other hand, the process $\mathbf{x}(t)$ constructed in the proof of (c) provides an example of a process having in $t = 0$ a right discontinuity point of the innovation spectrum (even of infinite multiplicity), which is nevertheless a continuity point of the process. This completes the proof of Lemma 2.

Innovation processes

6. From now on, it will be assumed that we are dealing with a vector process $\mathbf{x}(t)$ satisfying not only condition (A) of section 4, but also the following condition:

(B) $\mathbf{x}(t)$ *is a purely non-deterministic process, i.e. we have* $\mathfrak{H}(\mathbf{x}, -\infty) = 0$.

In the Hilbert space $\mathfrak{H}(\mathbf{x}) = \mathfrak{H}(\mathbf{x}, +\infty)$ of the process, we shall in general denote by $P_{\mathfrak{M}}$ the projection operator whose range is the subspace $\mathfrak{M}$. However, when $\mathfrak{M}$ is the particular subspace $\mathfrak{H}(\mathbf{x}, t)$, we write simply P_t instead of $P_{\mathfrak{H}(\mathbf{x}, t)}$.

As t increases from $-\infty$ to $+\infty$, the P_t form a never decreasing family of projections, with

$$P_{-\infty} = 0, \quad P_{+\infty} = I.$$

P_{t+0} is the projection on $\mathfrak{H}(\mathbf{x}, t+0)$, and similarly for P_{t-0}. The difference $P_{t_2} - P_{t_1}$, where $t_1 < t_2$, denotes the projection on the orthogonal complement $\mathfrak{H}(\mathbf{x}, t_2) - \mathfrak{H}(\mathbf{x}, t_1)$. It follows, in particular, that the projections $P_{t_2} - P_{t_1}$ and $P_{u_2} - P_{u_1}$ will be mutually orthogonal, as soon as the corresponding time intervals are disjoint.

Further, the points t of the innovation spectrum are characterized by the property

$$P_{t+h} - P_{t-h} > 0$$

for any $h > 0$, while the discontinuity points of that spectrum are characterized by the relation

$$P_{t+0} - P_{t-0} > 0.$$

Consider now any element z of the Hilbert space $\mathfrak{H}(\mathbf{x})$, and let us define a stochastic process $z(\lambda)$ by writing for any real λ

$$z(\lambda) = P_\lambda z. \tag{14}$$

It then follows from the above that $z(\lambda)$ is a process *with orthogonal increments*, such that

$$z(-\infty) = 0, \quad z(+\infty) = z.$$

We have $E\,z(\lambda) = 0$, and if we write

$$E\,|z(\lambda)|^2 = F_z(\lambda),$$

$F_z(\lambda)$ will be a never decreasing function of the real variable λ, such that

$$F_z(-\infty) = 0, \quad F_z(+\infty) = E|z|^2.$$

The points of increase of $z(\lambda)$, i.e. the points λ such that the increment $z(\lambda+h) - z(\lambda-h)$ does not reduce to zero for any $h > 0$, are identical with the points of increase of $F_z(\lambda)$, and form a subset of the innovation spectrum of the $\mathbf{x}(t)$ process. Similarly the left (right) discontinuities of $z(\lambda)$ are identical with the left (right) discontinuities of $F_z(\lambda)$, and form a subset of the set of all left (right) discontinuities of the innovation spectrum of $\mathbf{x}(t)$. Any increment $dz(\lambda)$ belongs to the subspace $d_\lambda \mathfrak{H}(\mathbf{x}, \lambda)$, and is thus built up by a certain part of the elements which enter as innovations into the $\mathbf{x}(t)$ process between the time points $t=\lambda$ and $t=\lambda+d\lambda$.

On account of these facts, we shall denote the (one-dimensional) $z(\lambda)$ process as a *partial innovation process* associated with the given vector process $\mathbf{x}(t)$.

7. For any $t \leqslant +\infty$, we shall denote by $\mathfrak{H}(z, t)$ the Hilbert space spanned by the random variables $z(\lambda)$ for all $\lambda \leqslant t$:

$$\mathfrak{H}(z, t) = \mathfrak{S}\{z(\lambda);\quad \lambda \leqslant t\}.$$

It follows from (14) that $z(\lambda)$ is always an element of $\mathfrak{H}(\mathbf{x}, \lambda)$, and consequently $\mathfrak{H}(z, t)$, is a subspace of $\mathfrak{H}(\mathbf{x}, t)$:

$$\mathfrak{H}(z, t) \subset \mathfrak{H}(\mathbf{x}, t).$$

Instead of $\mathfrak{H}(z, +\infty)$ we shall write briefly $\mathfrak{H}(z)$. Evidently $\mathfrak{H}(z, t)$ is the projection of $\mathfrak{H}(z)$ on $\mathfrak{H}(\mathbf{x}, t)$.

If no λ is at the same time a left and a right discontinuity of $z(\lambda)$, then $\mathfrak{H}(z, t)$ is, for every $t \leqslant +\infty$, identical[1] with the set $\mathfrak{H}^*(z, t)$ of all random variables y representable in the form

$$y = \int_{-\infty}^{t} g(\lambda)\, dz(\lambda) \tag{15}$$

with an F_z-measurable g such that

$$E|y|^2 = \int_{-\infty}^{t} |g(\lambda)|^2\, dF_z(\lambda) < \infty.$$

On the other hand, if in a certain point $\lambda < t$ the left and the right jumps of $z(\lambda)$, say u and v respectively, are both different from zero, all random variables $Au + Bv$ with constant A and B will belong to $\mathfrak{H}(z, t)$ while, with the usual definition of the integral (15), the discontinuity at λ will only provide the variables $A(u+v)$ as elements of $\mathfrak{H}^*(z, t)$.

We shall require the following Lemma, which is only a restatement of familiar facts concerning Hilbert space.

Lemma 3. *If y and z are elements of $\mathfrak{H}(\mathbf{x})$ such that $y \perp \mathfrak{H}(z)$, then $\mathfrak{H}(y) \perp \mathfrak{H}(z)$.*

Since $\mathfrak{H}(z)$ is the space spanned by all $z(\lambda) = P_\lambda z$, the relation $y \perp \mathfrak{H}(z)$ is equivalent to $y \perp P_\lambda z$ for all real λ. By the same argument, the assertion of Lemma 3 is equivalent to the relation

$$P_\lambda y \perp P_\mu z$$

[1] Cf., e.g., [2], p. 425–428. The integral in (15) should be so defined that, if the upper limit t is a discontinuity of $z(\lambda)$, a left jump of $z(\lambda)$ is included in the value of the integral, but not a right jump.

for all λ and μ. Now $y = P_\lambda y + w$, where $w \perp \mathfrak{H}(\mathbf{x}, \lambda)$, and thus in particular $w \perp P_\lambda z$. Hence for any λ

$$P_\lambda y = y - w \perp P_\lambda z.$$

Suppose now first $\lambda > \mu$. Then $P_\mu P_\lambda = P_\mu$, and thus

$$P_\lambda y = P_\mu y + w',$$

where $w' \perp \mathfrak{H}(\mathbf{x}, \mu)$. Hence in particular $w' \perp P_\mu z$. Since we have just proved that $P_\mu y \perp P_\mu z$, it follows that $P_\lambda y \perp P_\mu z$.

On the other hand, if $\lambda < \mu$ we have $P_\mu z = P_\lambda z + w''$, where $w'' \perp \mathfrak{H}(\mathbf{x}, \lambda)$, and thus $w'' \perp P_\lambda y$. Hence we obtain as before $P_\mu z \perp P_\lambda y$, and so Lemma 3 is proved.

Representation of $\mathbf{x}(t)$

8. We begin by proving the following Lemma, assuming as before that we are dealing with a given vector process $\mathbf{x}(t)$ satisfying conditions (A) and (B).

Lemma 4. *It is possible to find a finite or infinite sequence $z_1, z_2, \ldots$ of non-vanishing elements of $\mathfrak{H}(\mathbf{x})$ such that we have for every $t \leqslant +\infty$*

$$\mathfrak{H}(z_j, t) \perp \mathfrak{H}(z_k, t) \quad \text{for} \quad j \neq k, \tag{16}$$

$$\mathfrak{H}(\mathbf{x}, t) = \mathfrak{H}(z_1, t) + \mathfrak{H}(z_2, t) + \ldots, \tag{17}$$

where the second member of (17) *denotes the vector sum of the mutually orthogonal spaces involved.*

We first observe that it is sufficient to show that we can find a z_n sequence such that (16) and (17) hold for $t = +\infty$, since their validity for any finite t then easily follows.

By Lemma 1, the space, $\mathfrak{H}(\mathbf{x})$ is separable. Neglecting trivial cases, we may assume that $\mathfrak{H}(\mathbf{x})$ is infinite-dimensional. Thus a complete orthonormal system in $\mathfrak{H}(\mathbf{x})$ will form an infinite sequence, say $z_1^*, z_2^*, \ldots$. Starting from the sequence of the z_n^*, we shall now construct an infinite sequence $z_1, z_2, \ldots$ satisfying (16) and (17) for $t = +\infty$. Discarding any z_n which reduces to zero, we then obtain a finite or enumerable sequence of non-vanishing elements having the same properties, and so the Lemma will be proved.

We define the z_n sequence by the relations

$$\begin{aligned} z_1 &= z_1^*, \\ z_2 &= z_2^* - P_{\mathfrak{M}_1} z_2^*, \\ &\cdots\cdots \\ z_n &= z_n^* - P_{\mathfrak{M}_{n-1}} z_n^*, \\ &\cdots\cdots \end{aligned} \tag{18}$$

where $\mathfrak{M}_n$ denotes the vector sum

$$\mathfrak{M}_n = \mathfrak{H}(z_1) + \cdots + \mathfrak{H}(z_n).$$

Then for any $n > 1$ we have $z_n \perp \mathfrak{M}_{n-1}$, and consequently $z_n \perp \mathfrak{H}(z_j)$ for $j = 1, \ldots, n-1$. By Lemma 3 it then follows that we have $\mathfrak{H}(z_j) \perp \mathfrak{H}(z_k)$ for $j \neq k$, so that (16) is satisfied.

We further have

$$z_n^* = z_n + P_{\mathfrak{M}_{n-1}} z_n^*,$$

and since $z_n \in \mathfrak{H}(z_n)$, this shows that $z_n^* \in \mathfrak{H}(z_n) + \mathfrak{M}_{n-1} = \mathfrak{M}_n$. Now $\mathfrak{M}_n$ is a subset of the infinite vector sum $\mathfrak{H}(z_1) + \mathfrak{H}(z_2) + \ldots$. This sum, which is a subspace of $\mathfrak{H}(\mathbf{x})$, will thus contain the complete orthonormal system $z_1^*, z_2^*, \ldots$, and will consequently be identical with $\mathfrak{H}(\mathbf{x})$, so that (17) is also satisfied, and the Lemma is proved.

9. The sequence $z_1, z_2, \ldots$ considered in Lemma 4 is not uniquely determined, and we now proceed to show that it can be chosen in a way which will suit our purpose.

By Lemma 2, the discontinuities of the innovation spectrum of $\mathbf{x}(t)$ form an at most enumerable set. Let them be denoted by $\lambda_1. \lambda_2, \ldots$, and consider the subspaces

$$\mathfrak{U}_k = \mathfrak{H}(\mathbf{x}, \lambda_k) - \mathfrak{H}(\mathbf{x}, \lambda_k - 0),$$

$$\mathfrak{V}_k = \mathfrak{H}(\mathbf{x}, \lambda_k + 0) - \mathfrak{H}(\mathbf{x}, \lambda_k),$$

for $k = 1, 2, \ldots$. If h_k and j_k are the numbers of dimensions of $\mathfrak{U}_k$ and $\mathfrak{V}_k$ respectively, then by Lemma 2 we have $0 \leqslant h_k \leqslant q$, while j_k may be any finite non-negative integer or $\aleph_0$. According to the terminology of section 5, h_k is the left multiplicity of the point λ_k, while j_k is the right multiplicity. The number

$$N' = \sup_k (h_k + j_k)$$

will be called the *multiplicity of the discontinuous part* of the innovation spectrum. Let

$$u_{k1}, \ldots, u_{kh_k},$$

$$v_{k1}, \ldots, v_{kj_k}$$

be complete orthonormal systems in $\mathfrak{U}_k$ and $\mathfrak{V}_k$ respectively. If $\mathfrak{U}_k$ or $\mathfrak{V}_k$ reduces to zero, the corresponding system does of course not occur; however, when λ_k is a discontinuity, h_k and j_k cannot both be equal to zero, so that at least one of the u and v systems must contain a non-vanishing number of terms.

If z denotes any of the u_{kh} or v_{kj}, it will be seen that we have $P_\lambda z = z$ for $\lambda > \lambda_k$, and $P_\lambda z = 0$ for $\lambda < \lambda_k$. It follows that the space $\mathfrak{H}(z)$ will then be one-dimensional, and consist of all constant multiples of z. If y is any variable in $\mathfrak{H}(\mathbf{x})$ such that $y \perp z$, it then follows from Lemma 3 that $\mathfrak{H}(y) \perp \mathfrak{H}(z)$.

Suppose now that one of the orthonormal variables from which we started our proof of Lemma 4, say z_n^*, is identical with one of the u_{kh} or v_{kj}. By means of the remark just made, it then follows from the relations (18) that z_n^*

is orthogonal to $z_1, \ldots, z_{n-1}$, and thus also orthogonal to $\mathfrak{M}_{n-1}$. Consequently by (18) we obtain $z_n = z_n^*$.

Choosing the orthonormal system $z_1^*, z_2^*, \ldots$, so that all the variables u_{kh} and v_{kj} $(k=1, 2, \ldots, h=1, \ldots, h_k, j=1, \ldots, j_k)$ occur in it, we thus see that the same variables will also occur in the sequence $z_1, z_2, \ldots$ constructed according to (18), and satisfying the conditions of Lemma 4. Besides the u_{kh} and v_{kj}, there may be other elements in the z_n sequence. Let $w_1, w_2, \ldots$ be those elements, if any, which are different from all the u_{kh} and v_{kj}. We are going to show that, if

$$w(\lambda) = P_\lambda w$$

s the partial innovation process corresponding to any of the w_n, then $w(\lambda)$ *has no discontinuities.*

In fact, by a remark made in section 6, any discontinuity of $w(\lambda)$ would be a discontinuity of the innovation spectrum of $\mathbf{x}(t)$, i.e. equal to one of the λ_k. The corresponding jump of $w(\lambda)$, say w^*, would then be an element of the space $\mathfrak{H}(\mathbf{x}, \lambda_k+0) - \mathfrak{H}(\mathbf{x}, \lambda_k - 0)$. At the same time, w^* would belong to the space $\mathfrak{H}(w)$, and by Lemma 4 would thus be orthogonal to all the u_{kh} and v_{kj}. However, the latter variables form a complete orthonormal system in $\mathfrak{H}(\mathbf{x}, \lambda_k+0) - \mathfrak{H}(\mathbf{x}, \lambda_k - 0)$ so that we must have $w^*=0$, and it follows that $w(\lambda)$ has no discontinuities.

Let now the sequence of non-vanishing elements $z_1, z_2, \ldots$ considered in Lemma 4 be chosen in all possible ways that are consistent with the requirement that all the u_{kh} and v_{kj} should occur in it. Let, in each case, M denote the cardinal number of the corresponding sequence $w_1, w_2, \ldots$, formed by those elements which are different from all the u_{kh} and v_{kj}. The numbers M will then have a non-negative lower bound:

$$N'' = \inf M,$$

which we shall call the *multiplicity of the continuous part* of the innovation spectrum. Finally

$$N = \max(N', N'') \tag{19}$$

will be called the *spectral multiplicity* of the $\mathbf{x}(t)$ process. As soon as $\mathbf{x}(t)$ is not identically zero, N will be a finite positive integer, or equal to $\aleph_0$.

It follows from the definition of the multiplicity N'' of the continuous part that it is possible to find a sequence $z_1, z_2, \ldots$ satisfying the above requirements, and such that the corresponding set $w_1, w_2, \ldots$ will have precisely the cardinal number N''. We shall then say that these w_n form a *minimal w sequence.*[1]

In the sequel, $w_1, w_2, \ldots$ will denote the elements of a fixed minimal w sequence. We now propose to construct, by means of this given w sequence, a particular sequence $z_1, z_2, \ldots$ satisfying the conditions of Lemma 4, which will then be used for the proof of our representation theorem for $\mathbf{x}(t)$.

[1] By an adaptation of the proofs of theorem 7.5 and 7.6 of Stone [7] to the case considered here, it can be shown that a minimal w sequence can be chosen in such a way that the set of all points of increase of $w_1(\lambda) = P_\lambda w_1$ is identical with the continuous part of the innovation spectrum of $\mathbf{x}(t)$, and includes the corresponding set of any $w_n(\lambda)$ with $n > 1$ as a subset. As this property is not indispensable for the proof of the representation theorem given below, we shall restrict ourselves here to this remark.

We first observe that, owing to the way in which the w_j have been chosen, we have by Lemma 4 for every $t \leqslant +\infty$

$$\mathfrak{H}(\mathbf{x}, t) = \sum_{k,h} \mathfrak{H}(u_{kh}, t) + \sum_{k,j} \mathfrak{H}(v_{kj}, t) + \sum_{j} \mathfrak{H}(w_j, t), \tag{20}$$

where the sums denote vector addition, and all the $\mathfrak{H}$ spaces appearing in the second member are mutually orthogonal.

For any discontinuity point λ_k we now arrange the corresponding variables u_{kh} and v_{kj} into a single sequence

$$s_{k1}, \; s_{k2}, \; \ldots,$$

where the number of terms will be $h_k + j_k$. The new sequence $z_1, z_2, \ldots$ which we have in view is then defined by taking

$$z_n = w_n + \sum_k s_{kn}, \tag{21}$$

where the w_n are the elements of our fixed minimal w sequence. The summation is extended over all discontinuity points λ_k, and we take $s_{kn} = 0$ whenever $n > h_k + j_k$, and $w_n = 0$ whenever $n > N''$. The number of non-vanishing terms in the z_n sequence defined in this way will evidently be equal to the spectral multiplicity N of the $\mathbf{x}(t)$ process as defined by (19).

According to a remark made above, any space $\mathfrak{H}(s_{kn})$ is one-dimensional, and consists of all constant multiples of the variable s_{kn}. Hence it easily follows that we have for $t \leqslant +\infty$

$$\mathfrak{H}(z_n, t) = \mathfrak{H}(w_n, t) + \sum_k \mathfrak{H}(s_{kn}, t),$$

and further according to (20)

$$\begin{aligned} &\mathfrak{H}(\mathbf{x}, t) = \mathfrak{H}(z_1, t) + \mathfrak{H}(z_2, t) + \ldots, \\ &\mathfrak{H}(z_j t) \perp \mathfrak{H}(z_k, t) \quad \text{for} \quad j \neq k, \end{aligned} \tag{22}$$

so that the z_n defined by (21) satisfy the conditions of Lemma 4.

We observe that it follows from (21) that no $z_n(\lambda)$ can have a left and a right discontinuity in the same point λ. In fact, any discontinuity of $z_n(\lambda)$ will be either a left discontinuity with a jump u_{kh}, or a right discontinuity with a jump v_{kj}. By a remark made in section 6, the space $\mathfrak{H}(z_n, t)$ will then be identical with the set of all random variables y representable in the form (15).

In our given q-dimensional (column) vector process

$$\mathbf{x}(t) = (x_1(t), \ldots, x_q(t))$$

every component $x_j(t)$ is a random variable belonging to $\mathfrak{H}(\mathbf{x}, t)$. It then follows from (15) and (22) that we have for all real t

$$x_j(t) = \sum_{n=1}^{N} \int_{-\infty}^{t} g_{jn}(t, \lambda)\, d z_n(\lambda). \tag{23}$$

If the multiplicity N is infinite, the series in the second member will converge in the usual sense, so that we have

$$\sum_{n=1}^{N} \int_{-\infty}^{t} |g_{jn}(t, \lambda)|^2\, d F_{z_n}(\lambda) < \infty. \tag{24}$$

Introducing the $q \times N$ order matrix function

$$\mathbf{G}(t, \lambda) = \{g_{jn}(t, \lambda)\}, \tag{25}$$

$(j = 1, \ldots, q; n = 1, \ldots, N)$, and the N-dimensional (column) vector process

$$\mathbf{z}(\lambda) = (z_1(\lambda), \ldots, z_N(\lambda)), \tag{26}$$

it is seen that (23) may be written

$$\mathbf{x}(t) = \int_{-\infty}^{t} \mathbf{G}(t, \lambda)\, d\, \mathbf{z}(\lambda). \tag{27}$$

The vector process $\mathbf{z}(\lambda)$ defined by (26) has *orthogonal increments*, in the sense that two increments $\Delta z_j(\lambda)$ and $\Delta z_k(\mu)$ are always orthogonal if $j \neq k$, while for $j = k$ they are orthogonal if the corresponding time intervals are disjoint.

If we denote by $\mathfrak{H}(\mathbf{z}, t)$ the Hilbert space spanned by the variables $z_1(\lambda), \ldots, z_N(\lambda)$ for all $\lambda \leqslant t$:

$$\mathfrak{H}(\mathbf{z}, t) = \mathfrak{S}\{z_1(\lambda), \ldots, z_N(\lambda);\ \lambda \leqslant t\},$$

it follows from (22) that we have for every $t \leqslant +\infty$

$$\mathfrak{H}(\mathbf{x}, t) = \mathfrak{H}(\mathbf{z}, t). \tag{28}$$

According to the representation formula (27) and the property expressed by (28), it seems appropriate to call $\mathbf{z}(\lambda)$ a *total innovation process* associated with the given $\mathbf{x}(t)$. While $\mathbf{z}(\lambda)$ is not uniquely determined, its dimensionality N is uniquely determined by (19) as the spectral multiplicity of $\mathbf{x}(t)$. It is also seen that N is the smallest cardinal number for which there exists a representation of the form (27), with the properties specified by (22)–(28).

Summing up our results, we now have the following representation theorem.

Theorem 1. *Any stochastic vector process* $\mathbf{x}(t)$ *satisfying conditions* (A) *and* (B) *can be represented in the form* (27), *where* $\mathbf{G}(t, \lambda)$ *and* $\mathbf{z}(\lambda)$ *are defined by* (25) *and* (26). N *is the spectral multiplicity of the* $\mathbf{x}(t)$ *process. If* N *is infinite, the expansions* (23) *formally obtained for the components* $x_j(t)$ *are convergent in quadratic mean, as shown by* (24). $z_1, \ldots, z_N$ *are random variables in* $\mathfrak{H}(\mathbf{x})$ *satisfying* (22), *and such that no* $z_j(\lambda) = P_\lambda z_j$ *has a left and a right discontinuity in the same point* λ. *The vector process* $\mathbf{z}(\lambda)$ *has orthogonal increments and satisfies* (28). *No representation with these properties holds for any smaller value of* N.

If $\mathbf{x}(t) = (x_1(t), \ldots, x_q(t))$ is a vector process satisfying (A) and (B), each component $x_j(t)$, regarded as a one-dimensional process, has a certain spectral multiplicity N_j, and thus by Theorem 1 may be represented in the following form

$$x_j(t) = \sum_{n=1}^{N_j} \int_{-\infty}^{t} g_{jn}(t, \lambda)\, d z_{jn}(\lambda).$$

It can then be shown (although the proof is slightly more involved than may possibly be expected) that the spectral multiplicity N of $\mathbf{x}(t)$ satisfies the inequality

$$N \leqslant \sum_{j=1}^{q} N_j. \tag{29}$$

Consider, in particular, the case of a *stationary* vector process $\mathbf{x}(t)$, i.e. a process such that every second order covariance moment of the components is a function of the corresponding time difference:

$$E\left(x_j(t)\, \overline{x_k(u)}\right) = R_{jk}(t-u).$$

This process will satisfy (A) and (B) if and only if (a) the functions $R_{11}(t), \ldots, R_{qq}(t)$ are continuous at $t=0$, and (b) each component $x_j(t)$ is a purely non-deterministic stationary process. When these conditions are satisfied, each $x_j(t)$ has a representation of the form (7), and accordingly the spectral multiplicity of $x_j(t)$ is equal to one. It then follows from (29) that we have in this case $N \leqslant q$.

On the other hand, the process $\mathbf{x}(t)$ constructed in connection with the proof of Lemma 2, point (c), evidently provides an example of a vector process satisfying (A) and (B), and having an infinite spectral multiplicity N. As already observed, this example may be easily modified so as to yield a process with any given finite multiplicity.

If, in the relation (27), all components of the vectors on both sides are projected on the space $H(\mathbf{x}, u)$, where $u < t$, we finally obtain the following theorem, denoting by $P_u \mathbf{x}(t)$ the vector with the components $P_u x_j(t)$, for $j = 1, \ldots, q$.

Theorem 2. *The best linear (least squares) prediction of* $\mathbf{x}(t)$ *in terms of all variables* $x_j(\tau)$ *with* $j = 1, \ldots, q$ *and* $\tau \leqslant u$ *is given by the expression*

$$P_u \mathbf{x}(t) = \int_{-\infty}^{u} \mathbf{G}(t, \lambda)\, d\mathbf{z}(\lambda).$$

The square of the corresponding error of prediction for any component $x_j(t)$ *is*

$$E\left| x_j(t) - P_u x_j(t) \right|^2 = \sum_{n=1}^{N} \int_{u}^{t} \left| g_{jn}(t, \lambda) \right|^2 d F_{z_n}(\lambda).$$

REFERENCES

1. CRAMÉR, H., On some classes of non-stationary stochastic processes. To appear in: Proc. Fourth Berkeley Symp. on Mathematical Statistics and Probability.
2. DOOB, J. L., Stochastic processes. Wiley, New York, 1953.
3. HIDA, T., Canonical representations of Gaussian processes and their applications. Mem. Coll. Sci. Kyoto, A33 (1960).
4. LÉVY, P., A special problem of Brownian motion, and a general theory of Gaussian random functions. Proc. Third Berkeley Symp. on Mathematical Statistics and Probability, *2*, 133 (1956).
5. ——, Sur une classe de courbes de l'espace de Hilbert et sur une équation intégrale non linéaire. Ann. Ec. Norm. Sup. *73*, 121 (1956).
6. ——, Fonctions aléatoires à corrélation linéaire. Illinois J. Math. *1*, 217 (1957).
7. STONE, M. H., Linear transformations in Hilbert space. American Math. Soc. colloquium publication. New York, 1932.
8. WIENER, N., and MASANI, P., The prediction theory of multivariate stochastic processes. Acta Math. *98*, 111 (1957); and *99*, 93 (1958).
9. WOLD, H., A Study in the Analysis of Stationary Time Series. Thesis, University of Stockholm. Uppsala, 1938.

Tryckt den 22 februari 1961

Uppsala 1961. Almqvist & Wiksells Boktryckeri AB

Editors' Comments on Paper 8

At this point the benchmark paper of Kalman and Bucy on new results in filtering is presented. This paper does not depend on or bear directly upon the contemporarily obtained results on the multiplicity representation. However, it constitutes a fundamental advance in filtering theory and provides a model that is now standard in a number of engineering applications.

In addition to possessing the virtues of generality, implementability, and tractability, this model, in its linear versions, makes implicit use of the possibility of a causal, invertible prewhitening of either the state or the observation processes. On the other hand, the model applies only to systems that can be described by a finite number of differential equations. Thus more general impulse responses are not included.

It has been shown recently by Kailath that the Kalman–Bucy results may be obtained in some cases by the innovations approach, which is more directly related to the canonical decomposition of random processes.

This paper is reprinted with the permission of the authors and the American Society of Mechanical Engineers.

8

New Results in Linear Filtering and Prediction Theory[1]

R. E. KALMAN
Research Institute for Advanced Study,[2] Baltimore, Maryland

R. S. BUCY
The Johns Hopkins Applied Physics Laboratory, Silver Spring, Maryland

A nonlinear differential equation of the Riccati type is derived for the covariance matrix of the optimal filtering error. The solution of this "variance equation" completely specifies the optimal filter for either finite or infinite smoothing intervals and stationary or nonstationary statistics.

The variance equation is closely related to the Hamiltonian (canonical) differential equations of the calculus of variations. Analytic solutions are available in some cases. The significance of the variance equation is illustrated by examples which duplicate, simplify, or extend earlier results in this field.

The Duality Principle relating stochastic estimation and deterministic control problems plays an important role in the proof of theoretical results. In several examples, the estimation problem and its dual are discussed side-by-side.

Properties of the variance equation are of great interest in the theory of adaptive systems. Some aspects of this are considered briefly.

1 Introduction

At present, a nonspecialist might well regard the Wiener-Kolmogorov theory of filtering and prediction [1, 2][3] as "classical'—in short, a field where the techniques are well established and only minor improvements and generalizations can be expected.

That this is not really so can be seen convincingly from recent results of Shinbrot [3], Steeg [4], Pugachev [5, 6], and Parzen [7]. Using a variety of time-domain methods, these investigators have solved some long-standing problems in *nonstationary* filtering and prediction theory. We present here a unified account of our own independent researches during the past two years (which overlap with much of the work [3–7] just mentioned), as well as numerous new results. We, too, use time-domain methods, and obtain major improvements and generalizations of the conventional Wiener theory. In particular, our methods apply without modification to multivariate problems.

The following is the historical background of this paper.

In an extension of the standard Wiener filtering problem, Follin [8] obtained relationships between time-varying gains and error variances for a given circuit configuration. Later, Hanson [9] proved that Follin's circuit configuration was actually optimal for the assumed statistics; moreover, he showed that the differential equations for the error variance (first obtained by Follin) follow rigorously from the Wiener-Hopf equation. These results were then generalized by Bucy [10], who found explicit relationships between the optimal weighting functions and the error variances; he also gave a rigorous derivation of the variance equations and those of the optimal filter for a wide class of nonstationary signal and noise statistics.

Independently of the work just mentioned, Kalman [11] gave a new approach to the standard filtering and prediction problem. The novelty consisted in combining two well-known ideas:

(i) the "state-transition" method of describing dynamical systems [12–14], and

(ii) linear filtering regarded as orthogonal projection in Hilbert space [15, pp. 150–155].

As an important by-product, this approach yielded the *Duality Principle* [11, 16] which provides a link between (stochastic) filtering theory and (deterministic) control theory. Because of the duality, results on the optimal design of linear control systems [13, 16, 17] are directly applicable to the Wiener problem. Duality plays an important role in this paper also.

When the authors became aware of each other's work, it was soon realized that the principal conclusion of both investigations was identical, in spite of the difference in methods:

Rather than to attack the Wiener-Hopf integral equation directly, it is better to convert it into a nonlinear differential equation, whose solution yields the covariance matrix of the minimum filtering error, which in turn contains all necessary information for the design of the optimal filter.

2 Summary of Results: Description

The problem considered in this paper is stated precisely in Section 4. There are two main assumptions:

(A_1) A sufficiently accurate model of the message process is given by a linear (possibly time-varying) dynamical system excited by white noise.

(A_2) Every observed signal contains an additive white noise component.

Assumption (A_2) is unnecessary when the random processes in question are sampled (discrete-time parameter); see [11]. Even in the continuous-time case, (A_2) is no real restriction since it can be removed in various ways as will be shown in a future paper. Assumption (A_1), however, is quite basic; it is analogous to but somewhat less restrictive than the assumption of rational spectra in the conventional theory.

Within these assumptions, we seek the best linear estimate of the message based on past data lying in either a finite or infinite time-interval.

The fundamental relations of our new approach consist of five equations:

---

[1] This research was partially supported by the United States Air Force under Contracts AF 49(638)-382 and AF 33(616)-6952 and by the Bureau of Naval Weapons under Contract NOrd-73861.

[2] 7212 Bellona Avenue

[3] Numbers in brackets designate References at the end of paper.

Contributed by the Instruments and Regulators Division of The American Society of Mechanical Engineers and presented at the Joint Automatic Controls Conference, Cambridge, Mass., September 7–9, 1960. Manuscript received at ASME Headquarters, May 31, 1960. Paper No. 60—JAC-12.

(I) The differential equation governing the optimal filter, which is excited by the observed signals and generates the best linear estimate of the message.

(II) The differential equations governing the error of the best linear estimate.

(III) The time-varying gains of the optimal filter expressed in terms of the error variances.

(IV) The nonlinear differential equation governing the co-variance matrix of the errors of the best linear estimate, called the *variance equation.*

(V) The formula for prediction.

The solution of the variance equation for a given finite time-interval is equivalent to the solution of the estimation or prediction problem with respect to the same time-interval. The steady-state solution of the variance equation corresponds to finding the best estimate based on all the data in the past.

As a special case, one gets the solution of the classical (stationary) Wiener problem by finding the unique equilibrium point of the variance equation. This requires solving a set of algebraic equations and constitutes a new method of designing Wiener filters. The superior effectiveness of this procedure over present methods is shown in the examples.

Some of the preceding ideas are implicit already in [10, 11]; they appear here in a fully developed form. Other more advanced problems have been investigated only very recently and provide incentives for much further research. We discuss the following further results:

(1) The variance equations are of the Riccati type which occur in the calculus of variations and are closely related to the canonical differential equations of Hamilton. This relationship gives rise to a well-known analytic formula for the solution of the Riccati equation [17, 18]. The Hamiltonian equations have also been used recently [19] in the study of optimal control systems. The two types of problems are actually duals of one another as mentioned in the Introduction. The duality is illustrated by several examples.

(2) A sufficient condition for the existence of steady-state solutions of the variance equation (i.e., the fact that the error variance does not increase indefinitely) is that the information matrix in the sense of R. A. Fisher [20] be nonsingular. This condition is considerably weaker than the usual assumption that the message process have finite variance.

(3) A sufficient condition for the optimal filter to be stable is the dual of the preceding condition.

The preceding results are established with the aid of the "state-transition" method of analysis of dynamical systems. This consists essentially of the systematic use of vector-matrix notation which results in simple and clear statements of the main results independently of the complexity of specific problems. This is the reason why multivariable filtering problems can be treated by our methods without any additional theoretical complications.

The outline of contents is as follows:

In Section 3 we review the description of dynamical systems from the state point of view. Sections 4–5 contain precise statements of the filtering problem and of the dual control problem. The examples in Section 6 illustrate the filtering problem and its dual in conventional block-diagram terminology. Section 7 contains a precise statement of all mathematical results. A reader interested mainly in applications may pass from Section 7 directly to the worked-out examples in Section 11. The rigorous derivation of the fundamental equations is given in Section 8. Section 9 outlines proofs, based on the Duality Principle, of the existence and stability of solutions of the variance equation. The theory of analytic solutions of the variance equation is discussed in Section 10. In Section 12 we examine briefly the relation of our results to adaptive filtering problems. A critical evaluation of the current status of the statistical filtering problem is presented in Section 13.

3 Preliminaries

In the main, we shall follow the notation conventions (though not the specific nomenclature) of [11], [16], and [21]. Thus τ, t, t_0 refer to the time, α, β, . . ., x_1, x_2, . . ., ϕ_1, ϕ_2, . . ., a_{ij}, . . . are (real) scalars; $\mathbf{a}$, $\mathbf{b}$, . . ., $\mathbf{x}$, $\mathbf{y}$, . . ., $\boldsymbol{\phi}$, $\boldsymbol{\psi}$, . . . are vectors, $\mathbf{A}$, $\mathbf{B}$, . . ., $\boldsymbol{\Phi}$, $\boldsymbol{\Psi}$, . . . are matrices. The prime denotes the transposed matrix; thus $\mathbf{x}'\mathbf{y}$ is the scalar (inner) product and $\mathbf{x}\mathbf{y}'$ denotes a matrix with elements x_iy_j (outer product). $\|\mathbf{x}\| = (\mathbf{x}'\mathbf{x})^{1/2}$ is the euclidean norm and $\|\mathbf{x}\|^2_{\mathbf{A}}$ (where $\mathbf{A}$ is a nonnegative definite matrix) is the quadratic form with respect to $\mathbf{A}$. The eigenvalues of a matrix $\mathbf{A}$ are written as $\lambda_i(\mathbf{A})$. The expected value (ensemble average) is denoted by $\mathcal{E}$ (usually not followed by brackets). The covariance matrix of two vector-valued random variables $\mathbf{x}(t)$, $\mathbf{y}(\tau)$ is denoted by

$$\mathcal{E}\mathbf{x}(t)\mathbf{y}'(\tau) - \mathcal{E}\mathbf{x}(t)\mathcal{E}\mathbf{y}'(\tau) \quad \text{or} \quad \operatorname{cov}[\mathbf{x}(t), \mathbf{y}(\tau)]$$

depending on what form is more convenient.

Real-valued linear functions of a vector $\mathbf{x}$ will be denoted by $\mathbf{x}^*$; the value of $\mathbf{x}^*$ at $\mathbf{x}$ is denoted by

$$[\mathbf{x}^*, \mathbf{x}] = \sum_{i=1}^{n} x^*_i x_i$$

where the x_i are the co-ordinates of $\mathbf{x}$. As is well known, $\mathbf{x}^*$ may be regarded abstractly as an element of the *dual vector space* of the $\mathbf{x}$'s; for this reason, $\mathbf{x}^*$ is called a *covector* and its co-ordinates are the x^*_i. In algebraic manipulations we regard $\mathbf{x}^*$ formally as a row vector (remembering, of course, that $\mathbf{x}^* \neq \mathbf{x}'$). Thus the inner product is $\mathbf{x}^*\mathbf{y}^{*\prime}$ and we define $\|\mathbf{x}^*\|$ by $(\mathbf{x}^*\mathbf{x}^{*\prime})^{1/2}$. Also

$$\mathcal{E}[\mathbf{x}^*, \mathbf{x}]^2 = \mathcal{E}(\mathbf{x}^*\mathbf{x})^2 = \mathcal{E}\mathbf{x}^*\mathbf{x}\mathbf{x}'\mathbf{x}^{*\prime}$$
$$= \mathbf{x}^*(\mathcal{E}\mathbf{x}\mathbf{x}')\mathbf{x}^{*\prime} = \|\mathbf{x}^*\|^2_{\mathcal{E}\mathbf{x}\mathbf{x}'}$$

To establish the terminology, we now review the essentials of the so-called *state-transition method* of analysis of dynamical systems. For more details see, for instance, [21].

A linear dynamical system governed by an ordinary differential equation can always be described in such a way that the defining equations are in the *standard form:*

$$d\mathbf{x}/dt = \mathbf{F}(t)\mathbf{x} + \mathbf{G}(t)\mathbf{u}(t) \qquad (1)$$

where $\mathbf{x}$ is an n-vector, called the *state;* the *co-ordinates* x_i of $\mathbf{x}$ are called *state variables;* $\mathbf{u}(t)$ is an m-vector, called the *control function;* $\mathbf{F}(t)$ and $\mathbf{G}(t)$ are $n \times n$ and $n \times m$ matrices, respectively, whose elements are continuous functions of the time t.

The description (1) is incomplete without specifying the *output* $\mathbf{y}(t)$ of the system; this may be taken as a p-vector whose components are linear combinations of the state variables:

$$\mathbf{y}(t) = \mathbf{H}(t)\mathbf{x}(t) \qquad (2)$$

where $\mathbf{H}(t)$ is a $p \times n$ matrix continuous in t.

The matrices $\mathbf{F}$, $\mathbf{G}$, $\mathbf{H}$ can be usually determined by inspection if the system equations are given in block diagram form. See the examples in Section 5. It should be remembered that any of these matrices may be nonsingular. $\mathbf{F}$ represents the dynamics, $\mathbf{G}$ the constraints on affecting the state of the system by inputs, and $\mathbf{H}$ the constraints on observing the state of the system from outputs. For single-input/single-output systems, $\mathbf{G}$ and $\mathbf{H}$ consist of a single column and single row, respectively.

If $\mathbf{F}$, $\mathbf{G}$, $\mathbf{H}$ are constants, (3) is a *constant* system. If $\mathbf{u}(t) = 0$ or, equivalently, $\mathbf{G} = \mathbf{0}$, (3) is said to be *free.*

It is well known [21–23] that the general solution of (1) may be written in the form

$$\mathbf{x}(t) = \mathbf{\Phi}(t, t_0)\mathbf{x}(t_0) + \int_{t_0}^{t} \mathbf{\Phi}(t, \tau)\mathbf{G}(\tau)\mathbf{u}(\tau)d\tau \tag{3}$$

where we call $\mathbf{\Phi}(t, t_0)$ the *transition matrix* of (1). The transition matrix is a nonsingular matrix satisfying the differential equation

$$d\mathbf{\Phi}/dt = \mathbf{F}(t)\mathbf{\Phi} \tag{4}$$

(any such matrix is a *fundamental matrix* [23, Chapter 3]), made *unique* by the additional requirement that, for all t_0,

$$\mathbf{\Phi}(t_0, t_0) = \mathbf{I} = \text{unit matrix} \tag{5}$$

The following properties are immediate by the existence and uniqueness of solutions of (1):

$$\mathbf{\Phi}^{-1}(t_1, t_0) = \mathbf{\Phi}(t_0, t_1) \quad \text{for all} \quad t_0, t_1 \tag{6}$$

$$\mathbf{\Phi}(t_2, t_0) = \mathbf{\Phi}(t_2, t_1)\mathbf{\Phi}(t_1, t_0) \quad \text{for all} \quad t_0, t_1, t_2 \tag{7}$$

If $\mathbf{F}$ = const, then the transition matrix can be represented by the well-known formula

$$\mathbf{\Phi}(t, t_0) = \exp \mathbf{F}(t - t_0) = \sum_{i=0}^{\infty} [\mathbf{F}(t - t_0)]^i/i! \tag{8}$$

which is quite convenient for numerical computations. In this special case, one can also express $\mathbf{\Phi}$ analytically in terms of the eigenvalues of $\mathbf{F}$, using either linear algebra [22] or standard transfer-function techniques [14].

In some cases, it is convenient to replace the right-hand side of (3) by a notation that focuses attention on how the state of the system "moves" in the state space as a function of time. Thus we write the left-hand side of (3) as

$$\mathbf{x}(t) \equiv \boldsymbol{\phi}(t; \ \mathbf{x}, t_0; \ \mathbf{u}) \tag{9}$$

Read: The state of the system (1) at time t, evolving from the initial state $\mathbf{x} = \mathbf{x}(t_0)$ at time t_0 under the action of a *fixed* forcing function $\mathbf{u}(t)$. For simplicity, we refer to $\boldsymbol{\phi}$ as the *motion* of the dynamical system

4 Statement of Problem

We shall be concerned with the continuous-time analog of Problem I of reference [11], which should be consulted for the physical motivation of the assumptions stated below.

(A_1) The *message* is a random process $\mathbf{x}(t)$ generated by the *model*

$$d\mathbf{x}/dt = \mathbf{F}(t)\mathbf{x} + \mathbf{G}(t)\mathbf{u}(t) \tag{10}$$

The *observed signal* is

$$\mathbf{z}(t) = \mathbf{y}(t) + \mathbf{v}(t) = \mathbf{H}(t)\mathbf{x}(t) + \mathbf{v}(t) \tag{11}$$

The functions $\mathbf{u}(t)$, $\mathbf{v}(t)$ in (10–11) are independent random processes (white noise) with identically zero means and covariance matrices

$$\begin{aligned} \text{cov}\,[\mathbf{u}(t), \mathbf{u}(\tau)] &= \mathbf{Q}(t)\cdot\delta(t - \tau) \\ \text{cov}\,[\mathbf{v}(t), \mathbf{v}(\tau)] &= \mathbf{R}(t)\cdot\delta(t - \tau) \quad \text{for all} \quad t, \tau \\ \text{cov}\,[\mathbf{u}(t), \mathbf{v}(\tau)] &= \mathbf{0} \end{aligned} \tag{12}$$

where δ is the Dirac delta function, and $\mathbf{Q}(t)$, $\mathbf{R}(t)$ are symmetric, nonnegative definite matrices continuously differentiable in t.

We introduce already here a restrictive assumption, which is needed for the ensuing theoretical developments:

(A_2) The matrix $\mathbf{R}(t)$ is positive definite for all t. Physically, this means that no component of the signal can be measured exactly.

To determine the random process $\mathbf{x}(t)$ uniquely, it is necessary to add a further assumption. This may be done in two different ways:

(A_3) The dynamical system (10) has reached "steady-state" under the action of $\mathbf{u}(t)$, in other words, $\mathbf{x}(t)$ is the random function defined by

$$\mathbf{x}(t) = \int_{-\infty}^{t} \mathbf{\Phi}(t, \tau)\mathbf{G}(\tau)\mathbf{u}(\tau)d\tau \tag{13}$$

This formula is valid if the system (10) is uniformly asymptotically stable (for precise definition, valid also in the nonconstant case, see [21]). If, in addition, it is true that $\mathbf{F}$, $\mathbf{G}$, $\mathbf{Q}$ are constant, then $\mathbf{x}(t)$ is a stationary random process—this is one of the chief assumptions of the original Wiener theory.

However, the requirement of asymptotic stability is inconvenient in some cases. For instance, it is not satisfied in Example 5, which is a useful model in some missile guidance problems. Moreover, the representation of random functions as generated by a linear dynamical system is already an appreciable restriction and one should try to avoid making any further assumptions. Hence we prefer to use:

(A_3') The measurement of $\mathbf{z}(t)$ starts at some fixed instant t_0 of time (which may be $-\infty$), at which time $\text{cov}[\mathbf{x}(t_0), \mathbf{x}(t_0)]$ is known.

Assumption (A_3) is obviously a special case of (A_3'). Moreover, since (10) is not necessarily stable, this way of proceeding makes it possible to treat also situations where the message variance grows indefinitely, which is excluded in the conventional theory.

The main object of the paper is to study the

OPTIMAL ESTIMATION PROBLEM. *Given known values of $\mathbf{z}(\tau)$ in the time-interval $t_0 \leq \tau \leq t$, find an estimate $\hat{\mathbf{x}}(t_1|t)$ of $\mathbf{x}(t_1)$ of the form*

$$\hat{\mathbf{x}}(t_1|t) = \int_{t_0}^{t} \mathbf{A}(t_1, \tau)\mathbf{z}(\tau)d\tau \tag{14}$$

(where $\mathbf{A}$ is an $n \times p$ matrix whose elements are continuously differentiable in both arguments) with the property that the expected squared error in estimating any linear function of the message is minimized:

$$\mathcal{E}[\mathbf{x}^*, \mathbf{x}(t_1) - \hat{\mathbf{x}}(t_1|t)]^2 = \text{minimum for all } \mathbf{x}^* \tag{15}$$

Remarks. (*a*) Obviously this problem includes as a special case the more common one in which it is desired to minimize

$$\mathcal{E}\|\mathbf{x}(t_1) - \hat{\mathbf{x}}(t_1|t)\|^2$$

(*b*) In view of (A_1), it is clear that $\mathcal{E}\mathbf{x}(t_1) = \mathcal{E}\hat{\mathbf{x}}(t_1|t) = \mathbf{0}$. Hence $[\mathbf{x}^*, \hat{\mathbf{x}}(t_1|t)]$ is the *minimum variance linear unbiased estimate* of the value of any *costate* $\mathbf{x}^*$ at $\mathbf{x}(t_1)$.

(*c*) If $\mathcal{E}\mathbf{u}(t)$ is unknown, we have a more difficult problem which will be considered in a future paper.

(*d*) It may be recalled (see, e.g., [11]) that if $\mathbf{u}$ and $\mathbf{v}$ are gaussian, then so are also $\mathbf{x}$ and $\mathbf{z}$, and therefore the best estimate will be of the type (14). Moreover, the same estimate will be best not only for the loss function (15) but also for a wide variety of other loss functions.

(*e*) The representation of white noise in the form (12) is not rigorous, because of the use of delta "functions." But since the delta function occurs only in integrals, the difficulty is easily removed as we shall show in a future paper addressed to mathematicians. All other mathematical developments given in the paper are rigorous.

The solution of the estimation problem under assumptions (A_1), (A_2), (A_3') is stated in Section 7 and proved in Section 8.

5 The Dual Problem

It will be useful to consider now the dual of the optimal estimation problem which turns out to be the optimal regulator problem in the theory of control.

First we define a dynamical system which is the *dual* (or *adjoint*) of (1). Let

$$\left.\begin{aligned} t^* &= -t \\ \mathbf{F}^*(t^*) &= \mathbf{F}'(t) \\ \mathbf{G}^*(t^*) &= \mathbf{H}'(t) \\ \mathbf{H}^*(t^*) &= \mathbf{G}'(t) \end{aligned}\right\} \qquad (16)$$

Let $\mathbf{\Phi}^*(t^*, t_0^*)$ be the transition matrix of the dual dynamical system of (1):

$$d\mathbf{x}^*/dt^* = \mathbf{F}^*(t^*)\mathbf{x}^* + \mathbf{G}^*(t^*)\mathbf{u}^*(t^*) \qquad (17)$$

It is easy to verify the fundamental relation

$$\mathbf{\Phi}^*(t^*, t_0^*) = \mathbf{\Phi}'(t_0, t) \qquad (18)$$

With these notation conventions, we can now state the

OPTIMAL REGULATOR PROBLEM. *Consider the linear dynamical system* (17). *Find a "control law"*

$$\mathbf{u}^*(t^*) = \mathbf{k}^*(\mathbf{x}^*(t^*), t_0^*) \qquad (19)$$

with the property that, for this choice of $\mathbf{u}^*(t^*)$, *the "performance index"*

$$V(\mathbf{x}^*; t^*, t_0^*; \mathbf{u}^*) = \|\phi^*(t_0^*; \mathbf{x}, t^*; \mathbf{u}^*)\|^2_{\mathbf{P}_0} + \int_{t^*}^{t_0^*}\{\|\phi^*(\tau^*; \mathbf{x}^*, t^*; \mathbf{u}^*)\|^2_{\mathbf{Q}(\tau^*)} + \|\mathbf{u}^*(\tau^*)\|^2_{\mathbf{R}(\tau^*)}\}d\tau^* \qquad (20)$$

assumes its greatest lower bound.

This is a natural generalization of the well-known problem of the optimization of a regulator with integrated-squared-error type of performance index.

The mathematical theory of the optimal regulator problem has been explored in considerable detail [17]. These results can be applied directly to the optimal estimation problem because of the

DUALITY THEOREM. *The solutions of the optimal estimation problem and of the optimal regulator problem are equivalent under the duality relations* (16).

The nature of these solutions will be discussed in the sequel. Here we pause only to observe a trivial point: By (14), the solutions of the estimation problem are necessarily linear; hence the same must be true (if the duality theorem is correct) of the solutions of the optimal regulator problem; in other words, the optimal control law $\mathbf{k}^*$ must be a linear function of $\mathbf{x}^*$.

The first proof of the duality theorem appeared in [11], and consisted of comparing the end results of the solutions of the two problems. Assuming only that the solutions of both problems result in linear dynamical systems, the proof becomes much simpler and less mysterious; this argument was carried out in detail in [16].

Remark (*f*). If we generalize the optimal regulator problem to the extent of replacing the first integrand in (20) by

$$\|\mathbf{y}^*(\tau^*) - \mathbf{y}_d^*(\tau^*)\|^2_{\mathbf{Q}(\tau^*)}$$

where $\mathbf{y}_d^*(t^*) \not\equiv \mathbf{0}$ is the *desired output* (in other words, if the regulator problem is replaced by a servomechanism or follow-up problem), then we have the dual of the estimation problem with $\mathcal{E}\mathbf{u}(t) \not\equiv \mathbf{0}$.

6 Examples: Problem Statement

To illustrate the matrix formalism and the general problems stated in Sections 4–5, we present here some specific problems in the standard block-diagram terminology. The solution of these problems is given in Section 11.

Example 1. Let the model of the message process be a first-order, linear, constant dynamical system. It is not assumed that the model is stable; but if so, this is the simplest problem in the Wiener theory which was discussed first by Wiener himself [1, pp. 91–92].

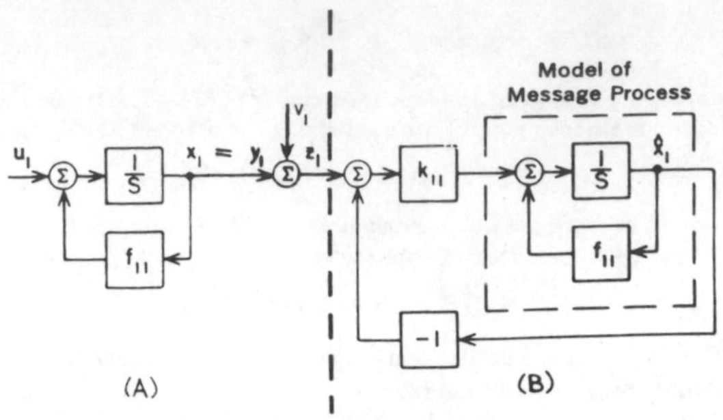

Fig. 1 Example 1: Block diagram of message process and optimal filter

The model of the message process is shown in Fig. 1(*a*). The various matrices involved are all defined by 1 × 1 and are

$$\mathbf{F}(t) = [f_{11}], \quad \mathbf{G}(t) = [1], \quad \mathbf{H}(t) = [1],$$
$$\mathbf{Q}(t) = [q_{11}], \quad \mathbf{R}(t) = [r_{11}].$$

The model is identical with its dual. Then the dual problem concerns the plant

$$dx^*_1/dt^* = f_{11}x^*_1 + u^*_1(t^*), \quad y^*_1(t) = x^*_1(t)$$

and the performance index is

$$\int_{t^*}^{t_0^*}\{q_{11}[x^*_1(\tau^*)]^2 + r_{11}[u^*_1(\tau^*)]^2\}d\tau^* \qquad (21)$$

The discrete-time version of the estimation problem was treated in [11, Example 1]. The dual problem was treated by Rozonoër [19].

Example 2. The message is generated as in Example 1, but now it is assumed that two separate signals (mixed with different noise) can be observed. Hence $\mathbf{R}$ is now a 2 × 2 matrix and we assume that

$$\mathbf{H} = \begin{bmatrix} 1 \\ 1 \end{bmatrix}$$

The block diagram of the model is shown in Fig. 2(*a*).

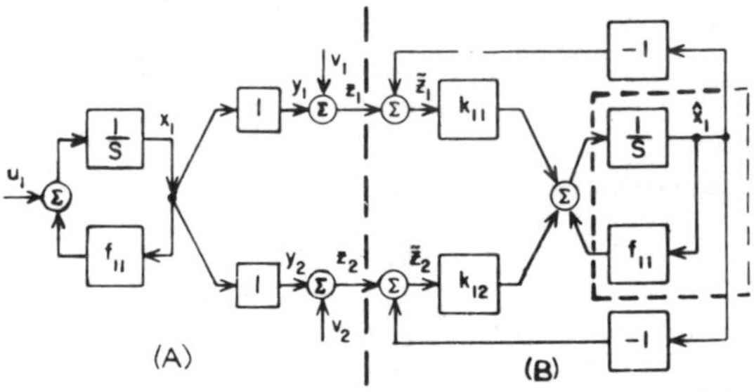

Fig. 2 Example 2: Block diagram of message process and optimal filter

Example 3. The message is generated by putting white noise through the transfer function $1/s(s+1)$. The block diagram of the model is shown in Fig. 3(*a*). The system matrices are:

$$\mathbf{F} = \begin{bmatrix} 0 & 0 \\ 1 & -1 \end{bmatrix} \quad \mathbf{G} = \begin{bmatrix} 1 \\ 0 \end{bmatrix} \quad \mathbf{H} = [0 \quad 1]$$

In the dual model, the order of the blocks $1/s$ and $1/(s+1)$ is interchanged. See Fig. 4. The performance index remains the same as (21). The dual problem was investigated by Kipiniak [24].

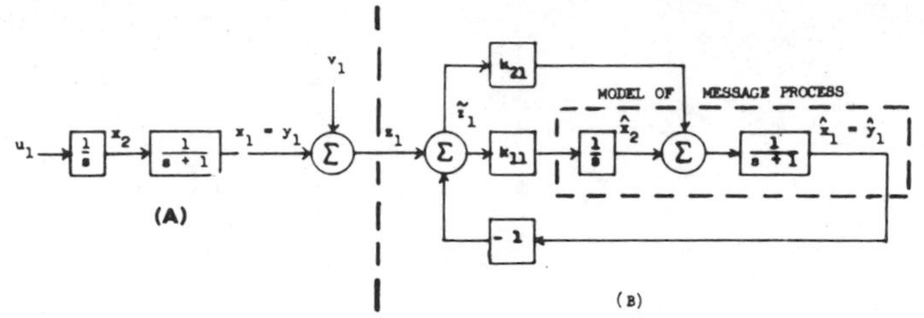

Fig. 3 Example 3: Block diagram of message process and optimal filter (x_1 and $\hat{x}_1$ should be interchanged with x_2 and $\hat{x}_2$.)

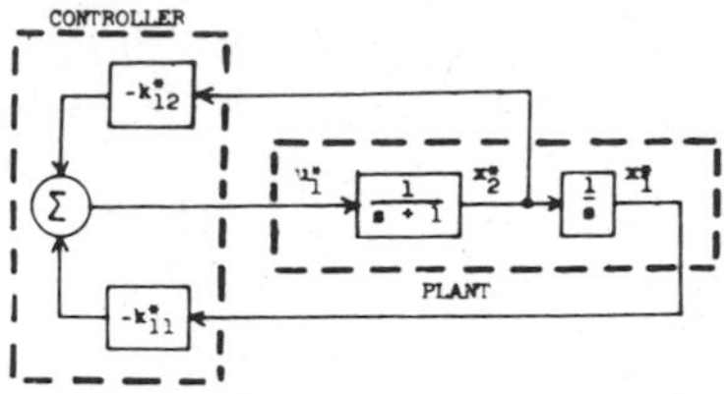

Fig. 4 Example 3: Block diagram of dual problem

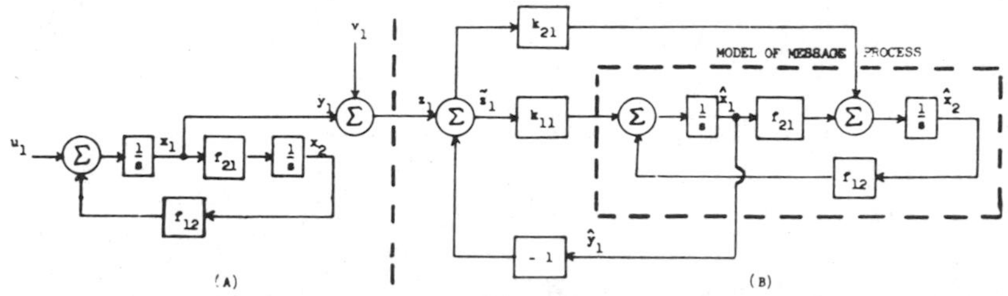

Fig. 5 Example 4: Block diagram of message process and optimal filter

Example 4. The message is generated by putting white noise through the transfer function $s/(s^2 - f_{12}f_{21})$. The block diagram of the model is shown in Fig. 5(a). The system matrices are:

$$\mathbf{F} = \begin{bmatrix} 0 & f_{12} \\ f_{21} & 0 \end{bmatrix} \quad \mathbf{G} = \begin{bmatrix} 1 \\ 0 \end{bmatrix} \quad \mathbf{H} = [1 \quad 0]$$

The transfer function of the dual model is also $s/(s^2 - f_{12}f_{21})$. However, in drawing the block diagram, the locations of the first and second state variables are interchanged, see Fig. 6. Evidently $f^*_{12} = f_{21}$ and $f^*_{21} = f_{12}$. The performance index is again given by (21).

The message model for the next two examples is the same and is defined by:

$$\mathbf{F} = \begin{bmatrix} 0 & 1 \\ 0 & 0 \end{bmatrix}$$

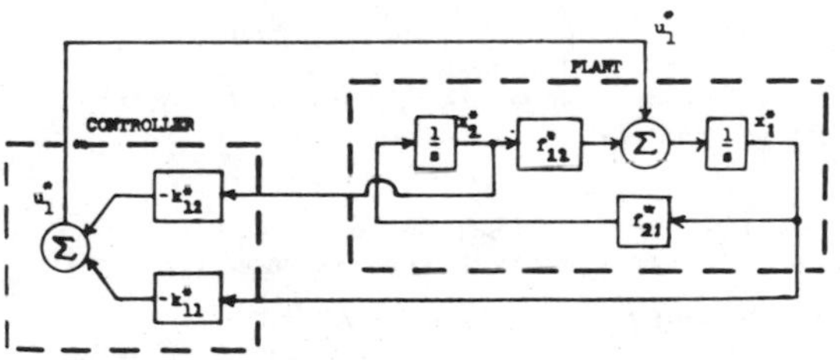

Fig. 6 Example 4: Block diagram of dual problem

The differences between the two examples lie in the nature of the "starting" assumptions and in the observed signals.

Example 5. Following Shinbrot [3], we consider the following situation. A particle leaves the origin at time $t_0 = 0$ with a fixed but unknown velocity of zero mean and known variance. The position of the particle is continually observed in the presence of additive white noise. We are to find the best estimator of position and velocity.

The verbal description of the problem implies that $p_{11}(0) = p_{12}(0) = 0$, $p_{22}(0) > 0$ and $q_{11} = 0$. Moreover, $\mathbf{G} = \mathbf{0}$, $\mathbf{H} = [1 \; 0]$. See Fig. 7(a).

The dual of this problem is somewhat unusual; it calls for minimizing the performance index

$$p_{22}(0)[\phi^*_2(0;\ \mathbf{x}^*, t^*;\ \mathbf{u}^*)]^2 + \int_{t^*}^{0} r_{11}[u^*_1(\tau^*)]^2 d\tau^* \quad (t^* < 0)$$

In words: We are given a transfer function $1/s^2$; the input u^*_1 over the time-interval $[t^*, 0]$ should be selected in such a way as to minimize the sum of (i) the square of the velocity and (ii) the control energy. In the discrete-time case, this problem was treated in [11, Example 2].

Example 6. We assume here that the transfer function $1/s^2$ is excited by white noise and that both the position x_1 and velocity x_2 can be observed in the presence of noise. Therefore (see Fig. 8a)

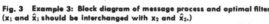

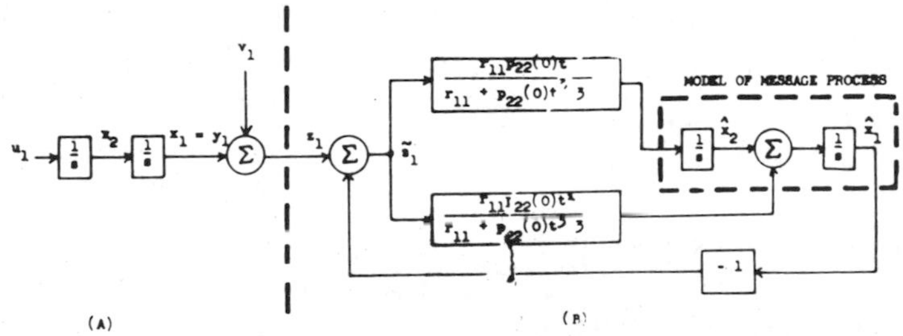

Fig. 7 Example 5: Block diagram of message process and optimal filter

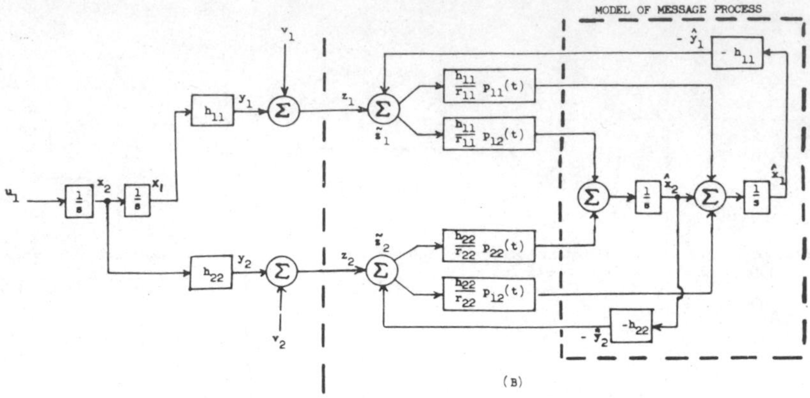

Fig. 8 Example 6: Block diagram of message process and optimal filter

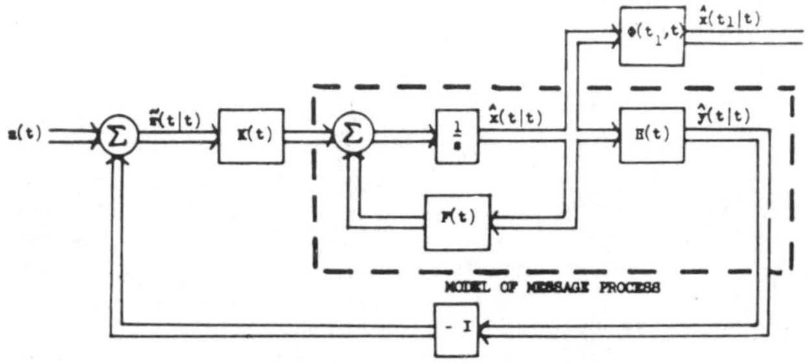

Fig. 9 General block diagram of optimal filter

$$\mathbf{G} = \begin{bmatrix} 0 \\ 1 \end{bmatrix} \quad \mathbf{H} = \begin{bmatrix} h_{11} & 0 \\ 0 & h_{22} \end{bmatrix}$$

This problem was studied by Hanson [9] and Bucy [25, 26]. The dual problem is very similar to Examples 3 and 4.

7 Summary of Results: Mathematics

Here we present the main results of the paper in precise mathematical terms. At the present stage of our understanding of the problem, the rigorous proof of these facts is quite complicated, requiring advanced and unconventional methods; they are to be found in Sections 8–10. After reading this section, one may pass without loss of continuity to Section 11 which contains the solutions of the examples.

(1) *Canonical form of the optimal filter.* The optimal estimate $\hat{\mathbf{x}}(t|t)$ is generated by a linear dynamical system of the form

$$\begin{aligned} d\hat{\mathbf{x}}(t|t)/dt &= \mathbf{F}(t)\hat{\mathbf{x}}(t|t) + \mathbf{K}(t)\tilde{\mathbf{z}}(t|t) \\ \tilde{\mathbf{z}}(t|t) &= \mathbf{z}(t) - \mathbf{H}(t)\hat{\mathbf{x}}(t|t) \end{aligned} \qquad \text{(I)}$$

The initial state $\hat{\mathbf{x}}(t_0|t_0)$ of (I) is zero.

For optimal extrapolation, we add the relation

$$\hat{\mathbf{x}}(t_1|t) = \mathbf{\Phi}(t_1, t)\hat{\mathbf{x}}(t|t) \qquad (t_1 \geqq t) \qquad \text{(V)}$$

No similarly simple formula is known at present for interpolation ($t_1 < t$).

The block diagram of (I) and (V) is shown in Fig. 9. The variables appearing in this diagram are vectors and the "boxes" represent matrices operating on vectors. Otherwise (except for the noncommutativity of matrix multiplication) such generalized block diagrams are subject to the same rules as ordinary block diagrams. The fat lines indicating direction of signal flow serve as a reminder that we are dealing with multiple rather than single signals.

The optimal filter (I) is a feedback system. It is obtained by taking a copy of the model of the message process (omitting the constraint at the input), forming the error signal $\tilde{\mathbf{z}}(t|t)$ and feeding the error forward with a gain $\mathbf{K}(t)$. Thus the specification of the optimal filter is equivalent to the computation of the optimal time-varying gains $\mathbf{K}(t)$. This result is general and does not depend on constancy of the model.

(2) *Canonical form for the dynamical system governing the optimal error.* Let

$$\tilde{\mathbf{x}}(t|t) = \mathbf{x}(t) - \hat{\mathbf{x}}(t|t) \qquad (22)$$

Except for the way in which the excitations enter the optimal error, $\tilde{\mathbf{x}}(t|t)$ is governed by the same dynamical system as $\hat{\mathbf{x}}(t|t)$:

$$d\tilde{\mathbf{x}}(t|t)/dt = \mathbf{F}(t)\tilde{\mathbf{x}}(t|t) + \mathbf{G}(t)\mathbf{u}(t) - \mathbf{K}(t)[\mathbf{v}(t) + \mathbf{H}(t)\tilde{\mathbf{x}}(t|t)] \qquad \text{(II)}$$

See Fig. 10.

(3) *Optimal gain.* Let us introduce the abbreviation:

$$\mathbf{P}(t) = \operatorname{cov}[\tilde{\mathbf{x}}(t|t), \tilde{\mathbf{x}}(t|t)] \qquad (23)$$

Then it can be shown that

$$\mathbf{K}(t) = \mathbf{P}(t)\mathbf{H}'(t)\mathbf{R}^{-1}(t) \qquad \text{(III)}$$

(4) *Variance equation.* The only remaining unknown is $\mathbf{P}(t)$. It can be shown that $\mathbf{P}(t)$ must be a solution of the matrix differential equation

$$d\mathbf{P}/dt = \mathbf{F}(t)\mathbf{P} + \mathbf{P}\mathbf{F}'(t) - \mathbf{P}\mathbf{H}'(t)\mathbf{R}^{-1}(t)\mathbf{H}(t)\mathbf{P} + \mathbf{G}(t)\mathbf{Q}(t)\mathbf{G}'(t) \qquad \text{(IV)}$$

This is the *variance equation;* it is a system of $n(n + 1)/2$[4] nonlinear differential equations of the first order, and is of the *Riccati* type well known in the calculus of variations [17, 18].

(5) *Existence of solutions of the variance equation.* Given any fixed initial time t_0 and a nonnegative definite matrix $\mathbf{P}_0$, (IV) has a unique solution

$$\mathbf{P}(t) = \mathbf{\Pi}(t;\ \mathbf{P}_0, t_0) \tag{24}$$

defined for all $|t - t_0|$ *sufficiently small,* which takes on the value $\mathbf{P}(t_0) = \mathbf{P}_0$ at $t = t_0$. This follows at once from the fact that (IV) satisfies a Lipschitz condition [21].

Since (IV) is nonlinear, we cannot of course conclude without further investigation that a solution $\mathbf{P}(t)$ exists for *all t* [21]. By taking into account the problem from which (IV) was derived, however, it can be shown that $\mathbf{P}(t)$ in (24) is defined for all $t \geqq t_0$.

These results can be summarized by the following theorem, which is the analogue of Theorem 3 of [11] and is proved in Section 8:

THEOREM 1. *Under Assumptions* (A_1), (A_2), (A_3'), *the solution of the optimal estimation problem with* $t_0 > -\infty$ *is given by relations* (I–V). *The solution* $\mathbf{P}(t)$ *of* (IV) *is uniquely determined for all* $t \geqq t_0$ *by the specification of*

$$\mathbf{P}_0 = \operatorname{cov}[\mathbf{x}(t_0), \mathbf{x}(t_0)];$$

knowledge of $\mathbf{P}(t)$ *in turn determines the optimal gain* $\mathbf{K}(t)$. *The initial state of the optimal filter is* $\mathbf{0}$.

(6) *Variance of the estimate of a costate.* From (23) we have immediately the following formula for (15):

$$\mathcal{E}[\mathbf{x}^*, \tilde{\mathbf{x}}(t|t)]^2 = \|\mathbf{x}^*\|^2_{\mathbf{P}(t)} \tag{25}$$

(7) *Analytic solution of the variance equation.* Because of the close relationship between the Riccati equation and the calculus of variations, a closed-form solution of sorts is available for (IV). The easiest way of obtaining it is as follows [17]:

Introduce the quadratic *Hamiltonian* function

$$\mathcal{H}(\mathbf{x}, \mathbf{w}, t) = -(1/2)\|\mathbf{G}'(t)\mathbf{x}\|^2_{\mathbf{Q}(t)} - \mathbf{w}'\mathbf{F}'(t)\mathbf{x} + (1/2)\|\mathbf{H}(t)\mathbf{w}\|^2_{\mathbf{R}^{-1}(t)} \tag{26}$$

and consider the associated *canonical* differential equations

$$\left.\begin{aligned} d\mathbf{x}/dt &= \partial\mathcal{H}/\partial\mathbf{w}^{5} = -\mathbf{F}'(t)\mathbf{x} + \mathbf{H}'(t)\mathbf{R}^{-1}(t)\mathbf{H}(t)\mathbf{w} \\ d\mathbf{w}/dt &= -\partial\mathcal{H}/\partial\mathbf{x} = \mathbf{G}(t)\mathbf{Q}(t)\mathbf{G}'(t)\mathbf{x} + \mathbf{F}(t)\mathbf{w} \end{aligned}\right\} \tag{27}$$

We denote the transition matrix of (27) by

$$\mathbf{\Theta}(t, t_0) = \begin{bmatrix} \mathbf{\Theta}_{11}(t, t_0) & \mathbf{\Theta}_{12}(t, t_0) \\ \mathbf{\Theta}_{21}(t, t_0) & \mathbf{\Theta}_{22}(t, t_0) \end{bmatrix} \tag{28}$$

[4] This is the number of distinct elements of the *symmetric* matrix $\mathbf{P}(t)$.

[5] The notation $\partial\mathcal{H}/\partial\mathbf{w}$ means the gradient of the scalar $\mathcal{H}$ with respect to the vector $\mathbf{w}$.

In Section 10 we shall prove

THEOREM 2. *The solution of* (IV) *for arbitrary nonnegative definite, symmetric* $\mathbf{P}_0$ *and all* $t \geqq t_0$ *can be represented by the formula*

$$\mathbf{\Pi}(t;\ \mathbf{P}_0, t_0) = [\mathbf{\Theta}_{21}(t, t_0) + \mathbf{\Theta}_{22}(t, t_0)\mathbf{P}_0]\cdot[\mathbf{\Theta}_{11}(t, t_0) + \mathbf{\Theta}_{12}(t, t_0)\mathbf{P}_0]^{-1} \tag{29}$$

Unless all matrices occurring in (27) are constant, this result simply replaces one difficult problem by another of similar difficulty, since only in the rarest cases can $\mathbf{\Theta}(t, t_0)$ be expressed in analytic form. Something has been accomplished, however, since we have shown that *the solution of nonconstant estimation problems involves precisely the same analytic difficulties as the solution of linear differential equations with variable coefficients.*

(8) *Existence of steady-state solution.* If the time-interval over which data are available is infinite, in other words, if $t_0 = -\infty$, Theorem 1 is not applicable without some further restriction.

For instance, if $\mathbf{H}(t) \equiv \mathbf{0}$, the variance of $\tilde{\mathbf{x}}$ is the same as the variance of $\mathbf{x}$; if the model (10–11) is unstable, then $\mathbf{x}(t)$ defined by (13) does not exist and the estimation problem is meaningless.

The following theorem, proved in Section 9, gives two sufficient conditions for the steady-state estimation problem to be meaningful. The first is the one assumed at the very beginning in the conventional Wiener theory. The second condition, which we introduce here for the first time, is much weaker and more "natural" than the first; moreover, it is almost a necessary condition as well.

THEOREM 3. *Denote the solutions of* (IV) *as in (24). Then the limit*

$$\lim_{t_0 \to -\infty} \mathbf{\Pi}(t;\ \mathbf{0}, t_0) = \bar{\mathbf{P}}(t) \tag{30}$$

exists for all t and is a solution of (IV) *if either*

(A_4) *the model* (10–11) *is uniformly asymptotically stable;* *or*

(A_4') *the model* (10–11) *is "completely observable"* [17], *that is, for all t there is some* $t_0(t) < t$ *such that the matrix*

$$\mathbf{M}(t_0, t) = \int_{t_0}^{t} \mathbf{\Phi}'(\tau, t)\mathbf{H}'(\tau)\mathbf{H}(\tau)\mathbf{\Phi}(\tau, t)d\tau \tag{31}$$

is positive definite. (*See* [21] *for the definition of uniform asymptotic stability.*)

Remarks. (*g*) $\bar{\mathbf{P}}(t)$ is the covariance matrix of the optimal error corresponding to the very special situation in which (i) an arbitrarily long record of past measurements is available, and (ii) the initial state $\mathbf{x}(t_0)$ was known exactly. When all matrices in (10–12) are constant, then so is also $\bar{\mathbf{P}}$—this is just the classical Wiener problem. In the constant case, $\bar{\mathbf{P}}$ is an equilibrium state of (IV) (i.e., for this choice of $\mathbf{P}$, the right-hand side of (IV) is zero). In general, $\bar{\mathbf{P}}(t)$ should be regarded as a moving equilibrium point of (IV), see Theorem 4 below.

(*h*) The matrix $\mathbf{M}(t_0, t)$ is well known in mathematical statistics. It is the *information matrix* in the sense of R. A. Fisher [20] corresponding to the special estimation problem when (i) $\mathbf{u}(t) \equiv \mathbf{0}$ and (ii) $\mathbf{v}(t)$ = gaussian with unit covariance matrix. In this case, the variance of any unbiased estimator $\mu(t)$ of $[\mathbf{x},^* \mathbf{x}(t)]$ satisfies the well-known Cramér-Rao inequality [20]

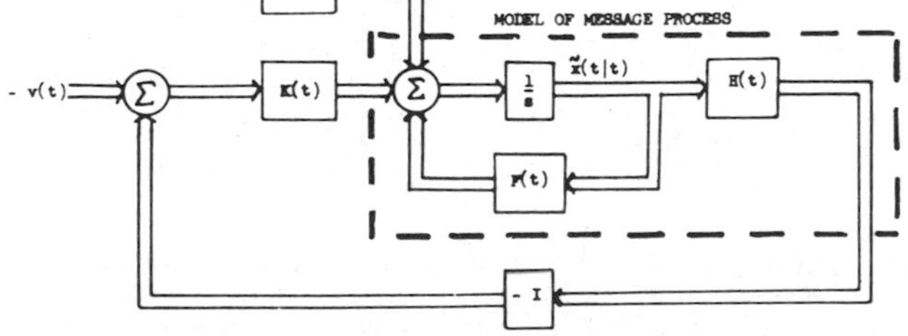

Fig. 10 General block diagram of optimal estimation error

$$\mathcal{E}[\mu(t) - \mathcal{E}\mu(t)]^2 \geqq \|\mathbf{x}^*\|^2_{\mathbf{M}^{-1}(t_0,\, t)} \tag{32}$$

Every costate $\mathbf{x}^*$ has a minimum-variance unbiased estimator for which the equality sign holds in (32) if and only if $\mathbf{M}$ is positive definite. This motivates the use of condition (A_4') in Theorem 3 and the term "completely observable."

(*i*) It can be shown [17] that in the constant case complete observability is equivalent to the easily verified condition:

$$\text{rank}[\mathbf{H}', \mathbf{F}'\mathbf{H}', \ldots, (\mathbf{F}')^{n-1}\mathbf{H}'] = n \tag{33}$$

where the square brackets denote a matrix with n rows and np columns.

(9) *Stability of the optimal filter.* It should be realized now that the *optimality* of the filter (I) does not at the same time guarantee its *stability*. The reader can easily check this by constructing an example (for instance, one in which (10–11) consists of two non-interacting systems). To establish weak sufficient conditions for stability entails some rather delicate mathematical technicalities which we shall bypass and state only the best final result currently available.

First, some additional definitions.

We say that the model (10–11) is *uniformly completely observable* if there exist fixed constants, α_1, α_2, and σ such that

$$\alpha_1\|\mathbf{x}^*\|^2 \leqq \|\mathbf{x}^*\|^2_{\mathbf{M}(t-\sigma,\, t)} \leqq \alpha_2\|\mathbf{x}^*\|^2 \quad \text{for all} \quad \mathbf{x}^* \text{ and } t.$$

Similarly, we say that a model is *completely controllable* [*uniformly completely controllable*] if the dual model is completely observable [uniformly completely observable]. For a discussion of these motions, the reader may refer to [17]. It should be noted that the property of "uniformity" is always true for constant systems.

We can now state the central theorem of the paper:

THEOREM 4. *Assume that the model of the message process is*

(A_4'') *uniformly completely observable;*
(A_5) *uniformly completely controllable;*
(A_6) $\alpha_3 \leqq \|\mathbf{Q}(t)\| \leqq \alpha_4, \quad \alpha_5 \leqq \|\mathbf{R}(t)\| \leqq \alpha_6 \quad$ for all $\;t$;
(A_7) $\|\mathbf{F}(t)\| \leqq \alpha_7$.

Then the following is true:

(i) *The optimal filter is uniformly asymptotically stable;*

(ii) *Every solution $\mathbf{\Pi}(t;\ \mathbf{P}_0,\ t_0)$ of the variance equation* (IV) *starting at a symmetric nonnegative matrix $\mathbf{P}_0$ converges to $\bar{\mathbf{P}}(t)$ (defined in Theorem 3) as $t \to \infty$.*

Remarks. (*j*) A filter which is not *uniformly* asymptotically stable may have an unbounded response to a bounded input [21]; the practical usefulness of such a filter is rather limited.

(*k*) Property (ii) in Theorem 4 is of central importance since it shows that the variance equation is a "stable" computational method that may be expected to be rather insensitive to roundoff errors.

(*l*) The speed of convergence of $\mathbf{P}_0(t)$ to $\bar{\mathbf{P}}(t)$ can be estimated quite effectively using the second method of Lyapunov; see [17].

(10) *Solution of the classical Wiener problem.* Theorems 3 and 4 have the following immediate corollary:

THEOREM 5. *Assume the hypotheses of Theorems 3 and 4 are satisfied and that* $\mathbf{F}, \mathbf{G}, \mathbf{H}, \mathbf{Q}, \mathbf{R}$, *are constants.*

Then, if $t_0 = -\infty$, the solution of the estimation problem is obtained by setting the right-hand side of (IV) *equal to zero and solving the resulting set of quadratic algebraic equations. That solution which is nonnegative definite is equal to* $\bar{\mathbf{P}}$.

To prove this, we observe that, by the assumption of constancy, $\bar{\mathbf{P}}(t)$ is a constant. By Theorem 4, all solutions of (IV) starting at nonnegative matrices converge to $\bar{\mathbf{P}}$. Hence, if a matrix $\mathbf{P}$ is found for which the right-hand side of (IV) vanishes and if this matrix is nonnegative definite, it must be identical with $\bar{\mathbf{P}}$. Note, however, that the procedure may fail if the conditions of Theorems 3 and 4 are not satisfied. See Example 4.

(11) *Solution of the Dual Problem.* For details, consult [17]. The only facts needed here are the following: The optimal control law is given by

$$\mathbf{u}^*(t^*) = -\mathbf{K}^*(t^*)\mathbf{x}(t^*) \tag{34}$$

where $\mathbf{K}^*(t^*)$ satisfies the duality relation

$$\mathbf{K}^*(t^*) = \mathbf{K}'(t) \tag{35}$$

and is to be determined by duality from formula (III). The value of the performance index (20) may be written in the form

$$\min_{\mathbf{u}^*} V(\mathbf{x}^*;\ t^*, t_0^*, \mathbf{u}^*) = \|\mathbf{x}^*\|^2_{\mathbf{\Pi}^*(t^*;\ \mathbf{x}^*,\, t_0^*)}$$

where $\mathbf{\Pi}^*(t^*;\ \mathbf{x}^*,\ t_0^*)$ is the solution of the dual of the variance equation (IV).

It should be carefully noted that the hypotheses of Theorem 4 are invariant under duality. Hence essentially the same theory covers both the estimation and the regular problem, as stated in Section 5.

The vector-matrix block diagram for the optimal regulator is shown in Fig. 11.

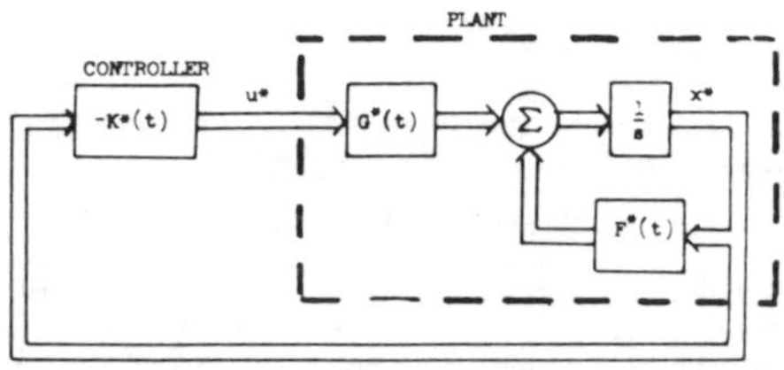

Fig. 11 General block diagram of optimal regulator

(12) *Computation of the covariance matrix for the message process.* To apply Theorem 1, it is necessary to determine cov $[\mathbf{x}(t_0), \mathbf{x}(t_0)]$. This may be specified as part of the problem statement as in Example 5. On the other hand, one might assume that the message model has reached steady state (see (A_3)), in which case from (13) and (12) we have that

$$\mathbf{S}(t) = \text{cov}\,[\mathbf{x}(t), \mathbf{x}(t)] = \int_{-\infty}^{t} \mathbf{\Phi}(t, \tau)\mathbf{G}(\tau)\mathbf{Q}(\tau)\mathbf{G}'(\tau)\mathbf{\Phi}'(t, \tau)d\tau$$

provided the model (10) is asymptotically stable. Differentiating this expression with respect to t we obtain the following differential equation for $\mathbf{S}(t)$

$$d\mathbf{S}/dt = \mathbf{F}(t)\mathbf{S} + \mathbf{S}\mathbf{F}'(t) + \mathbf{G}(t)\mathbf{Q}(t)\mathbf{G}'(t) \tag{36}$$

This formula is analogous to the well-known lemma of Lyapunov [21] in evaluating the integrated square of a solution of a linear differential equation. In case of a constant system, (36) reduces to a system of linear algebraic equations.

8 Derivation of the Fundamental Equations

We first deduce the matrix form of the familiar Wiener-Hopf integral equation. Differentiating it with respect to time and then using (10–11), we obtain in a very simple way the fundamental equations of our theory.

Much cumbersome manipulation of integrals can be avoided by recognizing, as has been pointed out by Pugachev [27], that the Wiener-Hopf equation is a special case of a simple geometric principle: *orthogonal projection.*

Consider an abstract space $\mathfrak{X}$ such that an inner **product** (X, Y) is defined between any two elements X, Y of $\mathfrak{X}$. **The norm** is defined by $\|X\| = (X, X)^{1/2}$. Let $\mathfrak{U}$ be a subspace of $\mathfrak{X}$. We seek a vector U_0 in $\mathfrak{U}$ which minimizes $\|X - U\|$ with respect to any U in $\mathfrak{U}$. If such a minimizing vector exists, it may be characterized in the following way:

ORTHOGONAL PROJECTION LEMMA. $\|X - U\| \geq \|X - U_0\|$ *for all* U *in* $\mathfrak{U}$ (i) *if and* (ii) *only if*

$$(X - U_0, U) = 0 \text{ for all } U \text{ in } \mathfrak{U} \quad (37)$$

(iii) *Moreover, if there is another vector* U_0' *satisfying* (37), *then* $\|U_0 - U_0'\| = 0$.

Proof. (i), (iii) Consider the identity

$$\|X - U\|^2 = \|X - U_0\|^2 + 2(X - U_0, U_0 - U) + \|U - U_0\|^2$$

Since $\mathfrak{U}$ is a linear space, it contains $U - U_0$; hence if Condition (37) holds, the middle term vanishes and therefore $\|X - U\| \geq \|X - U_0\|$. Property (iii) is obvious.

(ii) Suppose there is a vector U_1 such that $(X - U_0, U_1) = \alpha \neq 0$. Then

$$\|X - U_0 - \beta U_1\|^2 = \|X - U_0\|^2 + 2\alpha\beta + \beta^2\|U_1\|^2$$

For a suitable choice of β, the sum of the last two terms will be negative, contradicting the optimality of U_0. Q.E.D.

Using this lemma, it is easy to show:

WIENER-HOPF EQUATION. *A necessary and sufficient condition for* $[\mathbf{x}^*, \hat{\mathbf{x}}(t_1|t)]$ *(where* $\hat{\mathbf{x}}(t_1|t)$ *is defined by* (14)) *to be a minimum variance estimator of* $[\mathbf{x}^*, \mathbf{x}(t_1)]$ *for all* $\mathbf{x}^*$, *is that the matrix function* $\mathbf{A}(t_1, \tau)$ *satisfy the relation*

$$\operatorname{cov}[\mathbf{x}(t_1), \mathbf{z}(\sigma)] - \int_{t_0}^{t} \mathbf{A}(t_1, \tau) \operatorname{cov}[\mathbf{z}(\tau), \mathbf{z}(\sigma)]d\tau = \mathbf{0} \quad (38)$$

or equivalently,

$$\operatorname{cov}[\tilde{\mathbf{x}}(t_1|t), \mathbf{z}(\sigma)] = \mathbf{0} \quad (39)$$

for all $t_0 \leq \sigma < t$.

COROLLARY. $\operatorname{cov}[\tilde{\mathbf{x}}(t_1|t), \hat{\mathbf{x}}(t_1|t)] = \mathbf{0}$ (40)

Proof. Let $\mathbf{x}^*$ be a fixed costate and denote by $\mathfrak{X}$ the space of all scalar random variables $[\mathbf{x}^*, \mathbf{x}(t_1)]$ of zero mean and finite variance. The inner product is defined as $(X, Y) = \mathcal{E}[\mathbf{x}^*, \mathbf{x}(t_1)] \cdot [\mathbf{x}^*, \mathbf{y}(t_1)]$. The subspace $\mathfrak{U}$ is the set of all scalar random variables of the type

$$U = [\mathbf{x}^*, \mathbf{u}(t_1)] = \left[\mathbf{x}^*, \int_{t_0}^{t} \mathbf{B}(t_1, \tau)\mathbf{z}(\tau)d\tau\right]$$

(where $\mathbf{B}(t_1, \tau)$ is an $n \times p$ matrix continuously differentiable in both arguments). We write U_0 for the estimate $[\mathbf{x}^*, \hat{\mathbf{x}}(t_1|t)]$.

We now apply the orthogonal projection lemma and find that condition (37) takes the form

$$(X - U_0, U) = \mathcal{E}\,[\mathbf{x}^*, \tilde{\mathbf{x}}(t_1|t)][\mathbf{x}^*, \mathbf{u}(t_1)]$$
$$= \mathbf{x}^* \operatorname{cov}[\tilde{\mathbf{x}}(t_1|t), \mathbf{u}(t_1)]\mathbf{x}^{*\prime}$$

Interchanging integration and the expected value operation (permissible in view of the continuity assumptions made under (A_1), see [28]), we get

$$(X - U_0, U) = \mathbf{x}^* \left\{\int_{t_0}^{t} \operatorname{cov}[\tilde{\mathbf{x}}(t_1|t), \mathbf{z}(\sigma)]\mathbf{B}'(t_1, \sigma)d\sigma\right\} \mathbf{x}^{*\prime}$$

This expression must vanish for all $\mathbf{x}^*$. Sufficiency of (39) is obvious. To prove the necessity, we take $\mathbf{B}(t_1, \sigma) = \operatorname{cov}[\tilde{\mathbf{x}}(t_1|t), \mathbf{z}(\sigma)]$. Then $\mathbf{BB}'$ is nonnegative definite. By continuity, the integral will be positive for some $\mathbf{x}^*$ unless $\mathbf{BB}'$ and therefore also $\mathbf{B}(t_1, \sigma)$ vanishes identically for all $t_0 \leq \sigma < t$. The Corollary follows trivially by multiplying (39) on the right by $\mathbf{A}'(t_1, \sigma)$ and integrating with respect to σ. Q.E.D.

Remark. (*m*) Equation (39) does not hold when $\sigma = t$. In fact, $\operatorname{cov}[\tilde{\mathbf{x}}(t|t), \mathbf{z}(t)] = (1/2)\mathbf{K}(t)\mathbf{R}(t)$.

For the moment we assume for simplicity that $t_1 = t$. Differentiating (38) with respect to t, and interchanging $\partial/\partial t$ and $\mathcal{E}$, we get for all $t_0 \leq \sigma < t$,

$$\frac{\partial}{\partial t}\operatorname{cov}[\mathbf{x}(t), \mathbf{z}(\sigma)] = \mathbf{F}(t)\operatorname{cov}[\mathbf{x}(t), \mathbf{z}(\sigma)] + \mathbf{G}(t)\operatorname{cov}[\mathbf{u}(t), \mathbf{z}(\sigma)] \quad (41)$$

and

$$\frac{\partial}{\partial t}\int_{t_0}^{t} \mathbf{A}(t, \tau)\operatorname{cov}[\mathbf{z}(\tau), \mathbf{z}(\sigma)]d\tau = \frac{\partial}{\partial t}\int_{t_0}^{t} \mathbf{A}(t, \tau)\operatorname{cov}[\mathbf{y}(\tau), \mathbf{y}(\sigma)]d\tau + \frac{\partial}{\partial t}\mathbf{A}(t, \sigma)\mathbf{R}(\sigma)$$
$$= \int_{t_0}^{t} \frac{\partial}{\partial t}\mathbf{A}(t, \tau)\operatorname{cov}[\mathbf{z}(\tau), \mathbf{z}(\sigma)]d\tau + \mathbf{A}(t, t)\operatorname{cov}[\mathbf{y}(t), \mathbf{y}(\sigma)] \quad (42)$$

The last term in (41) vanishes because of the independence of $\mathbf{u}(t)$ of $\mathbf{v}(\sigma)$ and $\mathbf{x}(\sigma)$ when $\sigma < t$. Further,

$$\operatorname{cov}[\mathbf{y}(t), \mathbf{y}(\sigma)] = \mathbf{H}(t)\operatorname{cov}[\mathbf{x}(t), \mathbf{z}(\sigma)] - \operatorname{cov}[\mathbf{y}(t), \mathbf{v}(\sigma)] \quad (43)$$

As before, the last term again vanishes. Combining (41–43), we get, bearing in mind also (38),

$$\int_{t_0}^{t}\left[\mathbf{F}(t)\mathbf{A}(t, \tau) - \frac{\partial}{\partial t}\mathbf{A}(t, \tau) - \mathbf{A}(t, t)\mathbf{H}(t)\mathbf{A}(t, \tau)\right]\operatorname{cov}[\mathbf{z}(\tau), \mathbf{z}(\sigma)]d\tau = \mathbf{0} \quad (44)$$

for all $t_0 \leq \sigma < t$. This condition is certainly satisfied if the optimal operator $\mathbf{A}(t, \tau)$ is a solution of the differential equation

$$\mathbf{F}(t)\mathbf{A}(t, \tau) - \frac{\partial}{\partial t}\mathbf{A}(t, \tau) - \mathbf{A}(t, t)\mathbf{H}(t)\mathbf{A}(t, \tau) = \mathbf{0} \quad (45)$$

for all values of the parameter τ lying in the interval $t_0 \leq \tau \leq t$.

If $\mathbf{R}(\tau)$ *is positive definite in this interval, then condition* (45) *is necessary.* In fact, let $\mathbf{B}(t, \tau)$ denote the bracketed term in (44). If $\mathbf{A}(t, \tau)$ satisfies the Wiener-Hopf equation (38), then $\hat{\mathbf{x}}(t|t)$ given by (14) is an optimal estimate; and the same holds also for

$$\hat{\mathbf{x}}(t|t) + \int_{t_0}^{t}\mathbf{B}(t, \tau)\mathbf{z}(\tau)d\tau$$

since by (45) $\mathbf{A}(t, \tau) + \mathbf{B}(t, \tau)$ also satisfies the Wiener-Hopf equation. But by the lemma, the norm of the difference of two optimal estimates is zero. Hence

$$\mathbf{x}^*\left\{\int_{t_0}^{t}\int_{t_0}^{t}\mathbf{B}(t, \tau)\operatorname{cov}[\mathbf{z}(\tau), \mathbf{z}(\tau')]\mathbf{B}'(t, \tau')d\tau d\tau'\right\}\mathbf{x}^{*\prime} = 0 \quad (46)$$

for all $\mathbf{x}^*$. By the assumptions of Section 4, $\mathbf{y}(\tau)$ and $\mathbf{v}(\tau)$ are uncorrelated and therefore

$$\operatorname{cov}[\mathbf{z}(\tau), \mathbf{z}(\tau')] = \mathbf{R}(\tau)\delta(\tau - \tau') + \operatorname{cov}[\mathbf{y}(\tau), \mathbf{y}(\tau')]$$

Substituting this into the integral (46), the contribution of the second term on the right is nonnegative while the contribution of the first term is positive unless (45) holds (because of the positive definiteness of $\mathbf{R}(\tau)$), which concludes the proof.

Differentiating (14), with respect to t we find

$$d\hat{\mathbf{x}}(t|t)/dt = \int_{t_0}^{t}\frac{\partial}{\partial t}\mathbf{A}(t, \tau)\mathbf{z}(\tau)d\tau + \mathbf{A}(t, t)\mathbf{z}(t)$$

Using the abbreviation $\mathbf{A}(t, t) = \mathbf{K}(t)$ as well as (45) and (14), we obtain at once the differential equation of the optimal filter:

$$d\hat{\mathbf{x}}(t|t)/dt = \mathbf{F}(t)\hat{\mathbf{x}}(t|t) + \mathbf{K}(t)[\mathbf{z}(t) - \mathbf{H}(t)\hat{\mathbf{x}}(t|t)] \quad \text{(I)}$$

Combining (10) and (I), we obtain the differential equation for the error of the optimal estimate:

$$d\tilde{\mathbf{x}}(t|t)/dt = [\mathbf{F}(t) - \mathbf{K}(t)\mathbf{H}(t)]\tilde{\mathbf{x}}(t|t) + \mathbf{G}(t)\mathbf{u}(t) - \mathbf{K}(t)\mathbf{v}(t) \quad \text{(II)}$$

To obtain an explicit expression for $\mathbf{K}(t)$, we observe first that (39) implies that following identity in the interval $t_0 \leqq \sigma < t$:

$$\text{cov}\,[\mathbf{x}(t), \mathbf{y}(\sigma)] - \int_{t_0}^{t} \mathbf{A}(t, \tau)\,\text{cov}\,[\mathbf{y}(\tau), \mathbf{y}(\sigma)]d\tau = \mathbf{A}(t, \sigma)\mathbf{R}(\sigma) \quad (39')$$

Since both sides of (39′) are continuous functions of σ, it is clear that equality holds also for $\sigma = t$. Therefore

$$\mathbf{K}(t)\mathbf{R}(t) = \mathbf{A}(t, t)\mathbf{R}(t) = \text{cov}[\tilde{\mathbf{x}}(t|t), \mathbf{y}(t)] = \text{cov}\,[\tilde{\mathbf{x}}(t|t), \mathbf{x}(t)]\mathbf{H}'(t)$$

By (40), we have then

$$= \text{cov}\,[\tilde{\mathbf{x}}(t|t), \tilde{\mathbf{x}}(t|t)]\mathbf{H}'(t) = \mathbf{P}(t)\mathbf{H}'(t)$$

Since $\mathbf{R}(t)$ is assumed to be positive definite, it is invertible and therefore

$$\mathbf{K}(t) = \mathbf{P}(t)\mathbf{H}'(t)\mathbf{R}^{-1}(t) \quad \text{(III)}$$

We can now derive the variance equation. Let $\mathbf{\Psi}(t, \tau)$ be the common transition matrix of (I) and (II). Then

$$\mathbf{P}(t) - \mathbf{\Psi}(t, t_0)\mathbf{P}(t_0)\mathbf{\Psi}'(t, t_0) = \mathcal{E}\int_{t_0}^{t} \mathbf{\Psi}(t, \tau)[\mathbf{G}(\tau)\mathbf{u}(\tau) - \mathbf{K}(\tau)\mathbf{v}(\tau)]d\tau \times \int_{t_0}^{t} [\mathbf{u}'(\sigma)\mathbf{G}'(\sigma) - \mathbf{v}'(\sigma)\mathbf{K}'(\sigma)]\mathbf{\Psi}'(t, \sigma)d\sigma$$

Using the fact that $\mathbf{u}(t)$ and $\mathbf{v}(t)$ are uncorrelated white noise, the integral simplifies to

$$= \int_{t_0}^{t} \mathbf{\Psi}(t, \tau)[\mathbf{G}(\tau)\mathbf{Q}(\tau)\mathbf{G}'(\tau) + \mathbf{K}(\tau)\mathbf{R}(\tau)\mathbf{K}'(\tau)]\mathbf{\Psi}'(t, \tau)d\tau$$

Differentiating with respect to t and using (III), we obtain after easy calculations the variance equation

$$d\mathbf{P}/dt = \mathbf{F}(t)\mathbf{P} + \mathbf{P}\mathbf{F}'(t) - \mathbf{P}\mathbf{H}'(t)\mathbf{R}^{-1}(t)\mathbf{H}(t)\mathbf{P} + \mathbf{G}(t)\mathbf{Q}(t)\mathbf{G}'(t) \quad \text{(IV)}$$

Alternately, we could write

$$d\mathbf{P}/dt = d\,\text{cov}\,[\tilde{\mathbf{x}}, \tilde{\mathbf{x}}]/dt = \text{cov}\,[d\tilde{\mathbf{x}}/dt, \tilde{\mathbf{x}}] + \text{cov}\,[\tilde{\mathbf{x}}, d\tilde{\mathbf{x}}/dt]$$

and evaluate the right-hand side by means of (II). A typical covariance matrix to be computed is

$$\text{cov}\,[\tilde{\mathbf{x}}(t|t), \mathbf{u}(t)] = \text{cov}\left[\int_{t_0}^{t} \mathbf{\Psi}(t, \tau)[\mathbf{G}(\tau)\mathbf{u}(\tau) - \mathbf{K}(\tau)\mathbf{v}(\tau)]d\tau, \mathbf{u}(t)\right] = (1/2)\mathbf{G}(t)\mathbf{Q}(t)$$

the factor $1/2$ following from properties of the δ-function.

To complete the derivations, we note that, if $t_1 > t$, then by (3)

$$\mathbf{x}(t_1) = \mathbf{\Phi}(t_1, t)\mathbf{x}(t) + \int_{t_0}^{t} \mathbf{\Phi}(t_1, \tau)\mathbf{u}(\tau)d\tau$$

Since $\mathbf{u}(\tau)$ for $t < \tau \leqq t_1$ is independent of $\mathbf{x}(\tau)$ in the interval $t_0 \leqq \tau \leqq t$, it follows by (38) that the optimal estimator for the right-hand side above is $\mathbf{0}$. Hence

$$\hat{\mathbf{x}}(t_1|t) = \mathbf{\Phi}(t_1, t)\hat{\mathbf{x}}(t|t) \quad (t_1 \geqq t) \quad \text{(V)}$$

The same conclusion does *not* follow if $t_1 < t$ because of lack of independence between $\mathbf{x}(\tau)$ and $\mathbf{u}(\tau)$.

The only point remaining in the proof of Theorem 1 is to determine the initial conditions for (IV). From (38) it is clear that

$$\hat{\mathbf{x}}(t_0|t_0) = \mathbf{0}$$

Hence

$$\mathbf{P}_0 = \mathbf{P}(t_0) = \text{cov}[\tilde{\mathbf{x}}(t_0|t_0), \tilde{\mathbf{x}}(t_0|t_0)] = \text{cov}[\mathbf{x}(t_0), \mathbf{x}(t_0)]$$

In case of the conventional Wiener theory (see (A_3)), the last term is evaluated by means of (36).

This completes the proof of Theorem 1.

9 Outline of Proofs

Using the duality relations (16), all proofs can be reduced to those given for the regulator problem in [17].

(1) The fact that solutions of the variance equation exist for all $t \geqq t_0$ is proved in [17, Theorem (6.4)], using the fact that the variance of $\mathbf{x}(t)$ must be finite in any finite interval $[t_0, t]$.

(2) Theorem 3 is proved by showing that there exists a particular estimate of finite but not necessarily minimum variance. Under (A_4'), this is proved in [17; Theorem (6.6)]. A trivial modification of this proof goes through also with assumption (A_4).

(3) Theorem 4 is proved in [17; Theorems (6.8), (6.10), (7.2)]. The stability of the optimal filter is proved by noting that the estimation error plays the role of a Lyapunov function. The stability of the variance equation is proved by exhibiting a Lyapunov function for $\mathbf{P}$. This Lyapunov function in the simplest case is discussed briefly at the end of Example 1. While this theorem is true also in the nonconstant case, at present one must impose the somewhat restrictive conditions ($A_6 - A_7$).

10 Analytic Solution of the Variance Equation

Let $\mathbf{X}(t)$, $\mathbf{W}(t)$ be the (unique) matrix solution pair for (27) which satisfy the initial conditions

$$\mathbf{X}(t_0) = \mathbf{I}, \quad \mathbf{W}(t_0) = \mathbf{P}_0 \quad (47)$$

Then we have the following identity

$$\mathbf{W}(t) = \mathbf{P}(t)\mathbf{X}(t), \quad t \geqq t_0 \quad (48)$$

which is easily verified by substituting (48) with (IV) into (27). On the other hand, in view of (47–48), we see immediately from the first set of equations (27) that $\mathbf{X}(t)$ is the transition matrix of the differential equation

$$d\mathbf{x}/dt = -\mathbf{F}'(t)\mathbf{x} + \mathbf{H}'(t)\mathbf{R}^{-1}(t)\mathbf{H}(t)\mathbf{P}(t)\mathbf{x}$$

which is the adjoint of the differential equation (IV) of the optimal filter. Since the inverse of a transition matrix always exists, we can write

$$\mathbf{P}(t) = \mathbf{W}(t)\mathbf{X}^{-1}(t), \quad t \geqq t_0 \quad (49)$$

This formula may not be valid for $t < t_0$, for then $\mathbf{P}(t)$ may not exist!

Only trivial steps remain to complete the proof of Theorem 2.

11 Examples: Solution

Example 1. If $q_{11} > 0$ and $r_{11} > 0$, it is easily verified that the conditions of Theorems 3–4 are satisfied. After trivial substitutions in (III-IV) we obtain the expression for the optimal gain

$$k_{11}(t) = p_{11}(t)/r_{11} \quad (50)$$

and the variance equation

$$dp_{11}/dt = 2f_{11}p_{11} - p_{11}^2/r_{11} + q_{11} \quad (51)$$

By setting the right-hand side of (51) equal to zero, by virtue of the corollary of Theorem 4 we obtain the solution of the stationary problem (i.e., $t_0 = -\infty$, see (A$_3$)):

$$\bar{p}_{11} = \left[f_{11} + \sqrt{f_{11}^2 + q_{11}/r_{11}}\right] r_{11} \tag{52}$$

Since $\bar{p}_{11}$ and r_{11} are nonnegative, it is clear that only the positive sign is permissible in front of the square root.

Substituting into (50), we get the following expressions for the optimal gain

$$\bar{k}_{11} = f_{11} + \sqrt{f_{11}^2 + q_{11}/r_{11}} \tag{53}$$

and for the infinitesimal transition matrix (i.e., reciprocal time constant)

$$\bar{f}_{11} = f_{11} - \bar{k}_{11} = -\sqrt{f_{11}^2 + q_{11}/r_{11}} \tag{54}$$

of the optimal filter. We see, in accordance with Theorem 4, that the optimal filter is always stable, irrespective of the stability of the message model. Fig. 1(*b*) shows the configuration of the optimal filter.

It is easily checked that the formulas (52–54) agree with the results of the conventional Wiener theory [29].

Let us now compute the solution of the problem for a finite smoothing interval ($t_0 > -\infty$). The Hamiltonian equations (27) in this case are:

$$\left.\begin{aligned} dx_1/dt &= -f_{11}x_1 + (1/r_{11})w_1 \\ dw_1/dt &= q_{11}x_1 + f_{11}w_1 \end{aligned}\right\}$$

Let $\mathbf{T}$ be the matrix of coefficients of these equations.

To compute the transition matrix $\mathbf{\Theta}(t, t_0)$ corresponding to $\mathbf{T}$, we note first that the eigenvalues of $\mathbf{T}$ are $\pm\bar{f}_{11}$. Using this fact and constancy, it follows that

$$\mathbf{\Theta}(t, t_0) = \exp \mathbf{T}(t - t_0) = \mathbf{C}_1 \exp (t - t_0)\bar{f}_{11} + \mathbf{C}_2 \exp [-(t - t_0)\bar{f}_{11}]$$

where the constant matrices $\mathbf{C}_1$ and $\mathbf{C}_2$ are uniquely determined by the requirements

$$\mathbf{\Theta}(t_0, t_0) = \mathbf{C}_1 + \mathbf{C}_2 = \mathbf{I} = \text{unit matrix}$$

$$d\mathbf{\Theta}(t, t_0)/dt\big|_{t=t_0} = \mathbf{T}\mathbf{\Theta}(t, t_0)\big|_{t=t_0} = \bar{f}_{11}\mathbf{C}_1 - \bar{f}_{11}\mathbf{C}_2$$

After a good deal of algebra, we obtain

$$\mathbf{\Theta}(t_0 + \tau, t_0) = \begin{bmatrix} \cosh \bar{f}_{11}\tau - \dfrac{f_{11}}{\bar{f}_{11}} \sinh \bar{f}_{11}\tau & \dfrac{1}{r_{11}\bar{f}_{11}} \sinh \bar{f}_{11}\tau \\ \dfrac{q_{11}}{\bar{f}_{11}} \sinh \bar{f}_{11}\tau & \cosh \bar{f}_{11}\tau + \dfrac{f_{11}}{\bar{f}_{11}} \sinh \bar{f}_{11}\tau \end{bmatrix} \tag{55}$$

Knowledge of $\mathbf{\Theta}(t, t_0)$ can be used to derive explicit solutions to a variety of nonstationary filtering problems.

We consider only one such problem, which was treated by Shinbrot [3, Example 2]. He assumes that $f_{11} < 0$ and that the message process has reached steady-state. From (36) we see that

$$\mathcal{E}x_1^2(t) = -q_{11}/2f_{11} \quad \text{for all } t$$

We assume that the observations of the signal start at $t = 0$. Since the estimates must be unbiased, it is clear that $\hat{x}_1(0) = 0$. Therefore

$$p_{11}(0) = \mathcal{E}\tilde{x}_1^2(0) = \mathcal{E}x_1^2(0) = -q_{11}/2f_{11}$$

substituting this into (55), we get Shinbrot's formula:

$$p_{11}(t) = q_{11}\left[\frac{(f_{11} - \bar{f}_{11})e^{\bar{f}_{11}t} - (f_{11} + \bar{f}_{11})e^{-\bar{f}_{11}t}}{-(f_{11} - \bar{f}_{11})^2e^{\bar{f}_{11}t} + (f_{11} + \bar{f}_{11})^2e^{-\bar{f}_{11}t}}\right]$$

Since $\bar{f}_{11} < 0$, we see that as $t \to \infty$, $p_{11}(t)$ converges to

$$\bar{p}_{11} = -q_{11}/(f_{11} + \bar{f}_{11}) = (f_{11} - \bar{f}_{11})r_{11}$$

which agrees with (52).

To understand better the factors affecting convergence to the steady-state, let

$$\delta p_{11}(t) = p_{11}(t) - \bar{p}_{11}$$

The differential equation for δp_{11} is

$$d\delta p_{11}/dt = 2\bar{f}_{11}\delta p_{11} - (\delta p_{11})^2/r_{11} \tag{56}$$

We now introduce a *Lyapunov function* [21] for (56)

$$V(\delta p_{11}) = (\delta p_{11}/\bar{p}_{11})^2$$

The derivative of V *along motions* of (51) is given by

$$\dot{V}(\delta p_{11}) = \frac{\partial V(\delta p_{11})}{\partial \delta p_{11}} \cdot \frac{d\delta p_{11}}{dt} = -2[p_{11}/r_{11} + q_{11}/\bar{p}_{11}]V(\delta p_{11}) \tag{57}$$

This shows clearly that the "equivalent reciprocal time constant" for the variance equation depends on two quantities: (i) the message-to-noise ratio p_{11}/r_{11} at the input of the optimal filter, (ii) the ratio of excitation to estimation error $q_{11}/\bar{p}_{11}$.

Since the message model in this example is identical with its dual, it is clear that the preceding results apply without any modification to the dual problem. In particular, the filter shown in Fig. 1(*b*) is the same as the optimal regulator for a plant with transfer function $1/(s - f_{11})$. The Hamiltonian equations (27) for the dual problem were derived by Rozonoër [19] from Pontryagin's maximum principle.

Let us conclude this example by making some observations about the nonconstant case. First, the expression for the derivative of the Lyapunov function given by (57) remains true without any modification. Second, assume $\bar{p}_{11}(t_0)$ has been evaluated somehow. Given this number, $\bar{p}_{11}(t)$ can be evaluated for $t \geqq t_0$ by means of the variance equation (51); the existence of a Lyapunov function and in particular (57) shows that this computation is stable, i.e., not adversely affected by roundoff errors. Third, knowing $\bar{p}_{11}(t)$, equation (57) provides a clear picture of the transient behavior of the optimal filter, even though it might be impossible to solve (51) in closed form.

Example 2. The variance equation is

$$dp_{11}/dt = 2f_{11}p_{11} - p_{11}^2(1/r_{11} + 1/r_{22}) + q_{11}$$

If $q_{11} > 0$, $r_{11} > 0$, and $r_{22} > 0$, the conditions of Theorems 3–4 are satisfied. Therefore the minimum error variance in the steady-state is

$$\bar{p}_{11} = \frac{f_{11} + \sqrt{f_{11}^2 + q_{11}/r_{11} + q_{11}/r_{22}}}{1/r_{11} + 1/r_{22}}$$

and the optimal steady-state gains are

$$\bar{k}_{1i} = \bar{p}_{11}/r_{ii}, \qquad i = 1, 2$$

The same problem has been considered also by Westcott [30, Example]. A glance at his calculations shows that ours is the simpler and more natural approach.

Example 3. The variance equation is

$$\left.\begin{aligned} dp_{11}/dt &= -p_{12}^2/r_{11} + q_{11} \\ dp_{12}/dt &= p_{11} - p_{12} - p_{12}p_{22}/r_{11} \\ dp_{22}/dt &= 2(p_{12} - p_{22}) - p_{22}^2/r_{11} \end{aligned}\right\} \tag{58}$$

If $q_{11} > 0$, $r_{11} > 0$, the conditions of Theorems 3–4 are satisfied. Setting the right-hand side of (58) equal to zero, we get the solution of the stationary problem:

$$\bar{k}_{11} = \sqrt{q_{11}/r_{11}}$$

$$\bar{k}_{21} = -1 + \sqrt{1 + 2\sqrt{q_{11}/r_{11}}}$$

See Fig. 3(*b*).

The infinitesimal transition matrix of the optimal filter in the steady-state is:

$$\mathsf{F} = \begin{bmatrix} 0 & -\sqrt{q_{11}/r_{11}} \\ 1 & -\sqrt{1 + 2\sqrt{q_{11}/r_{11}}} \end{bmatrix}$$

The natural frequency of the filter is $(q_{11}/r_{11})^{1/4}$ and the damping ratio is $(1/2)[2 + (r_{11}/q_{11})^{1/2}]^{1/2}$. Even for such a very simple problem, the parameters of the optimal filter are not at all obvious by inspection.

The solution of the dual problem in the steady-state (see Fig. 4) is obtained by utilizing the duality relations

$$\bar{k}^*_{11} = \bar{k}_{11}, \qquad \bar{k}^*_{12} = \bar{k}_{21}$$

The same result was obtained by Kipiniak [24], using the Euler equations of the calculus of variations.

Example 4. The variance equation is

$$\left.\begin{aligned} dp_{11}/dt &= 2f_{12}p_{12} - p_{11}^2/r_{11} + q_{11} \\ dp_{12}/dt &= f_{21}p_{11} + f_{12}p_{22} - p_{11}p_{12}/r_{11} \\ dp_{22}/dt &= 2f_{21}p_{12} - p_{12}^2/r_{11} \end{aligned}\right\} \qquad (59)$$

If $f_{12} \neq 0$, $f_{21} \neq 0$, and $r_{11} > 0$, the conditions of Theorems 3–4 are satisfied. There are then two sets of possibilities for the right-hand side of (59) to vanish for nonnegative $\bar{p}_{22}$:

(A) $\bar{p}_{11} = \sqrt{q_{11}r_{11}}$, $\bar{p}_{12} = 0$, $\bar{p}_{22} = -(f_{21}/f_{12})\sqrt{q_{11}r_{11}}$

(B) $\bar{p}_{11} = \sqrt{(q_{11} + 4f_{12}f_{21}r_{11})r_{11}}$, $\bar{p}_{12} = 2f_{21}r_{11}$, $\bar{p}_{22} = (f_{21}/f_{12})\sqrt{(q_{11} + 4f_{12}f_{21}r_{11})r_{11}}$

The expression for $\bar{p}_{22}$ shows that Case (A) applies when $f_{12}f_{21}$ is negative (the model is stable but not asymptotically stable) and Case (B) applies when $f_{12}f_{21}$ is positive (the model is unstable).

The optimal filter is shown in Fig. 5(*b*). The optimal gains are given by

$$\bar{k}_{11} = \bar{p}_{11}/r_{11}, \qquad \bar{k}_{21} = \bar{p}_{12}/r_{11}$$

If $f_{12} \neq 0$ but $f_{21} = 0$, the model is completely observable but not completely controllable. Hence the steady-state variances exist but the optimal filter is not necessarily asymptotically stable since Theorem 4 is not applicable. As a matter of fact, the optimal filter in this case is partially "open loop" and it is not asymptotically stable.

If $f_{12} = 0$, then not even Theorem 3 is applicable. In this case, if $f_{21} \neq 0$, equations (59) have no equilibrium state; if $f_{21} = 0$, then equations (59) have an infinity of positive definite equilibrium states given by:

$$\bar{p}_{11} = \sqrt{q_{11}/r_{11}}, \qquad \bar{p}_{12} = 0, \qquad \bar{p}_{22} > 0$$

Thus if $f_{12} = 0$, the conclusions of Theorems 3–4 are false.

Example 5. The variance equation is

$$\left.\begin{aligned} dp_{11}/dt &= 2p_{12} - p_{11}^2/r_{11} \\ dp_{12}/dt &= p_{22} - p_{11}p_{12}/r_{11} \\ dp_{22}/dt &= -p_{12}^2/r_{11} \end{aligned}\right\}$$

We assume that $r_{11} > 0$; this assures that Theorem 3 is applicable. We then find that the steady-state error variances are all zero. The matrix of coefficients of the Hamiltonian equations (27) is:

$$\mathsf{T} = \begin{bmatrix} 0 & 0 & 1/r_{11} & 0 \\ -1 & 0 & 0 & 0 \\ 0 & 0 & 0 & 1 \\ 0 & 0 & 0 & 0 \end{bmatrix}$$

and the corresponding transition matrix is (here (4) is a *finite* series!)

$$\mathbf{\Theta}(t_0 + \tau, t_0) = \begin{bmatrix} 1 & 0 & \tau/r_{11} & \tau^2/2r_{11} \\ -\tau & 1 & -\tau^2/2r_{11} & -\tau^3/6r_{11} \\ 0 & 0 & 1 & \tau \\ 0 & 0 & 1 & 1 \end{bmatrix}$$

Using (29), we find ($t_0 = 0$):

$$\mathsf{P}(t) = \frac{r_{11}p_{22}(0)}{r_{11} + p_{22}(0)t^3/3}\begin{bmatrix} t^2 & t \\ t & 1 \end{bmatrix}$$

This formula, obtained here with little labor, is identical with the results of Shinbort [3, Example 1].

The optimal filter is shown in Fig. 7(*b*). The time-varying gains tend to 0 as $t \to \infty$; in other words, the filter pays less and less attention to the incoming signals and relies more and more on the previous estimates of x_1 and x_2.

Since the conditions of Theorem 4 are not satisfied, one might suspect that the optimal filter is *not* uniformly (and hence exponentially [21]) asymptotically stable. To check this conjecture, we calculate the transition matrix of the optimal filter. We find, for $t, \tau \geqq 0$,

$$\mathbf{\Psi}(t, \tau) = \frac{1}{\alpha(t)}\begin{bmatrix} \alpha(t) - \beta(t, \tau)t & -\alpha(t)\tau + \alpha(\tau)t + \beta(t, \tau)\tau t \\ -\beta(t, \tau) & \alpha(\tau) + \beta(t, \tau) \end{bmatrix}$$

where

$$\alpha(t) = t^3/3 + r_{11}/p_{22}(0)$$

$$\beta(t, \tau) = (t^2 - \tau^2)/2$$

Since $\psi_{11}(t\ \tau)$ does not converge to zero with $t - \tau \to \infty$, it is clear that the optimal filter is not even stable, let alone asymptotically stable.

From the transition matrix of the optimal filter, we can obtain at once its impulse response with respect to the input $z_1(t)$ and output $\hat{x}_1(t)$:

$$\psi_{11}(t, \tau)k_{11}(\tau) + \psi_{12}(t, \tau)k_{21}(\tau) = \frac{t\tau}{t^3/3 + r_{11}/p_{22}(0)}$$

This agrees with Shinbrot's result [3].

Example 6. The variance equation is:

$$\left.\begin{aligned} dp_{11}/dt &= 2p_{12} - h_{11}^2p_{11}^2/r_{11} - h_{22}^2p_{12}^2/r_{22} \\ dp_{12}/dt &= p_{22} - h_{11}^2p_{11}p_{12}/r_{11} - h_{22}^2p_{12}p_{22}/r_{22} \\ dp_{22}/dt &= -h_{11}^2p_{12}^2/r_{11} - h_{22}^2p_{22}^2/r_{22} + q_{11} \end{aligned}\right\} \qquad (60)$$

If $h_{11} \neq 0$, $q_{11} > 0$, $r_{11} > 0$, $r_{22} > 0$, then the conditions of Theorems 3–4 are satisfied. Setting the right-hand side of (60) equal to zero leads to a very complicated algebraic problem. We introduce first the abbreviations:

$$\alpha = |h_{11}|\sqrt{q_{11}/r_{11}}$$

$$\beta^2 = h_{22}^2q_{11}/r_{22}$$

It follows that

$$h_{11}\bar{k}_{11} = \frac{h_{11}^2}{r_{11}} p_{11} = \alpha \frac{\sqrt{2\alpha + \beta^2}}{\alpha + \beta^2}$$

$$h_{11}\bar{k}_{21} = \frac{h_{11}^2}{r_{11}} p_{12} = \frac{\alpha^2}{\alpha + \beta^2}$$

$$h_{22}\bar{k}_{12} = \frac{h_{22}^2}{r_{22}} p_{12} = \frac{\beta^2}{\alpha + \beta^2}$$

$$h_{22}\bar{k}_{21} = \frac{h_{22}^2}{r_{22}} p_{22} = \beta^2 \frac{\sqrt{2\alpha + \beta^2}}{\alpha + \beta^2}$$

It is easy to verify that the right-hand side of (60) vanishes for this set of p_{ij}'s; by Theorem 5, this cannot happen for any other set. Hence the solution of the stationary Wiener problem is complete. It is interesting to note that the conventional procedure would require here the spectral factorization of a two-by-two matrix which is very much more difficult algebraically than by the present method.

The infinitesimal transition matrix of the optimal filter is given by

$$\mathbf{F}_{\text{opt}} = \begin{bmatrix} -\alpha \dfrac{\sqrt{2\alpha + \beta^2}}{\alpha + \beta^2} & \dfrac{\alpha}{\alpha + \beta^2} \\ -\dfrac{\alpha^2}{\alpha + \beta^2} & -\beta^2 \dfrac{\sqrt{2\alpha + \beta^2}}{\alpha + \beta^2} \end{bmatrix}$$

The natural frequency of the optimal filter is

$$\omega = |\lambda(\mathbf{F}_{\text{opt}})| = \sqrt{\alpha}$$

and the damping ratio is

$$\zeta = |\text{Re}\, \lambda(\mathbf{F}_{\text{opt}})|/\omega = \frac{1}{\sqrt{2}} \sqrt{1 + \frac{\beta^2}{2\alpha}}$$

The quantities α and β can be regarded as signal-to-noise ratios. Since all parameters of the optimal filter depend only on these ratios, there is a possibility of building an adaptive filter once means of experimentally measuring α and β are available. An investigation of this sort was carried out by Bucy [31] in the simplified case when $h_{22} = \beta = 0$.

12 Problems Related to Adaptive Systems

The generality of our results should be of considerable usefulness in the theory of adaptive systems, which is as yet in a primitive stage of development.

An *adaptive system* is one which changes its parameters in accordance with measured changes in its environment. In the estimation problem, the changing environment is reflected in the time-dependence of $\mathbf{F}$, $\mathbf{G}$, $\mathbf{H}$, $\mathbf{Q}$, $\mathbf{R}$. Our theory shows that such changes affect only the values of the parameters but not the structure of the optimal filter. This is what one would expect intuitively and we now have also a rigorous proof. Under ideal circumstances, the changes in the environment could be detected instantaneously and exactly. The adaptive filter would then behave as required by the fundamental equations (I–IV). In other words, our theory establishes a basis of comparison between actual and ideal adaptive behavior. It is clear therefore that *a fundamental problem in the theory of adaptive systems is the further study of properties of the variance equation* (IV).

13 Conclusions

One should clearly distinguish between two aspects of the estimation problem:

(1) *The theoretical aspect.* Here interest centers on:

(i) The general form of the solution (see Fig. 1).

(ii) Conditions which guarantee a priori the existence, physical realizability, and stability of the optimal filter.

(iii) Characterization of the general results in terms of some simple quantities, such as signal-to-noise ratio, information rate, bandwidth, etc.

An important consequence of the time-domain approach is that these considerations can be completely divorced from the assumption of stationarity which has dominated much of the thinking in the past.

(2) *The computational aspect.* The classical (more accurately, old-fashioned) view is that a mathematical problem is solved if the solution is expressed by a formula. It is not a trivial matter, however, to substitute numbers in a formula. The current literature on the Wiener problem is full of semirigorously derived formulas which turn out to be unusable for practical computation when the order of the system becomes even moderately large. The variance equation of our approach provides a practically useful and theoretically "clean" technique of numerical computation. Because of the guaranteed convergence of these equations, the computational problem can be considered solved, except for purely numerical difficulties.

Some open problems, which we intend to treat in the near future, are:

(i) Extension of the theory to include nonwhite noise. As mentioned in Section 2, this problem is already solved in the discrete-time case [11], and the only remaining difficulty is to get a convenient canonical form in the continuous-time case.

(ii) General study of the variance equations using Lyapunov functions.

(iii) Relations with the calculus of variations and information theory.

14 References

1 N. Wiener, "The Extrapolation, Interpolation, and Smoothing of Stationary Time Series," John Wiley & Sons, Inc., New York, N. Y., 1949.

2 A. M. Yaglom, "Vvedenie v Teoriya Statsionarnikh Sluchainikh Funktsii" (Introduction to the theory of stationary random Processes) (in Russian), *Ups. Fiz. Nauk.*, vol. 7, 1951; German translation edited by H. Göring, Akademie Verlag, Berlin, 1959.

3 M. Shinbrot, "Optimization of Time-Varying Linear Systems With Nonstationary Inputs," Trans. ASME, vol. 80, 1958, pp. 457–462.

4 C. W. Steeg, "A Time-Domain Synthesis for Optimum Extrapolators," *Trans. IRE*, Prof. Group on Automatic Control, Nov., 1957, pp. 32–41.

5 V. S. Pugachev, "Teoriya Sluchainikh Funktsii i Ee Primenenie k Zadacham Automaticheskogo Upravleniya" (Theory of Random Functions and Its Application to Automatic Control Problems) (in Russian), second edition, Gostekhizdat, Moscow, 1960.

6 V. S. Pugachev, "A Method for Solving the Basic Integral Equation of Statistical Theory of Optimum Systems in Finite Form," *Prikl. Math. Mekh.*, vol. 23, 1959, pp. 3–14 (English translation pp. 1–16).

7 E. Parzen, "Statistical Inference on Time Series by Hilbert-Space Methods, I," Tech. Rep. No. 23, Applied Mathematics and Statistics Laboratory, Stanford Univ., 1959.

8 A. G. Carlton and J. W. Follin, Jr., "Recent Developments in Fixed and Adaptive Filtering," Proceedings of the Second AGARD Guided Missiles Seminar (Guidance and Control) AGARDograph 21, September, 1956.

9 J. E. Hanson, "Some Notes on the Application of the Calculus of Variations to Smoothing for Finite Time, etc.," JHU/APL Internal Memorandum BBD-346, 1957.

10 R. S. Bucy, "Optimum Finite-Time Filters for a Special Nonstationary Class of Inputs," JHU/APL Internal Memorandum BBD-600, 1959.

11 R. E. Kalman, "A New Approach to Linear Filtering and Prediction Problems," Trans. ASME, Series D, Journal of Basic Engineering, vol. 82, 1960, pp. 35–45.

12 R. E. Bellman, "Adaptive Control: A Guided Tour" (to be published), Princeton University Press, Princeton, N. J., 1960.
13 R. E. Kalman and R. W. Koepcke, "The Role of Digital Computers in the Dynamic Optimization of Chemical Reactions," Proceedings of the Western Joint Computer Conference, 1959, pp. 107–116.
14 R. E. Kalman and J. E. Bertram, "A Unified Approach to the Theory of Sampling Systems," *Journal of the Franklin Institute*, vol. 267, 1959, pp. 405–436.
15 J. L. Doob, "Stochastic Processes," John Wiley & Sons, Inc., New York, N. Y., 1953.
16 R. E. Kalman, "On the General Theory of Control Systems," Proceedings of the First International Congress on Automatic Control, Moscow, USSR, 1960.
17 R. E. Kalman, "Contributions to the Theory of Optimal Control," Proceedings of the Conference on Ordinary Differential Equations, Mexico City, Mexico, 1959; Bol. Soc. Mat. Mex., 1961.
18 J. J. Levin, "On the Matrix Riccati Equation," *Trans. American Mathematical Society*, vol. 10, 1959, pp. 519–524.
19 L. I. Rozonoër, "L. S. Pontryagin's Maximum Principle in the Theory of Optimum Systems, I," *Avt. i Telemekh.*, vol. 20, 1959, pp. 1320–1324.
20 S. Kullback, "Information Theory and Statistics," John Wiley & Sons, New York, N. Y., 1959.
21 R. E. Kalman and J. E. Bertram, "Control System Analysis and Design Via the 'Second Method' of Lyapunov. I. Continuous-Time Systems," Journal of Basic Engineering, Trans. ASME, series D, vol. 82, 1960, pp. 371–393.
22 R. E. Bellman, "Introduction to Matrix Analysis," McGraw-Hill Book Company, Inc., New York, N. Y., 1960.
23 E. A. Coddington and N. Levinson, "Theory of Ordinary Differential Equations," McGraw-Hill Book Company, Inc., New York, N. Y., 1955.
24 W. Kipiniak, "Optimum Nonlinear Controllers," Report 7793-R-2, Servomechanisms Lab., M.I.T., 1958.
25 R. S. Bucy, "A Matrix Formulation of the Finite-Time Problem," JHU/APL Internal Memorandum BBD-777, 1960.
26 R. S. Bucy, "Combined Range and Speed Gate for White Noise and White Signal Acceleration," JHU/APL Internal Memorandum BBD-811, 1960.
27 V. S. Pugachev, "General Condition for the Minimum Mean Square Error in a Dynamic System," *Avt. i Telemekh.*, vol. 17, 1956, pp. 289–295, translation, pp. 307–314.
28 M. Loève, "Probability Theory," Van Nostrand and Company, New York, N. Y., 1955, Chap. 10.
29 W. B. Davenport and W. L. Root, "An Introduction to the Theory of Random Signals and Noise," McGraw-Hill Book Company, Inc., New York, N. Y., 1956.
30 J. H. Westcott, "Design of Multivariable Optimum Filters," Trans. ASME, vol. 80, 1958, pp. 463–467.
31 R. S. Bucy, "Adaptive Finite-Time Filtering," JHU/APL Internal Memorandum BBD-645, 1959.

Editors' Postscript to Papers 1–8

The Kalman–Bucy filter model has been a basis for the analysis of a number of engineering problems during the last decade. Several authors since then have combined the ideas of Lévy and Cramér on unit multiplicity with the spectral factorization approach to the causal prewhitening procedure. The conditions on the filter's characteristics implied by the linear version of the Kalman–Bucy model have been generalized or modified in order to apply to non-white-noise cases.

The literature on the subjects of shaping filters and of the covariance factorization is extensive. Purely mathematical contributions are not included in this volume, since most of them appear to be rather abstract, with relatively obscure bearing on filtering applications. Gohberg and Krein (1964) should be cited for their significant work on the factorization of operators.

Most of the applications-oriented authors have dealt with the characterization of shaping filters and have usually focused on rather special cases; due to the fragmented nature of this literature, such contributions are also not included. E. B. Stear (1965), for example, gives an outline of representative results on the subject and draws an accurate and fairly complete picture of the state of shaping-filter research in the mid-sixties. Other typical contributions can be found in the reference list in Stear's paper. More recent results that should be mentioned include those of Brandenburg and Meadows (1971) and Anderson (1969).

References

Anderson, B. D. O. (1969). The inverse problem of stationary covariance generation. *J. Stat. Phys.*, **1**, 1, 133–147.

Brandenburg, L. H., and Meadows, H. E., (1971). Shaping filter representation of nonstationary colored noise. *IEEE Trans. Inform. Theory*, **IT-17**, 26–31.

Gohberg, I., and Krein, M. G., (1964). On the factorization of operators in Hilbert spaces. *Acta Sci. Math. Univ. Szeged*, **25**, 90–123. *Amer. Math. Soc. Transl., Ser. 2*, **51** (1966).

Stear, E. B. (1965). Shaping filters for stochastic processes. *In* "Modern Control Systems Theory" (C. T. Leondes, ed.), pp. 121–155. McGraw-Hill, New York.

Editors' Comments on Papers 9 and 10

In these two papers Professor V. Kallianpur of the University of Minnesota and Professor G. Mandrekar of Michigan State University have considered the extension of the Cramér–Hida representation to the case of random fields or multivariate random processes. The parameter set of the process is generalized from the standard time interval to an arbitrary Hausdorff space. The difficulties and subtleties involved in extending the canonical decomposition to this case are considerable.

The direction of the authors is, of course, not directly motivated by a consideration of any engineering application. However, the results are significant in a mathematical context, and the eventual potential for benefit to the applications-oriented reader should not be precluded.

Their first paper starts from O. Hanner's time-domain analysis and connects it to the multiplicity approach. It goes on to generalize some of his results and to present alternative proofs for some others.

Their second paper includes an abstract derivation of the multiplicity representation for random processes with generalized parameter sets. The paper is purely mathematical and constitutes a complete study of the problem.

These papers are reprinted with the permission of the authors, with the permission of the Department of Statistics of the University of North Carolina for the first paper, and with the permission of the Society for Industrial and Applied Mathematics for the second paper.

Reprinted from The *University of North Carolina Monograph Series in Probability and Statistics*, No. 3, 385–396 (1970)

9

On the Connection between Multiplicity Theory and O. Hanner's Time Domain Analysis of Weakly Stationary Stochastic Processes

G. KALLIANPUR AND V. MANDREKAR,
University of Minnesota

1. INTRODUCTION

In an early paper full of original ideas, O. Hanner [5] obtained a decomposition for a mean-continuous, purely non-deterministic, weakly stationary process depending on a continuous time parameter. Such a representation had already been derived by K. Karhunen in his 1947 paper [8] and is now generally known by his name. The interest of Hanner's work arises from the fact that his method is based on a time-domain analysis of the process itself and is entirely free of spectral considerations. In recent years, in the light of the extensive development of multivariate stationary processes, it has appeared desirable to separate the time-domain analysis from spectral studies, and the interest in the former has revived. As an example, we mention the paper of P. Masani and J. Robertson [9] whose approach considered the discretized process corresponding to the given continuous one and makes essential use of the Cayley transform associated with the unitary group of the process. The extension of this method to (finite-dimensional) multivariate stationary processes has been carried out by J. Robertson in his thesis [11].

We deduce in Section 2 as an easy corollary of the Karhunen representation theorem the fact that a purely non-deterministic stationary process has multiplicity one of Lebesgue type. A proof of this result does not seem to have been given in the literature. Our main concern, in this paper, however, is to show how a modification of Hanner's original proof directly leads to this result. We show how this is done in Theorem 1 (Section 3). The Karhunen representation is then derived as a consequence.

The purpose of studying this multiplicity question directly is to bring out in the analysis the separation of two issues (1) the form of the spectral type, (2) the multiplicity. The fact that the process has Lebesgue type is intimately connected with the stationarity and hence with the construction of the element $Z(I_a^b)$ of Hanner [5], (p. 166). On the other hand, the dimension of the process governs the determination of multiplicity. The modification of Hanner's technique which gives us both the above properties lies in recasting his arguments in the form of multiplicity theory. In so doing we can bring out the generality of the reasoning by which he solves the first problem. It will be shown that the univariate character of the process enters into the arguments only to establish the fact that the multiplicity is one. It is interesting to note that the reasoning of Hanner used in determining the multiplicity assumes the fact that every non-trivial process has multiplicity at least one. Hence the problem is to exploit the univariate character of the process to show that the multiplicity does not exceed one. Thus an extension of Hanner's method for multivariate stationary processes lies in the study of the spectral types and multiplicity. We have undertaken such a study in another paper in which using these ideas we are able to study the corresponding time domain problems for a large class of infinite-dimensional stationary processes [7]. It is now also obvious that the recent work of H. Cramér [1], [2], [3] and T. Hida [6] on the multiplicity theory of stochastic processes is an extension of Hanner's ideas to non-stationary processes.

2. THE MULTIPLICITY OF A WEAKLY STATIONARY STOCHASTIC PROCESS

Let $x_t(-\infty < t < \infty)$ be a one dimensional (complex valued) weakly stationary (w.s.) process with mean assumed zero and satisfying the following assumptions (A) :

A— (i). x_t is purely non deterministic, and

A—(ii). x_t is continuous in quadratic mean.

Following [5], let $L_2(x)$ be the Hilbert space of the proeess $\{x_t\}$, and let $L_2(x; a) = \mathfrak{S}\{x_t, t \leqslant a\}$ where $\mathfrak{S}\{\ldots\}$ denote the closed linear subspace of $L_2(x)$ spanned by the set of random variables in $\{\ldots\}$. We shall also write

$L_2(x; a, b) = L_2(x; b) \ominus L_2(x; a)(a \leqslant b)$. As we proceed further, whenever necessary, we shall draw on Hanner's notation and also on some of the notation and terminology of [6].

The following expresentation was obtained in [8] for a w.s. process $\{x_t\}$ satisfying (A).

$$x_t = \int_{-\infty}^{t} G(u-t)d\xi(u), \tag{2.1}$$

where ξ is a homogeneous random set function, i.e., a random set function with the property

$$\mathcal{E}[\xi(\Delta_1)\overline{\xi(\Delta_2)}] = \mu(\Delta_1 \cap \Delta_2),$$

μ being Lebesgue measure and Δ_1, Δ_2 any two sets of finite μ measure;

$L_2(x; t) = L_2(\xi; t) = \mathfrak{S}\{\xi(\Delta), \Delta \subset (-\infty, t)\}$ for every t and $\int_{-\infty}^{0} |G(u)|^2 d\mu(u)$ is finite.

It has been shown in Hida in [6] that every second order process $\{x_t\}$ satisfying (A) and not necessarily stationary has the representation

$$(2.2) \qquad x_t = \sum_{n=1}^{N} \int_{-\infty}^{t} F_n(t, u) dE(u) f^{(n)}$$

where (i) $\{E(u)\}$ $(-\infty < u < \infty)$ is the resolution of the identity determined by the projection operator with range $L_2(x; u)$, (ii) N is the multiplicity of x_t (N may be finite or ∞), (iii) $f^{(n)}$ $(n = 1, \ldots, N)$ are elements of $L_2(x)$ with the following properties: (a) the processes $Z_n(\Delta) = E(\Delta) f^{(n)}$, where $\Delta = (a, b]$ and $E(\Delta) = E(b) - E(a)$ have orthogonal increments and are mutually orthogonal, (b) the variance function of $Z_n(\Delta)$ is given by $\rho_{f(n)}(\Delta)$ where $\rho_{f(n)}(\Delta) = \| E(\Delta) f^{(n)} \|^2$, (c) $\rho_{f(1)} \gg \rho_{f(2)} \gg \cdots \rho_{f(N)}$, (d) $F_n(t, \cdot)$ is square integrable with respect to $\rho_{f(n)}$, and

(e) $\sum_{h=1}^{N} \int_{-\infty}^{+\infty} |F_n(t, u)|^2 d\rho_{f(n)}(u) < \infty$. The representation (2.2) can be chosen so as to satisfy

$$(2.3) \qquad L_2(x; t) = \sum_{n=1}^{N} \oplus L_2(z_n; t) \quad \text{for each } t.$$

It is convenient at this point to recall some of the terminology of multiplicity theory in a separable Hilbert space H. We shall introduce just those ideas that will be used in this paper. Let A be any self-adjoint operator with spectral measure function $E(\cdot)$. For any element f of H let ρ_f be the finite measure on the Borel sets of the line given by $\rho_f(\Delta) = \| E(\Delta) f \|^2$. The family of all finite measures on the line is divided into equivalence classes by the relation of equivalence between measures (equivalence here means mutual absolute continuity). If ρ is used to denote the equivalence class to which the measure ρ_f belongs, ρ will be called the spectral type of f with respect to A. ρ is also referred to as the spectral type belonging to A. If elements f and g are such that $\rho_f \equiv \rho_g$, they obviously have the same spectral type ρ. If for every measure ν belonging to ρ, $\nu \equiv \mu$, the Lebesgue measure then ρ is called the Lebesgue type. We shall say that the spectral type ρ dominates the spectral type $\sigma(\rho > \sigma, \sigma < \rho)$ if any (and thus every) measure belonging to σ is absolutely continuous with respect to any measure belonging to ρ. ρ and σ are said to be independent spectral types if for any spectral type ν such that $\nu < \rho$ and $\nu < \sigma$ we have $\nu = 0$. An element f is said to be of maximal spectral type ρ (with respect to A) if for every g in H, $\rho_g \ll \rho_f$. The subspace $\mathfrak{S}\{E(\Delta)f, \Delta$ over all finite intervals$\}$ is called the cyclic subspace with respect to A generated by f. If this space coincides with H, A is said to be cyclic and f is called a cyclic or generating element of A. A has multiplicity one in this case. Also, if f is a generating element of A then it is of maximal spectral type with respect to A. The spectral type of a generating element of A is referred to as the spectral type of A. The reader is referred to the article by A. I. Plessner and V. A. Rohlin [10] for further details.

Assuming the Karhunen representation (2.1) we shall show that the multiplicity of $\{x_t\}$ is one of Lebesgue type (Proposition, see also [3]). It might seem obvious that $N = 1$ on comparing (2.1) and (2.2). Although this is true it should be noted that in (2.1) the function ξ has for its variance function the Lebesgue measure μ, so that it is not possible to write $\xi(u) = E(u)f$ (for all u) for some f in $L_2(x)$.

Proposition. *The multiplicity of a weakly stationary process is one of Lebesgue type.*

Lemma 1 : *If x_t is a w.s. process, satisfying assumption A—(ii), then for any element $f \epsilon L_2(x)$ $\rho_f \ll \mu$.*

Proof : Let $\{T_h\}$ $(-\infty < h < +\infty)$, be the strongly continuous unitary group of the x_t-process ([2] (p. 55)). We recall from [5] that for every h, and any t real, we have

$$T_h E(t-h) = E(t) T_h. \tag{2.4}$$

Now $\rho_f(\Delta - h) = \|E(\Delta - h)f\|^2$ where $\Delta - h = \{u-h \mid u \epsilon \Delta\}$ and Δ is a Borel measurable set. Therefore, by (2.4) $\rho_f(\Delta - h) = \|E(\Delta)T_h f\|^2$. By strong continuity of the group,

$$\rho_f(\Delta - h) \to \rho_f(\Delta) \text{ as } h \to 0.$$

The assertion of the lemma is now an immediate consequence of a theorem due to N. Weiner and R. C. Young (see [12], p. 91).

Proof of the Proposition : By the property of the Karhunen representation and (2.3), we have

$$L_2(x; t) = L_2(\xi; t) = \sum_1^N \oplus L_2(z_n; t) \text{ for all } t.$$

Hence (see Doob [4], pp. 425–428), $z_n(t) = \int_{-\infty}^{t} h_n(t, u)d\xi(u)$. Since $\{z_n(t), -\infty < t < +\infty\}$ are mutually orthogonal processes with orthogonal increments, holding t fixed, we have for any finite interval $\Delta \subseteq (-\infty, t)$, $z_n(\Delta) = \int_\Delta h_n(t, u)d\xi(u)$ by (2.3), and hence $(n \neq 1)$, $\mathcal{E}[z_1(\Delta)\overline{z_n(\Delta)}] = \int_\Delta h_1(t, u)\overline{h_n(t, u)}d\mu(v) = 0$. This implies that

(2.5) $$\mu\{v\epsilon(-\infty, t] \,|\, h_1(t, v)\bar{h}_n(t, v) \neq 0\} = 0.$$

Define for $u\epsilon(-\infty, t]$, $S_n^{(t)} = \{u \,|\, h_n(t, u) \neq 0\}$ for $n = 1, 2, \ldots, N$. By (2.5), $\mu\{S_1^{(t)} \cap S_n^{(t)}\} = 0$, and therefore by Lemma 2.1,$\rho_{f(n)}$ $(S_1^{(t)} \cap S_n^{(t)}) = 0$ for all $n \neq 1$. Clearly, $\rho_{f(1)}([S_1^{(t)}]^c) = 0$ and $\rho_{f(n)}$ $([S_n^{(t)}]^c) = 0$ and hence by the maximality of $f^{(1)}$, $\rho_{f(n)}$ $([S_1^{(t)}]^c)$ $\cup [S_n^{(t)}]^c) = 0$ for $n \neq 1$ where $(\quad)^c$ denotes complementation with respect to $(-\infty, t]$. Thus $\rho_{f(n)}\{(-\infty, t]\} = 0$ for every t, so that $\rho_{f(n)} = 0$ $(n \neq 1)$. This gives $L_2(\xi; t) = L_2(z_1; t)$ i.e., $\xi(t) = \int_{-\infty}^{t} \nu_t(u)dz_1(u)$. From this and the proper canonical property, for each finite interval $\Delta \subseteq (-\infty, t]$, $\mu(\Delta) = \mathcal{E}\,|\xi(\Delta)|^2$ $= \int_\Delta |\nu_t(u)|^2 d\rho_{f(1)}(u)$. For each t, $\mu \ll \rho_{f(1)}$ on $(-\infty, t)$ and hence $\mu \ll \rho_{f(1)}$. This along with Lemma 2.1 proves the proposition.

3. A NEW PROOF OF THE MULTIPLICITY RESULT AND THE KARHUNEN REPRESENTATION

A new proof of the Karhunen representation using multiplicity arguments is presented in this section. The following main theorem is obtained essentially using the modified arguments of Hanner.

Theorem 1. *The self-adjoint operator A of a weakly stationary continuous parameter, purely non-deterministic univariate process continuous in q.m. is cyclic and has Lebesgue spectral type.*

Proof : Since $L_2(x)$ is separable there is an element $f^{(1)}$ of maximal spectral type with respect to A. Given $u > 0$ and $a < b$, let Δ_0 be a subinterval of $(0, u]$, $A < a-u$, $B > b$ and define

$$g_a^b = E(a, b] \int_A^B T_h E(\Delta_0) f^{(1)} dh. \tag{3.1}$$

Then g_a^b depsnds on u and Δ_0 but not on A and B. As a matter of fact let A' $(< A)$ then $\int_A^B T_h E(\Delta_0) f^{(1)} dh \in E(-\infty, a] L_2(x)$, since for each $(A' \leqslant h \leqslant A)$, $T_h E(\Delta_0) f^{(1)} = E(\Delta_0 + h) T_h f^{(1)} \in E(u+A)$ $L_2(x)$. Thus $E(a, b) \int_{A'}^{A} T_h E(\Delta_0) f^{(1)} dh = 0$. A similar argument holds if $(A' > A)$. Let $B' > B$, then $\int_B^{B'} T_h E(\Delta_0) f^{(1)} \in E(b, \infty)$ $L_2(x)$ by a similar consideration. As Hanner showed for $a < b < c$

$$\begin{cases} g_a^b + g_b^c = g_a^c,\ g_a^b \perp g_b^c \text{ and for any real } t \\ T_t\, g_a^b = g_{a+t}^{b+t}. \end{cases} \tag{3.2}$$

From these properties it follows that $\|g_a^b\|^2 = \tau(b-a)$, where τ is a non-negative number which does not depend on a and b. In fact, there is a subinterval, say Δ_0 of $(0, u)$ such that for all a, b $(a < b)$, $\|g_a^b\|^2 = (b-a)$. It can be shown that there exists a u such that $\|g_0^u\|^2 \neq 0$. The proof of this fact is on the same lines as that of Hanner's Proposition C and hence is omitted.

In the definition of g_a^b in (3.1), we shall choose this Δ_0. Now if for $0 < b < u$, we consider $g_0^b = E(0, b] \int_{A'}^{B'} T_h E(\Delta_0) f^{(1)} dh$ where $A' < -u$, $B' > b$, then by definition of g_0^b, $g_0^b = E(0, b] g_0^u$. Also $g_0^u = g_0^b + g_b^u$ with $g_b^u \perp g_0^b$. Hence $\|g_0^u\|^2 = \|g_0^b\|^2 + \|g_b^u\|^2$. If $g_0^b = 0$, we get $\tau(u-b) = \tau u$. The τ in the left hand side is positive by the choice of Δ_0. This however leads to a contradiction since $b < u$. For $b > u$ $\|g_0^b\|^2 \geqslant \|g_0^u\|^2 > 0$. If $b' < 0$, let $\beta = -b'$ and consider $T_\beta g_{b'}^0 = g_0^\beta$. But $\|g_0^\beta\|^2 \neq 0$ by previous arguments and hence $\|g_{b'}^0\|^2 \neq 0$. It is easy to deduce that $g_c^d \neq 0$ for all intervals $(c, d]$, c, d real. Thus by this choice of Δ_0, we get $\|g_a^b\|^2 = \tau(b-a)$ where τ is independent of a, b and $\tau > 0$. Without loss of generality we assume $\|g_a^b\|^2 = (b-a)$.

Our definition of g_a^b differs from Hanner's definition of $Z(I_a^b)$ in the following respect. Hanner defines $Z(I_a^b)=E(a, b] \int_A^B T_h\, zdh$, where z is any non-zero element of $L_2(x; 0, u)$. We choose a particular z of the form $E(\Delta_0)f^{(1)}$, $f^{(1)}$ being an element with maximal spectral type. The advantage of this consists in the fact that by bringing $f^{(1)}$ explicitly in the evidence in the construction of g_a^b we are able to prove that the maximal spectral type is Lebesgue.

Lemma 3.1. *The maximal spectral type of A is equivalent to μ, the Lebesgue measure.*

Proof : It suffices to show that $\rho_{f^{(1)}} \equiv \mu$ where $f^{(1)}$ is the element introduced above. Now for $I = (a, b]$, $\rho_{g_a^b}$ the spectral measure of g_a^b is given by $\rho_{g_a^b}(\Delta) = \mu^I(\Delta)$ for every finite interval Δ, where the measure μ^I is defined as $\mu^I(S) = \mu(I \cap S)$ for every measurable set S on the real line. Hence $\rho_{g_a^b} = \mu^I$. Since $f^{(1)}$ is an element of maximal spectral type, $\rho_{f^{(1)}} \gg \rho_{g_a^b} = \mu^I$ for every I. Hence $\mu \ll \rho_{f^{(1)}}$. An application of Lemma 1 (Section 2) now comyletes the proof.

Let us define $H_1(t) = \mathfrak{S}\{g_a^b, -\infty < a \leqslant b \leqslant t\}$ for each t, and $H_1 = H_1(\infty)$. With P_{H_1} denoting the projection onto H_1, let $x_t^{(1)} = P_{H_1} x_t$; then $x_t^{(1)}$ is weakly stationary, continuous in q.m. and purely non-deterministic. Further defining $y_t = x_t - x_t^{(1)}$ we have

$$L_2(x; t) = L_2(x^{(1)}; t) \oplus L_2(y; t) \quad \text{for every } t \text{ and} \tag{3.3}$$

$$L_2(x^{(1)}; +\infty) \perp L_2(y; +\infty)$$

Lemma 3.2. (i) $L_2(x^{(1)}; a, b) = \mathfrak{S}\{E(\alpha, \beta]g_a^b, a < \alpha \leqslant \beta \leqslant b\}$
$= H_1(a, b)$ $(a, b$ real)

(ii) *For all* a, b, $L_2(x^{(1)}; a, b)$ *is the cyclic subspace generated by* g_a^b.

(iii) A_I *has the spectral type* μ^I, *where for any interval* I, A_I *is the restriction of* A *to* $L_2(x^{(1)}; I)$.

Proof : For $a < t \leqslant b$, since $x_t \perp H_1 \ominus H_1(t)$ and by (3.3), $x_t^{(1)} - P_{L_2(x^{(1)}; a)} x_t^{(1)} = P_{H_1(t)} x_t - P_{L_2(x; a)} P_{H_1(t)} x_t$. But

$P_{L_2(x; a)} P_{H_1(t)} = P_{L_2(x; a)} (P_{H_1(a)} + P_{H_1(t) \ominus H_1(a)}) = P_{L_2(x; a)} P_{H_1(a)}$

$= P_{H_1(a)}$. Hence $x_t^{(1)} - P_{L_2(x^{(1)}; a)} x_t^{(1)} = (P_{H_1(t)} - P_{H_1(a)}) x_t$ and $L_2(x^{(1)}; a, b) \subseteq H_1(a, b)$. The other side of the equality is obvious since for all α, β, $E(\alpha, \beta]g_a^b \epsilon L_2(x; a, b)$ and $E(\alpha, \beta)g_a^b \perp L_2(y; \infty)$. This proves (i).

For any interval Δ, $E(\Delta)g_a^b = E(\Delta \cap (a, b])g_a^b = 0$ if Δ and $(a, b]$ are disjoint and $E(\Delta)g_a^b = E(\alpha, \beta]g_a^b$ if $\Delta \cap (a, b] = (\alpha, \beta]$. This fact and (i) together imply that $L_2(x^{(1)}; a, b) = \mathfrak{S}\{E(\Delta)g_a^b$, Δ any finite interval of the real line], which is assertion (ii) of the lemma. It is now obvious that A is reduced by $L_2(x^{(1)}; a, b)$. Conclusion (iii) follows from this and the fact that $\rho_{g_a^b} = \mu^I$.

Lemma 3.3. $L_2(x^{(1)})$ *reduces* A *and* $A^{(1)}$, *the restriction of* A *to* $L_2(x^{(1)})$ *is a cyclic operator of Lebesgue type.*

Proof : If $\omega \epsilon DA \cap L_2(x^{(1)})$ (which is non-empty) where D_A denotes the domain of A, then for $n = 0, \pm 1, \ldots$.

$$(3.2) \quad E(n-1, n]A\omega = AE(n-1, n]\omega = AP_{L_2(x^{(1)};\, n-1,\, n)} \omega \epsilon L_2(x^{(1)}: n-1, n)$$

Hence $A\omega \epsilon L_2(x^{(1)})$ i.e., A is reduced by $L_2(x^{(1)})$. Now if I_j $(j = 1, 2, \ldots)$ are disjoint finite intervals whose union is the real line we have $L_2(x^{(1)}) = \sum_j \oplus L_2(x^{(1)}; I_j)$. Hence

$$A^{(1)} = \sum_{j=1}^{\infty} A_{I_j}^{(1)} P_{L_2(x^{(1)}, I_j)}.$$

Each $A_{I_j}^{(1)}$ is cyclic and the corresponding spectral types μ_2 are independent (see definition given in Section 2). Hence ([10], p. 152) it follows that $A^{(1)}$ is cyclic with spectral type equivalent to μ. The following lemma gives multiplicity conditions under which a stationary process has Karhunen representation.

Lemma 3.4. *Let $\{u_t, -\infty < t < +\infty\}$ be a weakly stationary purely non-deterministic process continuous in q.m., and B be the self-adjoint operator of the process given by the resolution of the identity $\{\beta(t), -\infty < t < +\infty\}$ where $\beta(t)$ denotes the projection of $L_2(u)$ onto $L_2(u; t)$. If B is cyclic and has Lebesgue spectral type then $u_t = \int_{-\infty}^{t} G(u-t)d\xi(u)$ where $\{\xi(u), -\infty < u < +\infty\}$ is a process with stationary orthogonal increments such that $L_2(u; t) = \mathcal{S}\{\xi(\Delta), \Delta$ a sub-interval of $(-\infty, t]\} = L_2(\xi; t)$.*

Proof : Suppose that B is cyclic and let f be a generating element of A. Let $\rho_f^{(h)}(S) = \| \beta(S)T_h f \|^2$ for every measurable S in $(-\infty, t]$. Clearly by the generating property of f

$$T_h f = \int_{-\infty}^{+\infty} r(h, u)d\beta(u)f \text{ giving } \beta(\Delta)T_h f = \int_\Delta r(h, u)d\beta(u)f$$

for every finite interval Δ. Hence

$$d\beta(u)T_h f = r(h, u)d\beta(u)f. \tag{3.4}$$

This implies that $\rho_f^{(h)} \ll \rho_f$ for all h and $r(h, u) = \left[\dfrac{d\rho_{f(u)}^{(h)}}{d\rho_f}\right]^{\frac{1}{2}}$

where $\dfrac{d\rho_f^{(h)}}{d\rho_f}$ denotes the Radon-Nikodym derivative of $\rho_f^{(h)}$ with respect to ρ_f. Again, from the fact that f is a generating element and $u_0 \epsilon L_2(u)$ one has

$$u_0 = \int_{-\infty}^{0} F(0, u)d\beta(u)f \text{ with } \int_{-\infty}^{0} | F(0, u) |^2 d\rho_f(u) < \infty.$$

and therefore

$$u_t = T_t u_0 = \int_{-\infty}^{t} F(0, u-t)d\beta(u)T_t f \tag{3.5}$$

Now since $\rho_f \equiv \mu$, we can define an orthogonal random set function ξ by

$$\xi(\Delta) = \int_\Delta \left[\frac{d\rho_f}{d\mu}(u)\right]^{-\frac{1}{2}} d\beta(u)f \tag{3.6}$$

having Lebesgue measure as its variance function. Inverting (3.6) we get

$$d\beta(u)f = \left[\frac{d\rho_f}{d\mu}(u)\right]^{\frac{1}{2}} d\xi(u) \tag{3.7}$$

From (3.4) and (3.7), we can write (3.5) as

$$u_t = \int_{-\infty}^{t} F(0, u-t) \left[\frac{d\rho_f^{(t)}}{d\mu}\right]^{\frac{1}{2}} d\xi(u) \tag{3.8}$$

However, in view of (2.4)

$$\rho_f^{(t)}(\Delta) = \rho_f(\Delta - t) \text{ and thus } \frac{d\rho_f^{(t)}}{d\mu}(u) = \frac{d\rho_f}{d\mu}(u-t).$$

If we define $G(u-t) = F(0, u-t)\left[\frac{d\rho_f}{d\mu}(u-t)\right]^{\frac{1}{2}}$, (3.8) has the form (2.1), since by (2.3), (3.6) and (3.7) it follows that $L_2(u; t) = L_2(z; t) = L_2(\xi; t)$. Now in the above Lemma consider $x_t^{(1)}$ as u_t and $A^{(1)}$ as B then we obtain

$$x_t^{(1)} = \int_{-\infty}^{t} G_1(u-t)\, d\xi_1(u),$$

where $\xi_1(t)$-process has stationary orthogonal increments with $L_2(x^{(1)}; t) = L_2(\xi_1; t)$. Consider now the y_t-process. Our object is to show that $y_t = 0$ for each t[i.e., $L_2(y; t) = \{0\}$ for every t]. If this is not the case, since y_t is a weakly stationary, purely non-deterministic process continuous in q.m. we get by repeating previous arguments [for x_t] that $y_t = y_t^{(1)} + z_t$ where $L_2(z) \perp L_2(y^{(1)})$ and $y_t^{(1)} = \int_{-\infty}^{t} G_2(u-t)\, d\xi_2(u)$ with $L_2(y^{(1)}; t) = L_2(\xi_2; t)$.

Thus we have

$$\begin{cases} x_t = \int_{-\infty}^{t} G_1(u-t) d\xi_1(u) + \int_{-\infty}^{t} G_2(u-t) d\xi_2(u) + z_t, \text{ where} \\ L_2(x; t) = L_2(x^{(1)}; t) \oplus L_2(y^{(1)}; t) \oplus L_2(z; t) \\ \qquad = L_2(\xi_1; t) \oplus L_2(\xi_2; t) \oplus L_2(z; t) \end{cases} \tag{3.9}$$

and

$$L_2(x; 0) = \mathfrak{S}\{x_t, t \leqslant 0\}. \tag{3.10}$$

These relations enable us to obtain a contradiction to $y_t \neq 0$, essentially using the last arguments of Hanner's Proposition D (pp. 172–173), by construction an element in $L_2(x; 0)$ of the form,

$$W_s = \int_s^0 \overline{G_2(u-s)} d\xi_1(u) - \int_s^0 \overline{G_1(u-s)} d\xi_2(u).$$

Thus $y_t = 0$, $A = A^{(1)}$ and the theorem is proved.

As a corollary to the above result we can obtain the Karhunen representation.

Corollary. A weakly stationary continuous in q.m. continuous parameter purely non-deterministric process has the Karhunen representation.

The proof follows immediately from Theorem 1 and Lemma 3.4.

References

[1] Cramér, H. (1961). "On Some Classes of Non-Stationary Processes," *Proc. Fourth Berkeley Symp. Math. Statist. Prob.*, **2**, 57–77.

[2] Cramér, H. (1961). "On the Structure of Purely Non-Deterministic Processes," *Arkiv For Math.*, **4**, 249–266.

[3] Cramér, H. (1964). "Stochastic Processes as Curves in Hilbert Space," *Theoriya Veryatnosteii i ee Promeneniya IX*, **2**, 1964.

[4] Doob, J. L. (1953). *Stochastic Processes*, John Wiley and Sons, New York.

[5] Hanner, O. (1950). "Deterministic and Non-Deterministic Processes," *Arkiv For Math.*, **1**, 161–177.

[6] Hida, T. (1960). "Canonical Representations of Gaussian Processes and Their Applications," *Mem. Coll. Sci. Kyoto, Ser. A.*, **33**, 109–155.

[7] Kallianpur, G. and Mandrekar, V. (1965). "Multiplicity and Representation Theory of Purely Non-Deterministic Stochastic Processes," *Theoriya Veryatnosteii i ee Promeneniya*, 1965.

[8] Karhunen, K. (1947). "Uber lineare Methoden in der Wahrscheinlichkeitsrechnung," *Ann. Ac. Sci. Fennicae, Ser. A*, I, No. 37.

[9] Masani, P. and Robertson, J. (1962). "The Time-Domain Analysis of a Continuous Parameter Weakly Stationary Stochastic Process," *Pacific J. Math.*, **12**, 1361-1378.

[10] Plessner, A. I. and Rohlin, V. A. (1946). "Spectral Theory of Linear Operators II," (Russian) *Uspehi Mat. Nauk I*, No. 1, 71–191.

[11] Robertson, J. (1963). "Multivariate Continuous Parameter Weakly Stationary Stochastic Processes," Thesis, Indiana University, June.

[12] Saks, S. (1938) *Theory of the Integral* (translation by L. C. Young), New York.

(Received Nov. 12, 1965.)

THEORY OF PROBABILITY AND ITS APPLICATIONS

Volume X **Number 4**

1965

10

MULTIPLICITY AND REPRESENTATION THEORY OF PURELY NON-DETERMINISTIC STOCHASTIC PROCESSES [1]

G. KALLIANPUR, V. MANDREKAR

1. Introduction

In this paper we study the multiplicity theory of a wide class of purely non-deterministic weakly stationary processes and show how this theory provides a natural means of obtaining representations of continuous parameter processes that are extensions of the well known result due to K. Karhunen [10]. Our work can be described as a unified time domain analysis that applies equally to finite dimensional and infinite dimensional stationary processes. The earliest time domain analysis of a (univariate) continuous parameter weakly stationary process was made by O. Hanner in a remarkably original paper [6]. More recently, in the light of the extensive development of multidimensional stationary processes, it has appeared desirable to separate time domain studies from the spectral, and consequently, interest in the former has revived. As an example, we mention the paper of P. Masani and J. Robertson [11] whose approach makes extensive use of the Cayley transform associated with the unitary group of the process. The extension of this method to finite dimensional stationary processes has been carried out by J. Robertson in his thesis [14]. The earlier work of E. G. Gladyshev also belongs to the same order of ideas [5]. Hanner's paper, nevertheless, has remained an isolated piece of work and his method has apparently given the impression of being *ad hoc*. As a matter of fact, as we have shown elsewhere [9], Hanner's ideas reveal an intimate connection with multiplicity theory. Thus the generalization of Hanner's approach to multidimensional (even infinite dimensional) stationary processes is to be sought in the development of the multiplicity theory of the process, i.e., in the study of the self-adjoint operator A of the process and its spectral types. This is one of the central problems discussed in this paper, in Sections 4, 5, and 6.

In recent years a theory of representation of purely non-deterministic (possibly non-stationary) processes has been introduced by H. Cramér and also by T. Hida ([1], [2], [3], [7]). Following the technique of the latter author it is easy to extend the main representation theorem of [7] to the processes considered by us. This is done in Sections 2 and 3. Our purpose in doing so it to compare the representation theorem of the Hida–Cramér theory (Theorem 2.2 of this paper) with the result of Section 5 which is essentially independent of Sections 2 and 3. The generalized Hanner approach leads naturally to a definition of multiplicity which is seen to be identical with the con-

[1] Research supported by U. S. Army Research Office – Durham – under Grant DA–ARO–D–31–124–G562.

cept of the multiplicity of the process introduced by Hida. Indeed it is shown that every spectral type belonging to A has this multiplicity. Further discussion of this question is deferred to Section 5. Section 6 brings to light the natural role of multiplicity as a generalization of the rank of a stationary finite dimensional process.

We consider stochastic processes of the following kind.

Let Φ be a Hausdorff space satisfying the second countability axiom but otherwise arbitrary. We shall say that $\mathbf{x}_t$ $(-\infty < t < \infty)$ is a stochastic process on Φ if for each φ in Φ, $\mathbf{x}_t(\varphi)$ is a complex-valued random variable with mean zero and $\mathbf{E}|\mathbf{x}_t(\varphi)|^2$ finite. The process $\{\mathbf{x}_t\}$ $(-\infty < t < \infty)$ on Φ is called weakly stationary (or briefly, stationary) if for all φ, ψ in Φ and arbitrary real numbers s, t, and τ we have

$$\mathbf{E}[\mathbf{x}_{t+\tau}(\varphi)\overline{\mathbf{x}_{s+\tau}(\psi)}] = \mathbf{E}[\mathbf{x}_t(\varphi)\overline{\mathbf{x}_s(\psi)}].$$

The covariance function $\mathbf{E}[\mathbf{x}_t(\varphi)\overline{\mathbf{x}_s(\psi)}]$ of the process depends on $t-s$, φ and ψ. The definition of a discrete parameter process $\{\mathbf{x}_n\}$ is similarly given. It should be noted that the stationarity considered here is a temporal one and does not involve Φ. Nevertheless, it is sufficiently general and useful for our purpose since it includes as special cases many stationary random processes of practical interest. For instance, if Φ is a q-dimensional euclidean (or unitary) space and $\mathbf{x}_t(\varphi)$ is linear with respect to φ for each t, then the $\mathbf{x}_t$-process can be regarded as a q-vector stationary process (see [15]); if Φ is an infinite dimensional locally convex linear space and $\mathbf{x}_t(\varphi)$ is again supposed linear in φ (with probability one), then $\mathbf{x}_t$ is a weak stochastic process on Φ. On the other hand, stationary processes $\mathbf{x}_t$ as defined above include those that are not linear in φ (indeed Φ itself need not be a linear space). Such processes can serve as useful models for certain problems in meteorology (e.g. see [8]).

Associated with the $\mathbf{x}_t$ process (not assumed to be stationary) are the following spaces:

(a) the (Hilbert) space of the process $H(\mathbf{x})$, defined to be $\mathfrak{S}[\mathbf{x}_t(\varphi), t \in T, \varphi \in \Phi]$, the subspace of $L_2(\Omega, \mathbf{P})$ generated by the family of random variables $\mathbf{x}_t(\varphi)$ as t and φ vary respectively over T and Φ;

(b) the subspace $H(\mathbf{x}; t)$ of $H(\mathbf{x})$ given by $H(\mathbf{x}; t) = \mathfrak{S}[\mathbf{x}_\tau(\varphi), \tau \leqq t, \text{ and } \varphi \in \Phi]$ for every real t.

We say that $\mathbf{x}_t$ on Φ is purely non-deterministic if $H(\mathbf{x}; -\infty)$, the intersection of the subspaces $H(\mathbf{x}; t)$ for all $t \in T$, is trivial.

The process $\mathbf{x}_t$ is said to be deterministic if, for each t, $H(\mathbf{x}; t) = H(\mathbf{x}; -\infty)$.

In the concluding part of the paper we consider in greater detail Hilbert space valued processes since they are, perhaps, mathematically the simplest examples of infinite dimensional processes. If the process $\{\mathbf{x}_t\}$ is a weak process on the Hilbert space Φ, its representation is already given by Theorem 5.1. In Sections 7, 8 and 9 we make the stronger assumption that, for each t, $\mathbf{x}_t$ is a random element in the dual of a separable Hilbert space Φ, satisfying the further requirement that $\mathbf{E}||\mathbf{x}_t||^2$ is finite. Strengthened versions (involving random, Hilbert space valued integrals) of the representation theorems of Sections 2 and 5 are obtained in Section 9.

The problem of relating the multiplicity of a stationary process with spectral theory and of actually determining the multiplicity in concrete instances will be studied in a later paper.

The Hida — Cramér Theory

2. Representations of Stochastic Processes on Φ

Although our main interest will be in the study of continuous parameter weakly stationary processes we begin by considering representations of arbitrary second order purely non-deterministic processes $\mathbf{x}_t(\varphi)$ on Φ. It can be easily seen that the results stated in this section contain as special cases those of H. Cramér [2] and of T. Hida [7] (if Gaussian assumptions are made). They will, however, be stated without proof since they are proved by following essentially the method of the latter author. Our only reason for including them here is for the purpose of relating the representation and the definition of multiplicity given in this section with similar concepts for stationary processes obtained in Sections 5 and 6. For the sake of completeness we begin with the following 'Wold decomposition' of $\mathbf{x}_t$.

Proposition 2.1. *If* $\{\mathbf{x}_t, t \in T\}$ *is a stochastic process on* Φ, *then*

$$\mathbf{x}_t(\varphi) = \mathbf{x}_t^{(1)}(\varphi) + \mathbf{x}_t^{(2)}(\varphi) \text{ for each } \varphi \in \Phi, \text{ where}$$

(i) $\{\mathbf{x}_t^{(1)}\}$ *is a deterministic and* $\{\mathbf{x}_t^{(2)}\}$ *a purely non-deterministic process on* Φ; *and*

(ii) $H(\mathbf{x}^{(1)})$ *is orthogonal to* $H(\mathbf{x}^{(2)})$.

Observe that the topological assumptions concerning Φ in no way enter into the proof of this result.

Writing $J = T \times \Phi$, $\alpha = (t, \varphi)$, $\beta = (s, \psi)(\alpha, \beta \in J)$ define $K(\alpha, \beta) = \mathbf{E}[\mathbf{x}_t(\varphi)\overline{\mathbf{x}_s(\psi)}]$. Then, clearly, K is a covariance function on $J \times J$. Let us denote by $H(K)$ the reproducing kernel Hilbert-space of functions defined on J whose reproducing kernel is K. Let $H(K; t) = \mathfrak{S}[k(\cdot, \alpha). \alpha \in J_t]$, i.e. the subspace of $H(K)$ generated by $\{K(\cdot, \alpha), \alpha \in J_t\}$ where $J_t = \{(u, \varphi), u \leqq t \text{ and } \varphi \in \Phi\}$. It is well known that there exists an isometry, which we denote by V, from $H(K)$ to $H(\mathbf{x})$ taking functions $K(\cdot, \alpha)$ into the random variables $\mathbf{x}_t(\varphi)$ and such that $VH(K; t) = H(\mathbf{x}; t)$.

The following assumptions (A) will be basic for our purpose:

(A.1) The space $H(\mathbf{x})$ is separable;

(A.2) $H(\mathbf{x}; -\infty) = \{0\}$.

Condition (A.2) is equivalent to the process $\{\mathbf{x}_t\}$ being purely non-deterministic, while the following lemma gives sufficient conditions on the random variables $\mathbf{x}_t(\varphi)$ for (A.1) to hold.

Lemma 2.1 *Suppose that for each* t $(-\infty < t < \infty)$

(i) $\mathbf{x}_t(\varphi)$ *is continuous in quadratic mean relative to the topology of* Φ, *and*

(ii) *the random variables* $\mathbf{x}_{t-0}(\varphi)$ *and* $\mathbf{x}_{t+0}(\varphi)$ *exist (in quadratic mean) for each* $\varphi \in \Phi$.

Then $H(\mathbf{x})$ *is separable.*

This result is a generalization of a lemma due to Cramér [2] and takes as its starting point the fact, proved there, that for each φ, the set of all discontinuity points of the one-dimensional process $\{\mathbf{x}_t(\varphi), t \in T\}$ is at most denumerable.

Proof. It suffices to prove that there exists a countable dense set H_0 in $\{\mathbf{x}_t(\varphi), t \in T, \varphi \in \Phi\}$. Let $\Phi_0 = \{\varphi_k\}$ be a countable, everywhere dense set in Φ. The set D_k of discontinuities of the one-dimensional process $\mathbf{x}_t(\varphi_k)$ is at most denumerable. We shall show that

$$H_0 = \{\mathbf{x}_u(\varphi_k), \varphi_k \in \Phi_0, u \in \bigcup_k D_k, \text{ or } u \text{ rational}\}$$

is a dense subset of $\{\mathbf{x}_t(\varphi), t \in T, \varphi \in \Phi\}$. Since H_0 has at most denumerable elements, the proof of the lemma will be complete once we establish the preceding assertion. For τ and φ fixed consider an element $\mathbf{x}_\tau(\varphi)$ and let ε be an arbitrary positive number. By (i), there exists a $\varphi_k \in \Phi_0$ such that $\mathbf{E}|\mathbf{x}_\tau(\varphi)-\mathbf{x}_\tau(\varphi_k)|^2 < \varepsilon/2$. If τ is a discontinuity point of the one-dimensional process $\{\mathbf{x}_t(\varphi_k)\}(t \in T)$, then since $\mathbf{x}_\tau(\varphi_k) \in H_0$, the proof will be complete. On the other hand, if τ is not a discontinuity point of $\{\mathbf{x}_t(\varphi_k)\}$ then there exists a rational number r such that $\mathbf{E}|\mathbf{x}_\tau(\varphi_k)-\mathbf{x}_r(\varphi_k)|^2 < \varepsilon/2$. This implies that $\mathbf{E}|\mathbf{x}_t(\varphi)-\mathbf{x}_r(\varphi_k)|^2 < 2\varepsilon$ and since $\mathbf{x}_r(\varphi_k) \in H_0$ the proof is complete.

It might be remarked in passing that if $\mathbf{x}(t) = [x_1(t), \cdots, x_q(t)]$ is a q-dimensional process such that the random variables $x_i(t-0)$, $x_i(t+0)$ exist for $i = 1, \cdots, q$, the conditions of Lemma 2.1 are fulfilled if we take Φ to be q-dimensional Euclidean space and define $\mathbf{x}_t(\varphi) = \sum_{i=1}^q x_i(t)\varphi_i$, φ being the vector $(\varphi_1, \cdots, \varphi_q)$. In other words, Lemma 1 of [2] is a special case of Lemma 2.1. In view of the isometry V between $H(K)$ and $H(\mathbf{x})$, the assumptions (A) are equivalent to corresponding assumptions concerning the spaces $H(K)$ and $H(K; -\infty)$. Let us introduce the spaces

$$H^*(K; t) = \bigcap_{n=1}^{\infty} H(K; t+1/n).$$

We then have $H^*(K; -\infty) = \{0\}$ and $H(K) = H^*(K; \infty)$, the smallest subspace containing all the $H^*(K; t)$.

The spaces $H^*(\mathbf{x}; t)$ are similarly introduced. Let $\hat{E}(t)$ denote the projection operator from $H(K)$ onto $H^*(K; t)$ and $E(t)$ the projection from $H(\mathbf{x})$ onto $H^*(\mathbf{x}; t)$. It then follows easily that the families $\{\hat{E}(t), -\infty < t < \infty\}$ and $\{E(t), -\infty < t < \infty\}$ are right continuous resolutions of the identity in the respective Hilbert spaces $H(K)$ and $H(\mathbf{x})$.

The two results which follow are proved as in [7]. We omit the proof, which is essentially based on the Hellinger–Hahn decomposition of the self-adjoint operators $\hat{A}$ and A defined respectively on $H(K)$ and $H(\mathbf{x})$ by the resolutions of the identity introduced above. Observe that while the parameter set T of the process is always either the real line or the set of all integers, the resolution of the identity $\{E(t)\}$ determined by the process is defined for all real t.

Theorem 2.1. *Let assumptions* (A) *be satisfied. Then each element* $K(\cdot, \alpha)$ (α *in* J) *of* $H(K)$ *has the following representation*:

$$K(\cdot, \alpha) = \sum_{n=1}^{M_0} \int_{-\infty}^{t} G_n(\alpha, u)\, dE(u) f^{(n)} + \sum_{t_j \leq t} \sum_{l=1}^{M_j} a_{jl}(\alpha) g_{jl},$$

where the symbols introduced have the following meaning:

(a) $\{f^{(n)}\}$ *is a sequence of elements in* $H(K)$ *with the following properties*:

(i) *The inner product* $(E(\Delta_1)f^{(n)}, E(\Delta_2)f^{(m)}) = 0$ *whenever* Δ_1 *and* Δ_2 *are disjoint intervals or* $m \neq n$.

(ii) *For each* n, $G_n(\alpha, \cdot) \in L_2(\rho_n)$ *where* $\rho_n(\Delta) = \|E(\Delta)f^{(n)}\|^2$,

$$\sum_{n=1}^{M_0} \int |G_n(\alpha, u)|^2 d\rho_n(u) < \infty \text{ and } \rho_1 \gg \rho_2 \gg \cdots \text{ etc.}$$

and

(b) *For each* $j = 1, 2, \cdots$ *the sequence* $\{g_{jl}\}$ $(l = 1, \cdots, M_j)$ *are the eigenvectors of the self-adjoint operator* $\hat{A}$ *corresponding to the eigenvalue* t_j *and such that*

$$\sum_{j=1}^{\infty} \sum_{l=1}^{M_j} |a_{jl}(\alpha)|^2 \|g_{jl}\|^2 < \infty.$$

The elements $\{g_{jl}\}$, *further, form a complete orthonormal system in the subspace* $[E(t_j) - E(t_j - 0)]H(K)$ *with*

$$(g_{jl}, g_{im}) = 0 \quad \textit{if} \quad i \neq j.$$

For $\alpha = (t, \varphi)$ writing $\Gamma_n(\varphi; t, u) = G_n(\alpha, u)$ and $b_{jl}(\varphi; t) = a_{jl}(\alpha)$ we obtain the following representation for the process $\mathbf{x}_t$ on Φ.

Theorem 2.2. *If conditions* (A) *hold we have the following representation for* $\mathbf{x}_t$. *For each t and* φ, *with probability one*

$$\mathbf{x}_t(\varphi) = \sum_{n=1}^{M_0} \int_{-\infty}^{t} \Gamma_n(\varphi; t, u) dz_n(u) + \sum_{t_j \leqq t} \sum_{l=1}^{M_j} b_{jl}(\varphi; t)\xi_{jl}, \tag{2.1}$$

where

(a) $z_n(u)$ $(-\infty < u < \infty)$ *for each n, is an orthogonal random function with the further property that* $\mathbf{E}[z_m(u)\overline{z_n(v)}] = 0$ *for* $m \neq n$ *and* $\mathbf{E}|z_n(\Delta)|^2 = \rho_n(\Delta)$. *Further, the functions* Γ_n *and* ρ_n *satisfy the conditions stated in the preceding theorem.*

(b) *The random variables* ξ_{jl} $(l = 1, \cdots, M_j$ *and* $j = 1, 2, \cdots)$ *are mutually orthogonal with*

$$\sum_{j=1}^{\infty} \sum_{l=1}^{M_j} \sigma_{jl}^2 |b_{jl}(\varphi; t)|^2 \textit{ finite, where } \sigma_{jl}^2 = \mathbf{E}|\xi_{jl}|^2.$$

Definition. The cardinal number $M = \max\,[M_0, \sup_j M_j]$ is called the multiplicity of the stochastic process $\mathbf{x}_t$ on Φ.

It is to be noted that M can be infinite, in which case of course M is aleph null. The corresponding series that occur in our work are then to be treated as infinite series.

If T is the set of integers it is easy to see that M_0 is necessarily zero and $t_j = j$.

3. Canonical and Proper Canonical Representations.

The representation obtained in Theorem 2.2 has the following property. For $s < t$,

$$E(s)\mathbf{x}_t(\varphi) = \sum_{1}^{M_0} \int_{-\infty}^{s} \Gamma_n(\varphi; t, u) dz_n(u) + \sum_{t_j \leqq s} \sum_{l=1}^{M_j} b_{jl}(\varphi; t)\xi_{jl}. \tag{3.1}$$

A representation satisfying (3.1) will be called canonical. From the form of (2.1), it follows that $H(\mathbf{x}; t) \subset \mathfrak{S}[H(\mathbf{z}; t) \cup H(\xi; t)]$ where $H(\xi; t) = \mathfrak{S}[\xi_{jl}, l = 1, 2, \cdots, M_j, t_j \leqq t]$. For applications of the theory, however, it is more useful to consider canonical representations for which

$$\mathfrak{S}[H(\mathbf{z}; t) \cup H(\xi; t)] = H(\mathbf{x}; t) \text{ for all } t. \tag{3.2}$$

Following Hida, we refer to a representation with property (3.2) as proper canonical. In [7], Hida was concerned with proper canonical representations of multiplicity one. In order to be able to discuss the multiplicity theory of the more general processes considered by us it is necessary to establish the existence of a proper canonical representation of arbitrary multiplicity equivalent to the one given by Theorem 2.2. This we do in Theorem 3.1.

For the representation of Theorem 2.2. define the processes $B_n(u)$ as follows:

(i) If both M_0 and $\sup_j M_j$ are infinite, then

$$B_n(u) = z_n(u) + \sum_{t_j \leqq u} \xi_{jn} \qquad \text{for } n = 1, 2, \cdots \text{ ad inf.}$$

(ii) If M_0 is finite and $M_0 \leqq \sup_j M_j$, let

$$B_n(u) = \begin{cases} z_n(u) + \sum_{t_j \leqq u} \xi_{jn} & \text{for } n = 1, 2, \cdots, M_0, \\ \sum_{t_j \leqq u} \xi_{jn} & \text{for } M_0 < n \leqq \sup_j M_j. \end{cases}$$

(iii) In the remaining cases define

$$B_n(u) = \begin{cases} z_n(u) + \sum_{t_j \leqq u} \xi_{jn} & \text{for } n = 1, 2, \cdots, \sup_j M_j, \\ z_n(u) & \text{for } \sup_j M_j \leqq n \leqq M_0. \end{cases}$$

With the above notation we rewrite (2.1) as

$$\mathbf{x}_t(\varphi) = \sum_{n=1}^{M} \int_{-\infty}^{t} G_n(\varphi; t, u) dB_n(u), \tag{3.3}$$

where $M = \max(\sup_j M_j, M_0)$. What the functions G_n stand for is clear from the context.

Also, $H(B; t) = \mathfrak{S}[(\mathbf{z}; t) \cup H(\xi, t)]$. A representation of the form (3.3) will be denoted by $\{G_n, B_n\}_1^M$.

Theorem 3.3. *Let $\{G_n, B_n\}_1^M$ be a canonical representation. Then there exists a proper canonical representation $\{\tilde{G}_n, \tilde{B}_n\}_1^M$ such that for every (φ, t),*

$$\mathbf{x}_t(\varphi) = \sum_{n=1}^{M} \int_{-\infty}^{t} \tilde{G}_n(\varphi; t, u) d\tilde{B}_n(u)$$

with probability one.

PROOF. Let $\rho_n(\Delta) = \mathbf{E}|B_n(\Delta)|^2$. For each φ, t, and every measurable subset S of $(-\infty, t]$, define the measure $\mu_{(t,\varphi)}^{(n)}(S) = \int_S |G_n(\varphi; t, u)|^2 d\rho_n(u)$. Then for each n, the measure $\mu^{(n)}$ given by $\mu^{(n)}(S) = \bigvee_{(t,\varphi)} \mu_{(t,\varphi)}^{(n)}(S)$ (see [7]) is absolutely continuous with respect to ρ_n. Let $N_n = \{u | (d\mu^n/d\rho_n)(u) > 0\}$ and $\tilde{B}_n(S)$ be the random set function with variance function $\tilde{\rho}_n$ and defined by the stochastic integral $\tilde{B}_n(S) = \int_S I_{N_n}(u)\, dB_n(u)$. Further, set $\tilde{G}_n(\varphi; t, u) = G_n(\varphi; t, u)$ for all φ, t and u and consider the sum, $\mathbf{y}_t(\varphi) = \sum_1^M \int_{-\infty}^t \tilde{G}_n(\varphi; t, u) d\tilde{B}_n(u)$. If M is infinite, the right-hand side series is easily seen to be convergent in quadratic mean. From the fact that

$$\frac{d\mu_{(t,\varphi)}^{(n)}}{d\mu^{(n)}}(u) \frac{d\mu^{(n)}}{d\rho_n}(u) = |G_n(\varphi; t, u)|^2 \quad \text{for each } t, \varphi \text{ and } n,$$

it is easy to deduce that

$$\int_{-\infty}^{t} [1 - I_{N_n}(u)]^2 |G_n(\varphi; t, u)|^2 d\rho_n(u) = 0.$$

Thus for all t, φ,

$$\mathbf{E}|\mathbf{x}_t(\varphi) - \mathbf{y}_t(\varphi)|^2 = \sum_{n=1}^{M} \int_{-\infty}^{t} |1 - I_{N_n}(u)|^2 |G_n(\varphi; t, u)|^2 d\rho_n(u) = 0. \tag{3.4}$$

For (3.4) we find that for every t and φ

(3.5) $$\mathbf{x}_t(\varphi) = \mathbf{y}_t(\varphi) \text{ with probability one}$$

and that

(3.6) $$H(\mathbf{x}; s) = H(\mathbf{y}; s) \text{ for all } s \in T.$$

A similar argument also yields that, for every measurable subset S of $(-\infty, t]$,

(3.7) $$\sum_{n=1}^{M} \int_S |G_n(\varphi; t, u)|^2 \, d\rho_n(u) = \sum_{n=1}^{M} \int_S |\tilde{G}_n(\varphi; t, u)|^2 \, d\tilde{\rho}_n(u).$$

Since $\mathbf{E}[\tilde{B}_n(\Delta)\tilde{B}_m(\Delta')] = 0$ for $\Delta \neq \Delta'$ or $n \neq m$, we have

$$H(\tilde{\mathbf{B}}; t) = \sum_1^M \oplus H(\tilde{B}_n; t).$$

Therefore, to establish that $\{\tilde{G}_n, \tilde{B}_n\}$ is proper canonical, it suffices to show that $H(\tilde{B}_n; t) \subset H(\mathbf{x}; t)$ for all n and t. Now suppose that there is a t and an n, such that

$$H(\tilde{B}_n;\ t) \nsubseteq H(\mathbf{x}; t).$$

Then we can find a non-zero element $z \in H(\tilde{B}_n; t)$ which is orthogonal to $H(\mathbf{x}; t)$. Let $s'' \in T$ be arbitrary and $s \leqq s' \leqq t'$. By the canonical property of $\{G_n, B_n\}$, and (3.5)–(3.7), the projection of $\mathbf{x}_{s''}(\varphi)$ onto $H(\mathbf{x}; s')$ is given by

$$\textstyle\sum_1^M \int_{-\infty}^{s'} \tilde{G}_n(\varphi; s'', u) \, d\tilde{B}_n(u).$$

But $z \perp H(\mathbf{x}; t)$ and $z = \int_{-\infty}^{t} h(u) dB_n(u)$ with $h \in L_2(\tilde{\rho}_n)$(see [4], pp. 383–385). Hence

(3.8) $$\int_{-\infty}^{s'} \tilde{G}_n(\varphi; s'', u)\overline{h(u)}\, d\tilde{\rho}_n(u) = 0 \qquad \text{for all } s'', \varphi.$$

Using a similar argument with s we obtain

(3.9) $$\int_{s}^{s'} \tilde{G}_n(\varphi; s'', u)\overline{h(u)}\, d\tilde{\rho}_n(u) = 0 \qquad \text{for all } s'' \text{ and } \varphi.$$

Proceeding as in Theorem 1.2 of [7], it can be shown that (3.9) implies

$$\rho_n\{N(h) \cap N_n\} = 0 \text{ where } N(h) = \{u | h(u) \neq 0\}.$$

Hence

$$\mathbf{E}|z|^2 = \int_{-\infty}^{t} |h(u)|^2 \, d\tilde{\rho}_n(u) = \int_{-\infty}^{t} I_{N_n}(u)|h(u)|^2 \, d\rho_n(u) = \int_{N_n \cap N(h)} |h(u)|^2 \, d\rho_n(u) = 0,$$

contradicting the assumption that $z \neq 0$.

REMARKS. (i) The relation obtained in (3.5) is an equivalence relation. Hence we shall refer to $\{\tilde{G}_n, \tilde{B}_n\}_1^M$ as a proper canonical representation equivalent to $\{G_n, B_n\}_1^M$.

(ii) By definition of $\tilde{B}_n$ and the fact that $(d\tilde{\rho}_n/d\rho_n)(u) = I_{N_n}^2(u)$ if $\tilde{\rho}_n = 0$, we obtain $I_{N_n}(u) = 0$ a.e. ρ_n. But this will imply $\rho_n\{(d\mu^{(n)}/d\rho_n)(u) > 0\} = 0$. Hence $|G_n(\varphi; t, u)|^2$ which equals $(d\mu_{(t,\varphi)}^{(n)}/d\mu^{(n)})(u)(d\mu^{(n)}/d\rho_n)(u)$ vanishes almost everywhere $[\rho_n]$, i.e., for every φ and t $G_n(\varphi; t, u) = 0$ a.e. with respect to ρ_n, contradicting the fact that M is the multiplicity of $\{G_n, B_n\}_1^M$. Thus the representation $\{\tilde{G}_n, \tilde{B}_n\}$ also has multiplicity M.

(iii) Finally, from the definition of $\tilde{B}_n$ we have

$$\tilde{B}_n(S) = \int_S I_{N_n}(u)\,dz_n(u) + \sum_{t_j \in N_n \cap S} \xi_{jn} = \tilde{z}_n(S) + \sum_{t_j \in S} \tilde{\xi}_{jn},$$

say, where $\tilde{\xi}_{jn} = \xi_{jn}$ if $t_j \in N_n$, and 0 otherwise. Hence the proper canonical representation obtained can again be put in the form of (2.1).

Weakly Stationary Stochastic Processes on Φ

We now turn to the central task of this paper, the study of the multiplicity theory of weakly stationary processes on Φ. As we shall see, this theory applies also to a class of infinite dimensional stationary processes and shows that in the study of the latter the idea of multiplicity naturally supplants that of rank.

Before proceeding to the discrete parameter case whose results we shall need in Section 6 we make the following observations concerning the Wold decomposition of continuous parameter stationary processes on Φ. If for every real h, we define

$$T_h \mathbf{x}_t(\varphi) = \mathbf{x}_{t+h}(\varphi),$$

where t is an arbitrary real number and $\varphi \in \Phi$, it is easy to see that this definition can be extended so that T_h becomes a unitary operator. Indeed, $\{T_h\}$ $(-\infty < h < +\infty)$ is a group of unitary operators and for all real a and h

$$T_h E(a) = E(a+h) T_h.$$

Using this fact and Proposition 2.1 we are able to state the following proposition:

If $\{\mathbf{x}_t\}$ is a weakly stationary process on Φ then there exist weakly stationary processes on Φ, $\{\mathbf{x}_t^{(1)}\}$ and $\{\mathbf{x}_t^{(2)}\}$ such that

(1) $\mathbf{x}_t(\varphi) = \mathbf{x}_t^{(1)}(\varphi) + \mathbf{x}_t^{(2)}(\varphi)$ *for every* t,

(2) $\{\mathbf{x}_t^{(1)}\}$ *is deterministic*, $\{\mathbf{x}_t^{(2)}\}$ *is purely non-deterministic, and*

(3) $H(\mathbf{x}^{(1)})$, *and* $H(\mathbf{x}^{(2)})$ *are orthogonal.*

4. Discrete Parameter Processes

Let $\mathbf{x}_n$ $(n = 0, \pm 1, \cdots)$ be a purely nondeterministic stationary process on Φ. Since we want $H(\mathbf{x})$ to be separable, we shall assume that, for each n, $\mathbf{x}_n(\cdot)$ is continuous in quadratic mean in the Φ-topology. If in Theorem 2.2, T is the set of integers then the resolution of identity of the process is given by

$$E_t = \sum_{n \leq t} (p_n - p_{n-1}) \quad \text{where } p_n \text{ is the projection onto } H(\mathbf{x}; n).$$

The self-adjoint operator A then has a purely discrete spectrum, having each integer as an eigenvalue and $H(\mathbf{x}; n) \ominus H(\mathbf{x}; n-1)$ as the invariant subspaces. The multiplicity M of the process is therefore given by

$$M = \sup_n \left[\dim \{H(\mathbf{x}; n) \ominus H(\mathbf{x}; n-1)\}\right].$$

The following two lemmas show that $\dim \{H(\mathbf{x}; n) \ominus H(\mathbf{x}; n-1)\}$ is independent of n. Let $g_n(\varphi) = \mathbf{x}_n(\varphi) - p_{n-1}\mathbf{x}_n(\varphi)$.

Lemma 4.1. $H(\mathbf{x}; n) \ominus H(\mathbf{x}; n-1) = \mathfrak{S}[g_n(\varphi), \varphi \in \Phi]$ $(n = 0, +1, \cdots)$.

Lemma 4.2. *For arbitrary integers m and n, there exists a unitary operator T_m such that*

$$T_m \mathfrak{S}[g_n(\varphi), \varphi \in \Phi] = \mathfrak{S}[g_{m+n}(\varphi), \varphi \in \Phi].$$

To prove Lemma 4.1 it is enough to show that $H(\mathbf{x}; n) = H(\mathbf{x}; n-1) \oplus \mathfrak{S}[g_n(\varphi), \varphi \in \Phi]$. But this is true from the definition of $g_n(\varphi)$.

For the proof of Lemma 4.2, we consider the group $\{T_m\}$ of unitary operators given by

$$T_m \mathbf{x}_n(\varphi) = \mathbf{x}_{m+n}(\varphi) \text{ for all } n \text{ and } \varphi.$$

It can be easily verified that $T_m p_{n-1}\mathbf{x}_n(\varphi) = p_{m+n-1}\mathbf{x}_{m+n}(\varphi)$. Hence, $T_m g_n(\varphi) = g_{m+n}(\varphi)$ and the proof is complete.

For the process $\mathbf{x}_n$ of this section we now have the following result.

Theorem 4.1.

$$\mathbf{x}_n(\varphi) = \sum_{l=1}^{M} \sum_{m \leqq n} b_l(\varphi; m-n)\xi_l(m),$$

where

(i) $M = \dim[H(\mathbf{x}; n) \ominus H(\mathbf{x}; n-1)]$ *is the multiplicity of the process,*

(ii) *for each* l, $\{\xi_l(m)\}$ $(m = 0, \pm 1, \cdots)$ *has stationary orthogonal increments and* $\mathbf{E}[\xi_l(m)\overline{\xi_k(m')}] = 0$ *if* $k \neq l$. *Furthermore,*

$$\sum_{l=1}^{M} \sum_{m \leqq 0} |b_l(\varphi; m)|^2 \mathbf{E}|\xi_l(m)|^2$$

is finite and

(iii) $\sum_{i=1}^{M} \oplus H(\xi_i; n) = H(\mathbf{x}; n)$ *for all* n.

PROOF. From Theorems 2.2, 3.1 and the remarks preceding Lemma 4.1 about the resolution of the identity in $H(\mathbf{x})$, we have

$$\mathbf{x}_n(\varphi) = \sum_{m \leqq n} \sum_{l=1}^{M} b'_l(\varphi; n, m)\xi'_l(m) \quad \text{with } H(\mathbf{x}; n) = \sum_{1}^{M} \oplus H(\xi'_1; n). \tag{4.1}$$

By Lemma 4.1 and (4.1)

$$\mathfrak{S}[g_n(\varphi), \varphi \in \Phi, m \leqq n] = H(\mathbf{x}; n) = \sum_{1}^{M} \oplus H(\xi_i; n).$$

In particular

$$\mathfrak{S}[g_0(\varphi); \varphi \in \Phi] = \mathfrak{S}[\xi'_l(0), l = 1, 2, \cdots M].$$

Hence, if we define $\xi_l(m) = T_m \xi'_l(0)$, we have

$$\mathfrak{S}[\xi'_l(m), l = 1, 2, \cdots M] = \mathfrak{S}[\xi_l(m), l = 1, 2, \cdots M],$$

since

$$T_m \mathfrak{S}[\xi'_l(0), l = 1, 2, \cdots M] = T_m \mathfrak{S}[g_0(\varphi), \varphi \in \Phi] = \mathfrak{S}[g_m(\varphi), \varphi \in \Phi].$$

Therefore, $H(\mathbf{x}; n) = \sum_1^M \oplus H(\xi_i; n)$ and hence

$$\mathbf{x}_0(\varphi) = \sum_{l=1}^{M} \sum_{m \leqq 0} b_l(\varphi; m)\xi_l(m),$$

with

$$\sum_{l=1}^{M} \sum_{m \leqq 0} |b_l(\varphi; m)|^2 \mathbf{E}|\xi_l(m)|^2 < \infty,$$

$$\mathbf{x}_n(\varphi) = T_n \mathbf{x}_0(\varphi) = \sum_{l=1}^{M} \sum_{m \leqq 0} b_l(\varphi; m)\xi_l(m+n) = \sum_{l=1}^{M} \sum_{m \leqq n} b_l(\varphi; m-n)\xi_l(m).$$

5. Continuous Parameter, Weakly Stationary Processes.

We shall give in this section the generalization of what we believe to be the essence of Hanner's ideas underlying his time domain analysis of one-dimensional stationary processes. The desired generalization will turn out to be based on a study of the properties of the maximal spectral type of the operator A of the process and its multiplicity, thus effecting a unity with the work presented in Sections 2 and 3.

It is convenient to recall at this point some of the terminology of multiplicity theory in a separable Hilbert space H. Let A be any self-adjoint operator with spectral measure function $E(\cdot)$. For any element f in H let ρ_f be the finite measure on the Borel sets of line (sometimes also called the spectral function) given by $\rho_f(\Delta) = ||E(\Delta)f||^2$. The family of all finite measures on the line is divided into equivalence classes by the relation of equivalence between measures (equivalence here means mutual absolute continuity). If ρ is used to denote the equivalence class to which the measure ρ_f belongs, ρ will be called the spectral type of f with respect to A. ρ is also referred to as the spectral type belonging to A. If elements f and g are such that $\rho_f \equiv \rho_g$, they obviously have the same spectral type ρ. We shall say that the spectral type ρ dominates the spectral type σ ($\rho > \sigma$ or $\sigma < \rho$) if any (and thus every) measure belonging to σ is absolutely continuous with respect to any measure belonging to ρ. ρ and σ are said to be independent spectral types if for any spectral type ν such that $\nu < \rho$ and $\nu < \sigma$ we have $\nu = 0$. An element f is said to be of maximal spectral type ρ (with respect to A) if, for every g in H, $\rho_g \ll \rho_f$. The subspace $\mathfrak{S}\{E(\Delta)f,\ \Delta$ ranging over all finite intervals$\}$ is called the cyclic subspace (with respect to A) generated by f. If this subspace coincides with H, f is called a cyclic or generating element of A and A is called cyclic. Also if f is a generating element of A, f is of maximal spectral type and the latter is referred to as the spectral type of the (cyclic) operator A. It is to be noted that if A is any self-adjoint operator (since H is separable) there always exists a maximal spectral type belonging to A. Any system of mutually cyclic parts of A of type ρ is called an orthogonal system of type ρ relative to A. An orthogonal system of type ρ which cannot be enlarged by adding to it more cyclic parts of A is called maximal. It is a known result of this theory that all maximal systems of type ρ have the same cardinal number. This uniquely determined cardinal number is defined to be the multiplicity of the spectral type ρ with respect to A.

Finally we need the notion of a uniform spectral type. The spectral type $\rho(\neq 0)$ is said to be uniform if every non-zero type σ dominated by ρ has the same multiplicity as ρ itself. Most of the above definitions have been taken from the article by A. I. Plessner and T. A. Rohlin [12] to which the reader is also referred for further details.

When dealing with continuous parameter processes, we assume not only that $\mathbf{x}_t(\varphi)$ is continuous in q.m. in the topology of Φ but that for each $\varphi \in \Phi$, the complex valued univariate process $[\mathbf{x}_t(\varphi)]$ $(-\infty < t < +\infty)$ is continuous in q.m. in t. We shall refer to this as condition (C). It is easy to see that if (C) holds, the assumptions of Lemma 2.1 are valid so that the separability condition (A.1) is satisfied. In addition, it follows from condition (C) that the group $[T_h]$ introduced in Section 4 is strongly continuous. We recall from Section 4

$$T_h E(t) = E(t+h) T_h \tag{5.1}$$

for all real t, h. As in [9], (5.1) is the basic relation between the operator A and the unitary group of the process which we propose to exploit in our time domain analysis. We shall prove the central theorem on representation by means of a number of lemmas. The first group of lemmas concerns the properties of spectral types.

Lemma 5.1. *If f is any element of $H(\mathbf{x})$ then $\rho_f \ll \mu$, the Lebesgue measure.*

PROOF. Let us define for every real t, and every measurable set S of the real line $\rho_f^{(t)}(S) = \rho_f(S-t) = \|E(S-t)f\|^2$. From (5.1), however, $\rho_f^{(t)}(S)$ equals $\|E(S)\,T_t f\|^2$. Hence by the strong continuity of the group $\{T_t\}$, $\rho_f^{(t)}(S)$ converges to $\rho_f(S)$ as $t \to 0$. The assertion of the lemma now follows from a result due to N. Wiener and R. C. Young [see Saks [16], p. 140].

Let $f^{(1)}$ be a maximal element of A, i.e., an element of maximal spectral type with respect to A, and u any positive number. If we define

$$g_a^b = \{E(b)-E(a)\} \int_A^B T_h E(\Delta_0) f^{(1)}\, dh, \tag{5.2}$$

where $\Delta_0 \subset (0, u)$, $A < a-u$, $B > b$ and the integral is taken as in [6], we observe that g_a^b can be identified with Hanner's $Z(I_a^b)$ with $z = E(\Delta_0)f^{(1)}$ in the formula (3.2) of [6] (p. 166). We remark that g_a^b does not depend on A and B as long as these limits of integration satisfy the stated inequalities. We give here the properties of g_a^b which follow from those of $Z(I_a^b)$ [See [6], p. 167]. For $a < b < c$, we have

$$g_a^b + g_b^c = g_a^c, \tag{5.3}$$

$$g_a^b \text{ is orthogonal to } g_b^c, \tag{5.4}$$

and for arbitrary t,

$$T_t g_a^b = g_{a+t}^{b+t}. \tag{5.5}$$

It follows from (5.3), (5.4) and (5.5) that

$$\|g_a^b\|^2 = \tau(b-a), \tag{5.6}$$

where τ is a non-negative number thas does not depend on the interval $(a, b]$.

Lemma 5.2. *There exists a finite interval $\Delta_0 \subset (0, u]$ such that g_0^u as defined in (5.2) is different from zero.*

PROOF. We follow Hanner closely in proving this lemma ([6], Proposition C). Suppose $g_0^u = 0$, then for every $z' \in H(\mathbf{x})$ and every $\Delta_0 \subset (0, u]$, $\mathbf{E}[g_0^u \bar{z}'] = 0$. Hence, if $z = \omega(s_1, t_1)$ and $z' = \omega(s_2, t_2)$, where $\omega(s, t) = \{E(t)-E(s)\}f^{(1)}$ for $s < t$, then from the fact that $\mathbf{E}[g_0^u \bar{z}'] = 0$, we have

$$\int_{-u}^{u} \mathbf{E}[T_h \omega(s_1, t_1) \cdot \overline{\omega(s_2, t_2)}]\, dh = 0 \qquad (0 < s_1, t_1, s_2, t_2 \leqq u). \tag{5.7}$$

But for δ such that $0 < \delta < u/2$, $\mathbf{E}[T_h \omega(0, u)\overline{\omega(\delta, u-\delta)}] = \mathbf{E}|T_h \omega(\delta, u-\delta)|^2$ is a continuous function of h which converges, as $h \to 0$, to $\mathbf{E}|\omega(\delta, u-\delta)|^2$. Now, $\omega(\delta, u-\delta) = 0$ implies that $[E(\delta)-E(u-\delta)]f^{(1)} = 0$ and hence $[E(\delta)-E(u-\delta)]f = 0$ for all $f \in H(\mathbf{x})$, giving $H(\mathbf{x}; \delta) \ominus H(\mathbf{x}; u-\delta) = \{0\}$. This contradicts the fact that the $\mathbf{x}_t$-process is purely non-deterministic. Therefore we can find a γ $(0 < \gamma < u)$ such that

$$L = \int_{-\gamma}^{\gamma} \mathbf{E}[T_h \omega(0, u)\overline{\omega(\delta, u-\delta)}]\, dh \neq 0.$$

Let $t_0 = \delta < t_1 < \cdots < t_n = u-\delta$ be a finite subdivision of the interval $(\delta, u-\delta]$. Then

$$L = \sum_1^n \int_{-\gamma}^{\gamma} \mathbf{E}[T_h \omega(0, u) \cdot \overline{\omega(t_{i-1}, t_i)}]\, dh.$$

Let

$$M = \sum_1^n \int_{-\gamma-(t_i-t_{i-1})}^{\gamma+(t_i-t_{i-1})} \mathbf{E}[T_h \omega(t_{i-1}-\gamma, t_i+\gamma)\overline{\omega(t_{i-1}, t_1)}]\, dh$$

$$= \sum_1^n \int_{-u}^{+u} \mathbf{E}[T_h \omega(t_{i-1}-\gamma, t_i+\gamma)\overline{\omega(t_{i-1}, t_i)}]\, dh$$

which is zero from (5.7). Now $|M-L| \leqq 2u||\omega(0, u)|| \max_i ||\omega(t_{i-1}, t_i)||$. But

$$\omega(t_{i-1}, t_i) + P_{H(\mathbf{x}; t_i, u)} f^{(1)} = P_{H(\mathbf{x}; t_{i-1}, u)} f^{(1)}$$

and $\omega(t_{i-1}, t_i)$ is orthogonal to $H(\mathbf{x}; t_i, u)f^{(1)}$. Hence,

$$||\omega(t_{i-1}, t_i)||^2 = ||P_{H(\mathbf{x}; t_{i-1}, u)} f^{(1)}||^2 - ||P_{H(\mathbf{x}; t_i, u)} f^{(1)}||^2.$$

Since $||P_{H(\mathbf{x}; t, u)} f^{(1)}||^2$ is a continuous function of t, we make $||\omega(t_{i-1}, t_i)||$ as small as we please by taking a fine enough subdivision. Hence $M = L$. But $M = 0$ and $L \neq 0$. We arrive at a contradiction, thus proving the lemma.

Henceforth, Δ_0 will denote a fixed subinterval of $(0, u]$, such that $||g_0^u||^2 \neq 0$ where in (5.2) we take $(a, b] = (0, u]$.

Suppose that $0 < b < u$ and consider $g_0^b = [E(b)-E(0)] \int_{A'}^{B'} T_h E(\Delta_0) f^{(1)} dh$, where $A' < -u$, $B' > b$. Since the definition of g_0^b is independent of this particular choice of A', B', we have

$$g_0^b = [E(b)-E(0)]g_0^u = [E(b)-E(0)] \int_A^B T_h E(\Delta_0) f^{(1)} dh,$$

where $A < -u$ and $B > u$. Also from (5.3) and (5.4), $g_0^u = g_0^b + g_b^u$ with g_b^u orthogonal to g_0^b. Hence $||g_0^u||^2 = ||g_0^b||^2 + ||g_b^u||^2$. If $g_0^b = 0$, we have from (5.6) that $\tau u = \tau(u-b)$ where $\tau \neq 0$ by Lemma 5.2. Since u and b are distinct positive numbers, the above relation is absurd and thus $g_0^b \neq 0$. On the other hand, if $b > u$ then again (5.3) and (5.4) imply that $g_0^b = g_0^u + g_u^b$ with g_0^u being orthogonal to g_u^b. Therefore $||g_0^b||^2 = ||g_0^u||^2 + ||g_u^b||^2$ thus giving $g_0^b \neq 0$ for all positive b. Finally if $b' < 0$, then from (5.5), $T_\beta g_{b'}^0 = g_0^\beta$ where $\beta = -b'$. From previous arguments $g_0^\beta \neq 0$. Hence $g_{b'}^0 \neq 0$. Thus $g_0^b \neq 0$ if $b > 0$ and $g_{b'}^0 \neq 0$ if $b' < 0$. We therefore obtain $\tau \neq 0$ in (5.6), since for any $(c, d]$, $T_{-c} g_c^d = g_0^{d-c} \neq 0$.

Lemma 5.3. *The spectral measure* $\rho_{g_a^b} = \tau\mu^I(I = (a, b])$, *where* $\mu^I(S) = \mu(I \cap S)$ *for every measurable subset S of the real line.*

Proof. Let Δ be any finite interval. Then $\rho_{g_a^b}(\Delta) = ||E(\Delta)g_a^b||^2$. Therefore, from (5.2), $\rho_{g_a^b}(\Delta) - ||E(\Delta \cap I)g_a^b||^2$, which equals zero if $\Delta \cap I = \varnothing$ and, from (5.6), is equal to $\tau\mu(\Delta \cap I)$ if $\Delta \cap I \neq \varnothing$. The result follows immediately from the definition of μ^I.

The definition of g_a^b can obviously be adjusted to make $\tau = 1$. From now on we shall assume that this has been done.

Lemma 5.4. *If ρ is the maximal spectral type of A, then $\rho \equiv \mu$.*

PROOF. It suffices to prove that if $f^{(1)}$ is a maximal element then $\rho_{f(1)} \equiv \mu$. From the maximality of $f^{(1)}$ and the fact, shown in Lemma 5.3, that $\rho_{g^b_a} = \mu^I$ for an arbitrary interval $I = (a, b]$, it follows that $\mu \ll \rho_{f(1)}$. An appeal to Lemma 5.1 completes the proof.

We next define a complex-valued process $\xi_1(a)$ for all real a, as follows:

$$\xi_1(a) = -g^0_a \quad \text{if } a < 0,$$
$$\xi_1(0) = 0,$$
$$\xi_1(a) = g^a_0 \quad \text{if } a > 0.$$

If we set $\xi_1(I) = \xi_1(b) - \xi_1(a)$ for every interval $I = (a, b]$, it follows from (5.3) and (5.4) that

$$\xi_1(I) = g^b_a. \tag{5.8}$$

It is easy to see that $\{\xi_1(t)\}$ $(-\infty < t < +\infty)$ is a stochastic process with stationary orthogonal increments and $\mathbf{E}|\xi_1(\Delta)|^2 = \mu(\Delta)$. Let us write

$H(\xi_1) = \mathfrak{S}\{\xi_1(\Delta), \Delta$ ranging over all finite subintervals of real line$\}$ and

$H(\xi_1; t) = \mathfrak{S}\{\xi_1(\Delta), \Delta$ ranging over all finite intervals contained in $(-\infty, t]\}$.

Then by (5.5) it follows that, for every real t, $T_t P_{H(\xi_1)} = P_{H(\xi_1)} T_t$. If we now define

$$\mathbf{x}^{(1)}_t(\varphi) = P_{H(\xi_1)} \mathbf{x}_t(\varphi), \tag{5.9}$$

then the $\mathbf{x}^{(1)}_t$-process is stationary and $T_t \mathbf{x}^{(1)}_s(\varphi) = \mathbf{x}^{(1)}_{s+t}(\varphi)$ for all s and φ. Furthermore, since ξ_1 is a process with orthogonal increments, we have $H(\xi_1) = H(\xi_1; t) \oplus \mathfrak{S}\{\xi_1(\Delta), \Delta \subset (t, +\infty)\} = H(\xi_1; t) \oplus \mathfrak{S}\{g^b_a, t < a \leqq b < +\infty\}$ from (5.8). But, by definition of g^b_a, $\mathbf{x}_t(\varphi) \perp \mathfrak{S}\{g^b_a, t < a \leqq b < \infty\}$ so that $\mathbf{x}_t(\varphi) = P_{H(\xi_1; t)} \mathbf{x}_t(\varphi)$ for all t, φ. Since from (5.8) and (5.2), $\xi_1(\Delta) \in H(\mathbf{x}; t)$ for every finite interval Δ lying in $(-\infty, t]$, we have $H(\mathbf{x}^{(1)}; t) \subset H(\mathbf{x}; t)$. Hence the $\mathbf{x}^{(1)}_t$-process is purely non-deterministic.

Lemma 5.5. *For every real t and φ in Φ,*

$$\mathbf{x}^{(1)}_t(\varphi) = \int_{-\infty}^{t} F_1(\varphi; u-t) d\xi_1(u),$$

where

$$\int_{-\infty}^{0} |F_1(\varphi; u)|^2 d\mu(u) < \infty.$$

PROOF. Since $\mathbf{x}^{(1)}_0(\varphi) \in H(\xi_1;\ 0)$, it has the stochastic integral representation

$$\mathbf{x}^{(1)}_0(\varphi) = \int_{-\infty}^{0} F_1(\varphi; u) d\xi_1(u) \qquad \text{with} \int_{-\infty}^{0} |F_1(\varphi; u)|^2 d\mu(u) < \infty$$

(see [4], pp. 383–385). The $\mathbf{x}^{(1)}_t$-process is stationary and $T_t \xi_1(\Delta) = \xi_1(\Delta + t)$ from (5.5) and (5.8); hence

$$\mathbf{x}^{(1)}_t(\varphi) = T_t \mathbf{x}^{(1)}_0(\varphi) = \int_{-\infty}^{0} F_1(\varphi; u) d\xi_1(u+t) = \int_{-\infty}^{t} F_1(\varphi; u-t) d\xi_1(u).$$

For every $\varphi \in \Phi$ and t real, set $\mathbf{y}^{(1)}_t(\varphi) = \mathbf{x}_t(\varphi) - \mathbf{x}^{(1)}_t(\varphi)$. Then $T_t \mathbf{y}^{(1)}_s(\varphi) = \mathbf{y}^{(1)}_{s+t}(\varphi)$

and $H(\mathbf{y}^{(1)}; t) \subset H(\mathbf{x}; t)$. Hence the $\mathbf{y}_t^{(1)}$-process is also weakly stationary and purely non-deterministic. From (5.9) we have $\mathbf{y}_t^{(1)} = \mathbf{x}_t(\varphi) - P_{H(\xi_1)}\mathbf{x}_t(\varphi)$ which implies that for all t, φ, $\mathbf{y}_t^{(1)}(\varphi) \perp H(\xi_1)$. Since $H(\mathbf{x}^{(1)}) \subset H(\xi_1)$ it follows that for every t and s

$$H(\mathbf{y}^{(1)}: s) \perp H(\mathbf{x}^{(1)}; t). \tag{5.10}$$

Lemma 5.6.

$$H(\mathbf{x}; t) = H(\mathbf{x}^{(1)}; t) \oplus H(\mathbf{y}^{(1)}; t) \qquad \textit{for each } t.$$

PROOF. Since $H(\mathbf{x}^{(1)}; t) \oplus H(\mathbf{y}^{(1)}; t) \subset H(\mathbf{x}; t)$, we need to show only that $H(\mathbf{x}^{(1)}; t) \oplus H(\mathbf{y}^{(1)}; t) \subset H(\mathbf{x}; t)$. But this follows from the fact that, for $\varphi \in \Phi$, $\mathbf{x}_\tau(\varphi) = \mathbf{x}_\tau^{(1)}(\varphi) + \mathbf{y}_\tau^{(1)}(\varphi)$ which belongs to $H(\mathbf{x}^{(1)}; t) \oplus H(\mathbf{y}^{(1)}; t)$ for $t \geqq \tau$.

Lemma 5.7. *Let a and b be arbitrary real numbers. If we write*

$$H(\mathbf{x}^{(1)}; a, b) = H(\mathbf{x}^{(1)}; b) \ominus H(\mathbf{x}^{(1)}; a),$$

then

$$\begin{aligned} H(\mathbf{x}^{(1)}; a, b) &= \mathfrak{S}\{g_\alpha^\beta, a < \alpha \leqq \beta \leqq b\} \\ &= \mathfrak{S}\{(E(\beta) - E(\alpha))g_a^b, \alpha < \alpha \leqq \beta \leqq b\}. \end{aligned} \tag{5.11}$$

PROOF. The second half of relation (5.11) is obvious since $[E(\beta) - E(\alpha)]g_a^b = g_\alpha^\beta$ for $a < \alpha \leqq \beta \leqq b$. To prove the first part we proceed as follows: For $a < t \leqq b$ and $\varphi \in \Phi$,

$$\mathbf{x}_t^{(1)}(\varphi) - P_{H(\mathbf{x}^{(1)}; a)}\mathbf{x}_t^{(1)}(\varphi) = P_{H(\xi_1; t)}\mathbf{x}_t(\varphi) - P_{H(\mathbf{x}^{(1)}; a)}\mathbf{x}_t^{(1)}(\varphi).$$

From Lemma 5.6 and (5.10),

$$\mathbf{x}_t^{(1)}(\varphi) - P_{H(\mathbf{x}^{(1)}; a)}\mathbf{x}_t^{(1)}(\varphi) = P_{H(\xi_1; t)}\mathbf{x}_t(\varphi) - P_{H(\mathbf{x}; a)}P_{H(\xi_1; t)}\mathbf{x}_t(\varphi).$$

Furthermore, for $a \leqq t$, writing $H(\xi_1; a, t) = H(\xi_1; t) \ominus H(\xi_1; a)$,

$$H(\xi_1; t) = H(\xi_1; a) \oplus H(\xi_1; a, t) \text{ and } \mathbf{x}_\tau(\varphi) \perp H(\xi_1; a, t). \tag{5.12}$$

The latter assertion follows from (5.8) and the definition of g_a^b. Thus we have

$$P_{H(\mathbf{x}; a)}P_{H(\xi_1; t)} = P_{H(\mathbf{x}; a)}\{P_{H(\xi_1; a)} + P_{H(\xi_1; a, t)}\} = P_{H(\mathbf{x}; a)}P_{H(\xi_1; a)}.$$

Further since $H(\xi_1; a) \subset H(\mathbf{x}; a)$, we have

$$\mathbf{x}_t^{(1)}(\varphi) - P_{H(\mathbf{x}^{(1)}; a)}\mathbf{x}_t^{(1)}(\varphi) = P_{H(\xi_1; t)}\mathbf{x}_t(\varphi) - P_{H(\xi_1; a)}\mathbf{x}_t(\varphi).$$

Hence $H(\mathbf{x}^{(1)}; a, b) \subset H(\xi_1; a, b)$ which from (5.8) is the same as $\mathfrak{S}\{g_\alpha^\beta, a < \alpha \leqq \beta \leqq b\}$. To complete the proof we have only to observe, because of Lemma 5.5, that, for $a < \alpha \leqq \beta \leqq b$, g_α^β is in $H(\mathbf{x}; a, b)$ and is orthogonal to $H(\mathbf{y}^{(1)}; a, b)$.

Let $\hat{\mathbf{x}}_t^{(1)}(\varphi) = \mathbf{x}_t^{(1)}(\varphi) - P_{H(\mathbf{x}^{(1)}; a)}\mathbf{x}_t^{(1)}(\varphi)$. From Lemma 5.7, it follows that for $a < t \leqq b$ and $\varphi \in \Phi$

$$\hat{\mathbf{x}}_t^{(1)} = \int_a^t F(\varphi; t, u)\, dE(u)g_a^b, \qquad \text{where } \int_a^b |F(\varphi; t, u)|^2\, d\mu(u) < \infty. \tag{5.13}$$

We are now in a position to prove the following result.

Lemma 5.8. *The operator A is reduced by* $H(\mathbf{x}^{(1)}; a, b)$.

PROOF. It suffices to prove that for $a < t \leqq b$ and $\varphi \in \Phi$, $A\hat{\mathbf{x}}_t^{(1)}(\varphi) \in H(\mathbf{x}^{(1)}; a, b)$ since $H(\mathbf{x}^{(1)}; a, b) = \mathfrak{S}\{\mathbf{x}_t^{(1)}(\varphi), \varphi \in \Phi, a < t \leqq b\}$. From (5.13)

$$A\hat{\mathbf{x}}^{(1)}(\varphi) = \int_a^t uF(\varphi; t, u)dE(u)g_a^b, \qquad \text{where } F(\varphi; t, u) \in L_2(\mu^I).$$

Hence $A\hat{\mathbf{x}}_t^{(1)}(\varphi) \in \mathfrak{S}\{(E(\beta)-E(\alpha))g_a^b, a < \alpha < \beta \leqq b\}$ since $uF(\varphi; t, u) \in L_2(\mu^I)$. From the preceding lemma it now follows that $A\tilde{\mathbf{x}}_t^{(1)}(\varphi) \in H(\mathbf{x}^{(1)}; a, b)$.

Lemma 5.9. *$H(\mathbf{x}^{(1)})$ reduces the operator A.*

PROOF. From the properties of the resolution of the identity corresponding to A, we have

$$E(\Delta)A \equiv AE(\Delta) \tag{5.14}$$

for every finite subinterval $\Delta = (a, b]$. If w is any element belonging to $\mathscr{D}_A \cap H(\mathbf{x}^{(1)})$ (which is non-empty) where $\mathscr{D}_A$ is the domain of A, then from Lemma 5.8 we have

$$E(\Delta)Aw = AE(\Delta)w = AP_{H(\mathbf{x}^{(1)}; a, b)}w \in H(\mathbf{x}^{(1)}; a, b).$$

Now letting $a = n-1$, $b = n$ and $\Delta_n = (n-1, n]$ we obtain

$$Aw = \sum_{n=-\infty}^{\infty} E(\Delta_n)Aw \in \sum_{n=-\infty}^{\infty} \oplus H(\mathbf{x}^{(1)}; n-1, n) = H(\mathbf{x}^{(1)}).$$

Let $A^{(1)}$ be the reduction of A to $H(\mathbf{x}^{(1)})$. Then (Lemma 5.8) clearly $A^{(1)}$ is reduced by $H(\mathbf{x}^{(1)}; a, b)$. We denote this operator on $H(\mathbf{x}^{(1)}; a, b)$ by $A_I^{(1)} (I = (a, b])$. An immediate implication of Lemma 5.6 is that $A_I^{(1)}$ is a cyclic operator with generating element g_a^b. We recall from Lemma 5.3 that the spectral function of g_a^b is given by $\rho_{g_a^b} = \mu^I$.

Now let $I_j = (a_j, b_j]$ $(j = 1, 2, \cdots)$ be disjoint intervals whose union is the real line. If ρ_j denotes the spectral type of the operator $A_j^{(1)}$ (which we write here for $A_{I_j}^{(1)}$), then it is easy to verify that the ρ_j's are independent spectral types. For let j and m be arbitrary $(j \neq m)$ and suppose that σ is a measure whose spectral type is dominated by both ρ_j and ρ_m. Since $\mu^I j(I_k) = 0$, we have, for all $k \neq j$, $\sigma(I_k) = 0$. But $\sigma(I_j)$ is also equal to zero since $\mu^{I_m}(I_j) = 0$. Hence $\sigma = 0$. Summarizing all the above facts we find that we have a representation of $A^{(1)}$ as the orthogonal sum of cyclic operators $A_{I_j}^{(1)}$ whose corresponding spectral types ρ_j are independent. It then follows that ([12], p. 152) $A^{(1)}$ itself is cyclic and since the spectral function μ^{I_j} belongs to the type ρ_j for each j we can conclude moreover that the spectral type of $A^{(1)}$ is equivalent to μ. From Lemma 5.4 it follows that the spectral type of $A^{(1)}$ is equal to ρ, the maximal spectral type of A.

Let us recall that $H(\mathbf{x}) = H(\mathbf{x}^{(1)}) \oplus H(\mathbf{y}^{(1)})$ and the self-adjoint operator A is reduced by $H(\mathbf{x}^{(1)})$. Hence A can be written as the orthogonal sum of the reduced operators, $A = A_{H(\mathbf{x}^{(1)})} + A_{H(\mathbf{y}^{(1)})}$.

Now $A_{H(\mathbf{y}^{(1)})}$, a self-adjoint operator on $H(\mathbf{y}^{(1)})$, is the operator of the weakly stationary non-deterministic process $\{\mathbf{y}_t^{(1)}, -\infty < t < +\infty\}$. We may, therefore, apply the above analysis to this process replacing $H(\mathbf{x})$ by $H(\mathbf{y}^{(1)})$ and A by $A_{H(\mathbf{y}^{(1)})}$. We then have $H(\mathbf{y}^{(1)}) = H(\mathbf{x}^{(2)}) \oplus H(\mathbf{y}^{(2)})$, where the $\mathbf{x}_t^{(2)}$-process is constructed from the $\mathbf{y}_t^{(1)}$-process in the same way as the $\mathbf{x}_t^{(1)}$-process is obtained from the given $\mathbf{x}_t$-process. The $\mathbf{y}_t^{(2)}$-process is stationary and purely non-deterministic. We also have the orthogonal decomposition

$$A = A^{(1)} + A^{(2)} + A_{H(\mathbf{y}^{(2)})},$$

where $A^{(i)} = A_{H(\mathbf{x}^{(i)})}$. Continuing the above procedure we arrive at the following relations:

$$H(\mathbf{x}) = H(\mathbf{x}^{(1)}) \oplus H(\mathbf{x}^{(2)}) \oplus \cdots \oplus H(\mathbf{x}^{(M)}), \tag{5.15}$$

$$A = A^{(1)} + A^{(2)} + \cdots + A^{(M)}, \tag{5.16}$$

where $\mathbf{x}_t^{(i)}(\varphi) = P_{H(\xi_i)}\mathbf{x}_t(\varphi)$ and $\{\xi_i(u), -\infty < u < +\infty\}$ are mutually orthogonal processes with stationary orhogonal increments. The operators $A^{(i)}$ are cyclic, all having the same spectral type ρ (the maximal spectral type of A). Further M is a cardinal number at most equal to $\aleph_0$.

Also from Lemmas 5.5, 5.6 and 5.7, we have

$$\mathbf{x}_t^{(i)}(\varphi) = \int_{-\infty}^{t} F_i(\varphi; u-t)\,d\xi_i(u) \tag{5.17}$$

with

$$H(\mathbf{x}; t) = \sum_{i=1}^{H} \oplus H(\mathbf{x}^{(i)}; t) = \sum_{n=1}^{M} \oplus H(\xi_i; t). \tag{5.18}$$

Let $f^{(i)}$ be the generating element of $A^{(i)}$. Since $\{E(b)-E(a)\}f^{(i)} = P_{H(\mathbf{x}^{(i)}; a,b)}f^{(i)}$, clearly $H(\mathbf{x}^{(i)})$ is the cyclic subspace generated by $f^{(i)}$, i.e.,

$$H(\mathbf{x}^{(i)}) = \mathfrak{S}\{E(\Delta)f^{(i)}, \Delta \text{ ranging over all finite subintervals of the real line}\}. \tag{5.19}$$

We also have $\rho_{f^{(i)}} \equiv \mu$. From (5.15) and (5.19), we have

$$H(\mathbf{x}) = \sum_{i=1}^{M} \oplus \mathfrak{S}\{E(\Delta)f^{(i)}, \Delta \text{ ranging over all finite subintervals}\}$$

and

$$\rho_{f^{(1)}} \equiv \rho_{f^{(2)}} \equiv \cdots \equiv \rho_{f^{(M)}}. \tag{5.20}$$

Hence, it follows that M is the multiplicity of the $\mathbf{x}_t$-process (see Section 2 where this notion is defined). Assembling all the results of this section together we observe that we have established the following basic representation theorem.

Theorem 5.1. *Let $\mathbf{x}_t$ $(-\infty < t < +\infty)$ be a weakly stationary, purely non-deterministic process on Φ satisfying* (C). *Then*

$$\mathbf{x}_t(\varphi) = \sum_{i=1}^{M} \int_{-\infty}^{t} F_i(\varphi; u-t)\,d\xi_i(u), \tag{5.21}$$

where

(i) *M is the multiplicity of the process,*

(ii) *each $\xi_i(u)$ is a process with stationary orthogonal increments* (*homogeneous process*) *and the ξ_i's are mutually orthogonal. Furthermore,*

$$H(\mathbf{x}; t) = \sum_{i=1}^{M} \oplus H(\xi_i; t) \qquad \textit{for every real } t,$$

and

$$\sum_{1}^{M} \int_{-\infty}^{0} |F_i(\varphi; u)|^2\,d\mu(u) \qquad \textit{is finite.}$$

It can be easily seen that the homogeneous processes ξ_i $(i = 1, 2, \cdots M)$ of the representation (5.21) are uniquely determined up to a unitary equivalence.

The above theorem is a generalization of the Karhunen representation to stationary stochastic processes $\mathbf{x}_t$ on Φ. This result also generalizes the Rozanov–Gladyshev representation for q-dimensional stationary processes as will be seen in the next section. The reader will observe that (5.21) has been derived essentially independently of the Hida representation (2.1) and the latter is referred to at the end of the proof only for the purpose of identifying M as the multiplicity of the process. Indeed, the whole point of the problem is to study the maximal spectral type and to construct the homogeneous processes $\xi_i(u)$. Once (5.21) has been obtained, however, it is easy to discover the special properties that the representation possesses in this case, e.g. to see that all the elements $f^{(i)}$ occurring in it are equivalent, with a common spectral type equivalent to μ. Moreover, starting with the ξ_i's one can construct without difficulty a sequence $\{f^{(i)}\}$ for the representation (2.1) of Section 2. This can be done as follows: It is clear that the elements f_i $(i = 1, \cdots, M)$ occurring in the proof of Theorem 5.1 and with the property that they have all the same spectral type equivalent to μ (see (5.19) and (5.20)) can be chosen as the elements in the Hida representation of $\mathbf{x}_t$. If we now set

$$\xi_i(\Delta) = \int_\Delta \left[\frac{d\rho_{f_i}}{d\mu}(u)\right]^{-1/2} dE(u)f_i,$$

it is easy to verify that the ξ_i are mutually orthogonal random set functions each having μ as its measure function, and that (Δ being a finite interval)

$$E(\Delta)f_i = \int_\Delta \left[\frac{d\rho_{f_i}}{d\mu}(u)\right]^{1/2} d\xi_i(u).$$

If we now make the appropriate substitution in (2.1) and compare it with the representation (5.21) it follows that for each t and φ

$$F_i(\varphi; t, u) = F_i(\varphi; u-t)\left[\frac{d\rho_{f_i}}{d\mu}(u)\right]^{-1/2} \qquad (i = 1, \cdots, M)$$

a.e. with respect to μ.

Thus, for stationary processes, the generalization of the approach of Hanner given in Theorem 5.1 leads to a deeper analysis which includes the proof of (5.19) and (5.20) and yields directly the representation we seek. It is interesting to explore further the connection between ρ and M. The following discussion presents another aspect of the problem and provides additional information.

Theorem 5.2. *ρ is a uniform spectral type with (uniform) multiplicity M.*

PROOF. We use the ideas of Plessner and Rohlin [12]. It will first be shown that ρ has multiplicity M. Let $\{A'_\beta\}$ be an orthogonal system of type ρ and cardinality M', i.e., a system of orthogonal cyclic parts A'_β of the operator A, the spectral type of each cyclic operator A'_β being ρ. According to the terminology of [12] M is the multiplicity of ρ if we can prove that $M' \leqq M$. Observe that neither M nor M' can exceed $\aleph_0$, for otherwise we would arrive at a contradiction of the fact that $H(\mathbf{x})$ is separable. Furthermore, there is obviously nothing to prove if $M = \aleph_0$. Thus the only case to be considered is when M is a finite cardinal. If possible let $M' > M$. We shall show that this leads to a contradiction. Let h_i $(i = 1, \cdots, M)$ be a generating element of the subspace $H(\mathbf{x}^{(i)})$ and h'_β $(\beta = 1, \cdots, M')$ be similarly a generating element of the cyclic subspace corresponding to A'_β. Clearly, there is no loss of generality in supposing that all these elements have the same spectral function, say ρ'. From (5.15) and (5.19) it follows that for each β we have

$$h'_\beta = \sum_{i=1}^{M} \int F_{i\beta}(u)\,dE(u)h_i, \qquad \text{where } \sum_i \int |F_{i\beta}(u)|^2\,d\rho'(u) \text{ is finite.}$$

For every measurable set Δ we obtain

$$\mathbf{E}\{E(\Delta)h'_\beta \cdot h'_\gamma\} = \int_\Delta \sum_{i=1}^{M} F_{i\beta}(u)\overline{F_{i\gamma}(u)}\,d\rho'(u).$$

The left-hand side of the above relation is zero if $\beta \neq \gamma$ and equals $\rho'(\Delta)$ if $\beta = \gamma$. Hence for u not belonging to a set $N_{\beta\gamma}$ of zero ρ'-measure we have

$$\sum_{i=1}^{M} F_{i\beta}(u)\overline{F_{i\gamma}(u)} = \delta_{\beta\gamma}.$$

Since M' is at most $\aleph_0$ the set $N = \bigcup_{\beta,\gamma} N_{\beta\gamma}$ is measurable and $\rho'(N) = 0$. Choosing a fixed point u_0 in the complement of N we see that

$$\sum_{i=1}^{M} F_{i\beta}(u_0)\overline{F_{i\gamma}(u_0)} = \delta_{\beta\gamma} \qquad \text{for all } \beta, \gamma. \tag{5.22}$$

If we now set $a_\beta = \{F_{1\beta}(u_0), \cdots, F_{M\beta}(u_0)\}$, the relations (5.22) imply that the a_β are M' orthonormal vectors in M-dimensional unitary space. Hence M' cannot exceed M. In other words ρ has multiplicity M.

The proof that the spectral type ρ is uniform is achieved by a modification of the above argument. The reader will no doubt, observe that the conclusion about uniformity rests on the fact that the orthogonal system $[A^{(i)}, i = 1, \cdots, M]$ is not only maximal but that the orthogonal sum of the $A^{(i)}$ is equal to A (see (5.16)).

Let σ be any spectral type dominated by ρ. The only change we make in the proof given above is to let $\{A'_\beta\}$ be an orthogonal system of type σ and cardinality M'. Let h'_β be a generating element of the cyclic subspace of A'_β. Assuming, as we may, that the h'_i have all the same spectral funtion ρ' and that the h'_β have the same spectral function σ' we obtain the relations

$$\sum_{i=1}^{M} F_{i\beta}(u)\overline{F_{i\gamma}(u)} = \frac{d\sigma'}{d\rho'}(u)\delta_{\beta\gamma}, \tag{5.23}$$

where $u \notin N$ and $d\sigma'/d\rho'$ is the Radon–Nikodym derivative of σ' with respect to ρ'. Since the set $S = \{u\colon (d\sigma'/d\rho')(u) > 0\}$ has positive ρ'-measure we can choose u_0 in $S \cap N^c$ when as before N is the set of zero ρ'-measure. Substituting u_0 for u in the relations (5.23) we are again led to the conclusion that $M' \leqq M$. Thus it has been shown that the multiplicity of any spectral type dominated by ρ is equal to the multiplicity of ρ. Hence ρ is a uniform spectral type.

REMARK. It follows at once from the theorem just proved that every spectral type belonging to the operator A of the stationary process $\mathbf{x}_t$ has multiplicity M.

To find the functions F_i and the value of M in the representation (5.21) in specific instances one would have to consider, individually, concrete examples of spaces Φ and perhaps have to assume additional properties of the process $\mathbf{x}_t$ such as linearity in φ. The study of some of these questions we postpone to a later paper. However, since it is important to relate our work to recent developments in the theory of multidimensional stationary processes, we consider in the next section the case when Φ is a q-dimensional unitary space.

6. Multiplicity as a Generalization of Rank

In the theory of finite dimensional weakly stationary processes the notion of rank plays a conceptually essential role. Zasukhin, in 1941, was the first to define the rank of a q-dimensional, discrete parameter stationary process as the rank of the $(q \times q)$ 'error matrix' (see [18]). More recently, the definition of rank for a continuous parameter process has been given by Glasdyshev [5] to be the rank of the discrete parameter process associated with the process. This point of view has been further explored in the recent thesis of Robertson [14].

It is also well known in the literature that the rank of the process is equal to the rank of the spectral density matrix. (See [15] where the rank is defined this way and [14].)

We shall show in this section that multiplicity M occurring in the representation given in Theorem 5.1 constitutes a generalization of rank in the following sense: If $\mathbf{x}_t$ is a weakly stationary process on Φ where Φ may be infinite dimensional (and $\mathbf{x}_t(\varphi)$ itself may or may not be linear in φ), then M is equal to the multiplicity of the associated discrete process (Theorem 6.1). In the case where Φ is a q-dimensional unitary space and $\mathbf{x}_t(\varphi)$ is linear in φ, so that we are dealing with a q-dimensional stationary process, it is shown in Theorem 6.2 that the multiplicity equals the rank of the process and the representation of Theorem 5.1 coincides with that obtained in [5] and [14].

The connection between multiplicity and spectral theory for infinite dimensional stationary processes $\mathbf{x}_t$ will be considered in a later paper.

If $\{\mathbf{x}_t\}(-\infty < t < +\infty)$ is a given stationary stochastic process on Φ satisfying condition (C), then, for each φ, the one dimensional weakly stationary process $\{\mathbf{x}_t(\varphi)\}$ is continuous in q.m. and hence, for fixed φ, $\mathbf{x}_t(\varphi) = \int_{-\infty}^{+\infty} e^{it\lambda} d_\lambda G(\lambda)\mathbf{x}_0(\varphi)$, where $\{G(\lambda), -\infty < \lambda < +\infty\}$ is a resolution of the identity of the unitary group $\{T_h\}$ of the $\mathbf{x}_t$-process,

With the process $\{\mathbf{x}_t(\varphi)\}$ (for fixed φ) is associated a discrete parameter process

$$\tilde{\mathbf{x}}_n(\varphi) = \int_{-\pi}^{\pi} e^{in\lambda} d_\lambda G\left(\frac{1}{2\pi} \tan^{-1} \lambda\right) \mathbf{x}_0(\varphi) \qquad (n = 0, \pm 1, \cdots) \tag{6.1}$$

([4], [11]). Let us now write for each φ and t, $H_\varphi(x; t) = \mathfrak{S}\{\mathbf{x}_\tau(\varphi), \tau \leqq t\}$ and $H_\varphi(\tilde{x}; m) = \mathfrak{S}\{\tilde{\mathbf{x}}_n(\varphi), n \leqq m\}$ (m any integer). We have, for all φ, $H_\varphi(\mathbf{x}; +\infty) = H_\varphi(\tilde{\mathbf{x}}; +\infty)$ and $H_\varphi(\mathbf{x}; 0) = H_\varphi(\tilde{\mathbf{x}}; 0)$ (see [4], [11]). Therefore,

$$H(\mathbf{x}; +\infty) = H(\tilde{\mathbf{x}}; +\infty) \quad \text{and} \quad H(\mathbf{x}; 0) = H(\tilde{\mathbf{x}}; 0). \tag{6.2}$$

From stationarity and (6.2), the following lemma is immediate.

Lemma 6.1. $\{\mathbf{x}_t, -\infty < t < +\infty\}$ *is deterministic if and only if* $\{\tilde{\mathbf{x}}_n, n = 0, \pm 1, \cdots\}$ *is deterministic.*

We recall here two lemmas from [5] which will be frequently used in what follows. It should be observed that in Lemma (G_2) stated below the process can be infinite-dimensional. Its proof, however, involves no change and is an easy consequence of (6.2).

Lemma (G_1). *If $\{\eta_t\}$ is a one-dimensional weakly stationary, continuous in q.m., purely non-deterministic process, then the $\tilde{\eta}_n$-process is purely non-deterministic.*

Lemma (G_2). *If $\{\boldsymbol{\eta}_t\}$ and $\{\zeta_t\}$ are stationary processes on Φ satisfying condition* (C) *and such that $H(\boldsymbol{\eta}; t) \subset H(\zeta; t)$ for all t, then $H(\tilde{\boldsymbol{\eta}}; m) \subset H(\tilde{\zeta}; m)$ for every m and conversely.*

We shall now obtain from Theorem 5.1 a representation for the $\tilde{\mathbf{x}}_n$-process. The notation will be that of Section 5. Let us define, for each $i = 1, 2, \cdots, M$,

$$\mathbf{x}_t^{(i)}(\varphi) = \int_{-\infty}^{t} F_i(\varphi; u-t)\, d\xi_i(u), \tag{6.3}$$

where the right-hand side expression is the term appearing in the representation (5.21) of $\mathbf{x}_t(\varphi)$. Consider now the process $h^{(i)}(t) = \int_{-\infty}^{t} e^{s-t}\, d\xi_i(s)$ $(-\infty < t < +\infty)$. Then $\{h^{(i)}(t)\}$ is a one-dimensional stationary stochastic process with $T_t h^{(i)}(0) = h^{(i)}(t)$. Furthermore, since

$$\xi_i(t) - \xi_i(s) = \{h^{(i)}(t) - h^{(i)}(s)\} + \int_s^t h^{(i)}(u)\, du \qquad (s < t),$$

it follows that for all t

$$H(\xi_i; t) = H(h^{(i)}; t) \qquad (i = 1, 2, \cdots, M). \tag{6.4}$$

The $h_t^{(i)}$-process which is obviously continuous in q.m., is also purely non-deterministic, since from (6.4)

$$\bigcap_{-\infty}^{+\infty} H(h^{(i)}; t) = \bigcap_{-\infty}^{+\infty} H(\xi_i; t) \subset \bigcap_{-\infty}^{+\infty} H(\mathbf{x}; t).$$

The discrete parameter process $\{\tilde{h}^{(i)}(m)\}$ is thus purely non-deterministic and therefore has a moving average representation given by

$$\tilde{h}^{(i)}(m) = \sum_{l=0}^{\infty} b_i(l) u_i(m-l), \tag{6.5}$$

where

$$H(\tilde{h}^{(i)}; m) = \mathfrak{S}\{u_i(m-l), 0 \leqq l < +\infty\} \tag{6.6}$$

and $\{u_i(m)\}$ (for fixed i) is a process with stationary orthogonal increments. From (6.2), (6.4), (6.6) and the mutual orthogonality of $\{\xi_i\}$, it follows that the processes $\{u_i(n)\} (i = 1, 2, \cdots, M)$ are mutually orthogonal. Also from (6.3) and (6.4), $H(\mathbf{x}^{(i)}; t) \subset H(h^{(i)}; t)$ for each t. But from Lemma (G_2) and (6.6), $H(\tilde{\mathbf{x}}^{(i)}; m)$ is a subspace of $\mathfrak{S}(u_i(m-l), l = 0, 1, 2, \cdots\}$. Hence

$$\tilde{\mathbf{x}}_m^{(i)}(\varphi) = \sum_{l=0}^{\infty} C_i(\varphi; l) u_i(m-l). \tag{6.7}$$

From (6.3) the $\{x_t^{(i)}(\varphi)\}$ process is stationary and continuous in q.m. with $T_t \mathbf{x}_s^{(i)}(\varphi) = \mathbf{x}_{s+t}^{(i)}(\varphi)$. Hence $\mathbf{x}_t^{(i)}(\varphi) = \int_{-\infty}^{+\infty} e^{it\lambda} d_\lambda G(\lambda) \mathbf{x}_0^{(i)}(\varphi)$. Furthermore,

$$\mathbf{x}_t(\varphi) = \sum_1^M \mathbf{x}_t^{(i)}(\varphi) \text{ for every } t, \tag{6.8}$$

where the (possibly) infinite series converges in q.m., since $\sum_1^M \mathbf{E}|\mathbf{x}_t^{(i)}(\varphi)|^2$ is finite. Also,

$$\tilde{\mathbf{x}}_n^{(i)}(\varphi) = \int_{-\pi}^{\pi} e^{in\lambda} d_\lambda G\left(\frac{1}{2\pi}\tan^{-1}\lambda\right)\mathbf{x}_0(\varphi). \tag{6.9}$$

Since $S = \int_{-\pi}^{\pi} e^{i\lambda} d_\lambda G((1/2\pi)\tan^{-1}\lambda)$ is a bounded linear (in fact, unitary) operator on $H(\mathbf{x})$, from (6.8) [with $t = 0$], (6.9) and (6.1), we have

$$\tilde{\mathbf{x}}_n(\varphi) = \sum_1^M \tilde{\mathbf{x}}_n^{(i)}(\varphi). \tag{6.10}$$

From (6.7) and (6.10), $\tilde{\mathbf{x}}_n(\varphi) = \sum_{i=1}^M \sum_{l=-\infty}^n C_i(\varphi; n-l)u_i(l)$. From Theorem 5.1 and (6.4),

$$H(\mathbf{x}; t) = \sum_{i=1}^M \oplus H(\xi_i; t) = \sum_{i=1}^M \oplus H(h^{(i)}; t).$$

In other words,

$$H(\mathbf{x}; t) = \mathfrak{S}\{h^{(i)}(\tau), \tau \leqq t, i = 1, 2, \cdots, M\}. \tag{6.11}$$

From Lemma (G_2), (6.11) and (6.6) we have

$$H(\tilde{\mathbf{x}}; m) = \sum_{i=1}^M \oplus H(\tilde{h}^{(i)}; m) \doteq \sum_{i=1}^M \oplus \mathfrak{S}\{u_i(m-l), l = 0, 1, 2, \cdots\}. \tag{6.12}$$

(6.11) and (6.12) imply (see Theorem 4.1) that

$$M = \dim\{H(\tilde{\mathbf{x}}; n\} \ominus H(\tilde{\mathbf{x}}; n-1)\}. \tag{6.13}$$

We summarize the above results.

Theorem 6.1. *Let $\mathbf{x}_t(-\infty < t < +\infty)$ be a stationary, purely non-deterministic process satisfying condition* (C). *Then its multiplicity is equal to the common dimension of the subspaces $H(\tilde{\mathbf{x}}; n) \ominus H(\tilde{\mathbf{x}}; n-1)$.*

The above discussion pertaining to multiplicity is very general since we have been dealing with weakly stationary processes on an arbitrary Hausdorff space, satisfying the second countability axiom. It is instructive to consider the case when Φ is a finite-dimensional unitary space and the process $\mathbf{x}_t$ is linear on Φ. We have referred to the fact that some recent work of H. Cramér [2] can be regarded as a special case of the results of Section 2. In [2], Cramér also includes a brief discussion of the stationary case and shows that the multiplicity of the q-dimensional process does not exceed q. We shall now deduce from Theorem 6.1 that the multiplicity is actually equal to the rank of the process. This corollary (Theorem 6.2), incidentally, provides an alternative proof of a theorem due to Gladyshev (Theorem 1, [5]).

Suppose $\{e_i\}(i = 1, 2, \cdots, q)$ is an orthonormal basis in Φ. If $\{\mathbf{x}_t\}$ is a weakly stationary process linear in φ then, if $\varphi = \sum_{i=1}^q a_i e_i$, $\mathbf{x}_t(\varphi) = \sum_{i=1}^q a_i x_i(t)$ where $x_i(t) = \mathbf{x}_t(e_i)$. Now, $(x_1(t), x_2(t), \cdots, x_q(t))$ is a q-dimensional process which is weakly stationary. Since $\{\mathbf{x}_t\}$ satisfies condition (C), $\{x_i(t)\}(i = 1, 2, \cdots q)$ are continuous in q.m. Also if $(x_1(t), x_2(t), \cdots, x_q(t))$ is a q-dimensional weakly stationary process continuous in q.m., then there corresponds a stationary process $\{\mathbf{x}_t\}$ on the q-dimensional unitary space Φ which is linear in φ and satisfies conditon (C); [viz., $\mathbf{x}_t(\varphi) = \sum_{i=1}^{q} a_i x_i(t)$ if φ is the vector $(a_1, a_2, \cdots, a_q)$]. Furthermore $H(\mathbf{x}; t) = \mathfrak{S}[x_i(u), u \leqq t, i = 1, 2, \cdots q]$.

Theorem 6.2. *Let $(x_1(t), x_2(t), \cdots, x_q(t))$ be a continuous in q.m., purely non-deterministic, weakly stationary process. Then*

$$x_i(t) = \sum_{i=1}^{M} \int_{-\infty}^{t} F_{in}(u-t)\,d\xi_i(u),$$

where the ξ_i-processes and the number M are as introduced in Theorem 5.1,

$$\mathfrak{S}[x_i(u), u \leqq t, i = 1, 2, \cdots, q] = \sum_{i=1}^{M} \oplus H(\xi_i; t)$$

and M is the rank of the process.

Proof. All the assertions of the theorem follow immediately upon setting $\varphi = e_i$ in the representation obtained in Theorem 5.1. It remains only to show that M is the rank of the process. From Theorem 6.1 and Lemma 4.1 it follows that $M = \dim\,[H(\tilde{\mathbf{x}}; n) \ominus H(\tilde{\mathbf{x}}; n-1)] = \dim\{\mathfrak{S}[\tilde{g}_n(\varphi), \varphi \in \Phi]\}$. Writing $\tilde{\mathfrak{S}}_n = \mathfrak{S}[\tilde{\mathbf{x}}_i(m), m \leqq n$, integer, $l = 1, 2, \cdots, q]$ and $\tilde{g}_i(n) = \tilde{x}_i(n) - P_{\tilde{\mathfrak{S}}_{n-1}}\tilde{x}_i(n)$ we find that $\tilde{g}_n(\varphi) = \sum_{i=1}^{q} a_i \tilde{g}_i(n)$. Therefore, $M = \dim \mathfrak{S}[g_i(n), i = 1, 2, \cdots, q]$. But the latter quantity is the rank of the $(q \times q)$ 'error matrix' with elements $\mathbf{E}\tilde{g}_i(0)\tilde{g}_j(0) (i, j = 1, 2, \cdots, q)$, i.e., the rank of the process $(\tilde{x}_i(n), \tilde{x}_2(n), \cdots, \tilde{x}_q(n))$, [18]. Hence the multiplicity M of the $\mathbf{x}_t$-process (Theorem 5.1) equals its rank.

Theorems 4.1, 5.1 and 6.1 apply to weakly stationary processes $\mathbf{x}_t$ on a Hausdorff space Φ. The only assumptions on the process is that it satisfies condition (C) and is purely non-deterministic, while no condition is imposed on Φ other than that its topology satisfy the second countability axiom. If, in particular, Φ is a locally convex linear space (e.g. if Φ is an infinite-dimensional separable Hilbert space) with a countable basis $\{e_i\}$ and if $\mathbf{x}_t(\varphi)$ is linear in φ(e.g. $\mathbf{x}_t$ is a weak process on Φ), then we may consider the $\mathbf{x}_t$-process as having an infinite number of components $x^i_{(t)} = \mathbf{x}_t(e_i)$ $(i = 1, 2, \cdots)$. Thus we may conclude from these results and Theorem 6.2 that for infinite-dimensional processes the representation given in Theorem 5.1 is a generalization of the Karhunen–Gladyshev representation and that the multiplicity is the appropriate generalization of rank.

Hilbert-Space Valued Processes

7. Preliminaries

In Theorem 2.2, and for the stationary case in Theorem 5.1 we obtained a representation of the purely non-deterministic process on an arbitrary Hausdorff space Φ. Suppose now that Φ is a locally convex linear space and that for each t, $\mathbf{x}_t$ is a random variable taking values in Φ', the dual space of Φ; i.e., for each t, there exists a mapping $\mathbf{x}_t$ from Ω to Φ' such that (1) $\langle x_t, \varphi\rangle (\langle \varphi', \varphi\rangle$ denotes the value of the functional φ' at φ) is a random variable on Ω, and (2) for all $\varphi \in \Phi$, $\mathbf{x}_t(\varphi)\,[\omega] = \langle \mathbf{x}_t(\omega), \varphi\rangle$ with probability one. As is well-known these assumptions are stronger than the ones made in the concluding paragraph of Section 6 dealing with weak processes. We shall call $\{\mathbf{x}_t\}$ defined as above a process in Φ'. The definitions of deterministic and purely non-deterministic processes in Φ' are the same as the one given in the introduction.

By a representation of a purely non-deterministic process $\{\mathbf{x}_t\}$ in Φ', we mean a process $\{\mathbf{y}_t\}$ in Φ' such that $\mathbf{x}_t = \mathbf{y}_t$ with probability one for each t and $\mathbf{y}_t$ represents a 'moving average' over the present and past of the $\mathbf{x}_t$-process analogous to what was obtained in Theorem 2.2. In this section we confine our attention to the case in which Φ is a real separable Hilbert space and refer to $\{\mathbf{x}_t\}$ as a process in Φ. Although this is the only case studied in detail here, we feel that a similar theory can be developed to cover more general situations, e.g., where Φ is a separable, reflexive Banach space or a nuclear space. The last mentioned problem could well have points of contact with recent work of K. Urbanik and others on the representation of purely non-deterministic homogeneous generalized random fields ([17]).

We shall also make the stronger assumption that $\mathbf{E}||\mathbf{x}_t||^2$ is finite for each t, with the help of which we are able to prove a strengthened form of the Wold decomposition stated in Section 2.

Proposition 7.1. *Let $\{\mathbf{x}_t\}$ be a process in Φ with $\mathbf{E}||\mathbf{x}_t||^2 < \infty$, for each t. Then, with probability one we have $\mathbf{x}_t = \mathbf{x}_t^{(1)} + \mathbf{x}_t^{(2)}$ and $\mathbf{x}_t^{(i)}$, $i = 1, 2$, which are defined except possibly for an ω-set of probability zero, have the following properties*:

(1) *$\{\mathbf{x}_t^{(1)}\}$ and $\{\mathbf{x}_t^{(2)}\}$ are processes in Φ with $\mathbf{E}||\mathbf{x}_t^{(i)}||^2 < \infty$ $(i = 1, 2)$,*

(2) *$H(\mathbf{x}^{(1)})$ is orthogonal to $H(\mathbf{x}^{(2)})$, and*

(3) *$\{\mathbf{x}_t^{(1)}\}$ is deterministic and $\{\mathbf{x}_t^{(2)}\}$ is purely non-deterministic.*

PROOF. The process $\tilde{\mathbf{x}}_t(\varphi) = \langle \mathbf{x}_t, \varphi \rangle$ is a stochastic process on Φ. Hence Proposition 2.1 gives us $\tilde{\mathbf{x}}_t(\varphi) = \tilde{\mathbf{x}}_t^{(1)}(\varphi) + \tilde{\mathbf{x}}_t^{(2)}(\varphi)$. It suffices to show that $\tilde{\mathbf{x}}_t^{(i)}(\varphi) = \langle \mathbf{x}_t^{(i)}, \varphi \rangle$ $(i = 1, 2)$ where $\{\mathbf{x}_t^{(i)}\}$ are processes in Φ with the above mentioned properties. This is achieved by means of the following lemma.

Lemma 7.1. *Let $\{\mathbf{x}_t\}$ be a process in Φ and let P be a projection operator onto an arbitrary subspace of $H(\mathbf{x}; t)$. Then there exists an almost everywhere weakly measurable mapping $\mathbf{x}_{t,P}$ from Ω to Φ such that with probability one $\langle \mathbf{x}_{t,P}, \varphi \rangle = P\langle \mathbf{x}_t, \varphi \rangle$ for every $\varphi \in \Phi$.*

PROOF. Let t be fixed. It is well-known that our assumptions on $\mathbf{x}_t$ imply that, for all φ_1, φ_2 in Φ, $\mathbf{E}\langle \mathbf{x}_t, \varphi_1 \rangle \cdot \langle \mathbf{x}_t, \varphi_2 \rangle] = \langle B_t\varphi_1, \varphi_2 \rangle$, where B_t is an S-operator (see [13]). Choosing a complete orthonormal (c.o.n) system of eigenelements corresponding to the eigenvalues $\{\lambda_n\}$ of B_t and observing that B_t has finite trace, we obtain $\sum_1^\infty [P\langle \mathbf{x}_t(\omega), \varphi_n \rangle]^2 < \infty$. This implies that there is an ω-set N of zero probability such that

$$\sum_1^\infty [P\langle \mathbf{x}_t(\omega), \varphi_n \rangle]^2 \text{ is finite, if } \omega \notin N. \tag{7.1}$$

For every $\varphi \in \Phi$ and $\omega \notin N$, define

$$\eta_{t,P}(\varphi)[\omega] = \sum_{n=1}^\infty \langle \varphi, \varphi_n \rangle [P\langle \mathbf{x}_t(\omega), \varphi_n \rangle]. \tag{7.2}$$

Then $\eta_{t,P}$ is an a.e. weakly measurable, bounded linear functional on Φ. Hence, $\eta_{t,P}(\varphi)[\omega] = \langle \eta_{t,P}(\omega), \varphi \rangle$ for $\omega \notin N$. Clearly, for each φ,

$$\mathbf{E}[P\langle \mathbf{x}_t(\omega), \varphi \rangle - \langle \eta_{t,P}(\omega), \varphi \rangle]^2 = 0$$

and from (7.1) $||\eta_{t,P}(\omega)||^2$ is finite. If $\{\chi_m\}$ is any other c.o.n. system, then following the above argument we obtain an a.e. weakly measurable function $\zeta_{t,P}$ from Ω to Φ such that

$$\langle\zeta_{t,P}(\omega), \varphi\rangle = P\langle \mathbf{x}_t(\omega), \varphi\rangle, \qquad ||\zeta_{t,P}(\omega)||^2 < \infty,$$

and

$$\mathbf{E}[P\langle \mathbf{x}_t(\omega), \varphi\rangle - \langle\zeta_{t,P}(\omega), \varphi\rangle]^2 = 0 \text{ for every } \varphi.$$

Thus we have

$$||\eta_{t,P}(\omega) - \zeta_{t,P}(\omega)||^2 = \sum_1^\infty \mathbf{E}[\langle\eta_{t,P}(\omega) - \zeta_{t,P}(\omega), \chi_m\rangle]^2 = 0, \tag{7.3}$$

since, for every φ, $\mathbf{E}[\langle\eta_{t,P}(\omega), \varphi\rangle - \langle\zeta_{t,P}(\omega), \varphi\rangle]^2 = 0$. Let $\mathscr{L}_2(\Omega, \mathbf{P})$ be the space of weakly measurable functions g from Ω to Φ satisfying $\mathbf{E}||g(\omega)||^2 < \infty$ (strictly speaking, equivalence classes of functions, see Section 8). From (7.3) we see that $\eta_{t,P}$ and $\zeta_{t,P}$ are elements of the same equivalence class, say, $\mathbf{x}_{t,P}$ belonging to $\mathscr{L}_2(\Omega, \mathbf{P})$. Identifying $\mathbf{x}_{t,P}$ with any of its elements we have $\langle \mathbf{x}_{t,P}, \varphi\rangle = P\langle \mathbf{x}_t, \varphi\rangle$.

Since $\tilde{\mathbf{x}}_t^{(1)}(\varphi) = P_{H(\mathbf{x};-\infty)}\langle \mathbf{x}_t, \varphi\rangle$ and $\tilde{\mathbf{x}}_t^{(2)}(\varphi) = P_{H(\mathbf{x};t)\cap H^{\perp}(\mathbf{x};-\infty)}\langle \mathbf{x}_t, \varphi\rangle$, it follows from the lemma that there exist processes $\{\mathbf{x}_t^{(1)}\}$, $\{\mathbf{x}_t^{(2)}\}$ in Φ, defined for each t, except possibly on a null ω-set such that $\tilde{\mathbf{x}}_t^{(i)}(\varphi) = \langle \mathbf{x}_t^{(i)}, \varphi\rangle$ for $i = 1, 2$. Obviously, $\{\mathbf{x}_t^{(i)}\}$ satisfy all the other desired properties.

Before proving the representation theorem for purely non-deterministic processes $\mathbf{x}_t$ in Φ, we need to introduce stochastic integrals taking values in Φ, which we shall call stochastic Pettis integrals.

8. Stochastic Pettis Integrals

Let $(A, \mathfrak{A}, \mu)$ be an arbitrary σ-finite measure space and $\mathscr{L}_2(A, \mu)$ be the set of all weakly measurable functions g from A to Φ such that $\int ||g(a)||^2 \, d\mu(a)$ is finite. It is well known that upon identifying functions which are equal almost everywhere $[\mu]$ (i.e., setting $f = g$ if $\int ||f(a) - g(a)||^2 \, d\mu(a) = 0$), $\mathscr{L}_2(A, \mu)$ becomes a Hilbert space with inner product given by

$$(g_1, g_2)_{\mathscr{L}_2(A,\mu)} = \int \langle g_1(a), g_2(a)\rangle \, d\mu(a).$$

The norm of g will be denoted by $||g||_{\mathscr{L}_2(A,\mu)}$. It is easy to show that $\mathscr{L}_2(A, \mu)$ is separable if the Hilbert space $L_2(A, \mu)$ of real functions square integrable with respect to μ is separable. In particular, if $A = T$, the real line, and μ is a σ-finite measure on Borel sets, then the Hilbert space $\mathscr{L}_2(T, \mu)$ is separable. In what follows we write $\mathscr{L}_2(\mu)$ for $\mathscr{L}_2(T, \mu)$.

Lemma 8.1. *Let z be a real orthogonal random set function with $\mathbf{E}[z(\Delta)]^2 = \rho(\Delta)$. If $g \in \mathscr{L}_2(\rho)$, then there exists an a. e. $[\rho]$ weakly measurable mapping $J(g)$ from Ω to Φ with the following properties*:

$$J(g) \in \mathscr{L}_2(\Omega, \mathbf{P}); \tag{8.1}$$

if g_1, g_2 are any elements of $\mathscr{L}_2(\rho)$ and c_1, c_2 are real numbers, then

$$J(c_1 g_1 + c_2 g_2) = c_1 J(g_1) + c_2 J(g_2), \tag{8.2}$$

the equality holding in the sense of $\mathscr{L}_2(\Omega, \mathbf{P})$;

for every $\varphi \in \Phi$,

$$(8.3) \qquad \langle J(g), \varphi \rangle = \int \langle g(t), \varphi \rangle \, dz(t)$$

with probability one, where the right-hand side integral is an ordinary stochastic integral.

The element $J(g)$ *is called the Stochastic Pettis integral of* $g(t)$ *with respect to* z *and is written* $\int g(t)\,dz(t)$. *We also have*

$$(8.4) \qquad \mathbf{E}\left[\left\langle \int g_1(t)\,dz(t), \int g_2(t)\,dz(t) \right\rangle\right] = \int \langle g_1(t), g_2(t)\rangle \, d\rho(t).$$

PROOF. Let $\{\varphi_k\}$ be a c.o.n. system in Φ and let g be any element of $\mathscr{L}_2(\rho)$. Strictly speaking, each g represents an equivalence class belonging to $\mathscr{L}_2(\rho)$ and it is clear that elements of this equivalence class give rise to the same stochastic integral $\int \langle g(t), \varphi_k \rangle \, dz(t)$ since the latter is itself defined up to an equivalence. Denoting it (more precisely, a random variable belonging to the equivalence class) by $L(g, \varphi_k)$, we have

$$\sum_{k=1}^{\infty} \mathbf{E}[L(g, \varphi_k)]^2 = \sum_{k=1}^{\infty} \int \langle g(t), \varphi_k \rangle^2 \, d\rho(t) < \infty,$$

so that $\sum_{k=1}^{\infty} [L(g, \varphi_k)[\omega]]^2 < \infty$ except possibley when ω is a set N of zero $\mathbf{P}$-measure. If for any φ, we now set

$$L(g, \varphi)[\omega] = \sum_{k=1}^{\infty} \langle \varphi, \varphi_k \rangle L(g, \varphi_k)[\omega] \qquad (\omega \notin N),$$

it follows that $L(g, \cdot)[\omega]$ is a bounded linear functional on Φ. Hence we obtain

$$L(g, \varphi)[\omega] = \langle J_1(g)[\omega], \varphi \rangle,$$

where $J_1(g)[\omega] \in \Phi$. It is further easy to see that $J_1(g)[\cdot]$ is a.e. weakly measurable and that $\mathbf{E}\|J_1(g)[\omega]\|^2$ is finite. It is evident that we have relied on the choice of a particular c.o.n. system in our definition of $J_1(g)$. However, if $\{\psi_m\}$ is any other c.o.n. system in Φ and $J_2(g)[\cdot]$ is the corresponding a.e. weakly measurable mapping, then we have

$$\mathbf{E}\|J_1(g)[\omega] - J_2(g)[\omega]\|^2 = 0, \qquad \text{i.e., } \|J_1(g) - J_2(g)\|_{\mathscr{L}_2(\Omega, \mathbf{P})} = 0.$$

In other words, $J_1(g)$ and $J_2(g)$ belong to the same equivalence class, say $J(g)$, of $\mathscr{L}_2(\Omega, \mathbf{P})$. Thus, the equivalence class $J(g)$ in $\mathscr{L}_2(\Omega, \mathbf{P})$ is unambiguously defined for each g in $\mathscr{L}_2(\rho)$ and further $\|g\|_{\mathscr{L}_2(\rho)} = \|J(g)\|_{\mathscr{L}_2(\Omega, \mathbf{P})}$. For every $g \in \mathscr{L}_2(\rho)$, the corresponding element $J(g)$ of $\mathscr{L}_2(\Omega, \mathbf{P})$ will be called the stochastic Pettis integral of g with respect to the orthogonal process z and will be denoted by $\int g(t)\,dz(t)$. The assertions (8.2)–(8.4) of the lemma are easy to verify.

If z_1, z_2 are orthogonal random set functions with measure functions ρ_1 and ρ_2 respectively and are further mutually orthogonal, then it can be shown that

$$\mathbf{E}\left[\left\langle \int g_1(t)\,dz_1(t), \int g_2(t)\,dz_2(t) \right\rangle\right] = 0 \quad \text{for} \quad g_1 \in \mathscr{L}_2(\rho_1), g_2 \in \mathscr{L}_2(\rho_2).$$

The proof follows by the definition of the Pettis integral.

The following result will be useful in the next section.

Lemma 8.2. *Let* z_k $(k = 1, 2, \cdots)$ *be mutually orthogonal processes with or-*

thogonal increments and with respective measure functions ρ_k. If $g_k \in \mathscr{L}_2(\rho_k)$ are such that

$$\sum_{k=1}^{\infty} \int \|g_k(t)\|^2 d\rho_k(t) \text{ is finite,} \tag{8.5}$$

then $\sum_{k=1}^{\infty} \int g_k(t) dz_k(t)$ is an element of $\mathscr{L}_2(\Omega, \mathbf{P})$ (the series of stochastic Pettis integrals converging in the $\mathscr{L}_2(\Omega, \mathbf{P})$ sense), and for every $\varphi \in \Phi$,

$$\left\langle \sum_{k=1}^{\infty} \int g_k(t) dz_k(t), \varphi \right\rangle = \sum_{k=1}^{\infty} \int \langle g_k(t), \varphi \rangle dz_k(t)$$

with probability one.

PROOF. If is clear from the definition of $\int g_k(t) dz_k(t)$ that $\{\zeta_m\}$ where $\zeta_m = \sum_1^m \int g_k(t) dz_k(t)$ is a Cauchy sequence of elements in $\mathscr{L}_2(\Omega, \mathbf{P})$, since $(m' > m)$

$$\|\zeta_{m'} - \zeta_m\|^2_{\mathscr{L}_2(\Omega, \mathbf{P})} = \sum_m^{m'} \int \|g_k(t)\|^2 d\rho_k(t) \to 0$$

by (8.5). Hence the limit (in $\mathscr{L}_2(\Omega, \mathbf{P})$ sense) of ζ_m exists which we denote by $\sum_1^{\infty} \int g_k(t) dz_k(t)$. The other conclusions of the lemma are similarly proved.

9. Representation Theorems For Purely Non-Deterministic Hilbert-Space-Valued Processes

In this section we consider a purely non-deterministic process $\{\mathbf{x}_t\}$ in Φ, with $\mathbf{E}\|\mathbf{x}_t\|^2$ finite. As in Section 2 we confine ourselves to the continuous parameter case. The representation we seek for $\mathbf{x}_t$ is obtained in terms of stochastic Pettis integrals. Since

$$\mathbf{E}[\langle \mathbf{x}_t, \varphi \rangle - \langle \mathbf{x}_t, \psi \rangle]^2 \leqq \mathbf{E}\|\mathbf{x}_t\|^2 \|\varphi - \psi\|^2,$$

it follows that the $\mathbf{x}_t$-process is continuous in the topology of Φ. Hence, from Lemma 2.1, the space $H(\mathbf{x})$ is separable provided the limits $\mathbf{x}_{t-0}(\varphi)$ and $\mathbf{x}_{t+0}(\varphi)$ exist for each $\varphi \in \Phi$. We shall refer to this condition as assumption (B).

Theorem 9.1. *Let $\{\mathbf{x}_t\}$ be a purely non-deterministic process in Φ with $\mathbf{E}\|\mathbf{x}_t\|^2$ finite and satisfying assumption* (B). *Then for each t, with probability one,*

$$\mathbf{x}_t = \sum_1^{M_0} \int_{-\infty}^t F_n(t, u) dz_n(u) + \sum_{t_j \leq t} \sum_{l=1}^{M_j} b_{jl}(t) \xi_{jl} \tag{9.1}$$

where M_0, M_j, the processes z_n and the random variables ξ_{jl} have the same meaning as in Theorem 2.2.

Furthermore, for each t,

$$F_n(t, \cdot) \in \mathscr{L}_2(\rho_n), \tag{9.2}$$

ρ_n being the measure function of z_n, and $b_{jl}(t) \in \Phi$ for every j, l;

$$\sum_{n=1}^{M_0} \int_{-\infty}^t \|F_n(t, u)\|^2 d\rho_n(u) < \infty; \tag{9.3}$$

$$\sum_{j=1}^{\infty} \sum_{l=1}^{M_j} \|b_{jl}(t)\|^2 \mathbf{E}(\xi_{jl}^2) < \infty; \tag{9.4}$$

and

(9.5) $$H(\mathbf{x};t) = \mathfrak{S}\{H(\mathbf{z};t) \bigcup H(\xi;t)\}$$

for every t, where

$$H(\mathbf{z};t) = \mathfrak{S}[z_n(u)|u \leqq t, n = 1, \cdots, M_0] \text{ and}$$

$$H(\xi;t) = \mathfrak{S}[\xi_{jl}[l = 1, \cdots, M_j, t_j \leqq t].$$

PROOF. Since $\langle \mathbf{x}_t, \varphi \rangle$, is a s. P. on Φ, Theorem 2.2 applies without any change to it. Furthermore, it has been shown in Section 3 that the representation for $\langle \mathbf{x}_t, \varphi \rangle$ can be chosen to be proper canonical without changing the numbers M_0 and M_j and hence without affecting the multiplicity M of the process. This accounts for the conclusion (9.5) of the theorem. In order to prove the remaining assertions we need to use the additional hypothesis in the present case, viz., that $\mathbf{E}||\mathbf{x}_t||^2 < \infty$.

From Theorem 2.2, we obtain

(9.6) $$\sum_{n=1}^{M_0} \sum_{k=1}^{\infty} \int_{-\infty}^{t} F_n^2(\varphi_k; t, u) d\rho_n(u) \leqq \mathbf{E}||\mathbf{x}_t||^2 < \infty,$$

where $\{\varphi_k\}$ is a c.o.n. system in Φ. A fortiori, there exists a set A_n of ρ_n-measure zero such that for $u \notin A_n$

(9.7) $$\sum_{k=1}^{\infty} F_n^2(\varphi_k; t, u) < \infty.$$

For $\varphi \in \Phi$ setting $c_k = \langle \varphi, \varphi_k \rangle$, we obtain from (9.7) that, for $u \notin A_n$, $\sum_k c_k F_n(\varphi_k; t, u)$ converges and is in fact equal to $F_n(\varphi; t, u)$ a.e. $[\rho_n]$. Hence $F_n(\varphi; t, u)$ is a bounded linear functional on Φ for $u \notin A_n$. We may therefore write $F_n(\varphi; t, u) = \langle F_n(t, u), \varphi \rangle$, where $F_n(t, u)$ is an element of Φ and, moreover, $F_n(t, \cdot)$ is an element of $\mathscr{L}_2(\rho_n)$. From (9.6) we have

(9.8) $$\sum_{n=1}^{M_0} \int_{-\infty}^{+\infty} ||F_n(t, u)||^2 d\rho_n(u) < \infty.$$

Since $\mathbf{E}||\mathbf{x}_t||^2$ is finite it follows that for all j and l there exists a bounded linear functional $b_{jl}(t)$ such that, for each t,

(9.9) $$b_{jl}(\varphi; t) = \langle b_{jl}(t), \varphi \rangle \quad \text{with} \quad \sum_{j,l} ||b_{jl}(t)||^2 \sigma_{jl}^2 < \infty.$$

By (9.8), Lemma 8.2 and (9.9), we have

$$\mathbf{x}_t = \sum_{n=1}^{M_0} \int_{-\infty}^{t} F_n(t, u) dz_n(u) + \sum_{t_j \leqq t} \sum_{l=1}^{M_i} b_{jl}(t) \xi_{jl}.$$

The corresponding results for weakly stationary (see Introduction for definition of stationarity) Φ-valued processes are stated below without proof.

Theorem 9.2. *A discrete parameter weakly stationary, purely non-deterministic, process in Φ, with $\mathbf{E}||\mathbf{x}_t||^2 < \infty$, has the following representation*:

$$\mathbf{x}_n = \sum_{m=-\infty}^{n} \sum_{l=1}^{M} b_l(n-m) \xi_l(m).$$

Here M is the multiplicity of $\{\mathbf{x}_n\}$,

(I) *the discrete parameter processes $\{\xi_l(m)\}$ $(l = 1, \cdots, M)$ have orthogonal increments and are mutually orthogonal;*

(II) $H(\mathbf{x}; n) = \sum_{i=1}^{M} \oplus H(\xi_i; n)$ *for each* n,

(III) $b_i(n-m) \in \Phi$ *with* $\sum_{m=-\infty}^{0} \sum_{i=1}^{M} ||b_i^2(m)||^2 \mathbf{E}[\xi_i^2(m)] < \infty$.

The number M *is the multiplicity associated with the stochastic process.*

Theorem 9.3. *Let* $\{\mathbf{x}_t\}$ *be a continuous parameter weakly stationary process with values in* Φ *satisfying the assumptions of Theorem* 9.1 *and condition* (C). *Then for each* t, *with probability one,*

$$\mathbf{x}_t = \sum_{n=1}^{M} \int_{-\infty}^{t} F_n(u-t)\, d\xi_n(u).$$

In this representation

(I) *the* ξ_n*'s are mutually orthogonal and each* ξ_n *is a homogeneous orthogonal random set function (with Lebesgue measure* μ *for its measure function),*

(II) $H(\mathbf{x}; t) = \sum_{i=1}^{M} \oplus H(\xi_i, t)$ *for every* t,

(III) M *is the multiplicity of the process, and*

(IV) $F_n(u-t) \in \mathscr{L}_2(\mu)$ $(n = 1, \cdots, M)$ *such that*

$$\sum_{n=1}^{M} \int_{-\infty}^{0} ||F_n(u)||^2 d\mu(u) < \infty.$$

University of Minnesota
Minneapolis, Minnesota

Received by the editors
December 12, 1964

REFERENCES

[1] H. Cramér, *On some classes of non-stationary processes*, Proc. 4th Berkeley Sympos. Math. Statist. and Prob., II (1961), pp. 57–77.

[2] H. Cramér, *On the structure of purely non-deterministic processes*, Arkiv for Math., 4, 2–3 (1961), pp. 249–266

[3] H. Cramér, *Stochastic processes as curves in Hilbert space*, Theory Prob. Applications, 9 (1964), pp 169–179. (In the English translation.)

[4] J. L. Doob, *Stochastic Processes*, Wiley, New York, 1953.

[5] E. G. Gladyshev, *On multi-dimensional stationary random processes*, Theory Prob. Applications, 3(1958), pp. 425–428. (English translation.)

[6] O. Hanner, *Deterministic and non-deterministic processes*, Arkiv for Math., 1 (1950), pp. 161–177.

[7] T. Hida, *Canonical representations of Gaussian processes and their applications*, Mem. Coll. Sci. Univ. Kyoto, ser. A, 33 (1960), pp. 109–155.

[8] R. H. Jones, *Stochastic processes on a sphere as applied to meteorological 500-millibar forecasts*, Proc. of the Sympos. on Time Series Analysis, Brown-University, 1962. (Ed. by M. Rosenblatt, John Wiley and Sons).

[9] G. Kallianpur and V. Manderkar, *On the connection between multiplicity theory and O. Hanner's time domain analysis of weakly stationary stochastic processes*, Tech. Report 49, University of Minnesota, 1964.

[10] K. Karhunen, *Über lineare Methoden in der Wahrscheinlichkeitrechnung*, Ann. Acad. Sci. Fennicae, ser. AI, 37 (1947), pp. 3–79.

[11] P. Masani and J. Robertson, *The time domain analysis of a continuous parameter weakly stationary stochastic process*, Pacif. J. Math., 12 (1962), pp. 1361–1378.

[12] A. I. Plesner and V. A. Rokhlin, *Spectral theory of linear operators*, II, Uspekhi Mat. Nauk, I, 1 (1946), pp. 71–191. (In Russian.)

[13] Yu. V. Prokhorov, *Convergence of random processes and limit theorems in probability theory*, Theory Prob. Applications, 1 (1956), pp. 157–214. (English translation.)

[14] J. Robertson, *Multivariate continuous parameter weakly stationary stochastic processes*, Thesis, Indiana University, June, 1963.

[15] Yu. A. Rozanov, *Spectral theory of multi-dimensional stationary random processes with discrete time*, Uspekhi Mat. Nauk, XIII, 2 (1958), pp. 93–142. (In Russian.)

[16] S. Saks, *Theory of the Integrals* (translated by L. C. Young), G. E. Steckert and Co., New York, 1937.
[17] K. Urbanik, *A contribution to the theory of generalized stationary random fields*, Trans. of the Second Prague Conference, Prague, 1960, pp. 667–679.
[18] V. N. Zasukhin, *To the theory of multi-dimensional stationary processes*, DAN SSSR, 33 (1941), pp. 435–437. (In Russian.)

Editors' Comments on Papers 11 and 12

Theoretical research on the elaboration of the original Cramér–Hida results on the multiplicity representations continued and expanded during the last decade. These two papers, both by Cramér, are representative of this work.

The first discloses new aspects of the canonical decomposition by focusing its attention on the Hilbert-space-curve nature of the appropriate class of random processes. The fundamental nature of the spectral types of the "white" generators of a process is pointed out. Furthermore, examples of random processes with arbitrary multiplicity are provided, including one concerning the class of harmonizable processes.

The second paper raises and examines, on a preliminary basis, the questions of random processes with unit multiplicity. The conclusion appears to be that in addition to the discrete-time and the wide-sense stationary continuous-time processes, many others characterized by appropriate smoothness properties may have multiplicity equal to unity. Conditions on the autocovariance function, which is the sole quantity that determines the value of multiplicity, may be difficult to obtain.

These papers are reprinted with the permission of the author. The first has the permission of the Society for Industrial and Applied Mathematics. The second was originally published by the University of California Press and is reprinted by permission of the Regents of the University of California.

THEORY OF PROBABILITY
Volume IX AND ITS APPLICATIONS Number 2
1964

11

STOCHASTIC PROCESSES AS CURVES IN HILBERT SPACE

HARALD CRAMÉR

1. In this paper the theory of spectral multiplicity in a separable Hilbert space will be applied to the study of stochastic processes $x(t)$, where $x(t)$ is a complex-valued random variable with a finite second-order moment, while the parameter t may take any real values.

For an account of multiplicity theory we may refer to Chapter 7 of Stone's book [13] which deals with the case of a separable space. The treatment of the subject found e.g. in the books by Halmos [6] and Nakano [10] is mainly concerned with the more general and considerably more intricate case of a non-separable space.

Our considerations will apply to certain classes of curves in a purely abstract Hilbert space, and it is only a question of terminology when, throughout this paper, we confine ourselves to that particular realization of Hilbert space which has proved useful in probability theory.

We finally observe that all our statements may be directly generalized to the case of vector processes of the form $\{x_1(t), \cdots, x_n(t)\}$.

2. In Sections 2—4 we shall now introduce some basic definitions and some auxiliary concepts.

Consider the set of all complex-valued random variables x defined on a fixed probability space and satisfying the relations

$$\mathbf{E}\{x\} = 0, \qquad \mathbf{E}\{|x|^2\} < \infty.$$

Two variables x and y will be regarded as identical if

$$\mathbf{E}\{|x-y|^2\} = 0.$$

The set of all these variables forms a Hilbert space H, if the inner product is defined in the usual way:

$$(x, y) = \mathbf{E}\{x\bar{y}\}.$$

Convergence of sequences of random variables will always be understood as strong convergence in the topology thus introduced, i.e. as convergence in quadratic mean according to probability terminology.

If for every real t a random variable $x(t) \in H$ is given, the set of variables $x(t)$ may be regarded as a stochastic process with continuous time t or, alternatively, as a curve C in the Hilbert space H. It is well known that various properties of stochastic processes have been studied by regarding them as curves in H (cf., e.g., [7], [8], [11], [12], [2], [3]). We shall give in this paper some further applications of this point of view.

We define certain subspaces of H associated with the $x(t)$ curve or process by writing

$$H(x) = S\{x(u), -\infty < u < \infty\},$$
$$H(x, t) = S\{x(u), u \leqq t\},$$
$$H(x, -\infty) = \bigcap_t H(x, t).$$

Here we denote by $S\{\cdots\}$ the subspace of H spanned by the random variables indicated between the brackets.

Evidently $H(x)$ is the smallest subspace of H which contains the whole curve C generated by the $x(t)$ process, while $H(x, t)$ is the smallest subspace containing the "arc" of C formed by all points $x(u)$ with $u \leqq t$. If the parameter t is interpreted as time, $H(x, t)$ will perresent the "past and present" of the $x(t)$ process from the point of view of the instant t, while $H(x, -\infty)$ represents the "infinitely remote past" of the process.

For the processes $y(t), z(t), \cdots$ we use in the sequel the analogous notations $H(y, t)$, $H(z, t), \cdots$ for subspaces defined in the corresponding way.

In the sequel we shall only consider stochastic processes $x(t) \in H$ which are assumed to satisfy the following conditions (A) and (B):

(A) The subspace $H(x, -\infty)$ contains only the zero element of H.

(B) For all t the limits $x(t \pm 0)$ exist and $x(t-0) = x(t)$.

The condition (A) implies that $x(t)$ is a *regular*, or *purely non-deterministic* process (cf. [**3**]). From (B) it follows, as shown in [**3**], that the space $H(x)$ is separable, and that the $x(t)$ curve has at most an enumerable number of discontinuities. The condition $x(t-0) = x(t)$ is not essential, and is introduced here only in order to avoid some trivial complications.

As t increases through real values, the subspaces $H(x, t)$ will obviously form a never decreasing family. For a fixed finite t the union of all $H(x, u)$ with $u < t$ may not be a closed set, but if we define $H(x, t-0)$ as the closure of this union, it is easily proved that we always have $H(x, t-0) = H(x, t)$. Similarly, the union of the $H(x, u)$ for all real u may not be closed, but if $H(x, +\infty)$ is defined as its closure, we shall have $H(x, +\infty) = H(x)$.

Suppose that a certain time point t is such that for all $h > 0$ we have $H(x, t-h) \neq H(x, t+h)$. Every time interval $t-h < u < t+h$ will then contain at least one $x(u)$ not included in $H(x, t-h)$. This may be expressed by saying that the process receives a new impluse, or an *innovation*, during the interval $(t-h, t+h)$ for every $h > 0$. The set of all time points t with this property will be called the *innovation spectrum* of the $x(t)$ process.

3. Suppose that we are given a stochastic process $x(t)$ satisfying the conditions (A) and (B). By P_t we shall denote the projection operator in the Hilbert space $H(x)$ with range $H(x, t)$ It then follows from the properties of the $H(x, t)$ family given above that we have

$$\begin{aligned} & P_t \leqq P_u \text{ for } t < u, \\ (1) \qquad & P_{t-0} = P_t \text{ for all } t, \\ & P_{-\infty} = 0, \; P_{+\infty} = 1, \end{aligned}$$

where 0 and 1 denote respectively the zero and the identity operator in $H(x)$.

It follows that the P_t form a *spectral family* of projections or a *resolution of the identity* according to Hilbert space terminology. As we have seen, the projections P_t are, for all real t, uniquely determined by the given $x(t)$ process. We shall return to the properties of the P_t family in Section 5.

For any random variable $z \in H(x)$ with $\mathbf{E}\{|z|^2\} = 1$ we now define a stochastic process by writing

$$z(t) = P_t z.$$

It is then readily seen that $z(t)$ will be a process with orthogonal increments satisfying (A) and (B). Writing

$$F_z(t) = \mathbf{E}\{|z(t)|^2\},$$

it follows that $F_z(t)$ is, for any fixed z, a distribution function of t such that $F_z(t-0) = F_z(t)$ for all t.

The points of increase of $z(t)$ coincide with the points of increase of $F_z(t)$ and form a subset of the innovation spectrum of $x(t)$, and, accordingly, we shall denote $z(t)$ as a *partial innovation process* associated with $x(t)$. The space $H(z, t)$ is spanned by the random variables $z(u)$ with $u \leq t$, and it is known (cf., e.g., [5], p. 425—428) that $H(z, t)$ is identical with the set of all random variables of the form

$$\int_{-\infty}^{t} g(t, u) dz(u),$$

where $g(t, u)$ is a non-random function such that the integral

$$\int_{-\infty}^{t} |g(t, u)|^2 dF_z(u)$$

is convergent.

4. Consider now the class Q of all distribution functions $F(t)$ determined such as to be continuous to the left for all t. We introduce a partial ordering in Q by saying that F_1 is *superior* to F_2, and writing $F_1 \succ F_2$, whenever F_2 is absolutely continuous with respect to F_1. If $F_1 \succ F_2$ and $F_2 \succ F_1$, we say that F_1 and F_2 are *equivalent*.

The set of all distribution functions equivalent to a given $F(t)$ forms an *equivalence class* R. A partial ordering is introduced in the set of all equivalence classes in the obvious way by writing $R_1 \succ R_2$ when the corresponding relation holds for any $F_1 \in R_1$ and $F_2 \in R_2$. A point t is called a point of increase for the equivalence class R whenever t is a point of increase for any $F \in R$.

In the sequel we shall be concerned with never increasing sequences of equivalence classes:

$$R_1 \succ R_2 \succ \cdots \succ R_N. \tag{2}$$

The number N of elements in a sequence of this form, which may be finite or infinite, will be called the *total multiplicity* of the sequence. We shall also define a *multiplicity function* $N(t)$ of the sequence (2) by writing for any real t

$N(t)$ = the number of those R_n in (2) for which t is a point of increase.

$N(t)$, like N, may be finite or infinite, and the total multiplicity N will obviously satisfy the relation

$$N = \sup N(t),$$

where t runs through all real values.

If, in particular, we have $N(t) = 0$ for all t in some closed interval $[a, b]$, all functions $F(t)$ belonging to any of the equivalence classes in the sequence (2) will be constant throughout $[a, b]$.

5. To any $x(t)$ satisfying the conditions (A) and (B) there corresponds according to Section **3** a uniquely determined spectral family of projections P_t satisfying (1). It then follows from the theory of spectral multiplicity in a separable Hilbert space ([**13**], Chapter **7**) that to the same $x(t)$ there corresponds a uniquely determined, never increasing sequence (2) of equivalence classes, having the following properties:

If N is the total multiplicity of the sequence (2), it is possible to find N orthonormal random variables $z_1, \cdots, z_N \in H(x)$ such that the corresponding processes with orthogonal increments defined in Section **3** satisfy the relations

$$\begin{aligned} &F_{z_n}(t) \in R_n, \\ (3) \qquad &H(z_m, t) \perp H(z_n, t), \qquad m \neq n, \\ &H(x, t) = \sum_1^N H(z_n, t), \end{aligned}$$

where the last sum denotes the vector sum of the mutually orthogonal subspaces involved.

Now $x(t)$ is always an element of $H(x, t)$ and from Section **3** we then obtain the following theorem, previously given in somewhat less precise form in [**2**] and [**3**].

Theorem 1. *To any stochastic process $x(t)$ satisfying* (A) *and* (B) *there corresponds a uniquely determined sequence* (2) *of equivalence classes such that $x(t)$ can be represented in the form*

$$x(t) = \sum_1^N \int_{-\infty}^t g_n(t, u) dz_n(u), \tag{4}$$

where the $z_n(u)$ are mutually orthogonal processes with orthogonal increments satisfying (3). *The $g_n(t, u)$ are non-random functions such that*

$$\sum_1^N \int_{-\infty}^t |g_n(t, u)|^2 dF_{z_n}(u) < \infty.$$

The number N, which is called the total spectral multiplicity of the $x(t)$ process, is the uniquely determined number of elements in (2) *and may be finite or infinite. No representation of the form* (4) *with these properties exists for any smaller value of N.*

The sequence (2) corresponding to a given $x(t)$ process will be said to determine the *spectral type* of the process.

The relation (4) gives a linear representation of $x(t)$ in terms of past and present innovation elements $dz_n(u)$. The *total innovation process* associated with $x(t)$ is an N-dimensional vector process $\{z_1(t), \cdots, z_N(t)\}$ where, as before, N may be finite or infinite.

It is interesting to compare this with the situation in the case of a regular process with *discrete* time ([2], Theorem 1) where a similar representation always holds with $N = 1$.

Also in the particular case of a *stationary* process with continuous time, satisfying (A) and (B), it follows from well-known theorems that we have $N = N(t) = 1$ for all t, and that the only element in the corresponding sequence (2) may be represented by any absolutely continuous distribution function $F(t)$ having an everywhere positive density function.

6. The best linear least squares prediction of $x(t+h)$ in terms of all $x(u)$ with $u \leqq t$ is obtained from (4) in the form

$$P_t x(t+h) = \sum_1^N \int_{-\infty}^t g_n(t+h, u) dz_n(u).$$

The error involved in this prediction is

$$x(t+h) - P_t x(t+h) = \sum_1^N \int_t^{t+h} g_n(t+h, u) dz_n(u). \tag{5}$$

Now consider the multiplicity function $N(t)$ associated with the sequence (2), as defined in Section 3. Suppose that in the closed interval $t \leqq u \leqq t+h$ we have $N(u) \leqq N_1 < N$. Then all terms with $n > N_1$ in the second member of (5) will reduce to zero, so that the innovation entering into the process during $[t, t+h]$ will only be of dimensionality N_1. Speaking somewhat loosely, we may say that the multiplicity function $N(t)$ determines for every t the dimensionality of the innovation element $\{dz_1(t), \cdots\}$.

If, in particular, $N(u) = 0$ for $t \leqq u \leqq t+h$, it follows that the process does not receive any innovation at all during this interval. Accordingly in this case the whole second member of (5) reduces to zero, so that exact prediction is possible over the interval considered.

7. We now introduce the correlation function of the $x(t)$ process:

$$r(s, t) = \mathbf{E}\{x(s)\overline{x(t)}\}.$$

As before we assume that all stochastic processes considered satisfy the conditions (A) and (B). We proceed to prove the following theorem, which shows that the spectral type of a process is uniquely determined by the correlation function.

Theorem 2. *Let $x(t)$ and $y(t)$ be two processes satisfying* (A) *and* (B) *and having the same correlation function $r(s, t)$. The sequences of equivalence classes, which correspond to $x(t)$ and $y(t)$ in the way described in Theorem* 1, *are then identical.*

$x(t)$ and $y(t)$ define two curves situated, respectively, in the spaces $H(x)$ and $H(y)$. We now define a transformation V from the x-curve to the y-curve by writing

$$Vx(t) = y(t),$$

and extend this definition by linearity to the linear manifold in $H(x)$ determined by all points $x(t)$. It is readily seen that this definition is unique, and that the transformation is isometric. It follows, in fact, from the equality of the correlation functions that any linear relation $\sum c_n x(t_n) = 0$ implies and is implied by the corresponding relation $\sum c_n y(t_n) = 0$, which shows that the transformation is unique, while the isometry follows from the identity

$$r(s, t) = (x(s), x(t)) = (Vx(s), Vx(t)).$$

The transformation can now be extended to an isometric transformation V defined in the whole space $H(x)$. If we consider the restriction of V to $H(x, t)$, it is immediately seen that we have for all t

$$VH(x, t) = H(y, t).$$

Denoting by $P_t^{(x)}$ and $P_t^{(y)}$ the spectral families of projections corresponding, respectively, to $x(t)$ and $y(t)$, we then obtain

$$VP_t^{(x)}V^{-1} = P_t^{(y)}.$$

Thus the two spectral families are isometrically equivalent, and the assertion of the theorem now follows directly from Hilbert space theory. In the particular case when $H(x) = H(y)$, the transformation V will be unitary.

On the other hand, two processes with isometrically equivalent spectral families do not necessarily have the same correlation function. In other words, *the correlation function is not uniquely determined by the spectral type.*

In order to see this, it is enough to consider the two processes $x(t)$ and $y(t) = f(t)x(t)$, where $f(t)$ is a non-random function such that $0 < m < |f(t)| < M$ for all t. It is clear that $H(x, t) = H(y, t)$ for all t, while the correlation functions differ by the factor $f(s)\overline{f(t)}$.

8. In this section it will be shown that we can always find a stochastic process possessing any given spectral type. We shall even prove the more precise statement contained in the following theorem.

Theorem 3. *Suppose that a sequence of equivalence classes of the form* (2) *is given. Then there exists a harmonizable process*

$$x(t) = \int_{-\infty}^{\infty} e^{it\lambda} dy(\lambda)$$

which has the spectral type defined by the given sequence.

Comparing this statement with the final remark in Section 5, it will be seen how restricted the class of stationary processes is in comparison with the class of harmonizable processes.

In order to prove the theorem we denote by $A_1, A_2, \cdots$ a sequence of disjoint sets of real points such that the measure of every A_n is positive in any non-vanishing interval.[1] If $\alpha_n(v)$ is the characteristic function of A_n, we thus have

$$\int_a^a \alpha_n(v)dv > 0$$

for all n and for any real $a < b$.

We further take in each equivalence class R_n appearing in the given sequence (2) a distribution function $F_n(t) \in R_n$. Obviously we can choose the functions $F_1, \cdots, F_N$ so that the integrals

$$k_n^2 = \int_{-\infty}^{\infty} e^{t^2} dF_n(t) \tag{6}$$

converge for all n. We then have $1 \leqq k_n < \infty$. Assuming that the basic probability field is not too restricted, we can then find N mutually orthogonal stochastic processes $z_1(t), \cdots, z_N(t)$ with orthogonal increments such that

$$F_{z_n}(t) = \mathbf{E}\{|z_n(t)|^2\} = F_n(t).$$

We now introduce the following definition:

$$g_n(t, u) = \begin{cases} \dfrac{1}{nk_n} e^{-t} \displaystyle\int_u^t (t-v)\alpha_n(v)dv, & u < t, \\ 0, & u \geqq t, \end{cases} \tag{7}$$

and

$$x_n(t) = \int_{-\infty}^{t} g_n(t, u)dz_n(u), \tag{8}$$

$$x(t) = \sum_1^N x_n(t).$$

We then have for $u < t$

$$0 < g_n(t, u) < \frac{1}{nk_n} e^{-t}(t-u)^2,$$

and hence by (6),

$$\mathbf{E}\{|x_n(t)|^2\} < \frac{1}{n^2 k_n^2} e^{-2t} \int_{-\infty}^{\infty} (t-u)^4 dF_n(u)$$

$$\leqq \frac{8}{n^2 k_n^2} e^{-2t} \int_{-\infty}^{\infty} (t^4 + u^4) dF_n(u) \leqq \frac{8(t^4 + k_n^2)}{n^2 k_n^2},$$

so that the series for $x(t)$ converges in quadratic mean if $N = \infty$. (We note that the $x_n(t)$, like the $z_n(t)$, are mutually orthogonal.)

[1] The use of the sets A_n for the construction of processes with given multiplicity properties goes back to a correspondence between Professor Kolmogorov and the present author (cf. [4]). A simple way of constructing the A_n is the following: Let $1 < n_1 < n_2 < \cdots$ be positive integers such that $\sum_1^\infty 1/n_k$ converges. Almost every real x then has a unique expansion $x = r_0 + \sum_1^\infty r_k/(n_1 \cdots n_k)$, where the r_k are integers and $0 \leqq r_k < n_k$ for $k \geqq 1$. If A_n is the set of those x for which the number of zeros among the r_k with $k \geqq 1$ is finite and of the form $2^n(2p+1)$ where p is a non-negative integer, then the sequence $A_1, A_2, \cdots$ has the required properties.

We now proceed to prove a) that the $x(t)$ process defined by (8) has the given spectral type, and b) that it is harmonizable.

It follows from the construction of the $z_n(t)$ and from (8) that we have

$$F_{z_n}(t) \in R_n,$$

$$H(z_m, t) \perp H(z_n, t), \qquad m \neq n,$$

$$H(x, t) \subset \sum_1^N H(z_n, t).$$

If we can show that the sign of equality holds in the last relation, the relations (3) will be satisfied and it then follows that the $x(t)$ process defined by (8) has the given spectral type. In order to prove this it is sufficient to show that we have

$$z_n(t) \in H(x, t)$$

for all n and t.

We have

$$e^t x(t) = \sum_1^N \int_{-\infty}^t g_n(t, u) dz_n(u) = \sum_1^N \frac{1}{nk_n} \int_{-\infty}^t (t-u)\alpha_n(u) z_n(u) du.$$

It is shown without difficulty that the derivative in q.m. of this random function exists for all t and has the expression

$$\frac{d}{dt}(e^t x(t)) = \sum_1^N \frac{1}{nk_n} \int_{-\infty}^t \alpha_n(u) z_n(u) du, \tag{9}$$

where the last sum converges in q.m. We now want to show that for almost all t (Lebesgue measure) we may differentiate once more in q. m., and so obtain

$$\frac{d^2}{dt^2}(e^t x(t)) = \sum_1^N \frac{1}{nk_n} \alpha_n(t) z_n(t). \tag{10}$$

In order to prove this we must show that the random variable

$$W = \sum_1^N \frac{1}{nk_n} \left(\frac{1}{h} \int_t^{t+h} \alpha_n(u) z_n(u) du - \alpha_n(t) z_n(t) \right)$$

converges to zero in q. m. for almost all t as $h \to 0$. We have $W = W_1 + W_2$, where

$$W_1 = \sum_1^N \frac{1}{nk_n h} \int_t^{t+h} \alpha_n(u) (z_n(u) - z_n(t)) du,$$

$$W_2 = \sum_1^N \frac{1}{nk_n} z_n(t) \left(\frac{1}{h} \int_t^{t+h} \alpha_n(u) du - \alpha_n(t) \right).$$

Now both W_1 and W_2 are sums of mutually orthogonal random variables and we have

$$\mathbf{E}|W_1|^2 = \sum_1^N \frac{1}{n^2 k_n^2 h^2} \int_t^{t+h} \int_t^{t+h} \alpha_n(u)\alpha_n(v)[F_n(\min(u, v)) - F_n(t)]\, du\, dv$$

$$\leq \sum_1^N \frac{2}{n^2 h^2} \int_t^{t+h} (t+h-u)[F_n(u) - F_n(t)]du \leq 2 \sum_1^N \frac{F_n(t+h) - F_n(t)}{n^2}$$

and

$$\mathbf{E}|W_2|^2 = \sum_1^N \frac{1}{n^2 k_n^2} F_n(t) \left[\frac{1}{h}\int_t^{t+h} \alpha_n(u)du - \alpha_n(t)\right]^2$$

$$\leq \sum_1^N \frac{1}{n^2} \left[\frac{1}{h}\int_t^{t+h} \alpha_n(u)du - \alpha_n(t)\right]^2.$$

However, all the F_n are continuous almost everywhere, and it follows that W_1 tends to zero in q. m. for almost all t. On the other hand, the metric density of any A_n exists almost everywhere and is equal to $\alpha_n(t)$ so that W_2 tends to zero in q. m. almost everywhere. Thus we have shown that (10) holds for almost all t.

Let now m be a given integer, $1 \leq m \leq N$. The sets A_n being disjoint, it then follows from (10) that for almost all $t \in A_m$

$$\frac{d^2}{dt^2}(e^t x(t)) = \frac{1}{mk_m} z_m(t).$$

The first member of the last relation is evidently an element of $H(x, t)$, so that we have $z_m(t) \in H(x, t)$ for almost all $t \in A_m$. Now A_m is of positive measure in every non-vanishing interval, while $z_m(t)$ is by definition everywhere continuous to the left in q. m. Thus $z_m(t) \in H(x, t)$ for all t and all $m = 1, \cdots, N$, and according to the above this proves that $x(t)$ has the given spectral type.

In order to prove also that $x(t)$ is harmonizable we introduce the Fourier transform $h_n(\lambda, u)$ of $g_n(t, u)$ with respect to t. From (7) we obtain

$$h_n(\lambda, u) = \int_{-\infty}^{\infty} g_n(t, u) e^{-it\lambda} dt = \frac{1}{nk_n} \int_u^{\infty} e^{-t(1+i\lambda)} dt \int_u^t (t-v)\alpha_n(v)dv.$$

The double integral is absolutely convergent and we have

$$h_n(\lambda, u) = \frac{1}{nk_n} \int_u^{\infty} \alpha_n(v)dv \int_v^{\infty} (t-v)e^{-t(1+i\lambda)} dt$$

$$= \frac{1}{nk_n(1+i\lambda)^2} \int_u^{\infty} \alpha_n(v) e^{-v(1+i\lambda)} dv,$$

(11) $$|h_n(\lambda, u)| < \frac{e^{-u}}{nk_n(1+\lambda^2)}.$$

Thus $h_n(\lambda, u)$ is, for any fixed u, absolutely integrable with respect to λ. On the other hand, it follows from (7) that $g_n(t, u)$ is everywhere continuous, so that we have the inverse Fourier formula

(12) $$g_n(t, u) = \frac{1}{2\pi} \int_{-\infty}^{\infty} h_n(\lambda, u) e^{it\lambda} d\lambda.$$

Now the correlation functions of $x_n(t)$ and $x(t)$ are, by (7) and (8),

$$r_n(s, t) = \mathbf{E}\{x_n(s)\overline{x_n(t)}\} = \int_{-\infty}^{\infty} g_n(s, u)\overline{g_n(t, u)}dF_n(u),$$

$$r(s, t) = \mathbf{E}\{x(s)\overline{x(t)}\} = \sum_1^N r^n(s, t).$$

Replacing the g_n by their expressions according to (12) we obtain

$$r_n(s, t) = \frac{1}{(2\pi)^2}\int_{-\infty}^{\infty}\int_{-\infty}^{\infty} e^{i(s\lambda - t\mu)}\, d\lambda\, d\mu \int_{-\infty}^{\infty} h_n(\lambda, u)\overline{h_n(\mu, u)}dF_n(u),$$

the inversion of the order of integration being justified by absolute convergence according to (6) and (11).

If we write

$$c_n(\lambda, \mu) = \frac{1}{(2\pi)^2}\int_{-\infty}^{\infty} h_n(\lambda, u)\overline{h_n(\mu, u)}dF_n(u),$$

$$C_n(\lambda, \mu) = \int_{-\infty}^{\lambda}\int_{-\infty}^{\mu} c_n(\rho, \sigma)d\rho\, d\sigma,$$

it follows from well-known criteria (cf., e.g., [9], p. 466—469) that $C_n(\lambda, \mu)$ is a correlation function. Further $C_n(\lambda, \mu)$ is of bounded variation over the whole (λ, μ)-plane, its variation being bounded by the expression

$$\int_{-\infty}^{\infty}\int_{-\infty}^{\infty} |c_n(\lambda,\mu)|\, d\lambda\, d\mu < \frac{1}{(2\pi)^2 n^2 k_n^2}\int_{-\infty}^{\infty}\frac{d\lambda}{1+\lambda^2}\int_{-\infty}^{\infty}\frac{d\mu}{1+\mu^2}\int_{-\infty}^{\infty} e^{-2u} dF_n(u)$$

$$< \frac{1}{4n^2 k_n^2}\int_{-\infty}^{\infty} e^{1+u^2} dF_n(u) < \frac{1}{n^2}$$

obtained from (6) and (11).

It now follows that we have

$$r_n(s, t) = \int_{-\infty}^{\infty}\int_{-\infty}^{\infty} e^{i(s\lambda - t\mu)}\, d_{\lambda,\mu} C_n(\lambda, \mu),$$

$$r(s, t) = \int_{-\infty}^{\infty}\int_{-\infty}^{\infty} e^{i(s\lambda - t\mu)}\, d_{\lambda,\mu} C(\lambda, \mu),$$

where

$$C(\lambda, \mu) = \sum_1^N C_n(\lambda, \mu)$$

is a correlation function which, according to the above, is of bounded variation over the whole (λ, μ)-plane. Hence we may conclude (cf. [1], and [9], p. 476) that $x(t)$ is a harmonizable process

$$x(t) = \int_{-\infty}^{\infty} e^{it\lambda} dy(\lambda),$$

where $y(\lambda)$ has the correlation function $C(\lambda, \mu)$. We note that $x(t)$ is a regular process and is everywhere continuous in quadratic mean. The proof is completed.

Received by the editors
November 27, 1963

REFERENCES

[1] H. Cramér, *A contribution to the theory of stochastic processes*, Proc. Second Berkeley Sympos. Math. Statist. and Prob., 1951, pp. 329–339.
[2] H. Cramér, *On some classes of non-stationary stochastic processes*, Proc. 4-th Berkeley Sympos. Math. Statist. and Prob., II, 1961, pp. 57–77.
[3] H. Cramér, *On the structure of purely non-deterministic stochastic processes*, Arkiv Math., 4, 1961, pp. 249–266.
[4] H. Cramér, *Décompositions orthogonales de certains procès stochastiques*, Ann. Fac. Sciences Clermont, 11, 1962, pp. 15–21.
[5] J. L. Doob, *Stochastic Processes*, Wiley, N.Y., 1953.
[6] P. R. Halmos, *Introduction to Hilbert Space and the Theory of Spectral Multiplicity*, Chelsea, N.Y., 2-nd ed. 1957.
[7] A. N. Kolmogorov, *Curves in Hilbert space which are invariant with respect to a one-parameter group of motions*, DAN SSSR, 26, 1940, pp. 6–9. (In Russian.)
[8] A. N. Kolmogorov, *Wiener's spiral and some other interesting curves in Hilbert space*, DAN SSSR, 26, 1940, pp. 115–118. (In Russian.)
[9] M. Loève, *Probability Theory*, 3-rd ed., Van Nostrand, Princeton, N.J., 1963.
[10] H. Nakano, *Spectral Theory in the Hilbert Space*, Jap. Soc. for the Promotion of Science, Tokyo, 1953.
[11] J. v. Neumann and I. J. Schoenberg, *Fourier integrals and metric geometry*, Trans. Amer. Math. Soc., 50, 1941, pp. 226–251.
[12] M. S. Pinsker, *Theory of curves in Hilbert space with stationary n-th increments*, Izv. AN SSSR, Ser. Mat., 19, 1955, pp. 319–344. (In Russian.)
[13] M. H. Stone, *Linear Transformations in Hilbert Space*, American Math. Soc., N.Y., 1932.

STOCHASTIC PROCESSES AS CURVES IN HILBERT SPACE

HARALD CRAMÉR *(STOCKHOLM)*

(Summary)

Regular complex-valued random processes $x(t)$ with finite moments of second order are studied by methods of Hilbert space geometry. A representation formula (4) is given for the process $x(t)$ in terms of "past and present innovations". The number N is called the complete spectral multiplicity of the process $x(t)$ and is the smallest number for which such a representation exists. It is shown that the multiplicity of $x(t)$ is uniquely determined by the corresponding correlation function and that one can always find a harmonizing process $x(t)$ which has the multiplicity prescribed in advance.

Reprinted from *Proceedings of the Fifth Berkeley Symposium on Statistics and Applied Probability*, **II**, 215–221 (1965)

A CONTRIBUTION TO THE MULTIPLICITY THEORY OF STOCHASTIC PROCESSES

12

HARALD CRAMÉR
UNIVERSITY OF STOCKHOLM

1. Introduction

In a paper [1] read before the Fourth Berkeley Symposium in 1960, I communicated the elements of a theory of spectral multiplicity for stochastic processes. A related theory was given about the same time by Hida [5]. Since then, I have developed the theory in some subsequent papers [2]–[4], the most recent of which contains the text of a lecture given at the Seventh All-Soviet Conference of Probability and Mathematical Statistics in Tbilisi 1963. Further important work in the field has been made by Kallianpur and Mandrekar [6]–[8].

Many interesting problems arising in connection with this theory are still unsolved. The object of this paper is to offer a small contribution to the investigation of one of these problems.

We shall begin by giving in section 2 a brief survey of the results of multiplicity theory so far known for the simplest case of one-dimensional processes. For proofs and further developments we refer to the papers quoted above. A major unsolved problem will be discussed in section 3, whereas section 4 is concerned with some aspects of the well-known particular class of stationary processes, which are relevant for our purpose. Finally, section 5 is concerned with the construction of a class of examples which may be useful in the further study of the problem stated in section 3.

2. Spectral multiplicity of stochastic processes

Consider a stochastic process $x(t)$, where $x(t)$ is a complex-valued random variable defined on a fixed probability space, while t is a real-valued parameter. In general we shall allow t to take any real values, and shall only occasionally consider the case when t is restricted to the integers. We shall always assume that the relations

$$Ex(t) = 0, \qquad E|x(t)|^2 < \infty \tag{2.1}$$

are satisfied for all t.

We denote by $H(x)$ the Hilbert space spanned in a well-known way by the random variables $x(t)$ for all t, while $H(x, t)$ is the subspace of $H(x)$ spanned only by the $x(u)$ with $u \leq t$. The tail space $H(x, -\infty)$ may be regarded as representing the "infinitely remote past" of the process. If $H(x, -\infty)$ only contains the zero element of $H(x)$, the $x(t)$ process is said to be purely nondeterministic.

All processes $x(t)$ considered in the sequel will be assumed to satisfy the following conditions (A) and (B):

(A) the process is purely nondeterministic;

(B) the limits in quadratic mean $x(t+0)$ and $x(t-0) = x(t)$ exist for every t. Under these conditions the space $H(x)$ will be separable. We note that, in the case of a parameter t taking only integral values, condition (B) is irrelevant.

Let us now for a moment consider the case when t is restricted to the integers, so that we are concerned with a sequence of random variables x_n, with $n = 0, \pm 1, \cdots$. Then there exists a sequence of mutually orthogonal random variables z_n with

$$\begin{aligned} &Ez_n = 0, \qquad E|z_n|^2 = 1 \text{ or } 0 \text{ for every } n, \\ &Ez_m\overline{z_n} = 0 \qquad \text{for} \quad m \neq n, \end{aligned} \tag{2.2}$$

such that

$$x_n = \sum_{k=-\infty}^{n} c_{nk}z_k, \tag{2.3}$$

where the series

$$\sum_{k=-\infty}^{n} |c_{nk}|^2 \tag{2.4}$$

converges for every n, so that the expression for x_n converges in quadratic mean. The variable z_n may then be regarded as a (normalized) *innovation* entering into the process at time $t = n$.

By analogy, we might expect to have in the case of a continuous paramter t a representation of the form

$$x(t) = \int_{-\infty}^{t} g(t, u)\, dz(u) \tag{2.5}$$

where $z(u)$ would be a process with orthogonal increments, the increment $dz(u)$ representing the *innovation element* entering into the $x(t)$ process during the time element $(u, u + du)$.

However, in general this is not true. The situation in the continuous case turns out to be more complicated than in the discrete case. In general the innovation associated with a given time element must be regarded as a multi-dimensional or even infinite-dimensional random variable, so that the representation (2.5) is definitely too simple.

In order to present the representation formula which in the general case takes the place of (2.5), we must first consider the class C of all real-valued and never decreasing, not identically constant functions $F(t)$ which are continuous to the left for all t. A subclass D of C is called an *equivalence class* if any two functions F_1 and F_2 in D are mutually absolutely continuous. If D_1 and D_2 are equivalence classes, D_1 is said to be *superior* to D_2, and we write $D_1 > D_2$, if any $F_2 \in D_2$ is absolutely continuous relative to any $F_1 \in D_1$. Evidently, the relation $D_1 > D_2$ does not exclude the case that the two classes are identical.

Consider now a finite or infinite never increasing sequence of equivalence classes

$$D_1 > D_2 > \cdots > D_N, \tag{2.6}$$

where N may have any of the values $1, 2, \cdots, \infty$. Then N will be called the *total multiplicity* of the sequence. Further, let $N(t)$ for every t denote the number of those classes in (2.6) for which t is a point of increase of the corresponding functions F. Then $N(t)$ is called the *multiplicity function* of the sequence (2.6). Like N, $N(t)$ may be finite or infinite, and we have

$$N = \sup N(t), \tag{2.7}$$

where t runs through all real values.

The fundamental proposition of multiplicity theory for stochastic processes is the following. To any $x(t)$ stochastic process satisfying conditions (A) and (B), there is a uniquely determined sequence of the form (2.6) such that the following properties hold. For every $n = 1, 2, \cdots, N$, there is a process $z_n(t)$ of orthogonal increments, such that

$$\begin{aligned} &Ez_n(t) = 0, \qquad E|z_n(t)|^2 = F_n(t) \in D_n, \\ &Ez_m(t)\overline{z_n(u)} = 0 \qquad \text{for } m \neq n \text{ and all } t, u, \\ &H(x, t) = \sum_1^N H(z_n, t) \qquad \text{for all } t, \end{aligned} \tag{2.8}$$

where the last sum denotes the vector sum of the orthogonal subspaces $H(z_n, t)$. We then have for every t the representation

$$x(t) = \sum_1^N \int_{-\infty}^t g_n(t, u)\, dz_n(u), \tag{2.9}$$

where the g_n are nonrandom functions such that

$$\sum_1^N \int_{-\infty}^t |g_n(t, u)|^2\, dF_n(u) < \infty. \tag{2.10}$$

It is important to observe that the D_n sequence (2.6) is uniquely determined by the $x(t)$ process. Thus, in particular, the multiplicity function $N(t)$ and the total multiplicity N are also uniquely determined by $x(t)$. Accordingly, we shall say that the D_n sequence, as well as $N(t)$ and N, are spectral multiplicity characteristics of the stochastic process $x(t)$.

On the other hand, the $g_n(t, u)$ and $z_n(u)$ occurring in the representation (2.9) are not uniquely determined by the $x(t)$ process. Thus, for a given $x(t)$ we may have different representations of the form (2.9), all satisfying the relations (2.8). However, the D_n sequence (2.6), as well as the multiplicity characteristics $N(t)$ and N, will be identical for all these representations.

According to the representation (2.9), we may say that the multiplicity function $N(t)$ determines the dimensionality of the innovation element $[dz_1(u), dz_2(u), \cdots]$ entering into the process during the time element $(u, u + du)$.

It has been shown that the multiplicity characteristics of a given stochastic process $x(t)$ are uniquely determined by the covariance function of the process

$$r(t, u) = Ex(t)\overline{x(u)}. \tag{2.11}$$

We finally remark that the multiplicity theory as outlined above can be directly generalized to stochastic vector processes of a very general kind. We shall, however, not deal with these generalizations in the present paper.

3. Processes of total multiplicity $N = 1$

According to the above, we know that any given stochastic process $x(t)$ satisfying (A) and (B) has multiplicity characteristics which are uniquely determined by the process, and even by the covariance function $r(t, u)$ of the process.

On the other hand, so far we know very little about those properties of the process, or of the corresponding covariance function, which determine the actual values of multiplicity characteristics like $N(t)$ and N.

In the discrete case it follows from the above that, by analogy, it can be said that the total multiplicity is always $N = 1$. In the continuous case, the important class of (second-order) stationary processes has even $N(t) = 1$ for all t, and consequently $N = 1$, as follows from well-known properties of these processes to be presently recalled.

In view of these examples, it might well be asked if there exist any stochastic processes with a total multiplicity exceeding unity. The answer to this question is that such processes do, in fact, exist. It can even be shown that, as soon as we proceed from the class of stationary processes to the more general class of harmonizable processes introduced by Loève, any prescribed multiplicity properties may occur. In fact, it has been shown in [4] that, given any D_n sequence (2.6), there exists a harmonizable process $x(t)$ associated with this given D_n sequence. However, the example of such a process given in [4] is of a very special kind, and the corresponding representation (2.9) contains functions $g_n(t, u)$ having rather pathological properties, not likely to occur in applications to any physical problems.

Accordingly, it seems to be a problem of some interest to study more closely those properties of a stochastic process which determine the actual values of the multiplicity characteristics. In particular, it would be interesting to be able to define some fairly general class of processes having total multiplicity $N = 1$.

A natural approach to this last problem might be to start from the class of stationary processes, which always have $N = 1$, and then try to generalize the definition, still keeping sufficiently near the property of stationarity to conserve the multiplicity characteristic $N = 1$. We propose to give in the sequel an example of a generalization of this type. In order to do this, we must first recall some of the relevant properties of stationary processes.

4. Stationary processes

Let $x(t)$ be a (second-order) stationary process, satisfying (A) and (B). It then follows that the covariance function

$$r(t) = Ex(t+h)\overline{x(h)} \tag{4.1}$$

is everywhere continuous, and has the spectral representation

$$r(t) = \int_{-\infty}^{\infty} e^{it\lambda} f(\lambda)\, d\lambda \tag{4.2}$$

with a spectral density $f(\lambda) > 0$ for almost all λ (Lebesgue measure), such that $f(\lambda) \in L_1(-\infty, \infty)$, and

$$\int_{-\infty}^{\infty} \frac{\log f(\lambda)}{1+\lambda^2}\, d\lambda > -\infty. \tag{4.3}$$

The random variable $x(t)$ has the corresponding spectral representation

$$x(t) = \int_{-\infty}^{\infty} e^{it\lambda}\, dw(\lambda), \tag{4.4}$$

where $w(\lambda)$ is a process with orthogonal increments such that

$$E\, dw(\lambda) = 0, \qquad E|\, dw(\lambda)\,|^2 = f(\lambda)\, d\lambda. \tag{4.5}$$

Further, there exists a complex-valued function $h(\lambda) \in L_2(-\infty, \infty)$ and a process $z(t)$ of orthogonal increments such that

$$E\, dz(t) = 0, \qquad E|\, dz(t)\,|^2 = dt, \qquad |h(\lambda)|^2 = f(\lambda), \tag{4.6}$$

while the Fourier transform $g(t)$ of $h(\lambda)$ reduces to zero for $t < 0$, and we have the representation

$$x(t) = \int_{-\infty}^{t} g(t-u)\, dz(u) \tag{4.7}$$

with

$$H(x, t) = H(z, t) \tag{4.8}$$

for all t. The functions $h(\lambda)$ and $g(t)$ are uniquely determined, up to a constant factor of absolute value 1. Comparing this with the general representation formula (2.9), it is seen that the stationary process $x(t)$ has the multiplicity characteristics $N = 1$ and $N(t) = 1$ for all t.

5. A class of harmonizable processes with $N = 1$

We shall now define a class of harmonizable processes containing the stationary process $x(t)$ given by (4.4) or (4.7) as a particular case, and such that the multiplicity characteristics are the same as for $x(t)$, that is $N = 1$ and $N(t) = 1$ for all t.

Let $Q(\rho)$ be a never-decreasing function of the real variable ρ such that Q has a jump of size 1 at $\rho = 0$, whereas $Q(-\infty) = 0$, $Q(+\infty) < 2$, and

$$Q(\rho) + Q(-\rho) = Q(+\infty) \tag{5.1}$$

in all continuity points ρ of Q. The Fourier-Stieltjes transform of Q will then be real and positive, so that we may define an everywhere positive, continuous and bounded function $q(u)$ by the relation

$$[q(u)]^2 = \int_{-\infty}^{\infty} e^{-i\rho u}\, dQ(\rho). \tag{5.2}$$

We now define a stochastic process $X(t)$ by writing

$$X(t) = \int_{-\infty}^{t} g(t-u)q(u)\, dz(u), \tag{5.3}$$

where $g(t)$ and $z(u)$ are the same as in (4.7). As $q(u)$ is bounded, and $g(t) \in L_2(0, \infty)$, the integral in (5.3) exists as a quadratic mean integral. When Q is identically constant except for the jump at $\rho = 0$, it is seen that $X(t)$ reduces to the stationary process $x(t)$ given by (4.7).

We shall now first show that $X(t)$ has the required multiplicity characteristics. According to (2.8) and (2.9), we have to show that $H(X, t) = H(z, t)$ for all t. As it evidently follows from (5.3) that $H(X, t) \subset H(z, t)$, it will be sufficient to show that the opposite inclusion relation is also true. If, for some t, this were not so, there would be a nonzero element in $H(z, t)$ orthogonal to $X(u)$ for all $u \leq t$. Now every nonzero element in $H(z, t)$ is of the form

$$\int_{-\infty}^{t} m(v)\, dz(v), \tag{5.4}$$

with a quadratically integrable $m(v)$ not almost everywhere equal to zero. If this is orthogonal to $X(u)$ for $u \leq t$, we have

$$\int_{-\infty}^{u} g(u-v)q(v)\overline{m(v)}\, dv = 0 \tag{5.5}$$

for all $u \leq t$. However, since $q(v)$ is bounded and positive, it would follow that there is a nonzero element in $H(z, t)$ orthogonal to $x(u)$ for all $u \leq t$, in contradiction with the relation (4.8). Thus our assertion is proved.

We now proceed to prove that $X(t)$ as defined by (5.3) is a harmonizable process, and to deduce an expression for its spectral distribution. From (5.3) we obtain for the covariance function $R(s, t)$ of $X(t)$ the expression

$$\begin{aligned} R(s, t) = EX(s)\overline{X(t)} &= \int_{-\infty}^{\infty} g(s-u)\overline{g(t-u)}\, [q(u)]^2\, du \\ &= \int_{-\infty}^{\infty} g(s-u)\, \overline{g(t-u)}\, du \int_{-\infty}^{\infty} e^{-i\rho u}\, dQ(\rho). \end{aligned} \tag{5.6}$$

As $g(t) = 0$ for $t < 0$, and $g(t) \in L_2(0, \infty)$, it follows that the double integral is absolutely convergent, so that

$$R(s, t) = \int_{-\infty}^{\infty} dQ(\rho) \int_{-\infty}^{\infty} e^{-i\rho u} g(s-u)\overline{g(t-u)}\, du. \tag{5.7}$$

By the Parseval formula, this gives

$$R(s, t) = \int_{-\infty}^{\infty}\int_{-\infty}^{\infty} e^{i[s\lambda - t(\lambda+\rho)]} h(\lambda)\overline{h(\lambda+\rho)}\, d\lambda\, dQ(\rho). \tag{5.8}$$

Substituting here μ for $\lambda + \rho$, it will be seen that this is the expression of a harmonizable covariance function. The corresponding spectral mass is distributed over the (λ, μ)-plane so that the infinitesimal strip between the lines $\mu = \lambda + \rho$ and $\mu = \lambda + \rho + d\rho$ contains the mass $dQ(\rho)$, whereas the distribution within the strip has the relative density $h(\lambda)\overline{h(\mu)}$. Again we see that, in the particular case when $Q(\rho)$ is identically constant except for the jump at $\rho = 0$, the whole spectral mass is situated on the diagonal $\lambda = \mu$, so that we have the covariance function of a stationary process with spectral density $|h(\lambda)|^2 = f(\lambda)$. As soon as $Q(\rho)$ has some variation outside the point $\rho = 0$, we have the two-dimensional spectral distribution of a harmonizable covariance.

Thus the covariance function of the $X(t)$ process, given by (5.3), is harmonizable, and it then follows from known properties of harmonizable processes that $X(t)$ itself is harmonizable; that is, we have

$$X(t) = \int_{-\infty}^{\infty} e^{it\lambda}\, dZ(\lambda), \tag{5.9}$$

where the covariance function $EZ(\lambda)\overline{Z(\mu)}$ is obtained from the expression (5.8) with $\mu = \lambda + \rho$. At the same time, we have seen that the harmonizable process $X(t)$ has the multiplicity characteristics $N = 1$ and $N(t) = 1$ for all t.

REFERENCES

[1] H. Cramér, "On some classes of non-stationary stochastic processes," *Proceedings of the Fourth Berkeley Symposium on Mathematical Statistics and Probability*, Berkeley and Los Angeles, University of California Press, 1960, Vol. II, pp. 57–78.

[2] ———, "On the structure of purely non-deterministic stochastic processes," *Ark. Mat.*, Vol. 4 (1961), pp. 249–266.

[3] ———, "Décompositions orthogonales de certains processus stochastiques," *Ann. Fac. Sci., Clermont*, Vol. 8 (1962), pp. 15–21.

[4] ———, "Stochastic processes as curves in Hilbert space," *Teor. Verojatnost. i Primenen.*, Vol. 9 (1964), pp. 195–204.

[5] T. Hida, "Canonical representations of Gaussian processes and their applications," *Mem. Coll. Sci., Kyoto, Ser. A*, Vol. 33 (1960), pp. 109–155.

[6] G. Kallianpur and V. Mandrekar, "On the connection between multiplicity theory and Hanner's time domain analysis of weakly stationary stochastic processes," Department of Statistics, University of Minnesota, Technical Report 49 (1964).

[7] ———, "Multiplicity and representation theory of purely non-deterministic stochastic processes," Department of Statistics, University of Minnesota, Technical Report 51 (1964).

[8] ———, "Semi-groups of isometries and the representation and multiplicity of weakly stationary stochastic processes," to appear in *Ark. Mat.* (1966).

Editors' Comments on Papers 13 and 14

These papers are the first two of a series by Professor T. Kailath of Stanford University and his co-workers. The series begins with a recognition of the relationship of the canonical decomposition of a random process to the requirement of such a process's equivalence to white noise in a number of detection and estimation problems.

The term "innovations process" is revived by Kailath to describe the white generator of a random process in its canonical representation. The scope of subsequent papers in the series has expanded to include additional aspects of filtering theory.

The concept of multiplicity enters directly in the linear versions of the problems in question. The second paper (co-authored by Frost) involves noncausal linear systems and, therefore, is not as directly related to multiplicity theory as the first.

These papers are reprinted with the permission of the authors and the Director of Editorial Services of the Institute of Electrical and Electronics Engineers.

Reprinted from IEEE TRANSACTIONS
ON *AUTOMATIC CONTROL*
Volume AC-13, Number 6, December, 1968
pp. 646-655

13

An Innovations Approach to Least-Squares Estimation Part I: Linear Filtering in Additive White Noise

THOMAS KAILATH, MEMBER, IEEE

***Abstract*—The innovations approach to linear least-squares approximation problems is first to "whiten" the observed data by a causal and invertible operation, and then to treat the resulting simpler white-noise observations problem. This technique was successfully used by Bode and Shannon to obtain a simple derivation of the classical Wiener filtering problem for stationary processes over a semi-infinite interval. Here we shall extend the technique to handle nonstationary continuous-time processes over finite intervals. In Part I we shall apply this method to obtain a simple derivation of the Kalman-Bucy recursive filtering formulas (for both continuous-time and discrete-time processes) and also some minor generalizations thereof.**

I. INTRODUCTION

IN THE EARLY 1940's, Kolmogorov [1] and Wiener [2] first discussed problems of linear least-squares estimation for stochastic processes, but by entirely different methods. Kolmogorov [1] studied only discrete-time problems and he solved them by using a simple representation of such processes that was suggested in a 1938 doctoral dissertation by Wold [3]. This representation, which is obtained by a recursive orthonormalization procedure, is known as the Wold decomposition. [The original papers of Kolmogorov and Wold are quite readable, but a more accessible and very readable reference is the monograph by Whittle [4] (especially sec. 3.7).]

On the other hand, Wiener [2] took an almost completely nonprobabilistic approach. He mainly studied continuous-time problems and reduced them to the problem of solving a certain integral equation, the so-called Wiener–Hopf equation, that Wiener and Hopf had solved in 1931 [5] by using some of Wiener's results on harmonic analysis. Though Wiener undertook this work in response to an engineering problem (the design of antiaircraft fire-control systems), his solution was beyond the reach of his engineering colleagues, and his yellow-bound report soon came to be labeled the "Yellow Peril."

In 1950, Bode and Shannon [6] published a different derivation of Wiener's results that was, quite successfully, intended to make them more accessible to engineers. This paper was based on ideas in a classified 1944 report by Blackman, Bode, and Shannon [7]. The same approach was independently discovered by Zadeh (cf. footnote 3 in Zadeh and Ragazzini [8]).

Manuscript received January 31, 1968. This work was supported by the Applied Mathematics Division of the Air Force Office of Scientific Research under Contract AF 49(638)1517, and by the Joint Services Electronics Program at Stanford University, Stanford, Calif., under Contract Nonr 225(83).
The author is with Stanford University, Stanford, Calif.

However, it is somewhat ironic that these more engineering approaches were found later to be just the continuous-time versions of the original Wold–Kolmogorov technique, which had been developed in a purely mathematical context.

The results in [1]–[7] were all obtained for stationary processes with infinite or semi-infinite observation intervals. The paper of Zadeh and Ragazzini [8] was the first significant attempt to extend the theory. Over the last two decades, various extensions and generalizations have been obtained and many of these have been documented in textbooks, as for example, those of Doob [9], Laning and Battin [10], Pugachev [11], Lee [12], Yaglom [13], Whittle [3], Deutsch [14], Liebelt [15], Balakrishnan [16], Bryson and Ho [17], and others.

In recent years, applications in orbital mechanics and spacecraft tracking have spurred interest in recursive estimation for nonstationary processes over finite-time intervals. Such algorithms were used by Gauss in his numerical calculations of the orbit of the asteroid Ceres, but the modern interest in them is due to Swerling [18] and especially Kalman [19], [20], and Bucy [21], [22]. The great interest in recursive algorithms because of their obvious computational advantages has stimulated a great number of papers on them, providing alternate forms and derivations showing their relationship to more classical parameter estimation techniques (see, for example, the discussions and references in Deutsch [14] and Liebelt [15]). Nevertheless, it seems to us that the original derivations of Kalman [19] and Kalman and Bucy [22] still provide the most insight.

In order to obtain recursive solutions, Kalman and Bucy had to confine themselves to a special class of processes, viz., those that could be generated by passing white noise through a (possibly time-variant) "lumped" linear dynamical system, i.e., a system composed of a finite number of (possibly time-variant) R, L, C elements. (Such processes are sometimes called projections of wide-sense Markov processes, but we shall in the rest of this paper call them "lumped" processes.) They also assumed complete knowledge of this system, thus sidestepping the difficult problem of spectral factorization that had been a stumbling block to the extension of Wiener's classic solution (for semi-infinite observations on a stationary process) to more general situations. In his first paper in 1959, Kalman [19] treated discrete-time processes and obtained a recursive solution by a technique that was essentially the same as Kolmogorov's. In a later paper [20], he extended these results to the continuous-time case by the use of a particular limiting technique. This technique, though useful, is somewhat tedious to carry out rigorously. A careful discussion has been given by Wonham [23]. In [22], Kalman and Bucy attacked the continuous-time problem directly. However, they did not use the Wold-Kolmogorov approach because the direct continuous-time analog of Kalman's discrete-time procedure in [19] was hard to see. They therefore returned to the Wiener–Hopf integral equation and showed that (under certain assumptions on the signal and noise processes) the solution to this equation could be expressed in terms of the solution to a nonlinear Riccati differential equation. It is also worth noting that Siegert [24] had carried out essentially the same steps in a different (but mathematically isomorphic) problem.

The chief purpose of Part I is to give a derivation of the Kalman–Bucy results by the Wold–Kolmogorov method, which, for reasons that will be clear later, we shall call the innovations method. Not only does this close a gap in the preceding circle of ideas, but the insight it provides into the proof has also suggested some new results. These include some slight generalizations in the types of processes for which recursive estimation formulas can be obtained, and a very simple and general solution of the so-called smoothing (or interpolation or noncausal filtering) problem. The smoothing problem is one that has been somewhat difficult to solve by the original techniques of Kalman and Bucy, and the solutions that have been obtained are in a somewhat complicated form (see the discussions in Part II [25][1]). Our technique also enables a completely parallel method of attack for the discrete- and continuous-time problems. A new approach to linear estimation with additive colored (nonwhite) noise also follows from the present ideas (Geesey and Kailath [26]).

More strikingly, the innovations technique can also be extended to a large class of nonlinear least-squares problems, viz., those where the observation process is the sum of a non-Gaussian process and additive white Gaussian noise (cf. Kailath and Frost [27] and Frost [28]). The ideas of the present paper have also yielded some general results on the detection of general non-Gaussian signals in additive Gaussian noise (Kailath [29], discrimination between two general Gaussian processes (Kailath and Geesey [30]), and also in certain modeling problems (Kailath and Geesey [31]).

Finally, we should say a word about the level of rigor in the present work. It is difficult to work directly with white noise in a completely satisfactory and rigorous manner—one has usually, especially in the nonlinear case, to work with the integrated white noise. However, in our opinion, the key ideas can always be presented, quite simply, in the white-noise formulation. Then, after some familiarity with the appropriate mathematics has been gained, one can translate the white-noise formulation into the more rigorous (stochastic differential) framework. We shall do this in later papers. The more informal presentation here will, we hope, bring the basic ideas to a wider audience.

[1] This issue, page 655.

II. The Innovations Approach to Linear Least-Squares Estimation

The innovations approach is first to convert the observed process to a white-noise process, to be called the innovations process, by means of a *causal and causally invertible* linear transformation. The point is that the estimation problem is very easy to solve with white-noise observations. The solution to this simplified problem can then be reexpressed in terms of the original observations by means of the inverse of the original "whitening" filter.

This program, used by Bode and Shannon [6] for the stationary process problem with semi-infinite observation time, will now be carried out when the observations are made over a finite-time interval on a continuous-time (possibly nonstationary) stochastic process. Several initial sets of assumptions and several corresponding classes of problems, of varying degrees of generality, can be formulated. For simplicity, however, we shall deal largely with the following additive white-noise problem.

The given observation is a record of the form

$$y(t) = z(t) + v(t), \qquad t \in [a, b) \tag{1}$$

where

$v(\cdot)$ = a sample function of zero-mean white noise with covariance function[2]

$$\overline{v(t)v'(s)} = R(t)\delta(t-s), \qquad R(t) > 0,$$

$z(\cdot)$ = a sample function of a zero-mean "signal" process that has finite variance

$$\operatorname{tr}[\overline{z(t)z'(t)}] < \infty, \qquad t \in [a, b)$$

$[a, b]$ = a finite interval[3] on the real line.

We also assume that the "future" noise $v(\cdot)$ is uncorrelated from the "past" signal $z(\cdot)$, i.e.,

$$\overline{v(t)z'(s)} = 0, \qquad a \leq s < t < b. \tag{2}$$

We shall be interested in the linear least-squares estimate of a related process $x(t)$. Let

$\hat{x}(t \mid b)$ = a linear function of all the data $\{y(s), a \leq s < b\}$ that minimizes the mean-square error $\operatorname{tr}[z(t) - \hat{z}(t \mid b)][z(t) - \hat{z}(t \mid b)]'$. (3)

The corresponding instantaneous estimation error will be written

$$\tilde{z}(t \mid b) = z(t) - \hat{z}(t \mid b), \; \tilde{z}(t \mid t) = z(t) - \hat{z}(t \mid t). \tag{4}$$

[2] Bars will be used to denote expectations.

[3] The case of an infinite interval requires certain additional assumptions on the signal process such as stationarity, observability of models generating it, etc. Some more specific comments on this point will be made later [after (35)].

When $b=t$, the estimate is usually called the *filtered* estimate, when $b>t$ it is usually called the *smoothed* estimate, and when $b<t$ it is called the *predicted* estimate.

The major tools for the calculation of these estimates will be the following two theorems.

Theorem 1—The Projection Theorem: The best estimate $\hat{z}(t|b)$ is unique and satisfies the conditions

$$\tilde{z}(t \mid b) \triangleq z(t) - \hat{z}(t \mid b) \perp y(s), \qquad a \leq s < b \quad (5)$$

where

$$u \perp v \quad \text{means that} \quad \overline{uv'} = 0. \quad (6)$$

In words, the instantaneous error is uncorrelated with the observations.

Proof: This theorem, which was used by Kolmogorov [1], is by now fairly well known to engineers and is used in several of the textbooks cited earlier. A brief discussion of the relevant geometric picture is given in Appendix I.

Theorem 2—The Innovations Theorem: The process $\boldsymbol{\nu}(\cdot)$ defined by

$$\boldsymbol{\nu}(t) = y(t) - \hat{z}(t \mid t) = \tilde{z}(t \mid t) + v(t), \qquad a \leq t < b, \quad (7)$$

and to be called the "innovation process" of $y(\cdot)$, is a *white-noise* process with the *same covariance* as $v(\cdot)$, i.e.,

$$\overline{\boldsymbol{\nu}(t)\boldsymbol{\nu}'(s)} = \overline{v(t)v'(s)}, \qquad a \leq t, \quad s < b. \quad (8)$$

Furthermore, $y(\cdot)$ and $\boldsymbol{\nu}(\cdot)$ can be obtained from the other by causal (nonanticipative) linear operations. Therefore, $y(\cdot)$ and $\boldsymbol{\nu}(\cdot)$ are "equivalent" (i.e., they contain the same statistical information) as far as linear operations are concerned.

Proof: The proof will be deferred to Appendix II: however, a few remarks on the theorem and its significance are appropriate here.

Remark 1: The quantity $\boldsymbol{\nu}(t) = y(t) - \hat{z}(t|t) = y(t) - \hat{y}(t|t-)$ may be regarded as defining the "new information" brought by the current observation $y(t)$, being given all the past observations $y(t)$, and the old information deduced therefrom. Therefore, the name "innovation process of $y(\cdot)$" came into being. This term was first used for such processes by Wiener and has since gained wide currency. (A significant generalization, due to Frost [28] and to Kailath [29], of this theorem is that when the white noise $v(\cdot)$ is also Gaussian, but the signal $z(\cdot)$ is non-Gaussian, the innovation process $\boldsymbol{\nu}(\cdot)$ is not only white with the same covariance as $v(\cdot)$, but it is also Gaussian. Applications of this surprising result are given in [27]–[29].)

Remark 2: The fact that $\boldsymbol{\nu}(\cdot)$ is white has been noted before in the special case of lumped signal processes. In this case, the result was probably first noticed by several people (cf. [23], [32]–[34], and unpublished notes of the author and others). However, all their arguments rely, to varying degrees, on the explicit Kalman–Bucy formulas for $\hat{z}(t|t)$. Here we first obtain the result more generally [and with less computation, since we rely only on the projection properties of $\hat{z}(t|t)$], and then use it to obtain the Kalman–Bucy formulas. We note also that the equivalence of $y(\cdot)$ and $\boldsymbol{\nu}(\cdot)$ does not seem to have been explicitly pointed out before, even though, assuming knowledge of the Kalman–Bucy formulas, a proof is immediate (cf. Appendix II-D).

Remark 3: One reason the fact that $\boldsymbol{\nu}(\cdot)$ is white (for lumped processes) may have been known for a long time is that in the discrete-time solution of Kalman [19], $\boldsymbol{\nu}(\cdot)$ is shown to be white (in discrete time) with, however, a different covariance from that of the original noise. The exact formula will be given later (39).

III. Some Applications

We turn now to some applications of these two theorems. First, we present a new derivation of the Kalman–Bucy formulas for filtering of lumped signal processes in white noise. This derivation shows clearly the step at which restriction to such processes is essential to get a recursive solution, and this insight easily yields several (slight) generalizations of the Kalman–Bucy results, including some recent ones due to Kwakernaak [35], Falb [36], Balakrishnan and Lions [37] and Chang [38]. The same techniques apply to discrete-time problems as well. Our new method of proof yields, very simply, a general formula for the smoothed estimate (Part II) [26] and also, more importantly, can be generalized to the nonlinear case (Part III) [27].

A. The Kalman–Bucy Formulas for Recursive Filtering and Prediction

The Kalman–Bucy results are for lumped processes; however, we shall not begin with this assumption, but shall try to see how far we can go without any special assumptions.

We are given $\{y(s) = z(s) + v(s),\ a \leq s < t\}$ and wish to calculate the linear least-squares estimate $\hat{x}(t|t)$ of a related random variable $x(t)$.

The first step is to obtain the innovations, which, by Theorem 2, are given by

$$\boldsymbol{\nu}(t) = y(t) - \hat{z}(t/t), \qquad \overline{\boldsymbol{\nu}(t)\boldsymbol{\nu}(s)} = R(t)\delta(t-s). \quad (9)$$

Because the innovations $\boldsymbol{\nu}(\cdot)$ are equivalent to the original observations $y(\cdot)$, we can express $\hat{x}(t|t)$ as

$$\hat{x}(t \mid t) = \int_a^t g(t, s)\boldsymbol{\nu}(s)ds \quad (10)$$

where the linear filter $g(t, \cdot)$ is to be chosen so that [again using the equivalence of $y(\cdot)$ and $\boldsymbol{\nu}(\cdot)$]

$$x(t) - \hat{x}(t \mid t) \perp \boldsymbol{\nu}(s), \qquad a \leq s \leq t. \quad (11)$$

Putting together (9)–(11), we obtain

$$\overline{x(t)\boldsymbol{\nu}'(s)} = \int_a^t g(t, \sigma)\overline{\boldsymbol{\nu}(\sigma)\boldsymbol{\nu}'(s)}d\sigma \quad (12)$$

$$= g(t, s)R(s), \qquad a \leq s \leq t. \quad (13)$$

It is the last step that justifies the use of the innovation process $\mathbf{v}(\cdot)$. In (10)–(12), we could equally well have used the original observations $y(\cdot)$, but now (12), instead of being trivial, becomes the Wiener–Hopf integral equation, which cannot be solved by inspection. Returning to (13), we can now write

$$\hat{x}(t \mid t) = \int_a^t \overline{x(t)\mathbf{v}'(s)} R^{-1}(s)\mathbf{v}(s)ds. \tag{14}$$

This is the general formula for the *linear* least-squares estimate of $x(t)$ from a white-noise process. (We may point out, in anticipation, that the *nonlinear* (nl) least-squares estimate is given by the remarkably similar formula

$$\hat{x}_{nl}(t \mid t) = \int_a^t \widehat{x(t)\mathbf{v}'(s)} R^{-1}(s)\mathbf{v}(s)ds \tag{15}$$

where

$$\widehat{x(t)\mathbf{v}'(s)} = E[x(t)\mathbf{v}'(s) \mid y(\tau), a \leq \tau < s]. \tag{16}$$

This result will be derived in Part III [27].

So far we have made no special assumptions on $x(t)$. Kalman and Bucy [22] assumed that $x(t)$ satisfies the differential equation

$$\dot{x}(t) = F(t)x(t) + u(t), \qquad t \geq a, \qquad x(a) = x_a \tag{17}$$

where $u(\cdot)$ is white noise with intensity matrix $Q(\cdot)$ and uncorrelated with the observation white noise $v(\cdot)$, i.e.,

$$\overline{u(t)u'(s)} = Q(t)\delta(t-s), \qquad \overline{u(t)v'(s)} \equiv 0 \tag{18}$$ [4]

and the initial value x_a is a zero-mean random variable with variance P_a and uncorrelated with $u(\cdot)$, i.e.,

$$\bar{x}_a = 0, \quad \overline{x_a x_a'} = P_a, \quad \overline{u(s)x_a'} \equiv 0, \quad a \leq s < b. \tag{19}$$

To exploit this structure of $x(t)$, we can differentiate the general estimate formula (14) to obtain

$$\dot{\hat{x}}(t \mid t) = \overline{x(t)\mathbf{v}'(t)}R^{-1}(t)\mathbf{v}(t) + \left[\int_a^t \frac{d}{dt}\overline{x(t)\mathbf{v}'(s)}R^{-1}(s)\mathbf{v}(s)ds\right] \tag{20}$$

$$= \overline{x(t)\mathbf{v}'(t)}R^{-1}(t)\mathbf{v}(t) + \left[F(t)\int_a^t \overline{x(t)\mathbf{v}'(s)}R^{-1}(s)\mathbf{v}(s)ds + \int_a^t \overline{u(t)\mathbf{v}'(s)}R^{-1}(s)\mathbf{v}(s)ds\right]. \tag{21}$$

Now the second term in (21) is equal to $F(t)\hat{x}(t \mid t)$ [cf. (14)], and thus but for the last term, (21) would be a differential equation for $\hat{x}(t \mid t)$.

However, this last term will be zero if we assume that the white noise $u(\cdot)$ that generates the signal process $x(\cdot)$ is uncorrelated with the past observations $y(\cdot)$ [and therefore with the equivalent observations $\mathbf{v}(\cdot)$].

That is, with the further assumption

$$\overline{u(t)y'(s)} \equiv 0, \qquad s < t \tag{22}$$

we shall have

$$\dot{\hat{x}}(t \mid t) = F(t)\hat{x}(t \mid t) + K(t)\mathbf{v}(t), \; \mathbf{v}(t) = y(t) - \hat{z}(t \mid t) \tag{23}$$

where we have defined

$$K(t) \triangleq \overline{x(t)\mathbf{v}'(t)}R^{-1}(t). \tag{24}$$

A block diagram for (23) is shown in Fig. 1(a) where the box yielding $\hat{z}(t \mid t)$ will have a detailed structure similar to that for $\hat{x}(t \mid t)$ if we assume that the $z(t)$ obey a differential relation similar to (17) for $x(t)$. We can be somewhat more explicit about $\hat{z}(t \mid t)$ [and about the $K(t)$ of (24)] if we assume some specific functional relationship between $z(\cdot)$ and (past)[5] $x(\cdot)$. The simplest is, of course, the linear relationship, used by Kalman and Bucy [22],

$$z(t) = H(t)x(t) \tag{25}$$

which immediately yields (by linearity)

$$\hat{z}(t \mid t) = H(t)\hat{x}(t \mid t). \tag{26}$$

This is very useful because now [the innovations $\mathbf{v}(t)$ can be obtained directly from $\hat{x}(t \mid t)$ and $\hat{y}(t)$] the estimate $\hat{x}(t \mid t)$ can be realized by the *feedback* structure of Fig. 1(b).

The (*gain*) function $K(t)$ can also be written in a simpler form under the assumption (25):

$$\begin{aligned} K(t) &= \overline{x(t)\mathbf{v}'(t)}R^{-1}(t) \\ &= \overline{x(t)[\tilde{x}'(t \mid t)H'(t) + v'(t)]}R^{-1}(t) \\ &= \overline{[\hat{x}(t \mid t) + \tilde{x}(t \mid t)]\tilde{x}'(t \mid t)}H'(t)R^{-1}(t) + 0 \qquad (27) \\ &= 0 + \overline{\tilde{x}(t \mid t)\tilde{x}'(t \mid t)}H'(t)R^{-1}(t) \\ &= P(t, t)H'(t)R^{-1}(t), \quad \text{say} \qquad (28)^{6} \end{aligned}$$

where

$P(t, t)$ = the covariance function of the error in the estimate at time t.

It is easy to derive a differential equation for $P(t, t)$ by first noting from (17) and (23) that $\tilde{x}(t \mid t)$ obeys the differential equation

$$\dot{\tilde{x}}(t \mid t) = [F(t) - K(t)H(t)]\tilde{x}(t \mid t) - K(t)v(t) + u(t), \qquad \tilde{x}(a \mid a) = x_a. \tag{29}$$

Now applying a standard formula (Appendix I-B) we can show that $P(t, t)$ satisfies the (nonlinear) matrix Riccati equation

$$\dot{P}(t, t) = F(t)P(t, t) + P(t, t)F'(t) - K(t)R(t)K'(t) + Q(t), \qquad P(a, a) = P_a. \tag{30}$$

[4] The assumption $\overline{u(t)v](s)} \equiv 0$ can be relaxed to $\overline{u(t)v'(s)} = C(t)\delta(t-s)$; cf. (31) and (32).

[5] If $z(\cdot)$ depended on future $x(\cdot)$, we could not satisfy the condition (22).

[6] Note that (28) is true for general $x(\cdot)$, not only those with the differential representation (17).

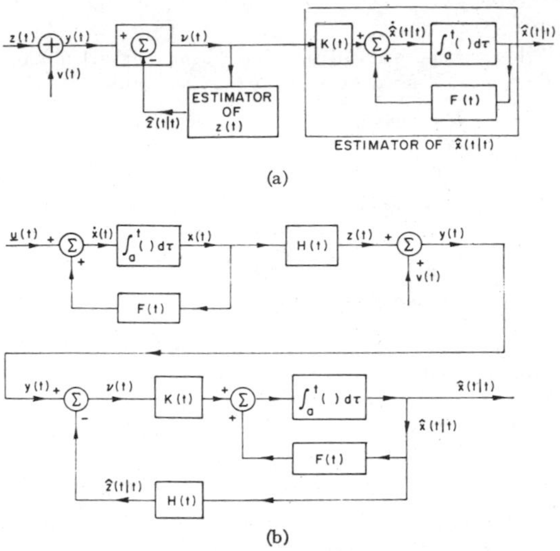

Fig. 1. (a) Filtered estimate of $x(t)$ from a related process $y(\tau)=z(\tau)+v(\tau)$, $a\leq\tau\leq t$. (b) The Kalman–Bucy filter; note that feedback of $\hat{x}(t|t)$ can be used to obtain $\nu(t)$ when $z(t)=H(t)x(t)$.

We now have obtained in formulas (23), (26), (28), and (30) the basic Kalman–Bucy formulas [for the problem defined by (1), (25), and (17)–(19)]. Our derivation is more direct than that of the original and reveals clearly the roles of the various assumptions in the Kalman-Bucy model. In Part III, we shall see the role that the corresponding assumptions play in the nonlinear problem. Our present proof also indicates some points at which the above arguments can be generalized. However, before doing this let us make a few supplementary remarks.

Correlated $u(\cdot)$ *and* $v(\cdot)$*:* We can, without violating the basic constraint (2) that $\overline{z(t)v'(s)}=0$, $s>t$, generalize the uncorrelatedness condition in (18) to

$$\overline{u(t)v'(s)} = C(t)\delta(t-s). \tag{31}$$

This will require minor changes in the above derivations,[7] which we shall leave for the reader's amusement. We shall only point out that finally the only change in the filter formulas [(23), (26), (28), (30)] will be that the gain function of (28) must be replaced by

$$K(t) = [P(t,t)'H'(t) + C(t)]R^{-1}(t). \tag{32}$$

The Prediction Problem: Suppose we are to estimate $x(t+\Delta)$, $\Delta>0$, given observations $\nu(\cdot)$ up to t. Then, by the innovations technique, we readily find

$$\hat{x}(t+\Delta\mid t) = \int_a^t \overline{x(t+\Delta)\nu'(s)}R^{-1}(s)\nu(s)ds. \tag{33}$$

[7] Notably that in (21) and (27) the terms that are zero will now be $(1/2)C(t)R^{-1}(t)$. The 1/2 arises from taking $\int_a^t\delta(t-s)\,ds=1/2$.

If $x(\cdot)$ is a lumped process described by the model (17)–(19), then it is easy to see that

$$\hat{x}(t+\Delta\mid t) = \Psi(t+\Delta, t)\hat{x}(t\mid t) \tag{34}$$

where $\Psi(t, s)$ is the fundamental (or state-transition) matrix of the differential equation (17) of the process $x(\cdot)$, i.e., $\Psi(t, s)$ is the solution of

$$\frac{d}{dt}\Psi(t,s) = F(t)\Psi(t,s), \qquad \Psi(s,s) = I. \tag{35}$$

The Steady-State Equation: In the preceding discussion, we have restricted the interval (a, t) to be finite. When the various matrices $F(\cdot)$, $H(\cdot)$, $Q(\cdot)$, and $R(\cdot)$, are time invariant, it is of interest to study the limiting behavior of the filter as the initial point a tends to $-\infty$. By a careful examination of the Riccati equation (30), Kalman and Bucy [22] have shown that when the model (17), (25) satisfies certain assumptions (stability, controllability, and observability, etc.), we can obtain a well-defined limiting solution by setting $\dot{P}=0$ in (30) and using the non-negative[8] solution of the resulting algebraic equations in the filter formulas (23) and (27). When the process $x(\cdot)$ has a rational spectral density, the above conditions are always met and the Kalman–Bucy solution reduces to the classical solution of Wiener. (The explicit equivalence has been shown by Leake [39].)

B. The Discrete-Time Problem

For discrete-time observations we will have similar results, with one rather trivial modification: the innovation process in the discrete-time case will have a different variance from that of the observation noise. Thus, let

$$y(k) = z(k) + v(k), \qquad k = 0, 1, 2, \cdots,$$
$$\overline{v(k)} = 0, \qquad \overline{v(k)v'(l)} = R(k)\delta_{kl} \tag{36}$$

with $\{z(k)\}$ a zero-mean finite-variance signal process. The innovation process will be defined by

$$\nu(k) \triangleq y(k) - \hat{z}(k\mid k-1) \tag{37}$$

where

$$\hat{z}(k\mid k-1) = \text{the linear least-squares estimate of } z(k) \text{ given } \{y(l), 0\leq l\leq k-1)\}. \tag{38}$$

Then it is easy to calculate (cf. Appendix II) that

$$\overline{\nu(k)} = 0, \qquad \overline{\nu(k)\nu'(l)} = [P_z(k) + R(k)]\delta_{kl} \tag{39}$$

where

$$P_z(k) = \text{covariance matrix of the error in the estimate } \tilde{z}(k\mid k-1) \tag{40}$$
$$= \overline{[z(k)-\hat{z}(k\mid k-1)][z(k)-\hat{z}(k\mid k-1)]'}.$$

Therefore, the innovation process is still (discrete-time)

[8] There are several solutions that are not non-negative definite.

white, but with a different variance. The estimation solution now proceeds essentially as in the continuous-time case; we shall rapidly outline the steps for a process $z(\cdot)$ of the form

$$z(k) = H(k)x(k),$$
$$x(k+1) = \Phi(k+1, k)x(k) + u(k), \tag{41}$$
$$\overline{u(k)u'(l)} = Q(k)\delta_{kl}, \quad \overline{u(k)v'(l)} = C(k)\delta_{kl}.$$

By the projection theorem, and assuming $[P_z(\cdot) + R(\cdot)]^{-1}$ exists,[9] we readily obtain the expression (42) for $\hat{x}(k+1|k)$ in terms of the $v(l)$, $l \le k$, which we can rearrange as

$$\hat{x}(k+1 \mid k)$$
$$= \sum_0^k \overline{x(k+1)v'(l)}[P_z(l) + R(l)]^{-1}v(l) \tag{42}$$
$$= \sum_0^{k-1} \overline{x(k+1)v'(l)}[P_z(l) + R(l)]^{-1}v(l)$$
$$+ \overline{x(k+1)v'(k)}[P_z(k) + R(k)]^{-1}v(k) \tag{43}$$
$$= \Phi(k+1, k)\hat{x}(k \mid k-1) + K(k)v(k), \quad \text{say} \tag{44}$$

where we have defined

$$K(k) \triangleq \overline{x(k+1)v'(k)}[P_z(k) + R(k)]^{-1}. \tag{45}$$

Now we note that

$$\overline{x(k)v'(k)}$$
$$= \overline{[\Phi(k+1, k)x(k) + u(k)][\tilde{x}'(k \mid k-1)H'(k) + v'(k)]} \tag{46}$$
$$= \Phi(k+1, k)\overline{x(k)\tilde{x}'(k \mid k-1)}H'(k) + C(k)$$
$$= \Phi(k+1, k)P(k)H'(k) + C(k) \tag{47}$$

where

$$P(k) \triangleq \overline{\tilde{x}(k \mid k-1)\tilde{x}'(k \mid k-1)}. \tag{48}$$

Therefore, using

$$P_z(k) \triangleq \overline{\tilde{z}(k \mid k-1)\tilde{z}'(k \mid k-1)} = H(k)P(k)H'(k) \tag{49}$$

we can write $K(k)$ as

$$K(k) = \Phi(k+1, k)[P(k)H'(k) + C(k)]$$
$$\cdot [H(k)P(k)H'(k) + R(k)]^{-1}. \tag{50}$$

Finally, with patience, we can derive a recursion relation for $P(k)$ which we quote without proof:

$$P(k+1) = \Phi(k+1, k)A(k)\Phi'(k+1, k) + Q(k),$$
$$A(k) = P(k) - K(k)[P_z(k) + R(k)]K'(k). \tag{51}$$

Equations (44) and (51) define the discrete-time Kalman filter, first derived in a slightly less direct way (but still using the innovations) in Kalman [19]. Our method here is exactly parallel to the one we used in the continuous-time case.

[9] If not, we use the Moore-Penrose pseudo-inverse, but we shall not pursue this refinement here.

C. Some Generalizations

The crucial step in our derivation of the Kalman–Bucy formulas was the use of the assumptions

$$\dot{x}(t) = F(t)x(t) + u(t), \quad \overline{u(t)v'(s)} \equiv 0$$

to write

$$\int_a^t \overline{\dot{x}(t)v'(s)}v(s)ds = F(t)\int_a^t \overline{x(t)v'(s)}v(s)ds$$
$$= F(t)\hat{x}(t \mid t).$$

However, suppose we had

$$\dot{x}(t) = F(t)x(t-1) + u(t), \quad \overline{u(t)y'(s)} \equiv 0. \tag{52}$$

Then we shall have

$$\int_a^t \overline{\dot{x}(t)v'(s)}v(s)ds = F(t)\int_a^t \overline{x(t-1)v'(s)}v(s)ds$$
$$= F(t)\hat{x}(t-1 \mid t). \tag{53}$$

Kwakernaak [35] was apparently the first to point out this result. More generally, suppose

$$\dot{x}(t) = \mathfrak{F} \circ x(\cdot) + u(t),$$
$$y(t) = \mathfrak{H} \circ x(\cdot) + v(t), \quad \overline{u(t)v'(t)} \equiv 0 \tag{54}$$

where $\mathfrak{F} \circ x(\cdot)$ and $\mathfrak{H} \circ x(\cdot)$ denote some linear operation on the "past" values $\{x(s), a \le s < t\}$ of the signal process. Then

$$\int_a^t \overline{\dot{x}(t)v'(s)}v(s)ds = \mathfrak{F} \circ \int_a^t \overline{x(\cdot)v'(s)}v(s)ds$$
$$= \mathfrak{F} \circ \hat{x}(\cdot \mid t) \tag{55}$$

and the obvious analogs of the Kalman–Bucy formulas (23)–(30) are again easily obtained (of course, suitable attention has to be paid to the proper topologies, etc.). Some problems of this type have been noted by Balakrishnan and Lions [36] and Falb [35] who use essentially an operator–theoretic analog of the Kalman–Bucy derivation. They given some specific examples with $\mathfrak{F}$ being a partial differential operator. For a different illustration, we note that $\mathfrak{H}$ may be a random sampling operation, a case that was recently studied in a less direct manner by Chang [37]. General representations of the form (39) often arise in describing stochastic process by evolution equations in abstract spaces and, in fact, some nonlinear processes may be made linear by such representations. We shall not explore this point further in the present elementary paper.

However, it may be of some value to point out that if $x(\cdot)$ obeys a nonlinear equation

$$\dot{x}(t) = f(x(s), s \le t, t) + u(t) \tag{56}$$

then the (linear) estimate $\hat{x}(t|t)$ obeys the equation

$$\dot{\hat{x}}(t \mid t) = \widehat{f(x(s), s \le t, t)} + \overline{x(t)v'(t)}v(t) \tag{57}$$

where $\widehat{f(x(s), s \le t, t)}$ is the best linear estimate of $f(x(\cdot), t)$ given $y(\tau)$, $a \le \tau \le t$. Such a problem was partially discussed by Chang [38].

Finally, we should make a brief comment about problems in which the additive observation noise is nonwhite. One solution is to apply a transformation that will whiten this noise and then use the Kalman–Bucy formulas. This method has been used by Bryson and Johansen [40]. However, a more powerful method is to whiten the whole observation process, the sum of the signal and the nonwhite noise; in other words, to obtain the innovations directly. This method is discussed in Geesey and Kailath [26]. It may be noted that the case of colored (finite-variance) noise plus white noise can be immediately treated by an obvious extension of Theorem 2—the observations can be whitened by subtracting out the estimates of the signal *and* the colored noise.

IV. Concluding Remarks

The main point of the innovations approach to statistical problems is that once we understand the basic probabilistic structure of the processes involved, many results can be obtained quite directly without resort to often more sophisticated (and analytical rather than probabilistic) tools like Wiener–Hopf techniques, Karhunen–Loève expansions, function space integrals, etc. In this paper we have illustrated this point for a class of nonstationary filtering problems.

In [25]–[31] applications are given for linear smoothing problems, nonlinear filtering and smoothing, covariance factorization, and detection problems.

Appendix I

A. The Projection Theorem

Formal proofs of the projection theorem are given in many textbooks. Here we shall make a few informal remarks that may aid in the understanding and application of the result. The projection theorem is probably quite familiar for linear (Hilbert) spaces of time function with inner product

$$\int_T u(t)v(t)dt \text{ or } \int_T u(t)v(t)p(t)dt, \text{ where } p(t) \geq 0. \quad (58)$$

Thus, the linear least-squares approximation to an unknown function $u(\cdot)$ in terms of a given function $v(\cdot)$ is obtained by projecting $u(\cdot)$ on $v(\cdot)$ with the given inner product (58). For our applications, we need to work with Hilbert spaces of random variables, these being values of a stochastic process $z(t)$, for different time instants $t \in [a, b]$, or linear combinations of such random variables. Now random variables are also functions not of t, but of a probability sample-space variable, say $\omega \in \Omega$. The inner product is (very heuristically) $\int_\Omega u(\omega)v(\omega)p(\omega)d\omega$ where $p(\omega)\,d\omega$ is a probability, or, as it is usually written, $\overline{uv}$. As long as we remember that ω, the probability variable, should replace time, all our intuitive notions of Hilbert function spaces (which are essentially generalizations of n-dimensional Euclidean space) carry over to random variables. The orthogonality relations of the projection theorem have a geometric setting in this space of random variables. In this context, there is often some initial confusion because the variable t is also present in the discussion of stochastic processes. However, it is essential to remember that in the Hilbert space of random variables, the elements are not functions of time but functions of ω; the variable t serves only to index some of the elements of the Hilbert space.

B. Covariance Relations for Lumped Processes

Let a random process $x(t)$ be obtained as the solution of the differential equation

$$\dot{x}(t) = F(t)x(t) + u(t), \quad x(a) = x_a, \quad t > a \quad (59)$$

where (the zero means are assumed for notational convenience)

$$\overline{u(t)} = 0, \quad \overline{u(t)u'(s)} = Q(t)\delta(t-s),$$
$$\bar{x}_a, \ \overline{u(t)x_a'} = 0, \quad t \geq a.$$

Then we can write

$$x(t) = \Psi(t, a)x(a) + \int_a^t \Psi(t, s)u(s)ds$$

where $\Psi(t, s)$ is the state-transition matrix defined as the (unique) solution of the equation

$$\frac{d\Psi(t, s)}{dt} = F(t)\Psi(t, s), \quad \Psi(a, a) = I, \ a \leq s \leq t. \quad (60)^{10}$$

By direct computation, we obtain $\bar{x}(t) \equiv 0$ and

$$\begin{aligned} R_x(t, t) &\triangleq \overline{[x(t) - \bar{x}(t)][x(t) - \bar{x}(t)]'} \\ &= \Psi(t, a)R_a\Psi'(t, a)' \\ &\quad + \int_a^t \Psi(t, s)Q(s)\Psi'(t, s)ds. \end{aligned} \quad (61)$$

Differentiating both sides of (61) with respect to t and using (60), we obtain

$$\frac{dR_x(t, t)}{dt} = F(t)R_x(t, t) + R_x(t, t)F'(t) + Q(t), \quad t \geq a \quad (62)$$

$$R_x(a, a) = R_a.$$

Furthermore, it follows by direct computation that

$$\begin{aligned} R_x(t, s) &\triangleq \overline{x(t)x'(s)} \\ &= \overline{\left[\Psi(t, s)x(s) + \int_s^t \Psi(t, \sigma)G(\sigma)u(\sigma)d\sigma\right]x'(s)} \quad (63) \\ &= \Psi(t, s)R_x(s, s) + 0 \qquad \text{for } t \geq s \\ &= R_x(t, t)\Psi'(s, t) \qquad \text{for } s \geq t \quad (64) \end{aligned}$$

where the last equation follows from the symmetry property $R_x(t, s) = R_x'(s, t)$. Equation (62), when applied to (29), yields the Riccati equation (30), as some

[10] When $F(\cdot)$ is time invariant, $\Psi(t,s) = e^{F(t-s)}$, $t \geq s$.

simple algebra will show. Equations (63) and (64) will be used for the smoothing problem in Part II. The above formulas are all well known.

Appendix II

The Innovation Process

If $y(\cdot)=z(\cdot)+v(\cdot)$, where $z(\cdot)$ is a second-order process and $v(\cdot)$ is white noise, we shall prove that the innovation process

$$\nu(t) = y(t) - \hat{z}(t \mid t), \qquad -\infty < a \leq t < b < \infty$$

is white with the same covariance as $v(\cdot)$, and that it is obtained from $y(\cdot)$ by a causal *invertible* linear operation. The first property follows easily by direct computation [and had been known for lumped process $z(\cdot)$]. The second property is more interesting and will be discussed first. (For simplicity, only the scalar case will be treated.)

A. The Relationship Between $y(\cdot)$ and $\nu(\cdot)$

Let $g_\nu(t, s)$ denote the optimum causal filter that operates on $\{y(s), s \leq t\}$ to give $\hat{z}(t|t)$, i.e.,

$$\hat{z}(t \mid t) = \int_a^t g_\nu(t, s) y(s) ds = \mathcal{G}_\nu y, \qquad \text{say} \tag{65}$$

where $\mathcal{G}_\nu$ denotes the integral operator with kernel $g_\nu(t, s)$.[11]

To make (65) well defined we need to assume that (cf. Doob [9], sec. 9.2)

$$\int_a^t g_\nu^2(t, s) ds < \infty \qquad \text{for every } t \in (a, b). \tag{66}$$

(If $g(t, \cdot)$ had delta functions in it, $\hat{z}(t|t)$ would have infinite variance.) From our assumption that $\int z^2(t)dt < \infty$, it can be shown that

$$\int_a^b \int_a^b g_\nu^2(t, s) dt ds < \infty, \tag{67}$$

a fact that will be useful presently. If we use I for the identity operator [the integral operator with kernel $\delta(t-s)$], then we can write, symbolically,

$$\nu = y - \hat{z} = y - \mathcal{G}_\nu y = (I - \mathcal{G}_\nu) y. \tag{68}$$

The problem, then, is to show that $(I-\mathcal{G}_\nu)$ is a causally invertible operator. The causality of $\mathcal{G}_\nu$ does the trick here because $\mathcal{G}_\nu$ is then what is called a Volterra kernel and it can be proved (see, e.g., Smithies [41], p. 34) that when $\mathcal{G}_\nu$ has a square-integrable kernel, then $(1-\mathcal{G}_\nu)^{-1}$ exists and is given by the Neumann (geometric) series

$$(1 - \mathcal{G}_\nu)^{-1} = 1 + \mathcal{G}_\nu + \mathcal{G}_\nu^2 + \mathcal{G}_\nu^3 + \cdots \tag{69}$$

where $\mathcal{G}_\nu^2 y = \mathcal{G}_\nu \mathcal{G}_\nu y$, and so on. The causality is obvious from (69).

In many applications, the signal process $z(\cdot)$ is continuous in the mean [which is equivalent to the continuity of the covariance function of $z(\cdot)$]. In this case, it can easily be shown that the kernel $g_\nu(t, s)$ is continuous in t and s (and, in this case, the arguments to establish (69) are even simpler (Riesz and Nagy [42], sec. 65).

[11] As an aside, we note that $\mathcal{G}_\nu$ can be regarded as an operator on L_2 (cf. Doob [9], sec. 9.2).

B. The Process $\nu(\cdot)$ is White

We shall establish by direct calculation that

$$\overline{\nu(t)\nu(s)} = \overline{v(t)v(s)} \quad \text{where} \quad \nu(t) = y(t) - \hat{z}(t \mid t).$$

First consider $t > s$. Then

$$\begin{aligned} \overline{\nu(t)\nu(s)} &= \overline{[\tilde{z}(t \mid t) + v(t)][\tilde{z}(s \mid s) + v(s)]} \\ &= \overline{v(t)v(s)} + \overline{v(t)\tilde{z}(s \mid s)} \\ &\quad + \overline{\tilde{z}(t \mid t)\tilde{z}(s \mid s)} + \overline{\tilde{z}(t \mid t)v(s)}. \end{aligned} \tag{70}$$

Now $\tilde{z}(s|s) = z(s) - \hat{z}(s|s)$ depends only on signal and noise up to time s. Since we have assumed that future noise is uncorrelated with past signal, the second term in (70) will be zero. Similarly, by the definition of $\hat{z}(t|t)$, $\overline{\tilde{z}(t|t)\tilde{z}(s|s)} = \overline{\tilde{z}(t|t)z(s)} - 0$ for $t > s$. Therefore, we can write (70) as

$$\begin{aligned} \overline{\nu(t)\nu(s)} &= \overline{v(t)v(s)} + \overline{\tilde{z}(t \mid t)z(s)} + \overline{\tilde{z}(t \mid t)v(s)} \\ &= \overline{v(t)v(s)} + \overline{\tilde{z}(t \mid t)[z(s) + v(s)]} \\ &= \overline{v(t)v(s)} + \overline{\tilde{z}(t \mid t)y(s)} \\ &= \overline{v(t)v(s)} + 0, \qquad t > s. \end{aligned} \tag{71}$$

A similar argument applies for $t < s$. Since $\overline{v(t)v(s)} = \delta(t-s) = 0$, $t \neq s$, we have $\overline{\nu(t)\nu(s)} = 0$, $t \neq s$. There remains only to examine the point $t = s$. Here we argue that $\overline{[\nu(t) - v(t)]^2} = \overline{\tilde{z}^2(t|t)} < \infty$, but $\overline{v^2(t)}$ is infinite (because $v(\cdot)$ is white), and therefore $\overline{\nu^2(t)}$ must be infinite (and we have just shown $\overline{\nu(t)\nu(s)} = 0$, $t \neq s$). This identifies $\nu(\cdot)$ as white noise. This argument seems shaky, but it is inevitable if we work with $v(\cdot)$ as an ordinary random process rather than as a generalized random process.

As long as we use the ordinary functional notation for $v(\cdot)$, all proofs, though they can be given in slightly different forms (especially with additional assumptions on $z(\cdot)$, e.g., that it is continuous in the mean or, more strongly, that it is a lumped process), must be essentially of the preceding form. A rigorous proof can be obtained by working with integrals of $v(\cdot)$ and $\nu(\cdot)$ (cf. [29]).

C. Discrete-Time Processes

Some insight is also shed on the preceding calculations by considering the discrete time

$$y(k) = z(k) + v(k), \qquad \overline{v(k)v(l)} = R(k)\delta_{kl},$$
$$\overline{v(k)z(l)} = 0, \qquad l \leq k.$$

In this case, the innovation process $\nu(\cdot)$ is defined as

$$\nu(k) = y(k) - \hat{z}(k \mid k-1) = \tilde{z}(k \mid k-1) + v(k) \tag{72}$$

and, by arguments similar to those in (71), we obtain

for $k>l$

$$\begin{aligned}\overline{\nu(k)\nu(l)} &= \overline{v(k)v(l)} + \overline{v(k)\tilde{z}(l \mid l-1)} \\ &\quad + \overline{\tilde{z}(k \mid k)-1)\tilde{z}(l \mid l-1)} + \overline{\tilde{z}(k \mid k-1)v(l)} \\ &= \overline{v(k)v(l)} + 0 + \overline{\tilde{z}(k \mid k-1)[z(l)+v(l)]} \\ &= \overline{v(k)v(l)}.\end{aligned}$$

Similarly, we can prove the equality for $k<l$. For $k=l$, we have

$$\begin{aligned}\overline{\nu^2(k)} &= \overline{v^2(k)} + \overline{2v(k)\tilde{z}(k \mid k-1)} + \overline{\tilde{z}^2(k \mid k-1)} \\ &= \overline{v^2(k)} + \overline{\tilde{z}^2(k \mid k-1)} = R(k) + P_z(k), \qquad \text{say.}\end{aligned}$$

Therefore

$$\overline{\nu(k)\nu(l)} = [R(k) + P_z(k)]\delta_{kl} \tag{73}$$

so that $\nu(\cdot)$, like $v(\cdot)$, is white but with a different variance. The continuous-time case can be approached by a limiting procedure in which $R(k)$ becomes indefinitely large while $P_z(k)$ remains finite, so that the variances of $\nu(\cdot)$ and $v(\cdot)$ are the same.

D. A Proof of the Equivalence of $\nu(\cdot)$ and $y(\cdot)$ Using the Kalman–Bucy Formulas

We noted in our discussion of Theorem 2 (cf. Remark 2) that the equivalence of $\nu(\cdot)$ and $y(\cdot)$ was obvious if the Kalman–Bucy result was assumed. The proof is trivial. Since $\nu(t)=y(t)-\hat{z}(t|t)$ and $\hat{z}(t|t)$ can be calculated from $y(s)$, $s\leq t$, $\nu(t)$ is completely determined by $y(s)$, $s\leq t$. Conversely, the Kalman–Bucy formula

$$\dot{\hat{x}}(t \mid t) = F(t)\hat{x}(t \mid t) + K(t)[y(t) - H(t)\hat{x}(t \mid t)],$$
$$\hat{x}(a \mid a) = 0$$

shows that $\hat{x}(t|t)$ is determined if $\{\nu(s),\ s\leq t\}$ is known, and then $y(t)$ can be obtained as $y(t)=H(t)\hat{x}(t|t)+\nu(t)$ since

$$\hat{z}(t|t)+\nu(t)=\hat{z}(t|t)+\tilde{z}(t|t)+v(t)=z(t)+v(t).$$

Therefore, $\nu(\cdot)$ and $y(\cdot)$ can each be obtained from the other by causal operations. This argument is due to R. Geesey.

Of course, the deeper result is that this fact is true without restriction to lumped processes and can indeed be used, as we have shown, to give a simple proof of the special formulas for lumped processes.

Acknowledgment

The author thanks R. Geesey, B. Gopinath, and P. Frost, former students at Stanford University, for many stimulating, enjoyable, and instructive conversations on various aspects of the innovations concept and, in particular, for teaching him various aspects of modern control theory.

References

[1] A. N. Kolmogorov, "Interpolation and extrapolation of stationary random sequences," *Bull. Acad. Sci. USSR, Math. Ser.* vol. 5, 1941. A translation has been published by the RAND Corp., Santa Monica, Calif., as Memo. RM-3090-PR.

[2] N. Wiener, *The Extrapolation, Interpolation, and Smoothing of Stationary Time Series with Engineering Applications.* New York: Wiley, 1949. Originally issued as a classified report by M.I.T. Radiation Lab., Cambridge, Mass., February 1942.

[3] H. Wold, *A Study in the Analysis of Stationary Time Series.* Uppsala, Sweden: Almqvist & Wiksell, 1938.

[4] P. Whittle, *Prediction and Regulation.* Princeton, N. J.: Van Nostrand, 1963.

[5] N. Wiener and E. Hopf, "On a class of singular integral equations," *Proc. Prussian Acad., Math.–Phys. Ser.*, p. 696, 1931.

[6] H. W. Bode and C. E. Shannon, "A simplified derivation of linear least square smoothing and prediction theory," *Proc. IRE*, vol. 38, pp. 417–425, April 1950.

[7] R. B. Blackman, H. W. Bode, and C. E. Shannon, "Data smoothing and prediction in fire-control systems," Research and Development Board, Washington, D.C., August 1948.

[8] L. A. Zadeh and J. R. Ragazzini, "An extension of Wiener's theory of prediction," *J. Appl. Phys.*, vol. 21, pp. 645–655, July 1950.

[9] J. L. Doob, *Stochastic Processes.* New York: Wiley, 1953.

[10] H. Laning and R. Battin, *Random Processes in Automatic Control.* New York: McGraw-Hill, 1958.

[11] V. S. Pugachev, *Theory of Random Functions and Its Applications in Automatic Control.* Moscow: Goztekhizdat, 1960.

[12] Y. W. Lee, *Statistical Theory of Communication.* New York: Wiley, 1960.

[13] A. M. Yaglom, *Theory of Stationary Random Functions*, R. A. Silverman, transl. Englewood Cliffs, N. J.: Prentice–Hall, 1966.

[14] R. Deutsch, *Estimation Theory.* Englewood Cliffs, N. J.: Prentice–Hall, 1966.

[15] P. B. Liebelt, *An Introduction to Optimal Estimation Theory.* Reading, Mass.: Addison–Wesley, 1967.

[16] A. V. Balakrishnan, "Filtering and prediction theory," in *Lectures on Communication Theory*, A. V. Balakrishnan, Ed. New York: McGraw–Hill, 1968.

[17] A. E. Bryson and Y. C. Ho, *Optimal Programming, Estimation and Control.* New York: Blaisdell, 1968.

[18] P. Swerling, "First-order error propagation in a stagewise smoothing procedure for satellite observations," *J. Astronautical Sci.*, vol. 6, no. 3, pp. 46–52, Autumn 1959. See also "A proposed stagewise differential correction procedure for satellite tracking and prediction," RAND Corp., Santa Monica, Calif., Rept. P-1292, January 1958.

[19] R. E. Kalman, "A new approach to linear filtering and prediction problems," *Trans. ASME, J. Basic Engrg.*, vol. 82, pp. 34–45, March 1960.

[20] ——, "New methods in Wiener filtering theory," *Proc. 1st Symp. on Engrg. Applications of Random Function Theory and Probability*, J. L. Bogdanoff and F. Kozin, Eds. New York: Wiley, 1963.

[21] R. S. Bucy, "Optimum finite-time filters for a special nonstationary class of inputs," Johns Hopkins University, Appl. Phys. Lab., Baltimore, Md., Internal Memo. BBD-600, 1959.

[22] R. E. Kalman and R. S. Bucy, "New results in linear filtering and prediction theory," *Trans. ASME, J. Basic Engrg.*, ser. D vol. 83, pp. 95–107, December 1961.

[23] W. M. Wonham, "Lecture notes on stochastic optimal control," Div. of Appl. Math., Brown University, Providence, R.I., Rept. 67–1.

[24] A. J. F. Siegert, "A systematic approach to a class of problems in the theory of noise and other random phenomena," Pt. 2 and 3, *IRE Trans. Information Theory*, vol. IT-3, pp. 38–43, March 1957; vol. IT-4, pp. 4–14, March 1958.

[25] T. Kailath and P. Frost, "An innovations approach to least-squares estimation—Part II: Linear smoothing in additive white noise," this issue, page 655.

[26] R. Geesey and T. Kailath, "An innovations approach to least-squares estimation—Part III: Estimation in colored noise" (to be published).

[27] ——, "An innovations approach to least-squares estimation—part IV: Nonlinear filtering and smoothing in white Gaussian noise" (to be published).

[28] P. A. Frost, "Estimation in continuous-time nonlinear systems," Ph.D. dissertation, Dept. of Elec. Engrg., Stanford University, Stanford, Calif., June 1968.

[29] ——, "A general likelihood ratio formula for random signals in Gaussian noise," *IEEE Trans. Information Theory*, to appear, 1969.

[30] T. Kailath, "An RKHS approach to detection and estimation—Part III: More on gaussian detection," to be submitted to *IEEE Trans. Information Theory.*

[31] T. Kailath and R. Geesey, "Covariance factoriztion—An explication via examples," *Proc. 2nd Asilomar Conference on Circuits and Systems*, Monterey, Calif., November 1968.

[32] H. J. Kushner, *Stochastic Stability and Control.* New York:, Academic Press, 1967.

[33] L. D. Collins, "Realizable whitening filters and state-variable realizations," *Proc. IEEE (Letters)*, vol. 56, pp. 100–101, January 1968.

[34] B. D. O. Anderson and J. B. Moore, "Whitening filters: A state-space viewpoint," Dept. of Elec. Engrg., University of Newcastle, Australia, Tech. Rept. EE 6707, August 1967. Also see *Proc. JACC*, (Michigan). 1968.
[35] H. Kwakernaak, "Optimal filtering in linear systems with time delays," *IEEE Trans. Automatic Control*, vol. AC-12, pp. 169–173, April 1967.
[36] P. Falb, "Kalman-Bucy filtering in Hilbert space," *Information and Control*, vol. 11, no. 1, pp. 102–137, August–September 1967.
[37] A. V. Balakrishnan and J. L. Lions, "State estimation for infinite-dimensional systems," *J. Computer and System Sciences*, vol. 1, no. 4, pp. 391–403, December 1967.
[38] S. S. L. Chang, "Optimum filtering and control of randomly sampled systems," *IEEE Trans. Automatic Control*, vol. AC-12, pp. 537–546, October 1967.
[39] R. J. Leake, "Duality condition established in the frequency domain," *IEEE Trans. Information Theory (Correspondence)*, vol. IT-11, p. 461, July 1965.
[40] A. E. Bryson, Jr., and D. E. Johansen, "Linear filtering for time-varying systems using measurements containing colored noise," *IEEE Trans. Automatic Control*, vol. AC-10, pp. 4–10, January 1965.
[41] F. Smithies, *Integral Equations*. London: Cambridge University Press, 1958.
[42] F. Riesz and B. S. Nagy, *Functional Analysis*. New York: Ungar, 1955.

Thomas Kailath (S'57–M'62) was born in Poona, India, on June 7, 1935. He obtained the Bachelor's degree in telecommunications engineering at the University of Poona, Poona, India, in 1956, and the S.M. and Sc.D. degrees at the Massachusetts Institute of Technology, Cambridge, in 1959 and 1961, respectively.

He worked at the Jet Propulsion Laboratories, Pasadena, Calif., until 1963, and since then has been at Stanford University, Stanford, Calif., where he is now Professor of Electrical Engineering. He was a visiting scholar in the Department of Electrical Engineering of the University of California, Berkeley, from January to June, 1963. His research interests are in communication through time-variant channels, feedback communication systems, continuous-time detection and estimation problems, and the analysis and structure of stochastic systems. He was coauthor of a paper on feedback systems that received the 1967 Information Theory Group Award. He is Consulting Editor for a Prentice-Hall series on information theory.

Dr. Kailath is a member of SIAM, the Institute of Mathematical Statistics, URSI, and Sigma Xi.

14

Reprinted from IEEE TRANSACTIONS
ON *AUTOMATIC CONTROL*
Volume AC-13, Number 6, December, 1968
pp. 655-660

An Innovations Approach to Least-Squares Estimation Part II: Linear Smoothing in Additive White Noise

THOMAS KAILATH, MEMBER, IEEE, AND PAUL FROST, MEMBER, IEEE

***Abstract*—The innovations method of Part I is used to obtain, in a simple way, a general formula for the smoothed (or noncausal) estimation of a second-order process in white noise. The smoothing solution is shown to be completely determined by the results for the (causal) filtering problem. When the signal is a lumped process, differential equations for the smoothed estimate can easily be derived from the general formula. In several cases, both the derivations and the forms of the solution are significantly simpler than those given in the literature.**

I. INTRODUCTION

IN THIS PAPER, we apply the innovations technique of Part I[1] to solve the smoothing problem. We shall show that the smoothing solution is completely determined in a simple way by the optimum causal filter and its adjoint. This result is valid for a general second-order (finite-variance) signal process in white noise, without restriction to lumped signal processes. For lumped processes, recursive solutions are available for the filtered estimate (Part I), and from these similar solutions can easily be found for the smoothed estimate. In the literature, recursive solutions to the smoothing problem have generally been obtained rather laboriously and often in less convenient form than ours (cf. Section III).

The problem we shall begin with is the following. We are given observations

$$y(t) = H(t)x(t) + v(t), \qquad a \le t < b \tag{1}$$

where

$$\overline{v(t)} = 0, \quad \overline{v(t)v'(s)} = R(t)\delta(t-s), \quad R(t) > 0 \tag{2}$$

$$\overline{x(t)} = 0, \quad \overline{x(t)x'(t)} < \infty, \quad \overline{x(t)v'(s)} \equiv 0, \quad s > t. \tag{3}$$

It is required to find the linear least-squares smoothed estimate

$$\hat{x}(t \mid b) = \text{the linear function of the data } \{y(s), a \le s < b\} \text{ that minimizes } \overline{[x(t) - \hat{x}(t \mid b)]' [x(t) - \hat{x}(t \mid b)]}. \tag{4}$$

Manuscript received January 31, 1968. This work was supported by the Applied Mathematics Division of the Air Force Office of Scientific Research under Contract AF 49(638)1517, by the Air Force Avionics Laboratory under Contract F 33615-67-C-1245, and by the Joint Services Electronics Program at Stanford University, Stanford, Calif., under Contract Nonr 225(83).
The authors are with Stanford University, Stanford, Calif.
[1] This issue, page 646.

We shall show, and the proof is trivial by the innovations method, that the smoothed estimate $\hat{x}(t|b)$ can be expressed

$$\hat{x}(t \mid b) = \int_a^b \overline{x(t)\nu'(s)}R^{-1}(s)\nu(s)ds \quad (5)$$

where

$$\nu(s) = y(s) - H(s)\hat{x}(s \mid s) \triangleq \text{the innovation process.}$$

By breaking up the range of integration, $\hat{x}(t|b)$ can be written as the sum of the filtered estimate $\hat{x}(t|t)$ and a correction term,

$$\hat{x}(t \mid b) = \hat{x}(t \mid t) + \int_t^b P(t, s)H'(s)R^{-1}(s)\nu(s)ds \quad (6)$$

where

$P(t, s) \triangleq \overline{\tilde{x}(t \mid t)\tilde{x}'(s \mid s)}$ = the covariance function of the error in the filtered estimate; viz., $\tilde{x}(\cdot) = x(\cdot) - \hat{x}(\cdot \mid \cdot)$.

We shall show that the covariances of the smoothed and filtered errors are related by a fairly simple equation that exhibits the reduction in estimation error due to the additional observations from t to b:

$$\overline{\tilde{x}(t \mid b)\tilde{x}'(t \mid b)} = \sum (t \mid b) = P(t \mid t) - \int_t^b P(t, s)H'(s)R^{-1}(s)H(s)P(s, t)ds. \quad (7)$$

Adjoint Filter Interpretation

When the signal and noise process $z(\cdot)$ and $v(\cdot)$ are *completely uncorrelated*, i.e.,

$$\overline{z(t)v'(s)} = 0 \qquad \text{for all } a \leq t, s \leq b, \quad (8)^2$$

we can express $\hat{x}(t|b)$ in a form that has a useful interpretation:

$$\hat{x}(t \mid b) = \int_a^b G_{y_m}(t, s)y_m(s)ds + \int^b G_{y_m}'(s, t)\nu_m(s)ds \quad (9)$$

where

$$G_{y_m}(t, s) \triangleq \begin{cases} P(t, s), & t \geq s \\ 0, & t < s \end{cases}$$

and

$\nu_m(s) \triangleq H'(s)R^{-1}(s)\nu(s) \triangleq$ modified innovation process

$y_m(s) \triangleq H'(s)R^{-1}(s)y(s) \triangleq$ modified observation process.

The first term on the right-hand side of (9) will be shown to be a representation of the filtered estimate, i.e.,

$$\hat{x}(t \mid t) = \int_a^t G_{y_m}(t, s)y_m(s)ds = \int_a^t P(t, s)H'(s)R^{-1}(s)y(s)ds. \quad (10)$$

Thus, $G_{y_m}(t, s)$ is the impulse response of the optimum causal filter with input $y_m(\cdot)$, and $G_{y_m}'(s, t)$ may be regarded as the impulse response of a filter that is *adjoint* to the optimum causal filter. It is clear from the definition of $G_{y_m}(\cdot, \cdot)$ that the adjoint filter is a completely noncausal filter (see, e.g., Zadeh and Desoer [1, sec. 8.3] for a discussion of adjoint filters). Equation (9) states that the smoothed estimate $\hat{x}(t|b)$ is obtained by adding a term that is acquired by passing the modified innovation $\nu_m(\cdot)$ through the adjoint of the optimum causal filter with $y_m(\cdot)$ [not $\nu_m(\cdot)$] as its input. The smoothing formula (9) was first obtained in a different way (see Kailath [2]) by using certain resolvent identities of Siegert [24'].[3] For *lumped* processes the formula (6) was obtained independently by Fraser [3], Kwakernaak [35'] and Frost, who all used different methods. The fact that the result holds for general second-order processes and the present simple proof, based on the innovations theorem of Part I, are due to Kailath. In the lumped case, other formulas for $\hat{x}(t|b)$ have been obtained by us and other authors, but as we shall show in Section III, these formulas are most easily deduced from the general formula (6).

The Discrete-Time Case

In the discrete-time case, we have the analog of the general formula (6) and (7)

$$\hat{x}(k \mid b) = \hat{x}(k \mid k) + \sum_{k+1}^{b} \overline{x(k)\nu(l)}[P_z(l) + R(l)]^{-1}\nu(l) \quad (11)$$

where

$$\nu(l) = y(l) - H(l)\hat{x}(l \mid l-1),$$
$$P_z(l) = \overline{\tilde{z}(l \mid l-1)\tilde{z}'(l \mid l-1)}.$$

But, somewhat surprisingly, there is no analog of the adjoint interpretation (9). The reason is that the adjoint interpretation really depends upon the resolvent identity used in [2] and, unfortunately, this identity does not hold in discrete time.

II. Derivation of General Smoothing Formulas

We start (as for $\hat{x}(t|t)$ in Part I) by noting that because of the equivalence of the observations $y(\cdot)$ and the innovations $\nu(\cdot)$, we can write $\hat{x}(t|b)$ as a linear

[2] For a lumped signal process, this condition can always be arranged by a special change of variables (cf. Kalman [20']).[3]

[3] Primes are used to denote references and equations that were given in Part I.

:$(t|t)$ and $P(t, s)$; of course, such deductions ,o have been made by Kwakernaak and Fraser, ependently obtained (6) for lumped processes, much less direct way. In fact, Fraser [3] made a ›eginning in this direction.

filtered estimate for the model (18)–(19) is

$$= F(t)\hat{x}(t|t) + K(t)\nu(t)$$
$$= [F(t) - K(t)H(t)]\,\hat{x}(t|t) + K(t)y(t), \quad t \geq a \tag{20}$$

$$= y(t) - H(t)\hat{x}(t|t),$$
$$= [P(t, t)H'(t) + C(t)]R^{-1}(t) \tag{21}$$

$$= F(t)P(t, t) + P(t, t)F'(t) - K(t)R(t)K'(t) + Q(t). \tag{22}$$

so note that we can write (cf. Appendix II of)

$$\triangleq \tilde{x}(t|t)\tilde{x}'(s|s) = P(t, t)\Phi'(s, t) \quad \text{for } s \geq t \tag{23}$$

$\Phi(t, s)$ is the fundamental (state-transition) x of the error equation

$$) = [F(t) - K(t)H(t)]\tilde{x}(t|t) + u(t) - K(t)v(t) \tag{24}$$

$\Phi(t, s)$ obeys the equation

$$(t, s) = [F(t) - K(t)H(t)]\Phi(t, s), \quad \Phi(s, s) = I. \tag{25}$$

later use, we note here the (well-known) formula[4]

$$\frac{d\Phi}{dt}(s, t) = -\Phi(s, t)[F(t) - K(t)H(t)], \qquad \Phi(s, s) = I. \tag{26}$$

w returning to the basic smoothing formula (6), we write, using (23),

$$|b)$$
$$= \hat{x}(t|t) + \int_t^b P(t, s)H'(s)R^{-1}(s)\nu(s)ds, \quad t \geq a \tag{27}$$
$$= \hat{x}(t|t) + P(t, t)\int_t^b \Phi'(s, t)H'(s)R^{-1}(s)\nu(s)ds$$
$$= \hat{x}(t|t) + P(t, t)\lambda(t), \quad \text{say} \tag{28}$$

here

$$\lambda(t) = \int_t^b \Phi'(s, t)\nu_m(s)ds \tag{29}$$

nd $\nu_m(s)$ is defined by (9). The function $\lambda(t)$ is closely elated to the adjoint system associated with (24), as we shall soon see. It will be useful to distinguish (as perhaps first done by Meditch [9]) three classes of smoothing problems, defined by the nature of the observation interval (a, b).

[4] This can be obtained by differentiation of the identity $\Phi(t,s)\Phi(s,t) = I$ and use of (25).

Fixed-Interval Smoothing: a and b Fixed

From (29),

$$\dot{\lambda}(t) = -\Phi'(t, t)\nu_m(t) + \int_t^b \frac{d}{dt}\Phi'(s, t)\nu_m(s)ds \tag{30}$$

which, by use of (26) and (27), can be written

$$\dot{\lambda}(t) = -[F'(t) - H'(t)K'(t)]\lambda(t) - \nu_m(t). \tag{31}$$

Equation (31) for $\lambda(\cdot)$ (with $\nu(\cdot)$ set equal to zero) is (cf. Zadeh and Desoer [1, ch. 6) the differential equation of the system that is *adjoint* to the systems of (20) or (24) (with driving terms $y(\cdot)$, $u(\cdot)$, $v(\cdot)$ set equal to zero). Therefore, the $\{\lambda(t)\}$ are often called the *adjoint* variables.

In any case, the formulas (20), (22), (28), and (31) define the smoothed estimate $\hat{x}(t|b)$ by a set of algebraic and differential equations. This set was the form in which the first recursive smoothing solutions were obtained (in a much less transparent manner) by Bryson and Frazier [4].

We shall now derive an alternative representation of $\hat{x}(t|t)$ which was first obtained by Rauch *et al.* [7], also in a more complicated way. We merely note that by differentiation of (28) we will have

$$\dot{\hat{x}}(t|b) = \dot{\hat{x}}(t|t) + \dot{P}(t, t)\lambda(t) + P(t, t)\dot{\lambda}(t) \tag{32}$$

which by use of (20), (22), (29), (31), and some simple algebra can be reduced to

$$\dot{\hat{x}}(t|b) = F(t)\hat{x}(t|b) + Q(t)\lambda(t) \tag{33}$$

and, by one more use of (28), this becomes

$$\dot{\hat{x}}(t|b) = F(t)\hat{x}(t|b) + Q(t)P^{-1}(t, t)[\hat{x}(t|b) - \hat{x}(t|t)]. \tag{34a}$$

By differentiating (7), it easily is shown that the smoothing error covariance can be represented as

$$\dot{\Sigma}(t|b) = (F(t) + Q(t)P^{-1}(t, t))\,\Sigma(t|b) + \Sigma(t|b)(F(t) + Q(t)P^{-1}(t|t))' - Q. \tag{34b}$$

The formulas (34a) and (34b) of Rauch *et al.* are noteworthy because they point out that knowledge of the filtering solution is sufficient for determining the smoothing solution. That is, once $\hat{x}(t|t)$ and $P(t, t)$ are determined for $a \leq t \leq b$, there is no further need to retain the original observations $\{y(t)\}$. A more detailed discussion of this sufficiency property of $\hat{x}(t|t)$ is give in Frost [28′, sec. II-K]. Practical implementatior (34a) and (34b) would dictate obtaining $P^{-1}(t, t)$ a solution of

$$\dot{P}^{-1}(t, t) = -P^{-1}(t, t)F(t) - F'(t)P^{-1}(t, t) - P^{-1}(t, t)QP^{-1}(t, t) + H'(t)R^{-1}($$

since inverting $P(t, t)$ for each t is not fe

There are various other representatic

Equ our

combination of the $\{\nu(s), a \le s < b\}$,

$$\hat{x}(t \mid b) = \int_a^b G_\nu(t, s)\nu(s)ds$$

where $G_\nu(\cdot, \cdot)$ is such that

$$\tilde{x}(t \mid b) = x(t) - \hat{x}(t \mid b) \perp \nu(s), \quad a \le s < b. \tag{12}$$

From these, we obtain

$$\begin{aligned} \overline{x(t)\nu'(s)} &= \int_a^b G_\nu(t, \sigma)\overline{\nu(\sigma)\nu'(s)}d\sigma \\ &= G_\nu(t, s)R(s), \qquad a \le t, \quad s < b. \end{aligned} \tag{13}$$

With this explicit expression for G_ν, we can rewrite $\hat{x}(t|b)$ as

$$\begin{aligned} \hat{x}(t \mid b) &= \int_a^b \overline{x(t)\nu'(s)}R^{-1}(s)\nu(s)ds \\ &= \int_a^t \overline{x(t)\nu'(s)}R^{-1}(s)\nu(s)ds \\ &\quad + \int_t^b \overline{x(t)\nu'(s)}R^{-1}(s)\nu(s)ds. \end{aligned} \tag{14}$$

By setting $b = t$ in the above equations, it is obvious that the first integral in (14) is just the filtered estimate $\hat{x}(t|t)$ [cf. (14′)].

To complete the proof of (6), we note that (the orthogonality property (12) is repeatedly used below)

$$\begin{aligned} &G_\nu(t, s)R(s) \\ &= \overline{x(t)\nu'(s)} = \overline{x(t)[\tilde{x}'(s \mid s)H'(s) + v'(s)]} \\ &= \overline{x(t)\tilde{x}'(s \mid s)}H'(s) = \overline{[\hat{x}(t \mid t) + \tilde{x}(t \mid t)]\tilde{x}'(s \mid s)}\, H'(s) \\ &= \overline{\tilde{x}(t \mid t)\tilde{x}'(s \mid s)}H'(s) \qquad \text{for } s > t \\ &= P(t, s)H'(s) \qquad \text{for } s > t. \end{aligned} \tag{15}$$

To obtain (7), we note that using (6) we can write

$$\tilde{x}(t \mid t) = \tilde{x}(t \mid b) + \int_t^b P(t, s)H'(s)R^{-1}(s)\nu(s)ds. \tag{16}$$

From (12), $\tilde{x}(t|b)$ is orthogonal to $\nu(s)$ for $a \le s < b$. Hence (7) follows immediately.

Finally, with the assumption (8), viz., $\overline{x(t)v'(s)} \equiv 0$, we shall prove (10) from which (9) easily follows. We note [by arguments similar to those for (15)] that

$$\begin{aligned} &P(t, s)H'(s) \\ &\triangleq \overline{\tilde{x}(t \mid t)\tilde{x}'(s \mid s)H'(s)} = \overline{\tilde{x}(t \mid t)[y(s) - H(s)\hat{x}(s \mid s) - v(s)]'} \\ &= -\overline{\tilde{x}(t \mid t)v'(s)}, \qquad \text{for } s < t \\ &\overline{\tilde{x}(t \mid t)v'(s)} = \int_a^t G_{y_m}(t, \sigma)H'(\sigma)R^{-1}(\sigma)\overline{y(\sigma)v'(s)}d\sigma \\ &G_{y_m}(t, s)H'(s), \qquad s < t. \end{aligned} \tag{17}$$

ation (10) follows easily from (17). This completes
erivation of the main smoothing formulas.

III. Recursive Sm
for Lumped

We shall now assume that
process described by

$$\dot{x}(t) = F(t)x($$

where

$$\overline{u(t)u'(s)} = Q(t)\delta(t - s), \quad \overline{u(t)v}$$

Some Earlier Results

Judging by the literature, rec
smoothing problem have been t
obtain than recursive-filtering
already mentioned the work of l
Fraser [3].

The first recursive solutions w
and Frazier [4] (see also Cox [
estimate that minimized a certai
dratic form.

Similar solutions and some alt
obtained (by discrete-time Bayes-
by Rauch [6], and Rauch, Tung,
discrete-time processes. Lee [8] d
using the calculus of variations. Ra
(all but one of) the forms to the co
by the formal limiting procedure
Meditch [9] has very recently done t
that had been omitted by Rauch *et*
earlier paper, Meditch [10] derived sm
for discrete-time processes by using
ments. A very direct projection metho
early report by Carlton [13].

We may also mention some recen
smoothing problem. Fraser [3] studied
for the smoothed estimate, expressing it
tion of two filtered estimates obtained
forward and backward Kalman–Bucy fil
the equivalence to earlier results and
computational aspects of the problem
merical examples. The two-filter solu
given by Mayne [14] and has been furt
Mehra [15]. Henrikson [16] gives a sol
discrete-time problem with colored obse
and Mehra [15] does the same in continuo
essentially combining the smoothing proo
and Frazier [4] and the colored noise t
Bryson and Johansen [40′]. Baggeroer [
proof similar to that of Kwakernaak [35
the Wiener-Hopf equation. A recent thesis b
[18] also obtains results similar to those of K
[35′], but by an ingenious discretization pro

In this paper, we shall derive all these ear
very directly from the recursive Kalman–Bu

las for
could al
who in
but in a
partial
The
$\dot{\hat{x}}(t|t)$

$\nu(t)$
$K(t)$
$\dot{P}(t, t)$

We a
Part
$P(t,$
wher
matr
$\dot{\hat{x}}(t|$
i.e.,
$d\Phi$
dt
For

No
ca
$\hat{x}($

las for $\hat{x}(t|t)$ and $P(t, s)$; of course, such deductions could also have been made by Kwakernaak and Fraser, who independently obtained (6) for lumped processes, but in a much less direct way. In fact, Fraser [3] made a partial beginning in this direction.

The filtered estimate for the model (18)–(19) is

$$\begin{aligned}\dot{\hat{x}}(t|t) &= F(t)\hat{x}(t|t) + K(t)\nu(t)\\ &= [F(t) - K(t)H(t)]\,\hat{x}(t|t) + K(t)y(t), \qquad t \geq a\end{aligned} \tag{20}$$

$$\begin{aligned}\nu(t) &= y(t) - H(t)\hat{x}(t|t),\\ K(t) &= [P(t,t)H'(t) + C(t)]R^{-1}(t)\end{aligned} \tag{21}$$

$$\begin{aligned}\dot{P}(t,t) &= F(t)P(t,t) + P(t,t)F'(t)\\ &\quad - K(t)R(t)K'(t) + Q(t).\end{aligned} \tag{22}$$

We also note that we can write (cf. Appendix II of Part I)

$$P(t,s) \triangleq \tilde{x}(t|t)\tilde{x}'(s|s) = P(t,t)\Phi'(s,t) \qquad \text{for } s \geq t \tag{23}$$

where $\Phi(t, s)$ is the fundamental (state-transition) matrix of the error equation

$$\dot{\tilde{x}}(t|t) = [F(t) - K(t)H(t)]\tilde{x}(t|t) + u(t) - K(t)v(t) \tag{24}$$

i.e., $\Phi(t, s)$ obeys the equation

$$\frac{d\Phi}{dt}(t,s) = [F(t) - K(t)H(t)]\Phi(t,s), \quad \Phi(s,s) = I. \tag{25}$$

For later use, we note here the (well-known) formula[4]

$$\begin{aligned}\frac{d\Phi}{dt}(s,t) &= -\Phi(s,t)[F(t) - K(t)H(t)],\\ \Phi(s,s) &= I.\end{aligned} \tag{26}$$

Now returning to the basic smoothing formula (6), we can write, using (23),

$$\begin{aligned}\hat{x}(t|b) &= \hat{x}(t|t) + \int_t^b P(t,s)H'(s)R^{-1}(s)\nu(s)ds, \qquad t \geq a\\ &= \hat{x}(t|t) + P(t,t)\int_t^b \Phi'(s,t)H'(s)R^{-1}(s)\nu(s)ds\end{aligned} \tag{27}$$

$$= \hat{x}(t|t) + P(t,t)\lambda(t), \qquad \text{say} \tag{28}$$

where

$$\lambda(t) = \int_t^b \Phi'(s,t)\nu_m(s)ds \tag{29}$$

and $\nu_m(s)$ is defined by (9). The function $\lambda(t)$ is closely related to the adjoint system associated with (24), as we shall soon see. It will be useful to distinguish (as perhaps first done by Meditch [9]) three classes of smoothing problems, defined by the nature of the observation interval (a, b).

[4] This can be obtained by differentiation of the identity $\Phi(t,s)\Phi(s,t) = I$ and use of (25).

Fixed-Interval Smoothing: a and b Fixed

From (29),

$$\dot{\lambda}(t) = -\Phi'(t,t)\nu_m(t) + \int_t^b \frac{d}{dt}\Phi'(s,t)\nu_m(s)ds \tag{30}$$

which, by use of (26) and (27), can be written

$$\dot{\lambda}(t) = -[F'(t) - H'(t)K'(t)]\lambda(t) - \nu_m(t). \tag{31}$$

Equation (31) for $\lambda(\cdot)$ (with $\nu(\cdot)$ set equal to zero) is (cf. Zadeh and Desoer [1, ch. 6) the differential equation of the system that is *adjoint* to the systems of (20) or (24) (with driving terms $y(\cdot)$, $u(\cdot)$, $v(\cdot)$ set equal to zero). Therefore, the $\{\lambda(t)\}$ are often called the *adjoint* variables.

In any case, the formulas (20), (22), (28), and (31) define the smoothed estimate $\hat{x}(t|b)$ by a set of algebraic and differential equations. This set was the form in which the first recursive smoothing solutions were obtained (in a much less transparent manner) by Bryson and Frazier [4].

We shall now derive an alternative representation of $\hat{x}(t|t)$ which was first obtained by Rauch *et al.* [7], also in a more complicated way. We merely note that by differentiation of (28) we will have

$$\dot{\hat{x}}(t|b) = \dot{\hat{x}}(t|t) + \dot{P}(t,t)\lambda(t) + P(t,t)\dot{\lambda}(t) \tag{32}$$

which by use of (20), (22), (29), (31), and some simple algebra can be reduced to

$$\dot{\hat{x}}(t|b) = F(t)\hat{x}(t|b) + Q(t)\lambda(t) \tag{33}$$

and, by one more use of (28), this becomes

$$\begin{aligned}&\dot{\hat{x}}(t|b)\\ &\quad = F(t)\hat{x}(t|b) + Q(t)P^{-1}(t,t)[\hat{x}(t|b) - \hat{x}(t|t)].\end{aligned} \tag{34a}$$

By differentiating (7), it easily is shown that the smoothing error covariance can be represented as

$$\begin{aligned}&\dot{\Sigma}(t|b)\\ &\quad = (F(t) + Q(t)P^{-1}(t,t))\,\Sigma(t|b)\\ &\qquad + \Sigma(t|b)(F(t) + Q(t)P^{-1}(t|t))' - Q.\end{aligned} \tag{34b}$$

The formulas (34a) and (34b) of Rauch *et al.* are noteworthy because they point out that knowledge of the filtering solution is sufficient for determining the smoothing solution. That is, once $\hat{x}(t|t)$ and $P(t, t)$ are determined for $a \leq t \leq b$, there is no further need to retain the original observations $\{y(t)\}$. A more detailed discussion of this sufficiency property of $\hat{x}(t|t)$ is given in Frost [28′, sec. II-K]. Practical implementation or (34a) and (34b) would dictate obtaining $P^{-1}(t, t)$ as the solution of

$$\begin{aligned}\dot{P}^{-1}(t,t) &= -P^{-1}(t,t)F(t) - F'(t)P^{-1}(t,t)\\ &\quad - P^{-1}(t,t)QP^{-1}(t,t) + H'(t)R^{-1}(t)H(t)\end{aligned} \tag{35}$$

since inverting $P(t, t)$ for each t is not feasible.

There are various other representations for the solu-

combination of the $\{ \mathbf{v}(s), a \leq s < b \}$,

$$\hat{\mathbf{x}}(t \mid b) = \int_a^b \mathbf{G}_\nu(t, s)\mathbf{v}(s)ds$$

where $\mathbf{G}_\nu(\cdot, \cdot)$ is such that

$$\tilde{\mathbf{x}}(t \mid b) = \mathbf{x}(t) - \hat{\mathbf{x}}(t \mid b) \perp \mathbf{v}(s), \quad a \leq s < b. \quad (12)$$

From these, we obtain

$$\begin{aligned} \overline{\mathbf{x}(t)\mathbf{v}'(s)} &= \int_a^b \mathbf{G}_\nu(t, \sigma)\overline{\mathbf{v}(\sigma)\mathbf{v}'(s)}d\sigma \\ &= \mathbf{G}_\nu(t, s)\mathbf{R}(s), \qquad a \leq t, \quad s < b. \end{aligned} \quad (13)$$

With this explicit expression for $\mathbf{G}_\nu$, we can rewrite $\hat{\mathbf{x}}(t|b)$ as

$$\begin{aligned} \hat{\mathbf{x}}(t \mid b) &= \int_a^b \overline{\mathbf{x}(t)\mathbf{v}'(s)}\mathbf{R}^{-1}(s)\mathbf{v}(s)ds \\ &= \int_a^t \overline{\mathbf{x}(t)\mathbf{v}'(s)}\mathbf{R}^{-1}(s)\mathbf{v}(s)ds \\ &\quad + \int_t^b \overline{\mathbf{x}(t)\mathbf{v}'(s)}\mathbf{R}^{-1}(s)\mathbf{v}(s)ds. \end{aligned} \quad (14)$$

By setting $b = t$ in the above equations, it is obvious that the first integral in (14) is just the filtered estimate $\hat{\mathbf{x}}(t|t)$ [cf. (14')].

To complete the proof of (6), we note that (the orthogonality property (12) is repeatedly used below)

$$\begin{aligned} &\mathbf{G}_\nu(t, s)\mathbf{R}(s) \\ &= \overline{\mathbf{x}(t)\mathbf{v}'(s)} = \overline{\mathbf{x}(t)[\tilde{\mathbf{x}}'(s \mid s)\mathbf{H}'(s) + \mathbf{v}'(s)]} \\ &= \overline{\mathbf{x}(t)\tilde{\mathbf{x}}'(s \mid s)}\mathbf{H}'(s) = \overline{[\hat{\mathbf{x}}(t \mid t) + \tilde{\mathbf{x}}(t \mid t)]\tilde{\mathbf{x}}'(s \mid s)}\,\mathbf{H}'(s) \quad (15) \\ &= \overline{\tilde{\mathbf{x}}(t \mid t)\tilde{\mathbf{x}}'(s \mid s)}\mathbf{H}'(s) \qquad \text{for } s > t \\ &= \mathbf{P}(t, s)\mathbf{H}'(s) \qquad \text{for } s > t. \end{aligned}$$

To obtain (7), we note that using (6) we can write

$$\hat{\mathbf{x}}(t \mid t) = \hat{\mathbf{x}}(t \mid b) + \int_t^b \mathbf{P}(t, s)\mathbf{H}'(s)\mathbf{R}^{-1}(s)\mathbf{v}(s)ds. \quad (16)$$

From (12), $\tilde{\mathbf{x}}(t|b)$ is orthogonal to $\mathbf{v}(s)$ for $a \leq s < b$. Hence (7) follows immediately.

Finally, with the assumption (8), viz., $\overline{\mathbf{x}(t)\mathbf{v}'(s)} \equiv 0$, we shall prove (10) from which (9) easily follows. We note [by arguments similar to those for (15)] that

$$\begin{aligned} &\mathbf{P}(t, s)\mathbf{H}'(s) \\ &\triangleq \overline{\tilde{\mathbf{x}}(t \mid t)\tilde{\mathbf{x}}'(s \mid s)\mathbf{H}'(s)} = \overline{\tilde{\mathbf{x}}(t \mid t)[\mathbf{y}(s) - \mathbf{H}(s)\hat{\mathbf{x}}(s \mid s) - \mathbf{v}(s)]'} \\ &= -\overline{\tilde{\mathbf{x}}(t \mid t)\mathbf{v}'(s)}, \qquad \text{for } s < t \\ &= \overline{\hat{\mathbf{x}}(t \mid t)\mathbf{v}'(s)} = \int_a^t \mathbf{G}_{y_m}(t, \sigma)\mathbf{H}'(\sigma)\mathbf{R}^{-1}(\sigma)\overline{\mathbf{y}(\sigma)\mathbf{v}'(s)}d\sigma \\ &= \mathbf{G}_{y_m}(t, s)\mathbf{H}'(s), \qquad s < t. \end{aligned} \quad (17)$$

Equation (10) follows easily from (17). This completes our derivation of the main smoothing formulas.

III. Recursive Smoothing Formulas for Lumped Processes

We shall now assume that the signal $\mathbf{x}(\cdot)$ is a lumped process described by

$$\dot{\mathbf{x}}(t) = \mathbf{F}(t)\mathbf{x}(t) + \mathbf{u}(t) \quad (18)$$

where

$$\overline{\mathbf{u}(t)\mathbf{u}'(s)} = \mathbf{Q}(t)\delta(t - s), \quad \overline{\mathbf{u}(t)\mathbf{v}'(s)} = \mathbf{C}(t)\delta(t - s). \quad (19)$$

Some Earlier Results

Judging by the literature, recursive solutions for the smoothing problem have been thought to be harder to obtain than recursive-filtering solutions. We have already mentioned the work of Kwakernaak [35'] and Fraser [3].

The first recursive solutions were given by Bryson and Frazier [4] (see also Cox [5]) who obtained an estimate that minimized a certain deterministic quadratic form.

Similar solutions and some alternative forms were obtained (by discrete-time Bayes-rule manipulations) by Rauch [6], and Rauch, Tung, and Striebel [7] for discrete-time processes. Lee [8] derived these results using the calculus of variations. Rauch *et al.* extended (all but one of) the forms to the continuous-time case by the formal limiting procedure of Kalman [20']. Meditch [9] has very recently done the same for the one that had been omitted by Rauch *et al.* In a somewhat earlier paper, Meditch [10] derived smoothing solutions for discrete-time processes by using projection arguments. A very direct projection method was given in an early report by Carlton [13].

We may also mention some recent theses on the smoothing problem. Fraser [3] studied a different form for the smoothed estimate, expressing it as the combination of two filtered estimates obtained by the use of forward and backward Kalman–Bucy filters. He showed the equivalence to earlier results and discussed the computational aspects of the problem in several numerical examples. The two-filter solution was first given by Mayne [14] and has been further studied by Mehra [15]. Henrikson [16] gives a solution for the discrete-time problem with colored observation noise and Mehra [15] does the same in continuous time, both essentially combining the smoothing proofs of Bryson and Frazier [4] and the colored noise techniques of Bryson and Johansen [40']. Baggeroer [17] gives a proof similar to that of Kwakernaak [35'] based on the Wiener-Hopf equation. A recent thesis by Lindquist [18] also obtains results similar to those of Kwakernaak [35'], but by an ingenious discretization procedure.

In this paper, we shall derive all these earlier results very directly from the recursive Kalman–Bucy formu-

tion to the fixed-interval smoothing problem. In particular, we note the two-filter solution introduced by Mayne [14] and Fraser [3]. For a fixed-time interval problem, the notion of a time direction is rather artificial and it is clear that recursive solutions can be defined for a time index running from $t=b$ to $t=a$. Thus, if we set $\tau=b-t$, the model equation (18) becomes

$$\frac{d}{d\tau}x(b-\tau) = -F(b-\tau)x(b-\tau) - u(b-\tau)$$

and the backward time filter equations are

$$\frac{d}{d\tau}\hat{x}_B(\tau) = -E(b-\tau)\hat{x}_B(\tau) + P_B(\tau,\tau)H'(b-\tau) \cdot R^{-1}(b-\tau)[y(b-\tau) - H(b-\tau)\hat{x}_B(\tau)]$$

$$\dot{P}_B(\tau,\tau) = -F(b-\tau)P_B(\tau,\tau) - P_B(\tau,\tau)F'(b-\tau) - P_B(\tau,\tau)H'(b-\tau)R^{-1}(b-\tau)H(b-\tau) \cdot P_B(\tau,\tau) + Q(b-\tau).$$

The backward filter equations expressed in terms of t are then

$$\dot{\hat{x}}_B(b-t) = F(t)\hat{x}_B(t) - P_B(t,t)H'(t)R^{-1}(t)[y(t) - H(t)\hat{x}_B(t)] \quad (36a)$$

$$\dot{P}_B(b-t,b-t) = -F(t)P_B(b-t,b-t) - P_B(b-t,b-t)F'(t) - P_B(t-b,t-b)H(t)R^{-1}(t)H(t) \cdot P_B(t-b,t-b) + Q(t). \quad (36b)$$

A simple calculation shows that the smoothed error covariance $\Sigma(t|b)$ can be expressed by

$$\Sigma^{-1}(t|b) = P^{-1}(t,t) + P_B^{-1}(b-t,b-t) \quad (37a)$$

provided we choose $P_B^{-1}(0,0)=0$.[5] Moreover, the smoothed estimate $\hat{x}(t|b)$ satisfies the relation[6]

$$\hat{x}(t|b) = \Sigma(t|b)[P^{-1}(t,t)\hat{x}(t|t) + P_B^{-1}(b-t,b-t)\hat{x}_B(b-t,b-t)]. \quad (37b)$$

It is useful to be able to express the solution to a smoothing problem in a number of different forms because of the various computational advantages that may accrue to each form. Such computational aspects have been partially considered by Fraser [3] and Mehra [16].

Fixed-Point Smoothing: t is Fixed, Say $t=a$, While b Increases

Our basic smoothing formula (6) yields, with $t=a$,

$$\hat{x}(a|b) = \hat{x}(a|a) + \int_a^b P(a,s)\nu_m(s)ds \quad (38)$$

where we recall that $P(t,s)$ is the covariance function of the error in the filtered estimate $\hat{x}(t|t)$, $a\leq t<b$. Moreover, our formula holds for second-order processes $x(\cdot)$. For lumped processes, Rauch *et al.* [7] and Meditch [11] have given different appearing but equivalent formulas. Thus, Meditch has[7]

$$\hat{x}(a|b) = \hat{x}(a|a) + \int_a^b B(s,a)P(s,s)\nu_m(s)ds \quad (39)$$

where $B(s,a)$ is the solution of the matrix differential equation

$$\frac{d}{ds}P(s,a) = -B(s,a)[F(s) + Q(s)P^{-1}(s,s)], \qquad B(a,a) = I. \quad (40)$$

The two solutions (38) and (39) are equivalent because the matrices $P(a,s)$ and $B(s,a)P(s,s)$ are equal. This can be shown [suggested by Rauch (private communication)] by simple differentiation because both matrices have the same initial value $[P(a,a)]$ and obey the same differential equations.

Fixed-Lag Smoothing: $b=t+\Delta$, $\Delta=A$ Fixed Positive Constant

We proceed as for fixed-interval smoothing, the only difference being that we have an additional term in the expression (31) for $\lambda(t)$. Because

$$\lambda(t) = \int_t^{t+\Delta} \Phi'(s,t)\nu_m(s)ds, \quad (41)$$

we now have

$$\dot{\lambda}(t) = -[F'(t) - H'(t)K'(t)]\lambda(t) - \nu_m(t) + \Phi'(t+\Delta,t)\cdot\nu_m(t+\Delta). \quad (42)$$

Substituting into (32), we have after some algebra

$$\dot{\hat{x}}(t|t+\Delta) = F(t)\hat{x}(t|t+\Delta) + Q(t)P^{-1}(t,t) \cdot (\hat{x}(t|t+\Delta) - \hat{x}(t|t)) + P(t,t)\Phi(t+\Delta,t)H'(t+\Delta) \cdot R^{-1}(t+\Delta)\nu(t+\Delta). \quad (43a)$$

The fixed-lag smoothing error covariance matrix $\Sigma_L(t)$ is given by

$$\dot{\Sigma}_L(t) = [F(t) + Q(t)P^{-1}(t,t)]\Sigma_L(t) + \Sigma_L(t)[F(t) + Q(t)P^{-1}(t,t)]' - (Q(t) + J(t)) \quad (43b)$$

where

$$J(t) = \Phi'(t+L,t)H'(t+L)R^{-1}(t+L)H(t+L)\Phi(t+L,t).$$

[5] Equation (37a) is trivially correct for $t=b$; equality for all t is simply verified by showing that the derivative of the right-hand side of (37a) is given by (34b).

[6] Equation (37b) is verified by showing that the derivative of the right-hand side is given by (34a).

[7] Very recently in another paper [12], Meditch has obtained the form (38) but in a less direct way (similar to that of Kwakernaak [35']).

We observe that the fixed-lag smoothing equations are essentially the same as those for a fixed interval [cf. (34a) and (34b)] except for an additional forcing term.

For this problem Meditch [9] obtains, by a tedious limiting argument on a discrete-time formula of Rauch [6], the formula

$$\begin{aligned}\dot{\hat{x}}(t \mid t+\Delta) &\\ = F(t)\hat{x}(t \mid t+\Delta) &+ Q(t)\lambda(t) + C(t+\Delta, t)\\ &\cdot P(t+\Delta, t+\Delta)\cdot \nu_m(t+\Delta) \end{aligned} \tag{44}$$

where the matrix $C(t+\Delta, t)$ is the solution of the differential equation

$$\begin{aligned}\dot{C}(t+\Delta, t) = [F(t) &+ Q(t)P^{-1}(t, t)]C(t+\Delta, t)\\ &- C(t+\Delta, t)[F(t+\Delta)\\ &+ Q(t+\Delta)P^{-1}(t+\Delta, t+\Delta)]\end{aligned} \tag{45}$$

with initial condition $C(a+\Delta, a) = B(a+\Delta, a)$ where $B(\cdot, \cdot)$ was defined in (40). The equivalence of (43a) and (44)–(45) can be proved in the same way as for (38) and (39), viz., it is easy to check that $P(t, t)\Phi'(t+\Delta, t)$ and $C(t+\Delta, t)P(t+\Delta, t+\Delta)$ have the same initial values and obey the same differential equations.

IV. Concluding Remarks

This paper gives a direct and simple derivation, by the innovations method of Part I, of all previously known smoothing formulas for lumped processes in additive white noise. It also gives a new general formula, (6), for arbitrary second-order signal processes and points out an interesting adjoint filter interpretation of the general smoothing solution.

References

[1] L. A. Zadeh and C. Desoer, *Linear System Theory—A State-Space Approach.* New York: McGraw-Hill, 1963.

[2] T. Kailath, "Application of a resolvent identity to a problem in linear smoothing," to appear in *J. SIAM on Control*, 1969.

[3] D. C. Fraser, "A new technique for the optimal smoothing of data," Sc.D. dissertation, Dept. of Aeronautical Engrg., Massachusetts Institute of Technology, Cambridge, Mass., January 1967.

[4] A. E. Bryson and M. Frazier, "Smoothing for linear and nonlinear dynamic systems," Aeronautical Systems Div., Wright-Patterson AFB, Ohio, Tech. Rept. ASD-TDR-63-119, February 1963.

[5] H. Cox, "On the estimation of state variables and parameters for noisy dynamic systems," *IEEE Trans. Automatic Control*, vol. AC-9, pp. 5–12, January 1964. Also, see Sc.D. dissertation, Dept. of Elec. Engrg., Massachusetts Institute of Technology, Cambridge, Mass., June 1963.

[6] H. E. Rauch, "Solutions to the linear smoothing problem," *IEEE Trans. Automatic Control (Short Papers)*, vol. AC-8, pp. 371–372, October 1963. Also, see Ph.D. dissertation, Dept. of Elec. Engrg., Stanford University, Stanford, Calif., 1962.

[7] H. E. Rauch, F. Tung, and C. T. Striebel, "Maximum likelihood estimates of linear dynamic systems," *AIAA J.*, vol. 3, no. 8, pp. 1445–1450, August 1965.

[8] R. C. K. Lee, *Optimal Estimation, Identification and Control.* Cambridge, Mass.: M.I.T. Press, 1964.

[9] J. S. Meditch, "On optimal linear smoothing theory," *J. Information and Control*, vol. 10, pp. 598–615, 1967.

[10] ——, "Orthogonal projection and discrete optimal linear smoothing," *J. SIAM on Control*, vol. 5, pp. 74–89, February 1967.

[11] ——, "Optimal fixed-point continuous linear smoothing," *Proc. 1967 Joint Automatic Control Conf.* (Philadelphia, Pa.).

[12] ——, "On optimal fixed-point linear smoothing," *Internat'l J. Control*, vol. 6, pp. 189–199, 1967.

[13] A. G. Carlton, "Linear estimation in stochastic processes," Johns Hopkins University, Appl. Phys. Lab., Baltimore, Md., Internal Rept., 1962.

[14] D. Q. Mayne, "A solution of the smoothing problem for linear dynamic systems," in *Automatica*, vol. 4. New York: Pergamon, 1966, pp. 73–92.

[15] R. K. Mehra, Ph.D. dissertation, Dept. of Appl. Phys. and Engrg., Harvard University, Cambridge, Mass., February 1968.

[16] J. Henrikson, Ph.D. dissertation, Dept. of Appl. Phys. and Engrg., Harvard University, Cambridge, Mass., September 1967.

[17] A. E. Baggeroer, Ph.D. dissertation, Dept. of Elec. Engrg., Massachusetts Institute of Technology, Cambridge, Mass., February 1968.

[18] A. Lindquist, "On optimal stochastic control with smoothed information," Royal Inst. Tech., Stockholm, Sweden, IOS Rept. R-25, May 1968.

Paul Frost (S'65–M'67) was born in Hartford, Conn., in October 1938. He received the B.S.E.E. and M.S. degrees from the University of Connecticut, Storrs, in 1962 and 1963, respectively, and the Ph.D. degree from Stanford University, Stanford, Calif., where he was a Howard Hughes Doctoral Fellow, in 1968.

He was an Instructor at the University of Connecticut for the academic year 1963–1964. He has been employed by the IBM Corporation and the Hughes Aircraft Company, and has been an Acting Assistant Professor at Stanford University since September 1967. His present research interests are concerned with applications of stochastic processes to control and communication theory.

Editors' Comments on Paper 15

Further work on the multiplicity representation of a random process has continued in recent years. Professor V. Mandrekar, known for his earlier work in the area, is the author of the following paper, which deals with a class of multivariate, wide-sense Markov processes.

For univariate, wide-sense Markov processes, i.e., for processes with autocovariance function of the form

$$R(t,s) = \Phi[\max(t,s)] \cdot \Theta[\min(t,s)]$$

it is known from earlier studies by Lévy (1955), Doob (1944), and others that, under some "smoothness" conditions on the functions Φ and Θ, the value of their multiplicity is unity.

In the next paper this result is generalized and connected to earlier versions. Furthermore a number of results are derived that touch on topics peripheral to the theory of multiplicity, as, for example, the problem of prediction of multivariate processes as studied by Beutler (1963), Wiener and Masani (1957), and others.

This paper is reprinted with the permission of the author and the Editor of the Nagoya Mathematical Journal.

References

Doob, J. L. (1944). The elementary Gaussian processes. *Ann. Math. Stat.*, **15,** 229–282.

Beutler, F. J. (1963). Multivariate wide-sense Markov processes and prediction theory. *Ann. Math. Stat.*, **34,** 424–438.

Wiener, N., and Masani, P. (1957). The prediction theory of multivariate stochastic processes I. *Acta Math.*, **98,** 111–149.

V. Mandrekar
Nagoya Math. J.
Vol. 33 (1968), 7-19

ON MULTIVARIATE WIDE-SENSE MARKOV PROCESSES*

15

V. MANDREKAR

1. **Introduction:** The idea of multivariate wide-sense Markov processes has been recently used by F.J. Beutler [1]. In his paper, he shows that the solution of a linear vector stochastic differential equation in a wide-sense Markov process. We obtain here a characterization of such processes and as its consequence obtain the conditions under which it satisfies Beutler's equation. Furthermore, in stationary Gaussian case we show that these are precisely stationary Gaussian Markov processes studied by J. Doob [5].

In their remarkable papers, T. Hida [6] and H. Cramér [2], [3] have studied the representation of a purely non-deterministic (not necessarily stationary) second order processes. We obtain such a representation for wide-sense Markov processes directly, by using their theory. The interesting part of our representation is that we are able to show that the multiplicity of q-dimensional wide-sense Markov processes does not exceed q, as, in general, even one-dimensional (not necessarily stationary) processes could have infinite multiplicity (see H. Cramér [2] and T. Hida [6]). We also show that the kernel splits (see Theorem 6. 1). As a consequence of this, we obtain the classical representation of Doob [5].

The paper is divided into 7 sections. The next section is devoted to the introduction of terminology and notation used in the rest of the paper.

2. **Direct-product Hilbert-spaces:** In this section we want to introduce the idea of direct-product Hilbert-spaces as in [10]. If H is a Hilbert-space we shall mean by $H^{(q)}$ the space of all vectors $\underline{h} = (h_1, h_2, \cdots, h_q)$ where for each i, $h_i \in H$. In $H^{(q)}$ is introduced a norm $\|\!| \underline{h} |\!\| = \sqrt{\sum_1^q \| h_i \|_H^2}$ and an inner product given by the Gramian matrix $[h, h^*] = \{< h_i, h_j >_H\}$.

Received November 8, 1967.

*) This research was supported by U.S. Army Research Office, Grant No. DA-ARO-D-31-124-G562 and U.S. Air Force Office of Scientific Research, Grant No. AFOSR-381-65. Sponsored by the Mathematics Research Center, United States Army, Madison, Wisconsin, under Contract No.: DA-31-124-ARO-D-462.

A linear manifold in $H^{(q)}$ is a non-void subset $\mathscr{M}$ of $H^{(q)}$ such that if $\underline{h}$, $\underline{h}' \in \mathscr{M}$ then $A\underline{h} + B\underline{h}' \in \mathscr{M}$ for all $q \times q$ matrices A, B. A subspace of $H^{(q)}$ is a linear manifold closed under the topology $|||\ \ |||$. We recall here a lemma due to N. Wiener and P. Masani [10] which proves the existence of the projection of an element $\underline{h}$ and gives its structure.

LEMMA WM (*Lemma* 5. 8 [10]). (a) *If $\mathscr{M}$ is a subspace of $H^{(q)}$ then there exists a subspace M of H such that $\mathscr{M} = M^{(q)}$, where $M^{(q)}$ denotes the Cartisian product $M \times \cdots \times M$ with q-factors. M is a set of all components of all elements in $\mathscr{M}$.*

(b) *If $\mathscr{M}$ is a subspace of $H^{(q)}$ and $\underline{h} \in H^{(q)}$, then there is a unique $\underline{h}' \in \mathscr{M}$ such that $|||\ \underline{h} - \underline{h}'\ ||| \leq |||\ \underline{h} - \underline{g}\ |||$ for all $\underline{g} \in \mathscr{M}$. For this $\underline{h}'$, $h'_i = P_M h_i$, M being as in* (a).. *An element $\underline{h}'$ satisfies preceding condition iff $[\underline{h} - \underline{h}', \underline{g}] = 0$ for all $\underline{g} \in \mathscr{M}$.*

The part (c), (d), and (e) of the original lemma are omitted since they won't be referred to here.

DEFINITION 2. 1. The unique element $\underline{h}'$ of Lemma WM (b) is called the orthogonal projection of $\underline{h}$ onto $\mathscr{M}$ and is denoted by $(\underline{h} | \mathscr{M})$.

Let (Ω, F, P) be a probability space. By a q-variate second-order stochastic process on (Ω, F, P), we mean a family of random vectors $\{\underline{x}(t), -\infty < t < +\infty\}$ where for each t, $\underline{x}(t) \in L_2^{(q)}(\Omega)$, $L_2(\Omega)$ denoting the Hilbert-space of complex-valued square-integrable random variables $L_2(\Omega)$. The past of the process up to s, $L_2(\underline{x}; s)$ is defined to be the subspace of $L_2(\Omega)$ generated by $\{x^{(i)}(\tau)\ \tau \leq s\ i = 1, z, \cdots, q\}$ with $\underline{x}(t) = \{x^{(1)}(t), \cdots, x^{(q)}(t)\}^*$. The following definition extends to q-variate case, the idea of wide-sense Markov process and that of wide-sense martingale [see Doob [4] pp. 90, 164].

DEFINITION 2. 2. (a) A q-variate process $\{\underline{x}(t)\}$ $(-\infty < t < +\infty)$ is a wide-sense martingale if for each t, $(\underline{x}(t) | L_2^{(q)}(\underline{x}; s)) = \underline{x}_s$ for $s < t$.

(b) A process $\{\underline{x}(t)\}$ is called wide-sense Markov if for each $s < t$, $(\underline{x}(t) | L_2^{(q)}(\underline{x}; s)) = A(t, s)\underline{x}(s)$.

3. Characterization of a wide-sense Markov process: The assumption (D) given below will be made through this paper.

(D. 1) $\underline{x}(t)$-process is continuous in q.m., i.e.,

$$\lim_{s \to t} |||\ \underline{x}(t) - \underline{x}(s)\ ||| = 0.$$

(D. 2) For all t, s real the covariance matrix $\Gamma(t, s)$ is non-singular.

The assumption (D. 2) and the definition of wide-sense Markov process imply $(\underset{\sim}{x}(t)|L_2^{(q)}(\underset{\sim}{x}; s)) = A(t,s)\underset{\sim}{x}_s$ where the matrix $A(t,s)$ is given by $A(t,s) = \Gamma(t,s)\Gamma^{-1}(s,s)$ for $s \leq t$. The function $A(t,s)$ is called a transition matrix function and is defined only for $s \leq t$. Observe that if $\underset{\sim}{x}(t)$ is wide-sense Markov, then for $s \leq t \leq u$ $A(u,s) = A(u,t)A(t,s)$.

The following is the main theorem of this section.

THEOREM 3. 1. *A q-variate second order continuous parameter process satisfying* (D) *is wide-sense Markov if and only if* $\underset{\sim}{x}(t) = \underline{\bar{\phi}}(t)\underline{u}_t$ *with probability one, where for each t,* $\underline{\bar{\phi}}(t)$ *is a non-singular* $q \times q$ *matrix and* $\underline{u}_t$ *is a q-dimensional wide-sense martingale such that* $L_2(\underline{u}; t) = L_2(\underset{\sim}{x}; t)$.

Proof. Sufficiency. Let $\underset{\sim}{x}(t) = \underline{\bar{\phi}}(t)\underline{u}_t$ where $\underline{\bar{\phi}}(t)$ and $\underline{u}_t$ are as described above. Then for $s \leq t$ $(\underset{\sim}{x}(t)|L_2^{(q)}(\underset{\sim}{x}: s)) = (\underline{\bar{\phi}}(t)\underline{u}_t|L_2^{(q)}(xs)) = (\underline{\bar{\phi}}(t)\underline{u}_t|L_2^{(q)}(\underline{u}; s)) = \underline{\bar{\phi}}(t)\underline{u}_s$. Since $\underline{u}_s = \underline{\bar{\phi}}^{-1}(s)\underset{\sim}{x}_s$ with probability one, we obtain that the transition matrix function $A(t,s) = \underline{\bar{\phi}}(t)\underline{\bar{\phi}}^{-1}(s)$.

Necessity. Let $\underset{\sim}{x}(t)$-process be wide-sense Markov. Then denoting by $A(t,s)$ the transition matrix function we have for $s \leq t$,

$$(\underset{\sim}{x}(t)|L_2^{(q)}(\underset{\sim}{x}; s)) = A(t,s)\underset{\sim}{x}_s \qquad \text{with probability one.} \tag{3. 1}$$

$$A(u,s) = A(u,t)A(t,s) \qquad \text{for } s \leq t \leq u. \tag{3. 2}$$

Let us now define, following Hida [6], for every real t, the function

$$\begin{aligned} \underline{\bar{\phi}}(t) &= A(t, s_0) \qquad \text{if } s_0 \leq t \\ &= A^{-1}(s_0, t) \qquad \text{if } t < s_0 \end{aligned}$$

where s_0 is a fixed real number. We now show that for all s, $t (s \leq t)$ real

$$A(t,s) = \underline{\bar{\phi}}(t)\underline{\bar{\phi}}^{-1}(s). \tag{3. 3}$$

First of all, if $s < s_0 \leq t$, (3. 3) is a restatement of (3. 2): i.e., $A(t,s) = A(t,s_0)A(s_0,s)$. If $s_0 \leq s < t$ from (3. 2) we have $A(t,s)A(s,s_0) = A(t,s_0)$, i.e., $A(t,s) = A(t,s_0)A^{-1}(s,s_0)$ giving (3. 3) again. Finally if $s < t \leq s_0$ we again get from (3. 3), $A(s_0,s) = A(s_0,t)A(t,s)$ and hence $A(t,s) = \underline{\bar{\phi}}(t)\underline{\bar{\phi}}^{-1}(s)$. $\underline{\bar{\phi}}(t)$ is non-singular since $A(t,s)$ is. Therefore from (3. 1) and (3. 3) we have for $s < t$,

$$(\underset{\sim}{x}(t)|L_2^{(q)}(\underset{\sim}{x}; s)) = \underline{\bar{\phi}}(t)\underline{\bar{\phi}}^{-1}(s)\underset{\sim}{x}_s \text{ with probability one.} \tag{3. 4}$$

Hence $\underline{u}_t = \underline{\bar{\phi}}^{-1}(t)\underset{\sim}{x}(t)$ is a martingale and $L_2(u; t) = L_2(\underset{\sim}{x}; t)$. The proof of the Theorem is now complete.

The characterization of Theorem 3. 1 will be used later to study purely non-deterministic wide-sense Markov processes and their multiplicity.

However, as a first application we show that if $\underline{x}_0 = 0$ and $\underline{\phi}(t)$ is differentiable, then it satisfies the following differential equation with probability one.

$$\overset{\circ}{\underline{x}}(t) = A(t)\underline{x}(t) + M(t)\underline{\eta}(t) \quad t \geq 0 \tag{3. 5}$$

where $\eta(\cdot)$ is a multivariate "white noise" random process and

$A(t) = \overset{\circ}{\underline{\phi}}(t)\underline{\phi}^{-1}(t)$; $M(t) = \underline{\phi}(t)$. The equation (3. 5) is to be interpreted as $\underline{x}(t) = \int A(t)\underline{x}(t)dt + \int M(t)d\underline{u}(t)$, η_t being the "fictitious derivative" of $\underline{u}_t$.

THEOREM 3. 2. *Let $\{\underline{x}(t),\ 0 \leq t < \infty\}$ be a wide-sense Markov process satisfying (D). If further $x_0 = 0$ and $\underline{\phi}(t)$ of Theorem 3. 1 is continuously differentiable then $\underline{x}(t)$ satisfies equation (3. 5) for $t \geq 0$ where η_t is a q-variate white noise process and the matrix function $A(t) = \overset{\circ}{\underline{\phi}}(t)\underline{\phi}^{-1}(t)$, $M(t) = \underline{\phi}(t)$.*

The proof of the Theorem follows by substituting in (3. 5) $\underline{x}_t = \underline{\phi}(t)\underline{u}_t$.

We now take up the study of covariance function of a stationary wide-sense Markov process.

DEFINITION 3. 2. We say that a q-variate second order process $\{\underline{x}(t),\ -\infty < t < +\infty\}$ is stationary if $\Gamma(t, s) = [\underline{x}(t),\ \underline{x}(s)] = R(t-s)$ for $s < t$.

By Theorem 3. 1 and the definition of wide-sense martingale we get for $h \geq 0$,

$$R(h) = [\underline{x}(t+h),\ \underline{x}(t)] = \underline{\phi}(t+h)J(t,t)\underline{\phi}^*(t), \text{ where } J(t,s) = [\underline{u}(t),\ \underline{u}(s)]_{L_2^{(q)}(\Omega)}. \tag{3. 6}$$

Let $h = 0$, we get

$$R(0) = \underline{\phi}(t)J(t,t)\underline{\phi}^*(t). \tag{3. 7}$$

With $t = 0$ in (3. 6), one has

$$R(h) = \underline{\phi}(h)J(0,0)\underline{\phi}^*(0). \tag{3. 8}$$

Relations (3. 6) and (3. 8) imply $h \geq 0$, $t \geq 0$

$$R(h) = R(t+h)R^{-1}(t)R(0). \tag{3. 9}$$

With $R_1(t) = R(t)R^{-1}(0)$, (3. 9) reduces to

$$R_1(t+h) = R_1(t)R_1(h). \tag{3. 10}$$

We prove now the following theorem.

THEOREM 3. 3. *Let* $\{\underline{x}(t)\}$ $(-\infty < t < \infty)$ *be a q-dimensional stationary process satisfying assumption (D). Then it is wide-sense Markov if and only if the transition matrix function* $B(t) = e^{tQ}$ *for* $t \geq 0$ *where* $B(t) = A(t,0)$ *and* Q *is uniquely determined constant* $q \times q$ *matrix none of whose eigenvalues has positive real part.*

Proof. Necessity. We have already shown that for $R_1(t) = R(t)R^{-1}(0)$ the equation (3. 10) holds. Further, from (D. 1) it follows that $R_1(t)$ is a continuous function and therefore $R_1(t) = e^{tQ}(t \geq 0)$ is the solution of (3. 10) where Q is a $q \times q$ constant matrix (see E. Hille and R.S. Phillips [11]). The assumption (D. 2) implies that $R_1(t)$ is non-singular and hence Q is uniquely determined by $R_1(t)$. Since $B(t) = R(t)R^{-1}(0)$ for $t \geq 0$ we have $B(t) = e^{tQ}$. Due to the fact that $\lambda(t) = \max_{j \leq q} \lambda_j(t)$ (where $\lambda_j(t)$ is j^{th} eigenvalue of $B(t)$) satisfies for all t, $|\lambda(t)| \leq \mathrm{tr}[R^{-1}(0)\,(R^{-1}(0))^*]\,(\sum_{i=1}^{q} E|x_i(0)|^2)^2$ it follows that the eigenvalues of $Q = \lim_{t \to 0} \frac{B(t) - I}{t}$ has non-negative real parts.

The above result was first proved by J.L. Doob [5] for Stationary Gaussian Markov processes. It was reproved by Beutler [1] for wide-sense Markov processes. We have proved it because our proof is based directly on the characterization of Theorem 3. 1. Furthermore it brings out the form of $\underline{\phi}(t)$ in stationary case which will be utilized in Theorem 5. 1. It is also interesting to note that the fact that $R(t-s) = \phi(t)J(s,s)\phi^*(s)$ could enable one to obtain a general form for the covariance function of stationary wide-sense Markov processes (see Kalmykov [8]).

4. Multiplicity of purely non-deterministic wide-sense Markov processes: A second order q-variate process is called purely non-deterministic if $\bigcap_t L_2(\underline{x}; t) = \{0\}$ where $L_2(\underline{x}; t)$ is as defined in Section 2. Let us denote by $E_x(t)$ the projection operators from $L_2(\underline{x})$ (the subspace generated by $\bigcup_t L_2(\underline{x}; t)$) onto $L_2(\underline{x}; t)$. Then under assumption (D. 1) of Section 3 and pure non-determinism, we obtain (see H. Cramér [3])

$$
\text{(4. 1)} \quad
\begin{aligned}
&\text{(i)} \quad L_2(\underline{x}) \text{ is separable}\\
&\text{(ii)} \quad E_x(+\infty) = I \quad E_x(-\infty) = 0\\
&\text{(iii)} \quad E_x(t)E_x(s) = E_x(s)E_x(t) = E_x(\min(s,t))\\
&\text{(iv)} \quad E_x(t+0) = \lim_{n\to\infty} E_x\left(t + \frac{1}{n}\right) = E_x(t) = E_x(t-0) = \lim_{n\to\infty} E_x\left(t - \frac{1}{n}\right).
\end{aligned}
$$

In other words $\{E_{\underline{x}}(t), -\infty < t < +\infty\}$ is a resolution of the identity in $L_2(\underline{x})$. A subset $A \subset L_2(\underline{x})$ is called a generating subset of $L_2(\underline{x})$ with respect to E if $L_2(\underline{x})$ is generated by $\{E(\Delta)f, f \in A$ and Δ a Borel subset of the real line$\}$. The idea of generating set is certainly not unique. However, it is known (see Yosida [12] p. 321), that such sets A can be ordered through their cardinality and there exists one with minimal cardinality. This minimal cardinal number, which because of 4. 1 (i) is at most countable, is called the multiplicity of E. Following H. Cramér [3] and T. Hida [6] we call this multiplicity the multiplicity of $\underline{x}(t)$. Our first result here is to show that under assumption (D) every q-variate wide-sense Markov process has multiplicity not exceeding q. For this purpose we need the following Lemma.

LEMMA 4. 1. *Let H be a separable Hilbert-space with H_1, H_2 be two subspaces of H such that $H_1 \perp H_2$ and $H = H_1 \oplus H_2$. Suppose $\{E_1(t)\}$ is a resolution of the identity in H_1 and $\{E_2(t)\}$ be a resolution of the identity in H_2 such that $E(t) = E_1(t) + E_2(t)$ is a resolution of the identity in H. If N_i is the multiplicity of $E_i (i = 1, 2)$ then multiplicity of E does not exceed $N_1 + N_2$.*

Proof. We are given $H_i = \mathfrak{S}\{E_i(\Delta)f, f \in A, \Delta$ a Borel subset of the real line$\}$, where $\mathfrak{S}$ denotes the "subspace generated by." Since card $A_1 = N_1$, card $A_2 = N_2$, and $E(\Delta)f = E_i(\Delta)f$ for $f \in A_i$ we get that $H = \mathfrak{S}\{E(\Delta)f, f \in A_1 \cup A_2, \Delta$ a Borel subset of real line$\}$. Thus multiplicity of $E \leq \text{card}(A_1 \cup A_2) \leq N_1 + N_2$ completing the proof.

LEMMA 4. 2. *Every purely non-deterministic univariate process $\{v(t); -\infty < t < +\infty\}$ with orthogonal increments has unit multiplicity.*

Proof. Let $\rho(\Delta) = \mathfrak{S}|v(t) - v(s)|^2$, $\Delta = [s, t]$. It is well-known (see Doob [4] Ch. IX) that $L_2(v) = \left\{\int_{-\infty}^{+\infty} f(t)v(dt), f \in L_2(\rho)\right\}$ where $\int_{-\infty}^{+\infty} f(t)v(dt)$ is a stochastic integral in the sense of Doob ([4] Ch. IX). Let $f \in L_2(\rho)$, where f is positive almost everywhere with respect to measure ρ. Then $\int_{-\infty}^{+\infty} f(t)v(dt) = f_0$ generates $L_2(v)$ completing the proof.

THEOREM 4. 1. *The multiplicity of a q-variate wide-sense Markov process satisfying assumption (D) does not exceed q.*

Proof. By Theorem 3. 1, $L_2(\underline{u}; t) = L_2(\underline{x}; t)$ for all t and hence in particular $L_2(\underline{x}) = L_2(\underline{u})$. Therefore $E_{\underline{x}}(t)$, the projection from $L_2(\underline{x})$ onto $L_2(\underline{x}; t)$ is the same operator as $E_{\underline{u}}(t)$ from $L_2(\underline{u})$ onto $L_2(\underline{u}; t)$. Therefore by definition of multiplicity, multiplicity of the process $\underline{x}(t)$ is the same as that of $\underline{u}(t)$. For the sake of simplicity, we shall establish that the multiplicity of a 2-variate wide-sense martingale does not exceed two. The general case being similar, this will conclude the proof. Define $v_1(t)=u_1(t)$, $v_2(t) = (I - P_{L_2(u_1)})u_2(t)$. Since $L_2(u_1) = L_2(u_1, t) \oplus \{u_1(\tau) - u_1(t)\ \tau \geq t\}$ by martingale property. But since $u_2(t) \perp \{u_1(\tau) - u_1(t),\ \tau \geq t\}$ we obtain that $v_2(t) = (I - P_{L_2(u_1\,;\,t)})u_2(t)$. Hence $L_2(\underline{u}; t) = L_2(v_1; t) \oplus L_2(v_2; t)$. It can be easily seen that $v_2(t) = P_{L_2(v_2\,;\,t)}u_2(t)$. This implies that both $\{v_1(t)\ \ -\infty < t < +\infty\}$ and $\{v_2(t)\ \ -\infty < t < +\infty\}$ are mutually orthogonal processes with orthogonal increments. Hence each has multiplicity one by Lemma 4. 2. But $E_u(t) = E_{v_1}(t) + E_{v_2}(t)$ and $L_2(\underline{u}) = L_2(v_1) \oplus L_2(v_2)$ and hence by Lemma 4. 1 we get multiplicity of $E_u \leq 2$. Q.E.D.

Before we conclude this section we want to recall here some ideas of Hida-Cramér theory. They are directly taken from G. Kallianpur and V. Mandrekar [7]. The following theorem of Hellinger-Hahn is well-known (see T. Hida [6]).

THEOREM H-H. *Let $L_2(\underline{x})$ be the separable Hilbert-space and $E(t)$ be any resolution of the identity in $L_2(\underline{x})$ (i.e., satisfies* 4. 1 (ii), (iii), (iv)) *then*

(i) $L_2(\underline{x}) = \sum_1^M \oplus \mathscr{M}_f(i)$ *where* $\mathscr{M}_f(i) = \mathfrak{S}\{E(\Delta)f^{(i)},\ \Delta$ *a Borel subset of the real line*$\}$.

(ii) *If* $\rho_f(i)$ *is the measure denoted by* $\rho_f(i)\,(\Delta) = \|E(\Delta)f^{(i)}\|^2$ *for each Borel set* Δ, *then* $\rho_f(1) \gg \rho_f(2) \gg \cdots$.

(iii) $\mathscr{M}_f(i) = \left\{\int_{-\infty}^{+\infty} f(u)Z_i(du);\ f \in L_2(\rho_f(i))\right\}$ *where* $Z_i(t) = E(t)f^{(i)}$ $(-\infty < t < +\infty,\ i = 1, 2, \cdots, M)$ *are mutually orthogonal processes with orthogonal increments.*

(iv) $\{f^{(1)}, \cdots, f^{(M)}\}$ *is the minimal generating sequence.*

The processes with orthogonal increments are defined in Doob [4] Chapter IX.

The above theorem is essentially the main theorem of Hida [6] and Cramér [3]. It is quoted here in the form as to bring out the connection of multiplicity as defined by us and the multiplicity of a representation as defined by Hida and Cramér.

Applying the above theorem we get

(4. 2) $$x^{(i)}(t) = \sum_{j=1}^{M} \int_{-\infty}^{+\infty} F_{ij}(t,u) Z_j(du) \text{ where } \sum_{1}^{M} \int_{-\infty}^{+\infty} |F_{ij}(t,u)|^2 \rho_f(j)(du) < \infty.$$

Equation (4. 2) gives the Hida-Cramér representation of a stochastic process where M is its multiplicity. It has the property $(s<t)$

(4. 3) $$E_x(s) x_i^{(i)}(t) = \sum_{1}^{M} \int_{-\infty}^{s} F_{ij}(t,u) Z_j(du).$$

A representation satisfying (4. 3) is called a canonical representation. A canonical representation is called proper canonical if

(4. 4) $$L_2(\underline{x};\, t) = \sum_{j=1}^{M} \oplus L_2(Z_j;\, t).$$

Note that $L_2(Z_j;\, t) = \mathscr{M}_f(i)(t) = \mathfrak{S}\{E(\Delta) f^{(i)},\ \Delta$ a Borel subset of $(-\infty, t)\}$. It is proved by Kallianpur and Mandrekar ([7] Theorem 3. 1) that every canonical representation can be assumed to be proper canonical.

Now by Theorem 4. 1 we get that for wide-sense Markov process $M \leq q$. Hence one can write representation (4. 3) in the form of a vector stochastic integral. In the next section we define this concept following M. Rosenberg [9] and obtain an analytic characterization so that a canonical representation be proper canonical.

5. Vector stochastic integrals and analytic characterization of proper canonical representations: Let P, Q be $q \times M$ $(M \leq q)$ matrix-valued functions of real numbers. We say that (P, Q) is integrable with respect to an $M \times M$ hermitian-matrix-valued function ρ if the matrix $P\rho' Q^*$ is integrable elementwise with respect to $\mathrm{tr}\rho$ where ρ' denotes the matrix of densities of elements of ρ, with respect to $\mathrm{tr}\rho$. We then define $\int P d\rho Q^* = \int P\rho' Q^* \mathrm{tr}\rho(du)$. P is said to be square-integrable $[\rho]$ if $\mathrm{tr}\left(\int P d\rho P^*\right)$ is finite. If we denote by $\mathscr{L}_2(\rho)$ the class of all measurable P square integrable with respect to $[\rho]$ where functions P, Q such that $\{P(u) - Q(u)\}\rho'(u) = 0$ a.e. $[\mathrm{tr}\rho]$ are identified. Then $\mathscr{L}_2(\rho)$ is a complete Hilbert space with gramian $[[P, Q]] = \mathrm{tr}\int P d\rho Q^*$ and $\mathrm{tr}[[P, P]] = \text{norm } P$. We shall call $\underline{\xi}$ an orthogonally scattered random vector-valued measure on the real line of dimension M if for each Borel set B, $\underline{\xi}(B) \in L^{(M)}(\Omega)$ and for Borel sets A and B $[\underline{\xi}(B), \underline{\xi}(A)] = \rho(A \cap B)$ where ρ is a Hermitian-matrix-valued measure on the real line. With this setup, Rosenberg [9] defined the vector stochastic integral $\int_{-\infty}^{+\infty} P(u)\underline{\xi}(du)$ for $P \in \mathscr{L}_2(\rho)$ in the same way as Doob does for the case $M=q$ ([4], p. 596). Further, if we denote by $\mathscr{L}_2(\underline{\xi})$ the subspace of $L_2^{(M)}(\Omega)$ generated by $\{\xi(B),\ B \in \mathscr{B}\}$ with $q \times M$ matrices as coefficients, then he has the following theorem, with $\mathscr{B}$ denoting the Borel subsets on the real line.

THEOREM R. *The correspondence* $P \rightarrow \int_{-\infty}^{+\infty} P(u)\underline{\xi}(du)$ *is an isomorphism from* $\mathscr{L}_2(\rho)$ *to* $\mathscr{L}_2(\underline{\xi})$.

In our context $\underline{Z}(B) = (Z_1(B), \cdots, Z_N(B))^*$ and $F(t,u)$ will be denoted by the matrix $\{F_{ij}(t,u)\}$. We then have from (4. 2) and Theorem 4. 1 that

$$\underline{x}(t) = \int_{-\infty}^{t} F(t,u)\underline{Z}(du);\ F \in \alpha_2(\rho) \text{ where } \rho(u) = \begin{bmatrix} \rho_f(1) & & 0 \\ & \ddots & \\ 0 & & \rho_f(M) \end{bmatrix}. \tag{5. 1}$$

If we denote by $\mathscr{L}_2(\underline{Z};\, t)$, the subspace of $\mathscr{L}_2(\underline{Z})$ generated by $\{Z(B),\ B$ a Borel subset of $(-\infty,\ t)\}$, then we trivially have

$$L_2^{(q)}(\underline{Z};\, t) = \mathscr{L}_2(\underline{Z};\, t). \tag{5. 2}$$

We now give an analytical characterization of a proper canonical representation. This is a direct generalization of Theorem 1. 7 of [6].

THEOREM 5. 1. *A canonical representation* (5. 1) *is proper canonical if and only if*

$$\int_{-\infty}^{t} P(u)d\rho(u)F^*(t,u) = 0 \ \textit{for}\ t \leq t_0 \ \textit{implies}\ P(u) = 0 \ \textit{a.e.}\ [\rho] \ \textit{on}\ (-\infty, t) \tag{5. 3}$$

where $P \in \mathscr{L}_2(\rho)$.

Proof. Sufficiency. Let (5. 3) hold and suppose that there is a t_0 with $L_2(\underline{Z};\, t_0) \neq L_2(\underline{x};\, t_0)$. Since by canonical property $L_2^{(q)}(\underline{x};\, t_0) \subseteq L_2(\underline{Z};\, t_0)$ we have a $\underline{V} \in \mathscr{L}_2(\underline{Z};\, t)$ (see 5. 2) such that $[\underline{V}, \underline{x}(t)] = 0$ for $t \leq t_0$. By Theorem R we have $\underline{V} = \int_{-\infty}^{t_0} P(u)Z(du)$ and $\neq 0$ and $\int_{-\infty}^{t} P(u)\underline{Z}(du)F^*(t,u) = 0$ for $t \leq t_0$. But (5. 3) this implies $P(u) = 0$ a.e. $[\rho]$ contradicting $\underline{V} \neq 0$.

Necessity. Suppose that $L_2(\underline{Z};\, t) = L_2(\underline{x};\, t)$ for each t and let t_0 be a real number such that

$$\int_{-\infty}^{t} P(u)\underline{Z}(du)F^*(t,u) = 0 \text{ for } t \leq t_0. \tag{5. 4}$$

Observe that from the proper canonical property $L_2^{(q)}(\underline{x}; t_0) = H^{(q)}(\underline{x}; t_0) = \mathscr{L}_2(\underline{Z}; t_0)$. Hence the vector $\underline{V} = \int_{-\infty}^{t} P(u)\underline{Z}(du)$ belongs to $L_2^{(q)}(\underline{x}; t_0)$. But (5. 4) implies that $[\underline{V}, \underline{x}(t)] = 0$ for all $t \leq t_0$. Hence $\underline{V} = 0$ giving $\underline{P}(u) = 0$ a.e. $[\rho]$. This proves the theorem.

In the next section we use this theorem to obtain the representation of purely non-deterministic processes.

6. Representation of a purely non-deterministic wide-sense Markov process and the result of Doob: In this section we obtain the representation of a purely-nondeterministic Markov process and as a consequence obtain the representation [(4. 3. 2 of [5]). The main theorem is as follows.

THEOREM 6. 1. *Let $\underline{x}(t)$ be a continuous parameter purely non-deterministic process satisfying (D). Then it is wide sense Markov if and only if*

(6. 1) $$\underline{x}(t) = \int_{-\infty}^{t} \underline{\phi}(t)G(u)\underline{Z}(du)$$

where

(i) *$\underline{\phi}(t)$ is as in Theorem* 3. 1,

(ii) *$\underline{Z}$ is an orthogonally scattered vector random function with*

$$[\underline{Z}(B), \ \underline{Z}(A)] = \begin{pmatrix} \rho_1(A \cap B) & & 0 \\ & \ddots & \\ 0 & & \rho_M(A \cap B) \end{pmatrix} = \rho(A \cap B)$$

for A, B Borel subsets of the real line,

(iii) $G \in \mathscr{L}_2(\rho)$

(iv) $L_2(\underline{Z}; t) = L_2(x; t)$.

Proof. Sufficiency. Define $\underline{u}(t) = \int_{-\infty}^{t} G(u)\underline{Z}(du)$. Then it suffices to prove that $\underline{u}(t)$ is a wide-sense Martingale. As then by Theorem 3. 1 the result will follow. Consider $s < t$ and $L_2^{(q)}(\underline{u}; s)$. Then

(6. 2) $$(\underline{u}(t) \,|\, L_2^{(q)}(\underline{u}; s)) = (\underline{u}(t) \,|\, L_2^{(q)}(\underline{x}; s)) = (\underline{u}(t) \,|\, L_2^{(q)}(\underline{Z}; s)),$$

where the first equality follows from non-singularity of $\phi(t)$ and the second (iv) of the hypothesis. Hence

(6. 3) $$(\underline{u}(t) | L_2^{(q)}(\underline{u}; s)) = \left(\int_{-\infty}^{t} G(u)\underline{Z}(du) | L_2^{(q)}(\underline{Z}; s)\right) = \int_{-\infty}^{s} G(u)d\underline{Z}(u).$$

The last equality in (6. 3) is a consequence of the fact that $\int_s^t G(u)d\underline{Z}(u) \perp L_2^{(q)}(Z; s)$. Hence $\underline{u}(t)$ is a wide-sense Martingale completing the sufficiency part.

Necessity. By wide-sense Markov hypothesis we obtain that $s < t$

(6. 4) $$\underline{x}(t) - A(t, s)\underline{x}_s \perp L_2^{(q)}(\underline{x}; \sigma) \text{ for } \sigma \leq s.$$

Equivalently (6. 4) gives

(6. 5) $$\int_{-\infty}^{\sigma} [F(t, u) - A(t, s)F(s, u)]\rho(du)F^*(\sigma, u) = 0 \text{ for } \sigma \leq s.$$

By Theorem (5. 1) we have

(6. 6) $$F(t, u) = A(t, s)\, F(s, u) \text{ a.e. } [\rho] \text{ on } (-\infty, s).$$

However, as in Theorem 3. 1 $A(t, s) = \underline{\phi}(t)\underline{\phi}^{-1}(s)$ and hence $F(t, u) = \underline{\phi}(t)\underline{\phi}^{-1}(s)F(s, u)$ a.e. $[\rho]$ on $(-\infty, s)$ i.e.,

(6. 7) $$\underline{\phi}^{-1}(t)F(t, u) = \underline{\phi}^{-1}(s)F(s, u) \text{ a.e. } [\rho] \text{ on } (-\infty, s).$$

From equation (6. 7) and the fact that $\| F(t, u) - F(s, u) \| \mathscr{L}_2(\rho) \to 0$ as $s \to t$. From (D. 1) we obtain that $G(u) = \underline{\phi}^{-1}(t)F(t, u)$ is independent of t. Hence $F(t, u) = \underline{\phi}(t)G(u)$ on $(-\infty, t)$ a.e. $[\rho]$. This completes the proof since (ii), (iii) and (iv) are consequences of the properties of proper canonical representation.

Now to obtain the result of Doob we appeal to the following theorem

THEOREM KM (*G. Kallianpur and V. Mandreker* [7] *Theorem* (5. 1)). *If $\underline{x}(t)$ is a q-variate purely non-deterministic stationary process satisfying* (D. 1), *then*

(6. 7) $$\underline{x}(t) = \int_{-\infty}^{t} K(t - u)\underline{\xi}(du) \text{ where } L_2(\underline{x}; t) = L_2(\underline{\xi}; t);$$

(i) $\underline{\xi}(\Delta) = (\xi_1(\Delta), \xi_2(\Delta), \cdots, \xi_M(\Delta))$ *with*

$$\xi_i(\Delta) = \int_\Delta \left[\frac{d\rho_f(i)}{d\mu} u\right]^{-\frac{1}{2}} Z_i(du)$$

for each Borel set Δ on the real line,

(ii) $K(t - \cdot) \in \alpha_2(\rho)$,

(iii) *M is the multiplicity of* $\underline{x}(t)$.*

We would like to remark that as a consequence of (i) $[\underline{\xi}(\Delta), \underline{\xi}(\Delta')] = \mu(\Delta \cap \Delta')I$ where μ is the Lebesgue measure on the real line and Δ, Δ' are Borel subsets. I denotes $M \times M$ identity matrix. From (6. 1) and (6. 7) we obtain that

$$K(t-u) = \underline{\Phi}(t)H(u) \qquad \text{a.e. } \mu \tag{6. 8}$$

where the equality is taken elementwise and $H(u) = G(u)\sum(u)$ where

$$\sum(u) = \begin{pmatrix} \left(\frac{d\rho_f(i)}{d\mu}\right)^{-\frac{1}{2}} & & 0 \\ & \ddots & \\ 0 \quad {}_{M\times M} & & \left(\frac{d\rho_f(i)}{d\mu}\right)^{-\frac{1}{2}} \end{pmatrix}.$$

Without loss of generality we can assume that (6. 8) holds at $u = 0$, otherwise the change is multiplication by a constant non-singular matrix. Putting $u = 0$ in (6. 8) we get

$$F(t) = \underline{\Phi}(t)H(0) \quad t \geq 0 \tag{6. 9}$$

i.e., $F(t) = e^{tQ}S$ where $S = \underline{\Phi}(0)H(0)$ from Theorem 3. 3. We have thus,

THEOREM 6. 2. *Let* $\underline{x}(t)$ *be a purely non-deterministic process satisfying* (D. 1). *It is widesense stationary Markov if and only if*

$$\underline{x}(t) = \int_{-\infty}^{t} e^{(t-u)Q}S\underline{\xi}(du) \tag{6. 10}$$

where Q *is as in Theorem* (3. 3), $[\underline{\xi}(\Delta), \underline{\xi}(\Delta')]_{L_2(q)(\underline{x})} = \mu(\Delta \cap \Delta')I$. I *is an* $M \times M$ *identity matrix where* M *is the rank of the process.*

The fact that M is the rank of the process from the representation (6. 10).

In comparing (6. 10) to Doob ([5]) we observe that Doob does not use Gaussian hypothesis. If we denote by $\zeta(t) = \int_{-\infty}^{t} e^{uQ}S\underline{\xi}(du)$ then $\underline{\xi}(t)$ is the ζ-process of equation (4. 3. 14) of Doob. M will then correspond to the number of ones occuring in his diagonal matrix U (see (4. 3. 18) In [5]).

7. Concluding Remark and Acknowledgements:

Remark. Theorem 3. 1 opens up the question of what processes can be represented at $\sum_{1}^{N}\underline{\Phi}_i(t)u_i(t)$ where $\underline{\Phi}_i(t)$ are some matrix functions and

$\underline{u}_i(t)$ are widesense martingales. It has been established by the author [AMS (1965) Abstract] that these lead under suitable conditions to continuous parameter N-ple Markov processes. Extension to such processes of the analytic questions studied here are being investigated and will be published later.

Acknowledgements. Some of the problems studied here were completed in the author's thesis written at Michigan State University under the guidance of Professor G. Kallianpur whom the author would like to thank for useful and interesting discussions. Finally, many thanks are due to the referee of this paper, without whose careful suggestions, the clarity in the writing could not have been achieved.

References

[1] Beutler, F.J. (1963), Multivariate wide-sense Markov processes and prediction theory. *Ann. Math. Statist.* **34,** 424–438.

[2] Cramér, H. (1961), On some classes of non-stationary processes. *Proc. 4th Berkeley Sympos. Math. Statist. and Prob.* **II** 57–77.

[3] Cramér, H. (1961), On the structure of purely non-deterministic processes. *Ark. Mat.* **4,** 2–3, 249–266.

[4] Doob, J.L. (1953), *Stochastic Processes* New York.

[5] Doob, J.L. (1944), The elementary Gaussian processes. *Annals of Math. Stat.* vol. **15,** 229–282.

[6] Hida, T. (1960), Canonical representations of Gaussian processes and their applications. *Mem. Coll. Sci.* A33 Kyoto. 109–155.

[7] Kallianpur, G. and Mandrekar, V. (1965), Multiplicity and representation theory of purely non-deterministic stochastic processes. *Teor. Veroyatnost. i Primenen.* **X,** USSR.

[8] Kalmykov, G.I. (1965), Correlation functions of a Gaussian Markov process. *Dokl. Akad. Nauk SSSR.*

[9] Rosenberg, M. (1964), Square integrability of matrix valued functions with respect to a non-negative definite Hermitian measure. *Duke Math. J.* **31.**

[10] Wiener, N. and Masani, P. (1957), The prediction theory of multivariate stochastic processes I. *Acta Math.* **98,** 111–149.

[11] Hille, E. and Phillips, R.S. (1957), Functional analysis and semigroups. *American Math. Soc. Coll.* Ch. X.

[12] Yosida, K. (1965), *Functional Analysis.* Springer-Verlag. 321–22.

University of Minnesota

Editors' Comments on Paper 16

Following his earlier work on the subject, Kailath offers in the following paper a comprehensive semitutorial presentation of the applicability of the innovations representation of a random process to problems of detection and estimation.

In particular, he considers applications to linear least-squares smoothing and prediction and to the detection of Gaussian and non-Gaussian signals in white Gaussian noise. Furthermore, he derives the Kalman–Bucy formulas, the solution to certain Fredholm integral equations, and the mutual information and channel capacity for white Gaussian noise channels.

This paper is reprinted with the permission of the author and the Director of Editorial Services of the Institute of Electrical and Electronics Engineers.

The Innovations Approach to Detection and Estimation Theory

16

THOMAS KAILATH, FELLOW, IEEE

Abstract—Given a stochastic process, its innovations process will be defined as a white Gaussian noise process obtained from the original process by a causal and causally invertible transformation. The significance of such a representation, when it exists, is that statistical inference problems based on observation of the original process can be replaced by simpler problems based on white noise observations. Seven applications to linear and nonlinear least-squares estimation. Gaussian and non-Gaussian detection problems, solution of Fredholm integral equations, and the calculation of mutual information, will be described. The major new results are summarized in seven theorems. Some powerful mathematical tools will be introduced, but emphasis will be placed on the considerable physical significance of the results.

I. Introduction

THE "innovations" approach to estimation and detection problems could also have been called the "whitening-filter" approach. However, this latter term has usually been used only in the context of linear processing of stationary and Gaussian processes; we use the term "innovations approach" to emphasize its applicability to nonlinear processing and non-Gaussian processes as well.

Given a stochastic process $\{y(t), t \in I\}$, where I is an interval on the real line, usually $I=[0, T]$, $T<\infty$, we shall define its *innovations* process $\{v(t), t \in I\}$ as a white Gaussian noise (WGN) process[1] such that $y(\cdot)$ can be calculated from $v(\cdot)$ by a causal (i.e., nonanticipative) and causally invertible transformation. If the transformation is linear, $y(\cdot)$ will be Gaussian, but this is not a necessary assumption. The point is that $v(\cdot)$ and $y(\cdot)$ contain the same "statistical information," since we can go back and forth in real time from one process to the other, but, of course, $v(\cdot)$ will generally be a much simpler statistical process than $y(\cdot)$. Moreover, since the values of $v(\cdot)$ at different instants of time are statistically independent of each other, each observation $v(t)$ brings "new information," unlike the observation $y(t)$ which is, in general, statistically related to past values of $y(\cdot)$. Therefore, following Wiener and Masani, the process $v(\cdot)$ will be called the "new information" or the "innovations" process of $y(\cdot)$.

The innovations approach was first used by Kolmogorov [1] in his solution of the linear least-squares prediction problem for stationary discrete-time stochastic processes. The technique was independently rediscovered, for continuous-time processes, by Bode and Shannon [2] and Zadeh and Ragazzini [3], and became known as the "whitening-filter" method In detection theory, the whitening-filter technique was proposed by Kotel'nikov [4] as a device for solving colored-noise problems by reduction to the white-noise case. Almost all of this pioneering work was for stationary second-order processes, though Zadeh and Ragazzini [3], Dolph and Woodbury [5], Darlington [6], and others attempted some extensions. However, the first really successful application to nonstationary second-order processes was made in 1960 by Kalman [7], with his recursive "state-space" algorithm for least-squares estimation of discrete-time Gauss–Markov processes. Though Kalman used (at least, implicity) the innovations method in [7], there were apparently some difficulties in extending the approach to continuous-time Gauss–Markov processes, for which Kalman and Bucy [8] (also [9]) used other methods.

The author's interest in the innovations method began with a desire to extend Kolmogorov's approach to obtain the result of Kalman and Bucy [8]. In doing this, however, it became clear that the method had greater scope and could be used without restriction to Gauss–Markov processes and "state-space" descriptions [10]–[13]. More important, the method could also be applied to several nonlinear least-squares estimation problems [14], [15] and to large classes of Gaussian and non-Gaussian detection problems [16]–[19]; some other applications, stochastic and nonstochastic as well, will be noted presently.

The innovations method has thus enabled us to significantly breach the "linear and/or Gaussian" barrier that seemed to have been reached in detection and least-squares estimation theory. But, even more important, in the author's opinion, is the fact that the innovations method has helped to bring forward a whole new set of powerful mathematical tools, including the Itô stochastic calculus and especially the martingale theory, including its newer aspects due to the French and Japanese schools of probability theory. A tutorial account of some of these new mathematical tools is given in [20], where further references may be found.

Manuscript received November 4, 1969; revised March 3, 1970. This work was supported by the Applied Mathematics Division of the Air Force Office of Scientific Research under Contract AF 44-620-69-C-0101 and by the JSEP at Stanford University under Contract Nonr 225(83). This paper was partially written at the Bell Telephone Laboratories, and, with the assistance of a Guggenheim Fellowship, at the Indian Institute of Science, Bangalore, India.

The author is with Stanford University, Stanford, Calif., but is presently a Visiting Professor at the Indian Institute of Science, Bangalore, India, and a Fellow of the John S. Guggenheim Memorial Foundation.

[1] This definition will suffice for the applications made in this paper; however, as briefly noted in Section V, and elsewhere, the notion of an innovations process can be formulated in a more general way.

Another important feature is that the innovations method has forced us to look deeper into the structural aspects of the detection and estimation problems we are interested in. In particular, it has brought out, as we shall illustrate in some detail later, the importance of processes of the form "signal" plus "white Gaussian noise" with a one-sided dependence between signal and past noise, but not necessarily a Gaussian signal (cf., Theorems 1–7). In obtaining these basic structural results, we are also forced to examine more closely the exact meaning of the integrals that arise in detection and estimation theory, even in the classical Gaussian detection problem. As discussed at some length in [16] and [18]–[20], this question of integrals is not merely a matter of rigor. Unless the proper interpretations are recognized and used, we will often obtain incorrect answers and we will neither be able to properly state the new general results for non-Gaussian problems nor to reduce the general results to previously known special cases.

These general remarks will be made more specific in later sections, but we do wish to make the point here that the general techniques most commonly used at present, e.g., the Karhunen–Loève and other sampling and series expansion methods, do not seem to lend themselves easily to use in nonlinear and non-Gaussian problems. On the other hand, the innovations approach works directly with the given discrete- or continuous-time processes; besides its greater applicability, the innovations method thus also provides a more direct derivation of earlier results, including those developed by state-space and Markov-process methods.

Finally, it may appear from the above remarks that the innovations method is rather abstruse and difficult to grasp physically; however, as we hope to illustrate below, one of the strengths of the innovations method is the great physical content of the basic results and the steps used in arriving at them. In fact, in this paper, we generally use only physical arguments to justify various statements, leaving the mathematical proofs to the references. In discussing this question of physical content, we should point out that one of the first papers using the innovations method, namely that of Bode and Shannon [2], set as its chief aim the derivation of Wiener's results by a method that would be more understandable and have more appeal to engineers. We are pleased that the further development of the innovations method has kept and, in fact, deepened the physical content of the method, while also bringing out its great mathematical content and wide applicability. In other words, we wish to stress that there need be no conflict between the greater mathematical power of a tool and its physical significance and interpretation. But now, as Bret Harte said, it is time to "cut the cackle and get to the hosses."

Outline of the Paper

We shall begin in Section II with some definitions and general problems associated with the existence and determination of innovation processes. The problem of existence is nontrivial, even for Gaussian processes. After the introductory discussions in Section II, we shall turn in the next two sections to two important classes of processes that have innovation representations. These basic processes are those that are the sum of a white Gaussian noise (WGN) process and a "smooth" signal process that may be Gaussian (Section III) or non-Gaussian (Section IV). Such processes are important because they are used to model many detection and estimation problems[2] and also because many more complicated processes can often be reduced by simple reversible transformations to one of the above special forms (Section V).

In Sections III and IV we shall see that the problem of finding the innovations is closely associated with certain causal least-squares estimation problems. This association will help to motivate and interpret several of our results. Such physical interpretations are especially important here, not only for their intrinsic value, but also because the rigorous mathematical proofs are rather deep and, by themselves, perhaps rather forbidding. As for the applications themselves, their wide range can be seen from the following partial list:

- Application I: The Kalman–Bucy formulas
- Application II: Linear least-squares smoothing
- Application III: Prediction in linear difference systems
- Application IV: Solution of certain Fredholm integral equations
- Application V: Detection of Gaussian signals in WGN
- Application VI: Detection of non-Gaussian signals in WGN
- Application VII: Mutual information and channel capacity for WGN channels.

The first five applications will be treated in Section III; the others in Section IV. For reasons of space, we have omitted several other applications, including some to nonlinear filtering [14], [15], to estimation in colored noise [12], to the so-called "separation theorem" of stochastic control [21], to adaptive estimation [22], to some modeling questions [23], and to system identification. However, the applications that we do treat should serve to illustrate well the power of the innovations method. Moreover, in Section V, we shall briefly discuss how to remove the restriction in Section III and IV that an additive WGN be present.

As stated previously, for several reasons, including space, all our discussion and proofs will be somewhat brief and largely heuristic, with detailed proofs being left to the references. We may also note that the sections dealing with the applications to estimation problems form a relatively self-contained unit and may be omitted, at least on a first reading, by a reader more interested in detection problems. The reader will note, however, that least-squares estimates enter in a fundamental way into our solution to the detection problem.

[2] It may be argued with some justice that WGN will always be present in communication and control problems, as thermal noise in analog systems and as round off noise in digital systems. However, we may note that much of our analyses for WGN can be generalized to certain more general non-Gaussian noises, see [17] and [20, secs. 10 and 11].

II. Innovation Processes and Innovation Representations

We stated in Section I that in this paper we would define[3] the innovations process of a given stochastic process, say, $y(\cdot)$, as a white Gaussian noise (WGN) process, and say, $v(\cdot)$, that is related to $y(\cdot)$ by a causal and causally invertible transformation. The representation of the process $y(\cdot)$ by such a transformation of a WGN process will be called the *innovations representation* (IR) of $y(\cdot)$. The term canonical representation (CR) will also sometimes be used.

The first general studies of such representations were made by Lévy in a series of papers in the 1950's; a summary of this work can be found in [24]. Lévy, who used the term *proper canonical representation* for the IR, obtained several interesting general results, but did not establish whether every process (under certain assumptions on nonperfect least-squares predictability of the process) has an IR. It turns out that, in general, the answer is no, even for Gaussian processes $y(\cdot)$, but innovations and innovation representations do exist for several important classes of processes. We shall first illustrate by example some of the general phenomena that can arise, and then proceed to a closer analysis of some processes that do have IR's.

Example 1: Stationary Gaussian Processes With Rational Power Spectral Densities. This was the class treated by Bode and Shannon [2] and admits of a simple solution (at least, for scalar processes). Thus, suppose the power spectral density of $y(\cdot)$ is

$$S_y(s) = \frac{s^2 - 4}{(s^2 - 1)(s^2 - 9)}, \qquad s = j\omega. \tag{1}$$

Then, it is easy to see that the (IR) of $y(\cdot)$ can be obtained by passing a WGN $v(\cdot)$ through a filter with transfer function

$$\mathscr{L}(s) = (s + 2)/(s + 1)(s + 3).$$

The general rule is to put into $\mathscr{L}(s)$ the left-half plane (LHP) poles and zeros of $S_y(s)$. The choice of LHP poles makes $\mathscr{L}(s)$ the transfer function of a causal filter and the choice of LHP zeros makes the filter causally invertible. Note that if we did not have this latter requirement, many filters could be found that would causally "whiten" $y(\cdot)$. However, there is a unique (up to sign) causal and causally invertible whitening filter. Note that $\mathscr{L}(s)$ yields the factorization

$$S_y(s) = \mathscr{L}(s)\mathscr{L}(-s)$$

which is why the problem of finding IR's (or equivalently, the innovations) for stationary Gaussian processes is often referred to as a spectral factorization problem. In nonstationary Gaussian processes, we have to "factor" the covariance function. We now examine a simple nonstationary process.

Example 2: Some Nonstationary Processes. The IR can be found rather trivially for a Wiener process; i.e., a Gaussian process with mean zero and covariance function

$$\overline{y(t)y(s)} = t\wedge s, \qquad 0 \le t, s \le T. \tag{2}$$

Then an invertible causal representation (i.e., the canonical representation) is

$$y(t) = \int_0^t v(\tau)\, d\tau \tag{3}$$

but there are many causal, but noncanonical (i.e., noncausally invertible) representations, e.g.,

$$y(t) = \int_0^t \left\{3 - \frac{12\tau}{t} + \frac{10\tau^2}{t^2}\right\} v(\tau)\, d\tau. \tag{4}$$

Equation (4) is a special case of a more general result of Lévy [24].

A less trivial example of an IR is obtained by considering Gauss–Markov processes, which have covariances of the form

$$\overline{y(t)y(s)} = a(t\vee s)b(t\wedge s), \qquad s \le t, s \le T \tag{5}$$[4]

where $b(t)>0$, $0\le t\le T$. Doob [25] showed that such processes have the IR

$$y(t) = b(t)\int_0^t p(\tau)\dot{v}(\tau)\, d\tau \tag{6}$$

where $p(\cdot)$ is defined by the relation

$$a(t)/b(t) = \int_0^t p^2(\tau)\, d\tau + a(0)/b(0). \tag{7}$$

Example 3: General Stationary Processes. A *nondeterministic stationary Gaussian process over* $(-\infty, \infty)$, viz., one whose power density $S(\omega)$ satisfies

$$\int \{|\ln S(\omega)|/(1 + \omega^2)\}\, d\omega < \infty,$$

has an IR. This was shown by Karhunen [26] and Hanner [27].

Example 4: Discrete-Time Processes. A discrete-time Gaussian process, stationary or not, over a half-line $t\ge n_0$, *always has an IR*, which can be obtained by a Gram–Schmidt orthonormalization. If the process is nondeterministic, we can have $n_0 = -\infty$.

Example 5: Processes Without an IR. However, *a (nondeterministic) nonstationary continuous-time process may not always have an IR.* Hida [28] gives the following example:

Consider two independent Wiener processes $w_1(\cdot)$ and $w_2(\cdot)$ on $[0, T]$. Then the process

$$y(t) = \begin{cases} w_1(t), & t\varepsilon \text{ rationals on } [0, T] \\ w_2(t), & t\varepsilon \text{ irrationals on } [0, T] \end{cases}$$

[3] The point is that instead of the WGN, we can use any process with independent values at every instant (cf. [20]).

[4] We shall use the compact notations $t\wedge s = \min(t, s)$ and $t\vee s = \max(t, s)$.

has no IR. To represent the process $y(\cdot)$ we will need to sum the outputs of two separate causal and causally invertible linear systems each driven by an independent white noise. In fact, Cramér [29] has shown that given any integer N, $0 \le N \le \infty$, there exists a process that will need N such filters, each driven by an independent white noise. For any process, the smallest number N that will suffice is called the multiplicity of the process; it is uniquely determined by the covariance function of the process.

As noted by Cramér [29], [30], this sharp difference in the possibilities for discrete-time and continuous-time nonstationary processes is rather surprising; one consequence is, of course, that care should be exercised in solving certain continuous-time problems by limiting arguments based on discrete-time approximations. However, we should note that the known examples of Gaussian processes of multiplicity greater than one are all rather pathological (e.g., the one above); less pathological examples can be given for non-Gaussian processes (cf., Motoo and Watanabe [30a]).

The important papers of Hida [28] and Cramér [29] establish several general properties, e.g., uniqueness of canonical representations for Gaussian processes, but they do not have many explicit examples. In fact, in a recent paper [30], Cramér poses as an important question the problem of identifying classes of nonstationary Gaussian processes that have multiplicity one. In Section III and IV we shall discuss two such classes: one Gaussian and one non-Gaussian. We shall also present several applications of the IR's for these classes of processes. In Section V, we shall briefly describe how to obtain the IR's for some more general classes of processes.

III. Innovations for a Class of Gaussian Processes and Some Applications

In this section we shall study processes of the form

$$y(t) = z(t) + v(t), \qquad 0 \le t \le T < \infty \tag{8}$$

where $v(\cdot)$ is WGN with unit spectral intensity

$$E[v(t)] = 0, \qquad E[v(t)v'(s)] = \delta(t-s) \qquad (9)^5$$

and $z(\cdot)$, which we shall often call the signal process, is Gaussian and such that

$$E[z(t)v'(s)] \equiv 0, \qquad \text{for } t < s. \tag{10}$$

In other words, $z(t)$ may depend upon past $\{v(s), 0 \le s < t\}$, as may happen in feedback communication problems, but future $v(\cdot)$ must be independent of past $z(\cdot)$. Furthermore, we shall define

$$E[z(t)z'(s) + z(t)v'(s) + v(t)z'(s)] = K(t, s), \quad 0 \le t, s \le T \tag{11}$$

and assume that

$$K(t, s) \text{ is continuous in } (t, s) \tag{12}$$

and

$$\int_0^T \operatorname{tr}[K(t, t)]\, dt < \infty. \tag{13}$$

We shall also assume that

$$E[z(t)] = 0, \qquad 0 \le t \le T. \tag{14}$$

Assumption (14) of zero mean-value is unnecessary, but is made here for notational convenience; (12)–(13) are more important, but they too can be somewhat relaxed, as we shall indicate later (Theorem 2). It will turn out that the problem of finding the innovations for processes of the form (8)–(14) depends intimately upon the problem of finding certain causal least-squares estimates. Consequently, Wiener–Hopf equations will play a key role in our solution. In fact, let us define

$$\hat{z}(t) = \text{the least-squares estimate of } z(t) \text{ given } \{y(s), 0 \le s < t\}, \qquad 0 \le t \le T. \tag{15}$$

That is, $\hat{z}(t)$ is the functional of past $y(\cdot)$ which yields

$$\operatorname{tr}\{E[z(t) - \hat{z}(t)][z(t) - \hat{z}(t)]'\} = \text{minimum}. \tag{16}$$

We shall also define

$$\tilde{z}(t) = z(t) - \hat{z}(t), \tag{17}$$

the instantaneous error in the estimate $z(t)$.

It is well known (e.g., Papoulis [31] and the comments in [10, Application I]) that $\hat{z}(t)$ is a linear functional of past $y(\cdot)$ that is uniquely determined by the *orthogonality* condition

$$\hat{z}(t) - z(t) \perp \{y(s), 0 \le s < t\} \tag{18}$$

where $a \perp b$ means $Eab' = 0$. From (18), we see easily that if

$$\hat{z}(t) = \int_0^t h(t, s) y(s)\, ds, \tag{19}$$

then the Volterra kernel $h(\cdot, \cdot)$ must satisfy

$$h(t, s) + \int_0^t h(t, \tau) K(\tau, s)\, d\tau = K(t, s), \qquad 0 \le s < t < T. \tag{20}$$

Because of the "causality constraint" that $s < t$, we shall call this equation the Wiener–Hopf equation for $h(t, s)$. To complete our definitions, let us also note that (19) and (20) can be more compactly written in an obvious operator notation as

$$\hat{z} = hy, \qquad h + \{hK\}_+ = K_+ \tag{21}$$

where

$$\{A(t, s)\}_+ = \text{the "causal" part of } A(t, s) = A(t, s), \qquad 0 \le s < t \le T. \tag{22}$$

We do not write $\{h_+\}$ because we shall always assume that $h(t, s)$ is causal; $K(t, s)$ must, of course, be symmetric because $\delta(t-s) + K(t, s)$ ($I + K$, in operator notation) is the covariance of $y(\cdot)$. Thus, we can now state our first basic result.

[5] The more general problem with $E[v(t)v'(s)] = r(t)\delta(t-s)$ can be easily reduced to the present one if $r(\cdot) > 0$, i.e., if $(\cdot)$ is strictly positive definite. Also note that we generally use matrix notation, and incidentally, thus also treat certain multivariable (vector) problems without special comment. Primes denote transpose and $tr[\cdot]$ stands for trace.

Theorem 1: Let $y(\cdot)$, $z(\cdot)$, and $v(\cdot)$ be defined as in (8)–(14), and $\hat{z}(\cdot)$ as in (14)–(20). Then the process

$$\nu(t) = y(t) - \hat{z}(t) = \tilde{z}(t) + v(t), \qquad 0 \le t \le T \tag{23}$$

is the innovations process of $y(\cdot)$, i.e., it is a WGN process with

$$E\nu(t) = 0, \qquad E\nu(t)\nu'(s) = \delta(t-s) \tag{24}$$

and $\nu(\cdot)$ is related to $y(\cdot)$ by a causal and causally invertible transformation. In fact, in operator notation.

$$\nu = (I-h)y, \qquad y = (I-h)^{-1}\nu. \tag{25}$$

Proof: Equations (23) and (24) have a nice interpretation. Since $\tilde{z}(t)$ is the portion of $z(t)$ that cannot be predicted from past $y(\cdot)$, and since $v(t)$, being WGN and also independent of past $z(\cdot)$, cannot be predicted from past $y(\cdot)$, it is reasonable to expect that $\nu(t) = \tilde{z}(t) + v(t)$ is the "new information" or the "innovation" in the observation $y(t)$, given all past observations $\{y(s), 0 \le s \le t\}$. Alternatively, since $\hat{v}(t) = 0$, $\hat{z}(t)$ is the least-squares estimate of $y(t)$ given past $\{y(s), s < t\}$ which also suggests that $\nu(t) = y(t) - \hat{z}(t)$ should be the "new information" in $y(t)$. Now if the names "innovation" or "new information" are really valid, the process $\nu(\cdot)$ should be white. And, in fact, this can be proved rigorously (see Application I of [16] or, for a somewhat less rigorous argument, Application II of [10]). However, from the second equality in (23), it seems surprising that the covariance of $\nu(\cdot)$ should be the same as that of $v(\cdot)$; the explanation, of course, is that $\hat{z}(\cdot)$ is correlated with $v(\cdot)$ in such a way that

$$\begin{aligned} E[\nu(t)\nu'(s)] &= E[v(t)\nu'(s)] + E[\tilde{z}(t)\nu'(s)] \\ &= E[v(t)v'(s)] + E[v(t)\tilde{z}(s)] + E[\tilde{z}(t)\nu'(s)] \\ &= E[v(t)v'(s)] + 0. \end{aligned}$$

To quickly check this, for example for $t > s$, note that by (10), the term $E[v(t)\tilde{z}'(s)]$ is zero and that by the orthogonality condition (18), so is the term $E[z(t)\nu'(s)]$! (Incidentally, most of the informal proofs using innovations are of this order of simplicity.)

To complete our argument that $\nu(\cdot)$ is the innovations process, we must show that $y(\cdot)$ can be recovered from $\nu(\cdot)$, or equivalently, that the operator $(I-h)$ in (24) has a causal inverse. The obvious candidate for the inverse is the Neumann series $I + h + h^2 + \cdots$, but, of course, such a series may not converge unless h is "sufficiently small." For example, if h is a scalar, $|h|$ must be less than one, while if h is a matrix, its largest eigenvalue must be less than one. Fortunately for us, in (25), h is a causal and Volterra operator which is well known to have no nonzero eigenvalues. (We may note that a Volterra operator is the analog of a triangular matrix with zeros *on* and above the main diagonal.) A more rigorous proof is given in [10].

Remark 1: Note the simplicity of the argument that we sketched to show $E[\nu(t)\nu'(s)] = E[v(t)v'(s)]$—the other proofs in [10], [11] are of the same general order of difficulty! Note also that no special assumptions have been needed on the structure of $z(\cdot)$. In the literature, the result (of the first half) of Theorem 1 has been known for the special case where process $z(\cdot)$ is generated by passing white noise through a lumped dynamical system; even in this special case the previous proofs are quite elaborate (see, e.g., Collins [32]).

Remark 2: The canonical representation of $y(\cdot)$ is

$$y = (I-h)^{-1}\nu = (I+k)\nu,$$

where $I + k = (I-h)^{-1}$ or equivalently $h + hk = k$. Note that if h is known $(I-h)^{-1} = (I+k)$ can be obtained by putting h into the return link of a simple unity forward gain feedback loop. Conversely, if the IR filter k is known, then h can be realized by putting k into the forward loop. Note that if by some means we had been able to first obtain the IR filter k, then the estimation problem can be solved by inspection. In other words, we reiterate that the filtering and IR problems are intimately related.

Remark 3: The conditions in Theorem 1 are only sufficient conditions; the innovations may be found as $y(\cdot) - \hat{z}(\cdot)$ under more general conditions. For example the classical Wiener filtering formula will show that the innovations are given by (23) when $z(\cdot)$ has a rational power-spectral-density function and the observation interval is semi-infinite (so that (13) is violated).

Remark 4: The filter h can be found by solving the Wiener–Hopf equation (20). When $K(t, s)$ has a "separable" form, we may expect from the work of Kalman and Bucy that the Wiener–Hopf equation can be replaced by a certain nonlinear Riccati differential equation, which is apparently often easier to solve on a computer than the Wiener–Hopf equation. This expectation will be realized later in Application IV. However, before that, let us note that Theorem 1 also enables us to obtain a simple derivation of the Kalman–Bucy formulas themselves.

Application I: The Kalman–Bucy Formulas

In the Kalman–Bucy problem, the signal process $z(\cdot)$ is modeled as the output of a dynamical system driven by white noise:

$$\dot{x}(t) = F(t)x(t) + G(t)u(t), \quad E[u(t)u'(s)] = Q(t)\delta(t-s) \tag{26}$$ [6]

$$\begin{aligned} z(t) = H(t)x(t), \quad E[x(0)] = 0 \quad E[x(0)x'(0)] = P_0, \\ E[u(t)x'(0)] \equiv 0, \qquad t \ge 0. \end{aligned} \tag{27}$$

The observations are

$$\begin{aligned} y(t) = z(t) + v(t), \qquad E[v(t)v'(s)] = I\delta(t-s), \\ E[v(t)u'(s)] \equiv 0. \end{aligned} \tag{28}$$

The "state vector" $x(\cdot)$ may represent the position and velocity (after linearization) of a satellite and the forcing term $u(\cdot)$ may represent random drag. With this model, Kalman and Bucy derived a differential equation for the estimates, $\hat{x}(\cdot)$, of the states; the estimate of $z(\cdot)$ can then be obtained by linearity as

$$\hat{z}(t) = H(t)\hat{x}(t). \tag{29}$$

To obtain the Kalman–Bucy equations by the innovations

[6] In general, $x(\cdot)$, $u(\cdot)$, $z(\cdot)$, $\dot{v}(\cdot)$, and $y(\cdot)$ may be column matrices and correspondingly F, G, H, Q, and P_0 will be matrices of appropriate dimensions. No special notation will be used to distinguish column matrices from more general matrices.

method, we first replace the given observations process $y(\cdot)$ by its innovations process $v(\cdot)$, which by Theorem 1 is given by

$$v(t) = y(t) - \hat{z}(t) = y(t) - H(t)\hat{x}(t). \tag{30}$$

Then the desired estimate $\hat{x}(t)$ is calculated as a linear functional of the innovations

$$\hat{x}(t) = \int_0^t g(t, s)v(s)ds. \tag{31}$$

By the orthogonality condition (18), $g(t, \cdot)$ is uniquely characterized by the condition

$$x(t) - \hat{x}(t) \perp v(\tau), \qquad 0 \le \tau < t \tag{32}$$

which yields

$$\overline{x(t)v'(\tau)} = \int_0^t g(t, s)\overline{v(s)v'(\tau)}ds, \qquad 0 \le \tau < t. \tag{33}$$

Now we can see the advantage of working with the innovations process $v(\cdot)$ rather than the original process $y(\cdot)$: if we had $y(\cdot)$ in (33) in place of $v(\cdot)$, (33) would reduce to a Wiener–Hopf integral equation of the form (20), which is, in general, difficult to solve. But when we work with $v(\cdot)$, we have a trivial equation since, by inspection, (33) yields

$$g(t, \tau) = \overline{x(t)v'(\tau)}, \qquad 0 \le \tau < t. \tag{34}$$

In any case with this expression for $g(t, \cdot)$, our estimate can be written as

$$\hat{x}(t) = \int_0^t \overline{x(t)v'(s)}\left[y(s) - H(s)\hat{x}(s)\right]ds. \tag{35}$$

We note that we have not yet obtained $x(t)$ explicitly, since the innovations process $v(\cdot)$ was itself expressed in terms of $\hat{x}(\cdot)$. But we have not yet used the additional structure that we have for $x(\cdot)$, viz., the differential equation (26). To exploit this structure, we differentiate $\hat{x}(t)$ as given by (35) to obtain

$$\begin{aligned}\dot{\hat{x}}(t) &= \overline{x(t)v'(t)}v(t) + \int_0^t \overline{\dot{x}(t)v'(s)}\,v(s)ds \\ &= K(t)v(t) + F(t)\int_0^t \overline{x(t)v'(s)}v(s)ds + G(t) \\ &\quad \cdot \int_0^t \overline{u(t)v'(s)}v(s)ds\end{aligned} \tag{36}$$

where we have set

$$K(t) = \overline{x(t)v'(t)}. \tag{37}$$

by (27) and (28) on $u(\cdot)$, $x(0)$, and $v(\cdot)$, it is easy to see that the last integral in (36) is zero; moreover, the second integral is [cf. (35)] equal to $\hat{x}(t)$. Therefore, (36) simplifies to yield the celebrated differential equation of Kalman and Bucy,

$$\dot{\hat{x}}(t) = F(t)\hat{x}(t) + K(t)\left[y(t) - H(t)\hat{x}(t)\right], \qquad \hat{x}(0) = 0. \tag{38}$$

The term $K(t)$, usually called the gain function, is related to the mean-square error in the estimate $\hat{x}(t)$:

$$\begin{aligned}K(t) &= \overline{x(t)v'(t)} = \overline{x(t)\tilde{x}'(t)}H'(t) + \overline{x(t)v'(t)} \\ &= \overline{\left[\hat{x}(t) + \tilde{x}(t)\right]\tilde{x}'(t)}H'(t) + 0 \\ &= \overline{\tilde{x}(t)\tilde{x}'(t)}H'(t) = P(t, t)H'(t).\end{aligned} \tag{39}$$

The mean-square error $P(t, t)$ does not depend upon the observations; moreover, direct calculation shows that it obeys the differential equation

$$\begin{aligned}\dot{P}(t, t) &= F(t)P(t, t) + P(t, t)F'(t) - K(t)K'(t) \\ &\quad + G(t)Q(t)G'(t), \qquad P(0, 0) = P_0.\end{aligned} \tag{40}$$ [7]

Equation (40) is a matrix nonlinear Riccati differential equation, whose numerical solution has been extensively investigated in the control-theory literature (e.g., Bucy and Joseph [33, pt. 2]).

For future use, let us note that we can generalize the model (26)–(28) slightly to permit correlation between $u(\cdot)$ and $v(\cdot)$ of the form

$$E\left[u(t)v'(s)\right] = C(t)\delta(t - s). \tag{41}$$

It is easy to check that allowing this form of correlation still preserves the requirement in the basic innovations Theorem 1 that the future white noise $v(\cdot)$ be independent of past signal $z(\cdot)$. We shall not repeat the derivations with the assumption (41), except to note that the only change in the Kalman–Bucy formulas (38) and (40) will be that now we will have

$$K(t) = P(t, t)H'(t) + G(t)C(t). \tag{42}$$

In most filtering problems $C(t)$ will be zero. However, we shall see that the case $C(t) \neq 0$ has important theoretical significance (cf. Theorems 2 and 4).

Remark 1: We have gone into somewhat more detail on the Kalman–Bucy formulas because of their importance in the now flourishing "state-variable" point of view in several communication problems. Moreover, the simplicity of the innovations derivation clearly brings out the role of the various assumptions and indicates various directions in which they may be generalized. In particular, we see that what is basic in the above arguments is a relation between $\dot{x}(\cdot)$ and past values of $x(\cdot)$, but that this relation can be more general (cf. [10]) than an ordinary differential equation, for example, there is no need to completely repeat, as is often done, the derivations for distributed parameter systems and then discover at the end their "surprising" similarity to the Kalman–Bucy equations. We stress this point because in many problems it is actually often easier to begin with a more general class of process and then to specialize

[7] Because of the form (26) for $x(\cdot)$, it is now well known and is easily shown that $\pi(t) = E[x(t)x'(t)]$, obeys the equation $\dot{\pi} = F\pi + \pi F' + GQG'$, $\dot{\pi}(0) = \pi_0$. Similarly because $v(\cdot)$ is white, (38) shows that $\Sigma(t) = E\,[\hat{x}(t)\hat{x}'(t)]$ obeys $\Sigma = F\Sigma + \Sigma F' + KK'$, $\Sigma(0) = 0$. Now (40) follows from the relation $\pi(t) = \Sigma(t) + P(t)$, which is a consequence of the orthogonality of $\hat{x}$ and $\tilde{x}$.

to state-variable processes. An illustration of this point was already given in Remark 1 following Theorem 1. Another quite striking example is provided by the so-called "linear smoothing" problem, which we shall now examine briefly.

Application II: The Linear Smoothing Problem

In their fundamental paper, Kalman and Bucy stated that the problem of smoothing (or noncausal estimation), viz., the determination of

$$\hat{x}(t|T) = \text{the least-squares estimate of } x(t) \text{ given } \{y(s), 0 \le s \le T\}, t \le T, \tag{43}$$

was more difficult than that of filtering (where $T=t$). This statement gave rise to a host of papers on the smoothing problem.

The innovations method yields very simply and directly, a general solution from which we can, by imposing further conditions (e.g., state-variable structure) on the general model, easily obtain all the previously derived special formulas (cf. [11]).

For the general smoothing formula we shall assume that the observations are of the form

$$y(t) = z(t) + v(t) = H(t)x(t) + v(t), \qquad 0 \le t \le T.$$

Now set

$$\hat{x}(t|T) = \int_0^t g(t, s)\nu(s)ds, \qquad \nu(s) = y(s) - H(s)\hat{x}(s) \tag{44}$$

where $g(t,\ )$ is to be determined by the orthogonality condition

$$x(t) - \hat{x}(t|T) \perp \nu(\tau), \qquad 0 \le \tau \le T. \tag{45}$$

Direct calculation, (33) and (34), yields

$$g(t, s) = \overline{x(t)\nu'(s)}, \qquad 0 \le s \le T. \tag{46}$$

The smoothed estimate can be written as

$$\begin{aligned}\hat{x}(t|T) &= \int_0^T \overline{x(t)\nu'(s)}\nu(s)ds \\ &= \int_0^t \overline{x(t)\nu'(s)}\nu(s)ds + \int_t^T P(t, s)H'(s)\nu(s)ds\end{aligned} \tag{47}$$

where

$$\begin{aligned}P(t, s)H'(s) &\triangleq \overline{x(t)\nu'(s)} = \overline{x(t)[\tilde{x}'(s)H'(s) + v'(s)]} \\ &= \overline{x(t)\tilde{x}'(s)}H'(s) = \overline{\tilde{x}(t)\tilde{x}'(s)}H'(s), \quad \text{for} \quad s > t.\end{aligned}$$

Therefore, we can identify $P(t, s)$ as

$$P(t, s) = \text{the covariance function of the error in the causal estimate } \tilde{x}(\cdot). \tag{48}$$

With this identification we can, from (47) [and (35)], express the smoothed estimate, $\hat{x}(t|T)$ entirely in terms of the filtered estimate $\hat{x}(t|t)=\hat{x}(t)$, viz.,

$$\hat{x}(t|T) = \hat{x}(t) + \int_t^T P(t, s)H'(s)[y(s) - H(s)\hat{x}(s)]ds. \tag{49}$$

Moreover, direct calculation shows that the smoothing error covariance can be expressed in terms of the filtered error covariance by the formula

$$\overline{\tilde{x}(t|T)\tilde{x}(t|T)} = P(t, t) - \int_t^T P(t, s)H'(s)H(s)P(s, t)ds. \tag{50}$$

It is now clear that if we have more information on $\tilde{x}(\cdot)$, e.g., a differential equation for it, then we can say more about $\tilde{x}(t|T)$, e.g., obtain differential equations for it. By these means, we can derive, as shown in [11], several special forms for $\tilde{x}(t|T)$, each of which may be most suited for a particular special application.

Application III: Prediction in Linear Difference System

As shown in [10] and [11], most of the results in Application I and II can be carried over to discrete-time problems with some slight changes. However, there are some situations where only the discrete-time model seems natural, and in such cases we can sometimes go beyond the results discussed in Applications I and II. We shall illustrate this point by applying the innovations method to a problem originally solved by Rissanen and Barbosa [34]. These authors dealt with the following difference equation model for a process $y(i)$, $i=0, 1, \cdots$

$$\begin{aligned}&y(i) = a_1(i)y(i-1) + \cdots + a_N(i)y_{i-N} + w(i), \qquad i \ge N \\ &y(i) = w(i), \qquad 0 \le i \le N-1\end{aligned} \tag{51}$$

where $w(i)$ is a colored-noise sequence with

$$E[w(i)] = 0, \qquad E[w(i)w'(j)] = r_{ij}. \tag{52}$$

Such high-order difference equation models often arise directly via identification algorithms. The problem is to calculate

$$\hat{y}(i|i-1) = \text{the least-squares linear estimate of } y(i) \text{ given } \{y(j), \quad 0 \le j \le i-1\}. \tag{53}$$

Rissanen and Barbosa [34] (see also [13]) discuss how this prediction problem may also be identified with certain state-variable filtering problems.

One method of calculating $\hat{y}(i|i-1)$ is by conversion of the given model (51) and (52) to a state-space model to which the Kalman–Bucy formulas can be applied. However this "two-step" method will involve several unnecessary computations since these formulas will not only yield the prediction $\hat{y}(i|i-1)$, but they will also give (the unwanted) estimates of the states in the equivalent state-space model for $y(\cdot)$. Therefore, Rissanen and Barbosa proposed a more direct solution, which requires first the factorization of the covariance matrix $[r_{ij}]$ of the $w(\cdot)$ into the product of a lower triangular matrix and its transpose. Later, Aasnaes [56] showed that the factorization of $[r_{ij}]$ could be avoided, thus somewhat simplifying the original algorithm. Here we shall obtain the same result by the innovations method. The first step is to determine the innovations for $\{y(\cdot)\}$. As we might expect from Theorem 1, these are

$$\nu(i) = \varepsilon(i)/\sqrt{\overline{\varepsilon^2(i)}}, \qquad \varepsilon(i) = y(i) - \hat{y}(i|i-1) \tag{54}$$

as can be verified by direct calculation. Note that the normalization of $\varepsilon(i)$ to $\nu(i)$ is not needed in the continuous-time case, which fact often makes certain continuous-time problems simpler than their discrete-time analogs.[8] Such simplifications, of which we shall encounter a very striking example in Theorem 5, help answer the often asked, but rather naive, questions of why continuous-time *models* are useful even in a purely digital environment. In our problem with $y(\cdot)$ described by (51), we will have

$$\hat{y}(i|i-1) = a_i(i)y(i-1) + \cdots + a_N(i)y(i-1) + \hat{w}(i|i-1). \tag{55}$$

Next, by direct calculation we obtain

$$\hat{w}(i|i-1) = \sum_0^{i-1} \psi_{ij}\left[\overline{\varepsilon(j)\varepsilon'(j)}\right]^{-1}\varepsilon(j) \tag{56}$$

where the coefficient is

$$\psi_{ij} = \overline{w(i)\varepsilon'(j)}. \tag{57}$$

Moreover, ψ_{ij} can be calculated by the simple recursive formula

$$\begin{aligned}\psi_{ij} &= \overline{w(i)\varepsilon'(j)} = \overline{w(i)w'(j)} - \overline{w(i)\hat{w}'(j|j-1)}\\ &= r_{ij} - E\left[w(i)\left(\sum_0^{j-1} \overline{w(j)\varepsilon'(k)}\left[\overline{\varepsilon(k)\varepsilon'(k)}\right]^{-1}\varepsilon(k)\right)\right] \qquad (58)\\ &= r_{ij} - \sum_{k=0}^{j-1} \psi_{ik}\left[\overline{\varepsilon(k)\varepsilon'(k)}\right]^{-1}\psi'_{jk}, \qquad 0 \le j \le i-1. \qquad (59)\end{aligned}$$

If $y(\cdot)$ is a p-dimensional vector, it can be shown [13] that the simple solution (55)–(59) requires for $N \gg 1$, about $4Np$ times fewer scalar multiplications than a solution based on a state-space model for $y(\cdot)$. We shall leave the problem here, referring the reader to [13] for further results on discrete-time processes, including non-Markovian processes of the "state-space" type.

Application IV: Solution of Certain Fredholm Integral Equations

The Fredholm integral equation of the second kind

$$H(t,s) + \int_0^T H(t,\tau)K(\tau,s)d\tau = K(t,s), \qquad 0 \le s, t \le T \tag{60}$$

arises in many problems in communication and control theory, e.g., in the detection of Gaussian signals in WGN [35, ch. 11], and in a functional analysis approach to the quadratic regulator problem of control theory. Several methods of solution have been developed for such equations, especially when $K(t, s)$ is symmetric and continuous in t and s, and is of the form

$$K(t,s) = \sum a_i e^{-b_i|t-s|} \tag{61}$$

or more generally when it is of the so-called "separable" form

$$\begin{aligned}K(t,s) = \sum_1^n m_i(t\vee s)n_i(t\wedge s) &= \begin{cases}\sum m_i(t)n_i(s), & t \ge s \quad (62)\\ \sum m_i(s)n_i(t), & t \le s \quad (63)\end{cases}\\ &= M(t\vee s)N(t\wedge s).\end{aligned}$$

The functions $M(\cdot)$ and $N(\cdot)$ may be matrix functions; if $K(t, s)$ is scalar, then $M(\cdot)$ will be a row matrix and $N(\cdot)$ a column matrix. Under certain additional assumptions on $K(t, s)$ of the form (63), Baggeroer [36] recently showed that the solution of the Fredholm equation (60) could be reduced to that of a matrix Riccati differential equation, similar to the one we encountered in the Kalman–Bucy problem. Baggeroer's assumptions are that $K(t, s)$ is the covariance function of a process generated by passing WGN through a known lumped linear dynamical system. Here, following [37], we shall use the innovations method first to provide an explanation of why a Riccati equation enters into this problem and then to show that there is no need to assume knowledge of a model[9] for the process whose covariance is $K(t, s)$. Furthermore we shall not even need to assume that $K(t, s)$ is a covariance: it will suffice to assume that its eigenvalues are greater than -1, which, of course, is the necessary and sufficient condition for (60) to have a (square-integrable) solution.

For further discussion it will be convenient to use an obvious integral operator notation under which (60) can be written as

$$(I + K)H = K \tag{64}$$

and in which the solution H can be expressed by

$$H = K(I+K)^{-1} = I - (I+K)^{-1} \tag{65}$$

which will be valid provided that

$$I + K > 0. \qquad (66)^{10}$$

However, to introduce the innovations method let us begin with the stronger assumption that $K \ge 0$. Then we can identify $I + K$ as the covariance function of a process $y(\cdot)$ of the form

[8] Thus even when digital implementation is necessary, it has often been found simpler to discretize (by some appropriate numerical-analysis technique) the continuous-time Kalman–Bucy equations (38)–(40) than to directly implement their more complicated exact discrete-time equivalents as given in [7]. Exclusive attention to the discrete-time case not only often introduces certain complicated but comparatively unimportant terms into many formulas, but, more important, usually forces us a priori to the possibly wasteful rectangular-grid methods of discretization. We should make clear that discrete-time models are undoubtedly often useful and unavoidable (see, in fact, our Application III); our objection is only to the cheap invocation of the computational necessity for ultimate discrete-time implementation to lightly dismiss and avoid the need for hard analysis of the continuous-time problem.

[9] Of course, if, as also assumed in the Kalman–Bucy theory, a model is known, then there is no need even to know $K(t, s)$ explicitly (see [38], [39]).

[10] $A \ge B$ means that $A(t, s) - B(t, s)$ is nonnegative definite. We stress that $K(t, s)$ will always be assumed symmetric and continuous, unless stated otherwise.

$$y(t) = z(t) + v(t), \qquad E[z(t)v'(s)] \equiv 0 \tag{67}$$

$$E[v(t)v'(s)] = \delta(t - s), \qquad Ez(t)z'(s)] = K(t, s). \tag{68}$$

The form of $y(\cdot)$ suggests that we can apply (the basic innovations) Theorem 1 to this process $y(\cdot)$ to obtain

$$v = y - \hat{z} = (I - h)y \tag{69}$$

where h is the estimating filter [cf., (21)]

$$\hat{z} = hy \quad \text{and} \quad h + \{hK\}_+ = K_+ \tag{70}$$

Representation (67) immediately suggests the calculations

$$\begin{aligned} I + K = \overline{yy'} &= (I - h)^{-1}\overline{vv'}(I - h^*)^{-1} \\ &= (I - h)^{-1}(I - h^*)^{-1} \end{aligned} \tag{71}$$

where

$$h^* \triangleq \text{the ``adjoint'' of the estimating filter } h, \text{ i.e.,} \tag{72}$$

$$h^*(t, s) = h'(s, t). \tag{73}$$

The point of these manipulations is that we can now write

$$\begin{aligned} H = I - (I + K)^{-1} &= I - (I - h^*)(I - h) \\ &= h + h^* - h^*h. \end{aligned} \tag{74}$$

We see from (74) that knowledge of the estimating filter h completely determines the solution H of the Fredholm equation (60). Furthermore, if the signal process $z(\cdot)$ can be modeled as the output of a known dynamical system driven by WGN then we know from the results of Kalman and Bucy that the estimating filter can be found via the solution of a certain Riccati differential equation, which explains the occurrence of a Riccati equation in Baggeroer's solution. However, note that to apply the Kalman–Bucy results we need to know the parameters of the model used to generate $z(\cdot)$ and, in general, unless such a model is known a priori, it is difficult to deduce a model for a given $K(t, s)$ of the form (63). ([38]–[40] discuss this problem, which is, of course, closely related to the problem of determining the IR for a process with this covariance, and give further references. Also see the remarks in Section V.) It is, therefore, somewhat surprising that the estimating filter can in fact be found via the solution of a Riccati equation determined directly by $K(t, s)$.[11] We refer the reader to [37] for the details of how this is done, and for references to related work. However, it may be useful to quote the appropriate Riccati equation here:

$$\Sigma(t) = (N - M')(N - \Sigma M')', \qquad \Sigma(0) = 0. \tag{75}$$

Not only is this equation completely determined by the given covariance $K(t, s)$ of (63), but, somewhat surprisingly, the same equation is appropriate even when $K(t, s)$ is not a covariance (i.e., $K \not> 0$) so long as $I + K > 0$, (i.e., whenever the integral equation has a solution.[12] The reason for this striking behavior is provided by the following interesting representation theorem. For later use in Application V, we shall state the theorem in a slightly more general form than is required for the present problem.

Theorem 2: A process $y(t)$, $0 \le t \le T$, has a covariance function of the form $\delta(t-s) + K(t, s)$, where $K(t, s)$ is square integrable on $[0, T] \times [0, T]$ and $I + K > 0$, if and only if it can be written in the form

$$y(t) = z(t) + v(t) \tag{76}$$

where

$$\begin{aligned} E[v(t)v'(s)] &= I\delta(t - s), K(t, s) \\ &= E[z(t)z'(s) + z(t)v'(s) + v(t)z'(s)] \\ E[v(t)z'(s)] &= 0, \qquad t > s \end{aligned} \tag{77}$$

and

$$\int_0^T \operatorname{tr} E[z(t)z'(t)]dt < \infty, \tag{78}$$

where $\operatorname{tr}[\cdot]$ denotes trace.

Proof. The proof involves a certain amount of analysis and we can only refer the reader to [18]. However, we may note that one proof of the "only if" part uses the innovations; in fact, our assumptions on $z(\cdot)$ immediately enable us to deduce the previously used chain of (69)–(71), from which the fact that $I + K > 0$ is obvious. The "if" part of the theorem is deeper and in [18] we used certain nonelementary (at least when $K(t, s)$ is only square-integrable and not also continuous[13]) results of Gohberg and Krein [41] on covariance factorization to obtain a proof.

As a first application of this theorem, we note that the "if" part immediately provides a signal-plus-noise model (now, however with a one-sided correlation between signal and noise) to which the arguments of (66)–(74) can again be applied. These arguments motivate the introduction of a Riccati equation, and some simple algebra (the details of which are in [37]) will now show that this equation can be reduced to (75).

A closely related application is to find the innovations for a Gaussian process $y(\cdot)$ with covariance $I + K$, $K > 0$. We shall leave to the interested reader the task of showing that

$$v = (I - h)y, \qquad h + \{hK\}_+ = K_+.$$

This calculation will give further insight into the significance of the "one-sided" correlation property.

To conclude this application, we note that the above ideas can be extended to calculate eigenvalues, Fredholm determinants, and eigenfunctions for symmetric and nonsymmetric separable kernels $K(t, s)$ and also for pairs of kernels (as arise in the so-called simultaneous diagonaliza-

[11] Note that we are thus directly obtaining a recursive solution for lumped signal processes in WGN without the need for an explicit state model. This is possible because in estimating the signal $z(\cdot)$ we are not directly concerned with the estimates of the states but only with a particular linear combination $\hat{z}(\cdot)$ of these state estimates. A similar circumstance was encountered in Application III.

[12] This case cannot be treated by the method of [36].

[13] A simple proof of the key relation (74) for continuous $K(t, s)$ can be found in [37a].

tion problem). However, we shall not go into these problems here, but refer the reader to [37] and [42], where a detailed bibliography of related work by Kalaba, Schumitzky, and others is also given. However, we shall discuss some applications to Gaussian detection problems.

Application V: Detection of Gaussian Signals

Though the problem of detecting Gaussian signals has been much studied, we shall see, following [16]–[18], that the innovations approach can still bring considerable simplification and new insights to the problem.

The most commonly studied Gaussian problem (e.g., [35] and [43]) is that of detecting a Gaussian signal process, say, $z(\cdot)$, with continuous covariance function $K(t, s)$ in independent additive WGN. The likelihood ratio (LR) for this problem is found in [35] and [43] by means of a simultaneous Karhunen–Loève (K–L) expansion of the signal and the independent noise.[14] The final expression is

$$\text{LR} = d^{-1/2}(1) \quad \exp \quad -\left\{\frac{1}{2}\sum_{1}^{\infty} y_i^2\lambda_i/(1+\lambda_i)\right\} \tag{79}$$

where

$$d(\lambda) = \prod_{1}^{\infty} (1 + \lambda\lambda_i), \quad \int_0^T K(t, \quad s)\psi_i(s)ds = \lambda_i\psi_i(t) \tag{80}$$

$$y_i = \int_0^T \psi_i(t)y(t)dt, \qquad i = 1, \; 2, \cdots \tag{81}$$

Since it is usually impractical to compute the quantities $\{\lambda_i\}$ and $\{y_i\}$, this formula is usually rewritten by introducing the function

$$H(t, s; \lambda) = \sum [\lambda_i/(1 + \lambda\lambda_i)]\psi_i(t)\psi_i(s) \tag{82}$$

in terms of which it is not hard to show that we can express the LR as

$$\text{LR} = \exp(\Lambda - B), \tag{83}$$

$$2\Lambda = \int_0^T\int_0^T H(t, s; 1)y(t)y(s)dt\, ds \tag{84}$$

$$2B = \sum \ln(1 + \lambda_i) = \int_0^1 d\lambda \int_0^T H(t, t; \lambda)dt. \tag{85}$$

Moreover, $H(t, s; \lambda)$ can be identified as the solution of the Fredholm integral equation

$$H(t, s; \lambda) + \lambda\int_0^T H(t, \tau; \lambda)K(\tau, s)d\tau = K(t, s). \tag{86}$$

This integral equation can be solved in a few cases, especially those where $K(t, s)$ is "separable," when we can, for example, use the Riccati equation methods presented in Application IV. In any case, numerical solutions of (86) can always be obtained with sufficient effort.[15] A basic question in detection theory is whether such an effort is worthwhile, both in terms of the difficulty in implementing a complicated expression for $H(t, s; 1)$ and also in view of the fact that in many problems our knowledge of $K(t, s)$ may be rather approximate and imperfect. It is with these thoughts in mind that the following rearrangement of (84), due to Price [45], has attracted so much attention. Price showed that we could write

$$2\Lambda = \int_0^T z_{e_1}(t)y(t)dt, \tag{87}$$

where

$$z_{e_1}(t) \triangleq \int_0^T H(t, s; 1)y(s)ds$$

= the noncausal (or smoothed) least-squares estimate of $z(t)$ given all the observations $\{y(s),\ 0 \le s \le T\}$ and assuming signal is present (which we shall designate as hypothesis h_1). (88)

The advantage of the interpretation (87) and (88) of Λ is that it enables some intelligent approximations to be made in actually computing Λ in a practical problem: it seems reasonable to replace $z_{e_1}(\cdot)$ by the best obtainable estimate, under the inevitable practical constraints of limited statistical knowledge and limited equipment complexity. The insight gained from the interpretation (87) and (88) was used by Price and Green to guide the development of a successful antimultipath communication system called the RAKE system [46].

However, despite its usefulness, we shall show that a different, but still related, interpretation of the IR can be more wide-ranging than the one given by (87) and (88). Thus, we note that (cf. [18]) the interpretation (88) fails both when the signal process $z(\cdot)$ has nonzero mean and, more important, when the signal and noise are correlated. There does not seem to be any simple explanation of this breakdown, and therefore there is a somewhat unsatisfying lack of "robustness" for the form (87) and (88).

However, by use of the innovations Theorem 1 and the representation Theorem 2, we shall be able to provide another form for the LR, which holds even when the signal process is not smooth, nor zero mean, nor independent of the WGN; the only requirement is that the detection problem be nonsingular, i.e., incapable of resolution with zero probability of error. Before proceeding to this most general problem, however, let us first consider an apparently simpler problem for which the innovations can be used to directly obtain a solution. Then we shall show (Theorem 4)

[14] Note that it is not quite obvious how this approach can handle correlated signal and noise, since it is not clear how such correlation will be reflected into the K–L coefficients of the signal and the noise. The K–L expansion can still be used for such problems, but in a different way (see [44] and [18, Application II].

[15] In fact, the likelihood ratio can always be evaluated by brute force as a ratio of probability densities on a sufficiently large grid of points.

that this apparently simpler problem is equivalent to the general problem.

Theorem 3: Consider a detection problem with hypotheses of the form

$$h_1 : y(t) = z(t) + w(t), \quad h_0 : y(t) = v(t), \quad 0 \le t \le T \quad (89)$$

where $v(\cdot)$ is zero-mean WGN and $z(\cdot)$ is a Gaussian process satisfying

$$E[z(t)] = m(t), \qquad \int Ez^2(t)dt < \infty \quad (90)$$

$$E[v(t)] = 0, \quad E[v(t)v(s)] = \delta(t-s), \quad E[v(t)z(s)] \equiv 0, \quad t > s. \quad (91)$$

This detection problem is nonsingular and the LR can be written

$$\mathrm{LR} = \exp \left\{ \int_0^T \hat{z}_1(t)y(t)dt - \frac{1}{2}\int_0^T \hat{z}_1^2(t)dt \right\} \quad (92)$$

where

$$\hat{z}_1(t) = E[z(t)|\{y(s), 0 \le s < t\}, h_1] = \text{the causal least-squares estimate of } z(t) \text{ given past } y(\cdot) \text{ and assuming } h_1 \text{ is true} \quad (93)$$

and $⨍$ denotes a special kind of integral called an Itô stochastic integral.

Proof: A rigorous proof is given in [17] and [18]. Here we shall give a plausibility argument. We first observe that if the signal process $z(\cdot)$ is completely known (i.e., it is deterministic) and is square-integrable then it is well known that the detection problem is nonsingular and that the LR is given by (92) with $\hat{z}_1(\cdot) = z(\cdot)$. Now observe that, by the innovations theorem, we can rewrite the hypothesis h_1 as

$$h_1 : y(t) = z(t) + v(t) = \hat{z}_1(t) + \nu(t)$$

where $\nu(\cdot)$ is zero-mean WGN with $E[\nu(t)\nu(s)] = \delta(t-s)$. Furthermore, since $v(\cdot)$ and $\nu(\cdot)$ are statistically indistinguishable, being Gaussian with the same mean and covariance, we can rewrite the hypotheses for our detection problem as

$$h_1 : y(t) = \hat{z}_1(t) + \nu(t), \quad h_0 : y(t) = \nu(t), \quad 0 \le t \le T. \quad (94)$$

In both forms of the detection problem, (89) and (94), the signal is random, but in (94) the signal $\hat{z}_1(\cdot)$ can be calculated given $y(\cdot)$. Therefore, though $\hat{z}_1(\cdot)$ is random, since it will change as $y(\cdot)$ changes, it is "conditionally known." Moreover, from the assumption $\int Ez^2(t)dt < \infty$ we can show that $\int E\hat{z}^2(t)dt < \infty$ and, therefore, that $\int \hat{z}^2(t)dt < \infty$ almost surely. Thus, it is plausible that the same LR formula will hold as for the known signal case, with the conditionally known signal used in place of the known signal.

This argument is not unreasonable (in fact, one of the rigorous proofs [17] is patterned on it) but, of course, some subtleties are left out. Most notably the question of the stochastic integral. This is an important point which we shall discuss in more detail a bit later. At the moment we wish to point out some important consequences of (92) and (93).

First we note that since $\hat{z}_1(\cdot)$ is a causal estimate, it can be recursively updated as more data come in. The LR formulas, (79), (84), and (87) do not share this important computational property, because if the interval $[0, T]$ is increased to, say, $[0, T_1]$ then the $\{\psi_i(\cdot)\}$ and $\{x_i\}$ for (79) and the $H(t, s; 1)$ for (84) and (87) will have to be completely recomputed over the larger interval and not just over the added interval $[T, T_1]$.

Secondly, (92) has a pleasing and useful interpretation: when the signal is not known, but is random, use an estimate of the signal in place of the unavailable actual signal. A somewhat similar argument was made earlier with the form (87), but in the present instance the argument is much more complete. For one thing, in (92) we use the estimate exactly as we would a known signal, which is not true of (87). Secondly, unlike (87), the formula continues to be valid when the signal process has a nonzero mean and also when the signal and noise are correlated. We may argue that the one-sided form (91) of the correlation is a restriction on the generality of the result. However, it is a somewhat surprising fact that the solution for the special model (89)–(91), with its one-sided correlation and very well-behaved signal variance [cf. (90)], is not really so special after all. This is shown by the next theorem.

Theorem 4: A Gaussian process $y(\cdot)$ is "equivalent" to a WGN, i.e., the problem of deciding whether an observation is a sample function of $y(\cdot)$ or of the WGN is nonsingular, if and only if we can express $y(\cdot)$ in the form

$$y(t) = z(t) + v(t)$$

where $z(\cdot)$ and $v(\cdot)$ obey the conditions (90) and (91) imposed in Theorem 3.

Proof: One rigorous proof is based on a result of Shepp [44] (also [18]) that equivalence holds if and only if the covariance of $y(\cdot)$ is of the form

$$E[y(t)y(s)] = \delta(t-s) + K(t,s), \quad \int_0^T\int_0^T K^2(t,s)dtds < \infty \quad (95)$$

and

$$I + K > 0. \quad (96)$$

Now our earlier representation of Theorem 2 yields the desired form (89)–(91) for $y(\cdot)$.

Another rigorous proof can be obtained by martingale theory (see Theorem 6). There does not seem to be any simple heuristic proof. However, the theorem has many striking aspects, and consequences, some of which are as follows.

1) The theorem shows the basic nature of the simple model (89)–(91) and stresses the importance of being able to handle correlated signal and noise (cf. footnote 12).

2) The theorem shows that the apparently more general "two-process" detection problem of Theorem 4 is really of the "signal-in-noise" type. Note also that the theorem enables us to say, under certain circumstances, how much WGN there is in a given process $y(\cdot)$. This is not a trivial problem even if $y(\cdot)$ is stationary, unless its power spectrum is bounded below by unity.

3) We see that (92) must continue to hold for the LR for the more general problem. How can we calculate the re-

quired $\hat{z}_1(\cdot)$ directly from the problem specifications? It can be shown [18] that the rule is

$$\hat{z}_1 = hy, \qquad h + \{hK\}_+ = K_+ \tag{97}$$

where K is defined by (95). We note also that for separable kernels the methods of Application IV can be used to obtain computationally efficient algorithms not only for $h(t, s)$, but also for the eigenvalues, eigenfunctions, and Fredholm determinants that arise in the calculation of the error probability and in signal design.

4) Note that the theorem does not say that there are representations of $y(\cdot)$ that are not of the form (89)–(91). For example, Shepp's result shows that the following process $y(\cdot)$ is equivalent to WGN:

$$y(t) = s(t) + n(t) \tag{98}$$

where

$$E[n(t)n(\tau)] = \delta(t - \tau), \qquad E[n(t)s(\tau)] \equiv 0 \tag{99}$$

and

$$E[s(t)s(\tau)] = K(t, \tau), \quad \int_0^T \int_0^T K^2(t)dtd\tau < \infty. \tag{100}$$

Now $K(t, \tau)$ can be square integrable on $T \times T$, but because a line has zero-measure in a square, we can have

$$\int_0^T Es^2(t)dt = \int_0^T K(t, t)dt = \infty. \tag{101}$$

which apparently violates (90). But note that Theorem 4 only says that there is a way of rewriting $y(\cdot)$ so that it obeys (89)–(91). We shall leave to the reader the simple exercise of finding the proper way for the present problem.

5) Note that for the problem (98)–(101), we cannot apply the usual formulas (83)–(85) because it turns out (cf. [18]) that each of the terms Λ and B is infinite but in such a way that $(\Lambda - B)$ is finite, so that the LR $= \exp(\Lambda - B)$ is still finite. Then how can we get a meaningful expression for the LR? We do not have space to pursue this interesting question here, but can refer the reader to [18, Application II] for a tutorial exposition. We should note here that the resolution of the problem requires careful attention to the definition and properties of the double integral in (84). Trouble may be expected here because the integral describes *nonlinear* operations on WGN.

6) In view of the above difficulties it may be somewhat surprising to note that our causal estimator formula (92)

$$\text{LR} = \exp \fint_0^T \hat{z}_1(t)y(t)dt - \frac{1}{2}\int_0^T \hat{z}_1^2(t)dt$$

is always valid. The explanation must lie, of course, in the special nature of the integral $\fint$. Unfortunately, we again do not have space here to explore and explain the special properties of this integral. Among these is the fact that $\fint$ does not obey the rules of the usual calculus, e.g., if $w(t) = \int_0^t v(s)ds$, then

$$\fint_0^T w(t)dw(t) = \frac{w^2(T) - w^2(0)}{2} - \frac{T}{2} = \int_0^T wdw - \frac{T}{2}$$

where $\int_0^T$ denotes the "ordinary" integral. A tutorial discussion of the Itô integral is given in [20]. We should mention that there is nothing mysterious about the Itô integral—its properties are well defined and the rules for manipulating it are easily remembered. The fact that the rules are different from the usual rules is curious, but this should not upset us. The special rules arise because we are setting up a special idealized model (with WGN) and we wish to make logically consistent deductions from the model. We could avoid the need for new rules by excluding WGN models, but this is too heavy and unnecessary a price to pay.[16] We should stress again how simple it is to adapt to the new rules; we may also point out that Itô integrals can be simulated by analog equipment as "ordinary" integrals can [16].

7) From another (pragmatic) point of view, the new rules are necessary if we are to show the equivalence of the general formula (92) to the formulas known for special cases. Thus, Stratonovich and Sosulin [47] showed that if $z(\cdot)$ was a Gaussian process, with a separable continuous covariance function, *independent of* $v(\cdot)$, then we could write

$$\text{LR} = \exp \int_0^T z_1(t)y(t)dt - \frac{1}{2}\int_0^T \widehat{z_1^2(t)}dt. \tag{102}$$[17]

If we define

$$\varepsilon^2(t) = E[z(t) - \hat{z}_1(t)]^2 \tag{103}$$

then some simple algebra will let us rewrite (102) as follows.

$$\begin{aligned}\text{LR} = &\left[\exp - \frac{1}{2}\int_0^T \varepsilon^2(t)dt\right] \\ &\cdot\left[\exp \int_0^T \hat{z}_1(t)y(t)dt - \frac{1}{2}\int_0^T \hat{z}_1^2(t)dt\right].\end{aligned} \tag{104}$$

The formula (104) was independently obtained by Schweppe [48], who also showed its equivalence to the usual formulas (83)–(85). Comparing (104) with (92), we see that they will not coincide unless the integrals $\fint$ and $\int$ are related as

$$\fint_0^T \hat{z}_1(t)y(t)dt = \int_0^T \hat{z}_1(t)y(t)dt - \frac{1}{2}\int_0^T \varepsilon^2(t)dt. \tag{105}$$

The first integral on the right-hand side of (105) is often called a Fisk–Stratonovich integral (cf. [20]). When it can be defined, the Fisk–Stratonovich integral obeys the usual rules of the calculus, but this integral exists only for a restricted class of problems. For example, it does not exist for the problem (98)–(101), whereas the Itô integral is still meaningful. For other reasons as well, discussed in [20], the

[16] Recall that WGN models are not of any particular interest in themselves. They are important because they provide mathematically simple and tractable models that can 1) often satisfactorily approximate many physical systems with wideband noise of unknown high-frequency behavior, and 2) play an important role in the analytical solution of problems with colored noise (see, e.g., [12], [18], [38], [44]). It is putting the cart before the horse, and misses the whole point of mathematical modeling of physical problems, to seek, as is sometimes done, to "rigorously" analyze WGN models by replacing the "nonphysical" WGN by some "physical" almost-white noise.

[17] The form (102) is invalid when $z(\cdot)$ and $v(\cdot)$ are correlated.

Itô form of integral is the most widely used type of stochastic integral. We shall encounter it again in the next section.

8) While on the topic of stochastic integrals we should mention a transformation of the double integral (84) that has often been proposed [35], [43] in order to obtain a "causal" implementation of Λ, viz., using the symmetry of $H(t, s; 1)$, we rewrite (84) as follows.

$$\begin{aligned} 2\Lambda &= \int_0^T\int_0^T H(t, s; 1)y(t)y(s)dtds \\ &= 2\int_0^T y(t)dt\int_0^T H(t, s; 1)y(s)ds. \end{aligned} \tag{106}$$

There is a difficulty here, however, since for the iterated integral we need to specify whether an Itô integral or (ordinary) Fisk–Stratonovich integral (if it exists) is to be used. The fact that such problems need attention has only been realized in recent years and since that is the point we wish to make here, we shall leave further discussion of this problem to [18].

We shall conclude by pointing out the somewhat unexpected fact that much of our discussion of (92) of the LR for the Gaussian problem continues to be valid for a very general class of non-Gaussian signals as well (Theorems 6 and 7). The fact that we can obtain such general results is rather surprising since most previous studies of detection formulas have stopped with Gaussian or at most Markov signals. These new results could not have been obtained without the help of the powerful new tools and ideas surrounding the Itô integral.

IV. Innovations for a Class of Non-Gaussian Processes and Some Applications

It is a striking fact that we shall be able to prove analogs of Theorems 1, 3, and 4 of the previous section for a large class of non-Gaussian processes as well. The processes we consider will be of the form

$$y(t) = z(t) + v(t) \tag{107}$$

where $v(\cdot)$ is WGN of unit intensity and $z(\cdot)$ is a *not necessarily Gaussian* process that obeys the following:

1) $\int_0^T Ez^2(t)dt < \infty$. (108)

2) The WGN $v(t)$ is independent of past

$$\{v(s), z(s), 0 \le s < t\}. \tag{109}$$

For example, $z(\cdot)$ could be a Poisson process or a random telegraph wave.

As in Section III, we shall find it useful to define the least-squares estimate

$\hat{z}(t)$ = the causal least-squares estimate, *not necessarily linear*, of $z(t)$ given past $\{y(s), 0 \le s < t\}$. (110)

Then, as in the Gaussian case, if we define the "new information" or "innovations" process

$$\nu(t) = y(t) - \hat{z}(t) = \tilde{z}(t) + v(t) \tag{111}$$

it is not unreasonable, and could be proved, that $\nu(\cdot)$ is a white-noise process,

$$\overline{\nu(t)\nu(s)} = \delta(t - s).$$

Unlike the Gaussian case, it does not immediately follow that the values of $\nu(\cdot)$ should be statistically independent. Despite the fact that $y(\cdot)$ is non-Gaussian and $\hat{z}(\cdot)$, being obtained by a nonlinear operation on $y(\cdot)$, may be expected to be even more non-Gaussian than $y(\cdot)$, it turns out that $\nu(\cdot)$ is Gaussian.

Theorem 5: Under the assumptions (107)–(110), the process

$$\nu(t) = y(t) - \hat{z}(t)$$

is a white Gaussian noise process with the same statistics as $v(\cdot)$.

Proof: A rigorous proof, based on certain martingale theorems, is given in [15] and [16]. Here we shall only explain why $\nu(\cdot)$ is Gaussian, assuming that the whiteness follows as in Theorem 1. As expected, a form of central limit theorem has to be used, but a subtlety arises that brings out the importance of the fact that $\hat{z}(\cdot)$ is a least-squares estimate. For the proof, let us consider the integral of $\nu(\cdot)$, say,

$$w(t) = \int_0^t \nu(\tau)d\tau = \int_0^t \tilde{z}(\tau)d\tau + \int_0^t v(\tau)d\tau$$

and we shall prove that $w(t)$ is Gaussian. First, note that the increments of $w(\cdot)$ over disjoint intervals must be uncorrelated because the white noise $\nu(\cdot)$ in the disjoint intervals is uncorrelated. Next, we observe that $w(\cdot)$ has continuous paths because the integral of the WGN $v(\cdot)$ is the so-called Wiener process which is known to have continuous paths[18] and the integral $\int_0^t \tilde{z}(\tau)d\tau$ is continuous, because by assumption (108), $\tilde{z}(\cdot)$ has finite variance. Now let $0 < t_1 < \cdots < t_i \cdots < t$ be a partition of the interval $[0, t]$ and let

$$w(t) = \sum_i [w(t_{i+1}) - w(t_i)].$$

The terms in this sum are uncorrelated and are easily seen to have variance $[t_{i+1} - t_i]$; moreover, because of the continuity of $w(\cdot)$ we can make the number of terms arbitrarily large. We might now try to appeal to the central limit theorem except that, at least as usually understood, that theorem applies to independent, not merely uncorrelated, random variables. Here is where the fact that $\hat{z}(\cdot)$ is a least-squares estimate enters the picture. It is known that least-squares estimates are conditional means and this fact makes the process $w(\cdot)$ a "martingale" process. Martingales have many remarkable properties and one of them is that the central limit theorem holds for them [25, p. 384].

Remark 1: In Theorem 1, we were able to prove that in the Gaussian case $\nu(\cdot)$ and $y(\cdot)$ were causally obtainable

[18] Note that this fact is not true for integrals of non-Gaussian, e.g., Poisson, white noise. Incidentally, it is important to note that unlike Theorem 1, there is no analog of Theorem 5 for discrete-time processes.

$$y(t) = \phi(t, m, \{y(s), 0 \le s < t\}) + v(t)$$

$(\cdot)$ is the actual "modulator" signal that is trans- ver the additive WGN channel. The signal $\phi(\cdot)$ end upon past $y(\cdot)$, as in a feedback communica- tem, but we shall assume that the messages m are dent of $v(\cdot)$ so that $\phi(\cdot)$ will be independent of $(\cdot)$. We shall assume also that there is an average onstraint on $\phi(\cdot)$,

$$E\int_0^T \phi^2(s)ds \le P_{av}. \tag{115}$$

as in Theorem 5,

$$\hat{\phi}_1(t) = E[\phi(t)|y(s), 0 \le s < t, h_1],$$

53] has recently shown that

$$I(y, m) = \frac{1}{2}\int_0^T E[\phi(t) - \hat{\phi}_1(t)]^2 dt. \tag{116}$$

ketching the proof of this result, let us note a simple tion [53], namely proving that the maximum mutual ation under the power constraint (115) is $P_{av}/2$ and s maximum value can actually be attained. For this at from (116) we have

$$I \le \frac{1}{2}\int_0^T E[\phi^2(t)] = \frac{P_{av}}{2}. \tag{117}$$

er, by taking $\phi(\cdot)$ to be a stationary Gaussian we can by increasing the bandwidth of its power density function, while keeping the total power nake $\hat{\phi}_1(\cdot) \to 0$. Therefore, the upper bound $P_{av}/2$ ually be achieved. It is well known [54] that $P_{av}/2$ is the capacity (maximum mutual information) of the y additive WGN channel, with only a power (and andwidth) constraint. But now we have also shown capacity with feedback is unchanged.

rove (116) heuristically, we note that

$$(y, m) = E\left[\ln \frac{p(y|m)}{p(y)}\right] = E\left[\ln \frac{p(y|m)/p(y|0)}{p(y)/p(y|0)}\right]. \tag{118}$$

umerator is the LR for deciding whether $(t, m\{y(s), 0 \le s < t\}) + v(t)$, with a particular m or $y(t)$ is WGN. The denominator has a similar inter- n, but now with m and consequently $\phi(\cdot)$ being . Now, using Theorem 6, we have

$$\ln \left[\frac{\exp \int_0^T \phi(t)y(t)dt - \frac{1}{2}\int_0^T \phi^2(t)dt}{\exp \int_0^T \hat{\phi}_1(t)y(t)dt - \frac{1}{2}\int_0^T \hat{\phi}_1^2(t)dt}\right]$$

$$E\int_0^T [\hat{\phi}_1^2(t) - \phi^2(t)]dt + E\int_0^T [\phi(t) - \hat{\phi}_1(t)]y(t)dt.$$

cond expectation is zero because of the independence of future v from past $\phi(\cdot)$. Th

$$I = \frac{1}{2}E\int_0^T [\hat{\phi}_1^2(t) - \phi^2(t)]dt = \frac{1}{2}$$

V. Some Extensions and Conclu

In the previous sections we have show innovations for processes with an additive nent. When there is no such component the pro difficult, but several results (cf. [38]–[40] and ences therein) have been obtained for Gaussian p As we might expect, these results are derived by re the covariance function of the given process, by differe tion, adjunction of initial conditions, etc., to covariances the form $I+K$, $I+K>0$, to which the results of Section I can be applied. We may refer to [12] for a summary of these results and also applications to detection and estimation in colored (non-white) Gaussian noise. For non-Gaussian processes, extensions can be obtained by allowing the innovations to be (formal) derivatives of independent increment processes and, even more generally, martingale processes (cf. [17], [20], [49]).

A more general approach to the colored Gaussian noise case can be obtained from some general results of Gohberg and Krein [41]. By combining these results with those obtained by the reproducing kernel Hilbert space (RKHS) approach to Gaussian detection problems, we have been able to solve the following problem: given two Gaussian processes y_1 and y_2 with covariances R_1 and R_2, find a causal and causally invertible transformation relating one to the other. As we might expect from Theorem 4 and the discussion following it, a sufficient condition is that the processes be equivalent, i.e., that the problem of deciding whether an observed sample function comes from y_1 or y_2 should be nonsingular. An interesting feature of these results is the close attention that most be paid to such basic concepts as causality, the distinction between Volterra and causal operators, etc. (cf. [19], [55]).

There are still other applications of the innovations method and philosophy that we have not been able to discuss here, but we hope that the theorems and applications we have described will illustrate the power and the scope of the method.

Acknowledgment

The author wishes to thank all his friends and colleagues who have greatly aided him in his work in this field, especially, P. Frost, R. Geesey, T. Duncan, D. Duttweiler, H. Aasnaes, B. Gopinath, M. Zakai, E. Parzen, J. M. Clark, L. Shepp, W. L. Root, T. Kadota, B. Anderson, T. Hida, M. Hitsuda, and E. Wong.

References

[1] A. N. Kolmogorov, "Stationary sequences in Hilbert spaces," *Bull. Math. Univ. Moscow* (in Russian), vol. 2, no. 6, p. 40, 1941.
[2] H. W. Bode and C. E. Shannon, "A simplified derivation of linear least square smoothing and prediction theory," *Proc. IRE*, vol. 38, pp. 417–425, April 1950.
[3] L. A. Zadeh and J. R. Ragazzini, "An extension of Wiener's theory of prediction," *J. Appl. Phys.*, vol. 21, pp. 645–655, July 1950.
[4] V. A. Kotel'nikov, *The Theory of Optimum Noise Immunity*. New

ch other, and, therefore, the process $\{v(s), 0 \leq s < t\}$ ed the same information as $\{y(s), 0 \leq s < t\}$. In the t non-Gaussian case, this property has not yet been ished in full generality. It is true for jump processes i.e., processes that can only assume a finite number of plitudes, and for uniformly bounded processes and for me other situations. However, the author conjectures that e result will be true under the assumptions (108) and (109) n $z(\cdot)$, and, in fact, under somewhat weaker conditions than (108). There is some support for this conjecture, but as yet, there is no complete proof. Therefore, here we shall only discuss applications that do not call upon this property of the "causal equivalence" of $y(\cdot)$ and $v(\cdot)$; this will exclude some interesting problems, e.g., nonlinear least-squares estimation (however, see [14]).

Our first application will be to a detection problem.

Application VI: Non-Gaussian Signals in WGN

Let us consider the hypotheses

$$h_1 : y(t) = z(t) + v(t), \quad h_0 : y(t) = v(t), \quad 0 \leq t \leq T \tag{112}$$

where $z(\cdot)$ and $v(\cdot)$ are as in (107)–(109). Then Theorem 3 can be generalized as follows.

Theorem 6: The LR for the detection problem (112) can be written as

$$\mathrm{LR} = \exp \left\{ \int_0^T \hat{z}_1(t) y(t) dt - \frac{1}{2} \int_0^T \hat{z}_1^2(t) dt \right\} \tag{113}$$

where

$$\hat{z}_1(t) = \text{the causal least-squares estimate of } z(t) \text{ given } y(s), 0 \leq s < t \text{ and assuming } h_1 \text{ is true} \tag{114}$$

and $\oint_0^T$ denotes the Itô stochastic integral.

Proof: A rigorous proof is given in [17], based on Theorem 5 and on some recent martingale theorems. Though we shall not pursue the topic here, we may note that these martingale theorems have recently yielded some more general results [49] than those in Theorems 5 and 6. As for a heuristic proof, the reader will note that the argument used for Theorem 3 will still hold. The basic reason for (113) is that, by Theorem 5, the detection problem can be reexpressed as

$$h_1 : y(t) = \hat{z}_1(t) + v(t), \quad h_0 : y(t) = v(t)$$

where $v(\cdot)$ is WGN and $\hat{z}_1(\cdot)$ is a conditionally known [given $y(\cdot)$] signal.

Remark 1: We have not explicitly mentioned here, as we did in Theorem 3, the nonsingularity of the detection problem. The reason is that a new "singularity" phenomenon arises in the non-Gaussian detection problem. To explain this consider the problem of determining whether an observation x comes from a rectangular distribution over [0, 2] (hypothesis h_1) or over [1, 2] (hypothesis h_0). Clearly, some of the time we will be able to affirm hypothesis h_1 with zero probability of error, but we cannot do this for h_0. More generally in non-Gaussian signal in noise problems of the type (112), even when (108) and (109) are met, there may be a

collection of $y(\cdot)$ with nonzerc
can affirm h_0 without error, but
we shall say that the process z
tinuous with respect to the W
problem, we cannot generally
in the special case that $z(\cdot)$ is
of results on absolute continuit
nical and we shall only refer th
which discuss such questions fo
However, we should mention he
Theorem 4, which was Gaussian
the theorem is motivated by e
model (112) and the formula (11
the next theorem, is that they are

Theorem 7: A process $y(\cdot)$ is
defined above) with respect to a
if it can be written in the form

$$y(t) = z(t, \{y(s), 0 \leq$$

where $v(\cdot)$ is WGN and $z(\cdot)$ is a
is square-integrable on [0, T].

Proof: The result is proved

As might be expected, many of
made after Theorems 3 and 4 ca
therefore we shall not repeat ther
note that several examples show
reduces to previously known forr
random phase, are worked out in
examples will show clearly the sig
tion of the integral and will also
with the Itô calculus.

As for previous work, the only
by a different method, of Straton
Duncan [52] for $z(\cdot)$ that are M
Our approach via the innovation
cial assumptions on $z(\cdot)$.

The generality of our formula (
previously known explicit formula
to stress that even more important
tion it provides on the ideal dete
cussed in the Gaussian case and
length in [16]. We cannot overem
qualitative aspects and would str
examine the discussion in [16]. V
Theorem 6 can also be found in [16
generalizations to certain classes
noise.

Our final application would seen
removed from the binary detection
studying, but we shall find that it
general LR formula.

Application VII: Mutual Inform Channels

We shall obtain a simple express
mation $I(y, m)$ between the mess
$y(\cdot)$ of a channel in which

where
mitted
may d
tion sy
indepe
future
power

Then i

Zakai

Before
applic
inform
that th
note th

Morec
proces
spectr
fixed,
can ac
actual
one-w
not a
that th
To

The
$y(t)=$
wheth
pretat
rando

$I =$

$=$

The s

from each other, and, therefore, the process $\{v(s), 0 \le s < t\}$ contained the same information as $\{y(s), 0 \le s < t\}$. In the present non-Gaussian case, this property has not yet been established in full generality. It is true for jump processes $z(\cdot)$, i.e., processes that can only assume a finite number of amplitudes, and for uniformly bounded processes and for some other situations. However, the author conjectures that the result will be true under the assumptions (108) and (109) on $z(\cdot)$, and, in fact, under somewhat weaker conditions than (108). There is some support for this conjecture, but as yet, there is no complete proof. Therefore, here we shall only discuss applications that do not call upon this property of the "causal equivalence" of $y(\cdot)$ and $v(\cdot)$; this will exclude some interesting problems, e.g., nonlinear least-squares estimation (however, see [14]).

Our first application will be to a detection problem.

Application VI: Non-Gaussian Signals in WGN

Let us consider the hypotheses

$$h_1 : y(t) = z(t) + v(t), \quad h_0 : y(t) = v(t), \quad 0 \le t \le T \qquad (112)$$

where $z(\cdot)$ and $v(\cdot)$ are as in (107)–(109). Then Theorem 3 can be generalized as follows.

Theorem 6: The LR for the detection problem (112) can be written as

$$\text{LR} = \exp \left\{ \int_0^T \hat{z}_1(t) y(t) dt - \frac{1}{2} \int_0^T \hat{z}_1^2(t) dt \right. \qquad (113)$$

where

$$\hat{z}_1(t) = \text{the causal least-squares estimate of } z(t) \text{ given } y(s),\ 0 \le s < t \text{ and assuming } h_1 \text{ is true} \qquad (114)$$

and $\oint_0^T$ denotes the Itô stochastic integral.

Proof: A rigorous proof is given in [17], based on Theorem 5 and on some recent martingale theorems. Though we shall not pursue the topic here, we may note that these martingale theorems have recently yielded some more general results [49] than those in Theorems 5 and 6. As for a heuristic proof, the reader will note that the argument used for Theorem 3 will still hold. The basic reason for (113) is that, by Theorem 5, the detection problem can be reexpressed as

$$h_1 : y(t) = \hat{z}_1(t) + v(t), \quad h_0 : y(t) = v(t)$$

where $v(\cdot)$ is WGN and $\hat{z}_1(\cdot)$ is a conditionally known [given $y(\cdot)$] signal.

Remark 1: We have not explicitly mentioned here, as we did in Theorem 3, the nonsingularity of the detection problem. The reason is that a new "singularity" phenomenon arises in the non-Gaussian detection problem. To explain this consider the problem of determining whether an observation x comes from a rectangular distribution over [0, 2] (hypothesis h_1) or over [1, 2] (hypothesis h_0). Clearly, some of the time we will be able to affirm hypothesis h_1 with zero probability of error, but we cannot do this for h_0. More generally in non-Gaussian signal in noise problems of the type (112), even when (108) and (109) are met, there may be a collection of $y(\cdot)$ with nonzero probability, for which we can affirm h_0 without error, but never h_1. When this is true we shall say that the process $z(\cdot)+v(\cdot)$ is *absolutely continuous* with respect to the WGN process $v(\cdot)$. In our problem, we cannot generally claim the converse, except in the special case that $z(\cdot)$ is also Gaussian. The proofs of results on absolute continuity become somewhat technical and we shall only refer the reader to [50] and [51], which discuss such questions for models of the type (112). However, we should mention here a result that generalizes Theorem 4, which was Gaussian signals. As in that context, the theorem is motivated by enquiring how general the model (112) and the formula (113) are. The answer, given in the next theorem, is that they are very general.

Theorem 7: A process $y(\cdot)$ is absolutely continuous (as defined above) with respect to a WGN process if and only if it can be written in the form

$$y(t) = z(t, \{y(s), 0 \le s < t\}) + v(t)$$

where $v(\cdot)$ is WGN and $z(\cdot)$ is a functional of past $y(\cdot)$ that is square-integrable on $[0, T]$.

Proof: The result is proved in [17] and [49].

As might be expected, many of the explanatory comments made after Theorems 3 and 4 can also be made here and therefore we shall not repeat them. It may be worthwhile to note that several examples showing how the LR formula reduces to previously known formulas, e.g., for signals with random phase, are worked out in some detail in [16]. These examples will show clearly the significance of the Itô definition of the integral and will also help familiarize the reader with the Itô calculus.

As for previous work, the only results are those, obtained by a different method, of Stratonovich and Sosulin [47] and Duncan [52] for $z(\cdot)$ that are Markov diffusion processes. Our approach via the innovations does not need such special assumptions on $z(\cdot)$.

The generality of our formula (113), in that it includes all previously known explicit formulas, is striking. But we wish to stress that even more important is the structural information it provides on the ideal detector. This point was discussed in the Gaussian case and is also treated at some length in [16]. We cannot overemphasize the value of such qualitative aspects and would strongly urge the reader to examine the discussion in [16]. Various generalizations of Theorem 6 can also be found in [16] [17], and [49], including generalizations to certain classes of non-Gaussian additive noise.

Our final application would seem a priori to be somewhat removed from the binary detection problems we have been studying, but we shall find that it is closely related to the general LR formula.

Application VII: Mutual Information in Additive WGN Channels

We shall obtain a simple expression for the mutual information $I(y, m)$ between the messages m and the outputs $y(\cdot)$ of a channel in which

$$y(t) = \phi(t, m, \{y(s), 0 \le s < t\}) + v(t)$$

where $\phi(\cdot)$ is the actual "modulator" signal that is transmitted over the additive WGN channel. The signal $\phi(\cdot)$ may depend upon past $y(\cdot)$, as in a feedback communication system, but we shall assume that the messages m are independent of $v(\cdot)$ so that $\phi(\cdot)$ will be independent of future $v(\cdot)$. We shall assume also that there is an average power constraint on $\phi(\cdot)$,

$$E\int_0^T \phi^2(s)ds \le P_{av}. \tag{115}$$

Then if, as in Theorem 5,

$$\hat{\phi}_1(t) = E[\phi(t)|y(s), 0 \le s < t, h_1],$$

Zakai [53] has recently shown that

$$I(y, m) = \frac{1}{2}\int_0^T E[\phi(t) - \hat{\phi}_1(t)]^2 dt. \tag{116}$$

Before sketching the proof of this result, let us note a simple application [53], namely proving that the maximum mutual information under the power constraint (115) is $P_{av}/2$ and that this maximum value can actually be attained. For this note that from (116) we have

$$I \le \frac{1}{2}\int_0^T E[\phi^2(t)] = \frac{P_{av}}{2}. \tag{117}$$

Moreover, by taking $\phi(\cdot)$ to be a stationary Gaussian process we can by increasing the bandwidth of its power spectral density function, while keeping the total power fixed, make $\hat{\phi}_1(\cdot) \to 0$. Therefore, the upper bound $P_{av}/2$ can actually be achieved. It is well known [54] that $P_{av}/2$ is actually the capacity (maximum mutual information) of the one-way additive WGN channel, with only a power (and not a bandwidth) constraint. But now we have also shown that the capacity with feedback is unchanged.

To prove (116) heuristically, we note that

$$I(y, m) = E\left[\ln \frac{p(y|m)}{p(y)}\right] = E\left[\ln \frac{p(y|m)/p(y|0)}{p(y)/p(y|0)}\right]. \tag{118}$$

The numerator is the LR for deciding whether $y(t) = \phi(t, m\{y(s), 0 \le s < t\}) + v(t)$, with a particular m or whether $y(t)$ is WGN. The denominator has a similar interpretation, but now with m and consequently $\phi(\cdot)$ being random. Now, using Theorem 6, we have

$$I = E \ln \left[\frac{\exp\left\{\int_0^T \phi(t)y(t)dt - \frac{1}{2}\int_0^T \phi^2(t)dt\right.}{\exp\left\{\int_0^T \hat{\phi}_1(t)y(t)dt - \frac{1}{2}\int_0^T \hat{\phi}_1^2(t)dt\right.}\right]$$

$$= \frac{1}{2}E\int_0^T [\hat{\phi}_1^2(t) - \phi^2(t)]dt + E\int_0^T [\phi(t) - \hat{\phi}_1(t)]y(t)dt.$$

The second expectation is zero because of the independence of future v from past $\phi(\cdot)$. Therefore

$$I = \frac{1}{2}E\int_0^T [\hat{\phi}_1^2(t) - \phi^2(t)]dt = \frac{1}{2}\int_0^T E[\phi(t) - \hat{\phi}_1(t)]^2 dt.$$

V. Some Extensions and Concluding Remarks

In the previous sections we have shown how to find the innovations for processes with an additive WGN component. When there is no such component the problem is more difficult, but several results (cf. [38]–[40] and the references therein) have been obtained for Gaussian processes. As we might expect, these results are derived by reducing the covariance function of the given process, by differentiation, adjunction of initial conditions, etc., to covariances of the form $I+K$, $I+K>0$, to which the results of Section III can be applied. We may refer to [12] for a summary of these results and also applications to detection and estimation in colored (non-white) Gaussian noise. For non-Gaussian processes, extensions can be obtained by allowing the innovations to be (formal) derivatives of independent increment processes and, even more generally, martingale processes (cf. [17], [20], [49]).

A more general approach to the colored Gaussian noise case can be obtained from some general results of Gohberg and Krein [41]. By combining these results with those obtained by the reproducing kernel Hilbert space (RKHS) approach to Gaussian detection problems, we have been able to solve the following problem: given two Gaussian processes y_1 and y_2 with covariances R_1 and R_2, find a causal and causally invertible transformation relating one to the other. As we might expect from Theorem 4 and the discussion following it, a sufficient condition is that the processes be equivalent, i.e., that the problem of deciding whether an observed sample function comes from y_1 or y_2 should be nonsingular. An interesting feature of these results is the close attention that most be paid to such basic concepts as causality, the distinction between Volterra and causal operators, etc. (cf. [19], [55]).

There are still other applications of the innovations method and philosophy that we have not been able to discuss here, but we hope that the theorems and applications we have described will illustrate the power and the scope of the method.

Acknowledgment

The author wishes to thank all his friends and colleagues who have greatly aided him in his work in this field, especially, P. Frost, R. Geesey, T. Duncan, D. Duttweiler, H. Aasnaes, B. Gopinath, M. Zakai, E. Parzen, J. M. Clark, L. Shepp, W. L. Root, T. Kadota, B. Anderson, T. Hida, M. Hitsuda, and E. Wong.

References

[1] A. N. Kolmogorov, "Stationary sequences in Hilbert spaces," *Bull. Math. Univ. Moscow* (in Russian), vol. 2, no. 6, p. 40, 1941.
[2] H. W. Bode and C. E. Shannon, "A simplified derivation of linear least square smoothing and prediction theory," *Proc. IRE*, vol. 38, pp. 417–425, April 1950.
[3] L. A. Zadeh and J. R. Ragazzini, "An extension of Wiener's theory of prediction," *J. Appl. Phys.*, vol. 21, pp. 645–655, July 1950.
[4] V. A. Kotel'nikov, *The Theory of Optimum Noise Immunity*. New

York: McGraw-Hill, 1959. (Ph.D. dissertation presented in January 1947 in Moscow, USSR.)

[5] C. I. Dolph and M. A. Woodbury, "On the relation between Green's functions and covariances of certain stochastic processes and its application to unbiased linear prediction," *Trans. Am. Math. Soc.*, vol. 72, pp. 519–550, 1952.

[6] S. Darlington, "Nonstationary smoothing and prediction using network theory concepts," *Trans. 1959 Internatl. Symp. on Circuit and Information Theory* (Los Angeles, Calif.), pp. 1–13, 1959.

[7] R. E. Kalman, "A new approach to linear filtering and prediction problems," *Trans. ASME, J. Basic Engrg.*, vol. 82, pp. 34–35, March 1960.

[8] R. E. Kalman and R. S. Bucy, "New results in linear filtering and prediction theory," *Trans. ASME, J. Basic Engrg., Ser. D*, vol. 83, pp. 95–107, December 1961.

[9] R. E. Kalman, "New methods in Wiener filtering theory," in *Proc. 1st Symp. on Engrg. Appl. of Random Function Theory and Probability*, J. L. Bogdanoff and F. Kozin, Eds. New York: Wiley, 1963.

[10] T. Kailath, "An innovations approach to least-squares estimation, part I: linear filtering in additive white noise," *IEEE Trans. Automatic Control*, vol. AC-13, pp. 646–655, December 1968.

[11] T. Kailath and P. Frost, "An innovations approach to least-squares estimation, part II: linear smoothing in additive white noise," *IEEE Trans. Automatic Control*, vol. AC-13, pp. 655–660, December 1968.

[12] R. Geesey and T. Kailath, "Applications of canonical representation to estimation and detection in colored noise," *Proc. 19th Polytech. Inst. of Brooklyn Symp. on Computer Processing in Communications.* Brooklyn, N. Y.: Polytechnic Press, April 1969.

[13] H. Aasnaes and T. Kailath, "An innovations approach to least-squares estimation—part V: more on discrete systems," Stanford University, Systems Theory Lab., Stanford, Calif., Tech. Rept., 1969.

[14] P. A. Frost and T. Kailath, "An innovations approach to least-squares estimation, part III: nonlinear filtering and smoothing in white Gaussian noise," to be published.

[15] P. A. Frost "Estimation in continuous-time nonlinear systems," Ph.D. dissertation, Dept. of Elec. Engrg., Stanford University, Stanford, Calif., June 1968.

[16] T. Kailath, "A general likelihood-ratio formula for random signals in Gaussian noise," *IEEE Trans. Information Theory*, vol. IT-15, pp. 350–361, May 1969.

[17] ——, "A further note on a general likelihood formula for random signals in Gaussian noise," *IEEE Trans. Information Theory*, vol. IF-16, pp. 393-396, July 1970.

[18] ——, "Likelihood ratios for Gaussian processes," *IEEE Trans. Information Theory*, May 1970 (to be published).

[19] D. Duttweiler, "Hilbert-space methods for detection and estimation theory," Ph.D. dissertation, Dept. of Elec. Engrg., Stanford University, Stanford, Calif., June 1970.

[20] T. Kailath and P. A. Frost, "Mathematical modeling and transformation theory of white noise processes," to be published.

[21] A. Lindquist, "An innovations approach to optimal control of linear stochastic systems with time delay," *Information Sciences*, vol. 1, no. 3, pp. 279–295, 1969.

[22] R. K. Mehra, "On the identification of variances and adaptive Kalman filtering," *1968 Proc. Joint Automatic Control Conf.*, (Boulder, Colo., August 1968), pp. 494–503.

[23] B. D. Anderson and T. Kailath, "The choice of signal-process models in Kalman–Bucy filtering," *IEEE Trans. Circuit Theory*, 1969 (to be published).

[24] P. Lévy, *Brownian Movement and Stochastic Processes.* Paris: Gauthier–Villars, 1964, Appendix II.

[25] J. L. Doob, *Stochastic Processes.* New York: Wiley, 1953.

[26] K. Karhunen, "Über Lineare Methoden in der Wahrscheinlichkeitsrechnung," *An. Acad. Sci. Fennicae*, Ser. A, I, vol. 37, pp. 3–79, 1947.

[27] O. Hanner, "Deterministic and non-deterministic processes," *Arkiv. Mat.*, vol. 1, pp. 161–177, 1950.

[28] T. Hida, "Canonical representations of Gaussian processes and their applications," *Mem. Coll. Sci. Univ. Kyoto*, Ser. A, vol. 33, pp 109–155, 1960.

[29] H. Cramér, "Stochastic processes as curves in Hilbert space," *Theory Probability Appl.*, vol. 9, pp. 199–204, 1964.

[30] ——, "A contribution to the multiplicity theory of stochastic processes," *Proc. 5th Symp. on Math. Statist. and Probability*, vol. 2. Berkeley, Calif.: University of California Press, 1967, pp. 215–221.

[30a] M. Motoo and S. Watanabee, "On a class of additive functionals of Markov process," *J. Math Kyoto Univer.*, vol. 4, pp. 429–469, 1965.

[31] A. Papoulis, *Probability, Random Variables and Stochastic Processes.* New York: McGraw-Hill, 1965.

[32] L. D. Collins, "Realizable whitening filters and state-variable realizations," *Proc. IEEE* (Letters), vol. 56, pp. 100–101, January 1968.

[33] R. S. Bucy and P. D. Joseph, *Filtering for Stochastic Processes with Applications to Guidance.* New York: Wiley, 1968.

[34] J. Rissanen and L. Barbosa, "Properties of infinite covariance matrices and stability of optimum predictors," *Information Sciences*, vol. 1, no. 3, pp. 221–236, July 1969.

[35] C. W. Helstrom, *Statistical Theory of Signal Detection*, 2nd. ed. London: Pergamon Press, 1968.

[36] A. Baggeroer, "A state-variable approach to the solution of Fredholm integral equations," *IEEE Trans. Information Theory*, vol. IT-15, pp. 557–570, September 1969.

[37] T. Kailath, "Fredholm resolvents, Wiener–Hopf equations and Riccati differential equations," *Trans. IEEE Information Theory*, vol. IT-15, pp. 665–672, November 1969.

[37a] ——, "Application of a resolvent identity to a linear smoothing problem," *SIAM J. Control*, vol. 7, pp. 68–74, February 1969.

[38] T. Kailath and R. Geesey, "Covariance factorization—an explication via examples," *Proc. 2nd Asilomar Conf. on Systems and Circuits* (Monterey, California), November 1968.

[39] R. Geesey, "Canonical representations of second-order processes with applications," Ph.D. dissertation, Stanford, Calif., December 1968.

[40] B. D. Anderson, J. B. Moore, and S. G. Loo, "Spectral factorization of time-varying covariance functions," *IEEE Trans. Information Theory*, vol. IT-15, pp. 550–557, September 1969.

[41] I. Gohberg and M. G. Krein. "On the factorization of operators in Hilbert spaces," *Acta Sci. Math. Univ. Szeged*, vol. 25, pp. 90–123, 1964. See also *Amer. Math Soc. Transl.*, ser. 2, vol. 51, 1966.

[42] B. D. Anderson and T. Kailath, "Some integral equations with nonsymmetric kernels," to be published.

[43] H. L. Van Trees, *Detection, Estimation and Modulation Theory*, vols. 1, 2. New York: Wiley, 1968, 1970.

[44] L. A. Shepp, "Radon–Nikodym derivatives of Gaussian measures," *Ann. of Math. Stat.*, vol. 37, pp. 321–354, 1966.

[45] R. Price, "Optimum detection of random signals in noise, with application to scatter-multipath communication—I," *IRE Trans. Information Theory*, vol. IT-2, pp. 125–135, December 1956.

[46] R. Price and P. E. Green, Jr., "A communication technique for multipath channels," *Proc. IRE*, vol. 46, pp. 555–570, March 1958.

[47] R. L. Stratonovich and Yu G. Sosulin, "Optimal detection of a Markov process in noise," *Engrg. Cybernetics*, vol. 6, pp. 7–19, October 1964.

[48] F. C. Schweppe, "Evaluation of likelihood functions for Gaussian signals," *IEEE Trans. Information Theory*, vol. IT-11, pp. 61–70, January 1965.

[49] T. Kailath, Radon Nikodym derivatives with respect to Weiner and related measures," *Proc. 2nd Mastech Conf. Probability and Statistics* (Madvas, India), to be published.

[50] T. T. Kadota and L. A. Shepp, "Conditions for the absolute continuity between a certain pair of probability measures," *Z. Wahrscheinlichkeitstheorie*, 1969.

[51] T. Kailath and M. Zakai, "Absolute continuity and Radon–Nikodym derivatives for certain measures relative to Wiener measure," Internal Memo., Bell Telephone Labs., N. J., 1969, to be published.

[52] T. E. Duncan, "Probability densities for diffusion processes with applications to nonlinear filtering theory and detection theory," Ph.D. dissertation, Dept. of Elec. Engrg., Stanford University, Stanford, Calif., May 1967. Also *Information and Control*, vol. 13, pp. 62–74, July 1968.

[53] M. Zakai, "On the mutual information of the white Gaussian noise channel with and without feedback," Internal Memo., Bell Telephone Labs., N. J., 1969.

[54] R. M. Fano, *Transmission of Information.* Cambridge, Mass.: MIT Press, 1961.

[55] T. Kailath and D. Duttweiler, "Generalised innovation processes and some applications," to be presented at the Internatl. Symp. Inform. Theory, Noordwijk, Netherlands, June 1970.

[56] H. Aasnaes, private communication.

Editors' Comments on Paper 17

This paper, by Professor H. Cramér, was presented as the first Samuel Stanley Wilks lecture at the dedication of the Jadwin Mathematics Building of Princeton University on March 17, 1970.

In addition to a brief survey of his earlier results on the multiplicity representation, the author presents new conditions for unit multiplicity, and he studies related problems of statistical inference. In particular, the equivalence between Gaussian probability measures induced by random processes on appropriate function spaces is related to the canonical decompositions of these processes.

The latter subject has been also studied by several authors in both its abstract version (Shepp; Clark; Rozanov et al.) and its applied version (Kadota; Kailath et al.) These works are presented later in this volume.

This material is reprinted with the permission of the author and the Princeton University Press.

17

STRUCTURAL AND STATISTICAL PROBLEMS FOR A CLASS OF STOCHASTIC PROCESSES

by
Harald Cramér

PRINCETON UNIVERSITY PRESS
PRINCETON, NEW JERSEY
1971

INTRODUCTION

by
Frederick Mosteller

MR. CHAIRMAN, LADIES AND GENTLEMEN:

The deed of gift of the Samuel Stanley Wilks Memorial Fund, provided by Professor Wilks's friends after his death in 1964, reads: "The income from this fund shall be used from time to time for a public lecture or conference on statistics, broadly conceived, the speaker and occasion to be chosen by the University in a manner deemed most likely to advance statistics in the service of the nation."

I am sure that we can all agree that this first Wilks Memorial Lecture is being held on a most appropriate occasion, since Fine Hall was a second home for many of us. The dedication of a new Fine Hall is certainly the end of an era and the start of a new one and thus a moment of national significance to statistics.

Our speaker today represents statistics in a manner that Wilks especially appreciated. Let me document this remark.

When Wilks returned from his National Scholarship year in England, he came to Princeton and there began a considerable tradition in mathematical statistics. (By the way his first statistics student was Churchill Eisenhart, whose father hired Sam.) Although Sam often worked on applications of statistics, he thoroughly enjoyed returning to mathematical statistics as his many papers show. This pleasure is made clear in a rare personal remark in the Preface to his book, entitled *Mathematical Statistics,* where, after saying that it gives, "a careful presentation of basic mathematical statistics and the underlying mathematical theory of a wide variety of topics in statistics with just enough discussion and examples to clarify the basic concepts," he continues, "I believe this approach to be prerequisite to a fuller understanding of statistical methodology, not to mention the one I find most satisfying." And so our speaker and his topic today are especially appropriate for the first Wilks Memorial Lecture.

Our speaker's contribution to education in our field through his great book *Mathematical Methods of Statistics,* published at both Princeton and Stockholm, gave new direction to the course work of students of statistics the world over.

As an applied statistician he has been statistical adviser to Swedish insurance companies; as a theoretician, he was a professor of mathematical statistics from 1929 to 1958. He took up the presidency of the University of Stockholm in 1950, continued to 1958 when he became Chancellor of the Swedish Universities. Naturally his honors are many and I shall mention but a few: after visiting here at Princeton in 1946, he was awarded an honorary D.Sc. degree in 1947. He is not just a member but an honorary member both of the International Statistical Institute and of the Royal Statistical Society. He is a member of the Royal Swedish Academy of Science and of the Royal Swedish Academy of Engineering and, coming back to the United States, of the American Academy of Arts and Sciences.

His contributions to the field of stochastic processes have been extensive and profound; indeed he was co-author of another book as recently as 1966. His title for the first Samuel Stanley Wilks Memorial Lecture is "Structural and Statistical Problems in a Certain Class of Stochastic Processes."

Ladies and Gentlemen: Professor Harald Cramér.

2

STRUCTURAL AND STATISTICAL PROBLEMS FOR A CLASS OF STOCHASTIC PROCESSES

S. S. WILKS MEMORIAL LECTURE
MARCH 17, 1970

by
Harald Cramér

1. Introduction

It is an honor and a privilege to have been asked to give the first S. S. Wilks Memorial Lecture.

Sam Wilks entered the field of statistical science at a time when it still had all the freshness of youth. Even during his early years he was able to make outstanding contributions to its development. Later, while always pursuing his scientific research, he became engaged in administrative work to an extent that rendered him the well-deserved name "A Statesman of Statistics." He organized the teaching of statistics in this university and elsewhere, always making it clear that this teaching should be built on rigorous mathematical foundations. And, last but not least, Sam was a friend of his many friends, always ready to go out of his way to help a friend, in every possible way.

The early statistical works of Sam Wilks belong to the classical theory of statistical inference. Accordingly, they are concerned with the probability distributions of groups of random variables, the number of which may be very large, but is always assumed to be finite. It is well known that, in many fields of applications, we nowadays encounter important problems that involve an infinite number of random variables, and so cannot be solved within the frames of the classical theory. Some of these problems, which come up in the theory of stochastic processes, may be regarded as straightforward generalizations of inference problems treated by Wilks and other authors. It seems to me that it will be in the spirit of Sam Wilks to choose

3

some of these generalized problems as the subject of a Wilks Memorial Lecture.

A systematic attempt to generalize classical statistical inference theory to stochastic processes first appeared in the 1950 Stockholm University thesis by Ulf Grenander [8]. Since then, a vigorous development has taken place, based on works by a large number of authors.

A group of probabilistic problems that seems to have been of fundamental importance throughout this development is concerned with finding convenient *types of representation* for the classes of stochastic processes under consideration, so as to enable us to investigate the properties of the processes, and, in particular, to work out methods of statistical inference. As examples of such types of representation, we may think of the spectral representation for stationary processes, the Karhunen-Loève representation in terms of eigenfunctions and eigenvalues of a symmetrical kernel, etc.

In the sequel we shall be concerned with one particular mode of representation, which is applicable to a large class of stochastic processes. After having introduced the general representation formula and discussed its properties, we shall consider the possibility of its application to some problems of statistical inference.

Let us first present a simple argument that suggests the type of representation formula in question. Consider the real-valued random variable $x(t)$ of a stochastic process with a continuous time parameter t. In various applications it seems natural to regard the value assumed by $x(t)$ at a given instant t as the accumulated effect of a stream of random *impulses* or *innovations* acting throughout the past. Suppose, e.g., that $x(t)$ represents the intensity at time t of an electric current generated by a stream of electrons arriving at an anode at randomly distributed time points. Let an electron arriving at time u bring an impulse $dz(u)$, while the effect at time t of a unit impulse acting at time u is measured by a *response function* $g(t, u)$. Assuming the effects to be simply additive, we should then be heuristically led to try a representation formula for $x(t)$ of the type

$$x(t) = \int_{-\infty}^{t} g(t, u)\, dz(u), \tag{1.1}$$

with some appropriate definition of the stochastic integral in the second member.

4

Similar arguments will apply to various other applied problems, and thus the following question arises in a natural way: *Is it possible to define some general class of stochastic processes that will admit a representation more or less similar to the formula* (1.1)?

It will appear that this is, in fact, possible if we modify (1.1) by replacing the second member of that formula by the sum of a certain number of terms of the same form. In some cases the number of terms will be equal to one, so that (1.1) will apply without modification.

It turns out that the most adequate approach to our question is supplied by the geometry of Hilbert space. In the following section we shall briefly recall some known facts from Hilbert space theory, and then show how this theory leads to a complete answer to our question.

2. The Representation Formula

We shall consider real-valued random variables $x = x(\omega)$ defined on a fixed probability space $(\Omega, \mathbf{F}, P)$, and satisfying the conditions

$$Ex = 0, \qquad Ex^2 < \infty.$$

It is well known that the set of all such random variables forms a real Hilbert space H, if addition and multiplication by real numbers are defined in the obvious way, while the inner product of two elements x and y is defined by

$$(x, y) = Exy.$$

The norm of an element x is then

$$\|x\| = (Ex^2)^{1/2}.$$

Two elements x and y are regarded as identical if the norm of their difference is zero. We then have $E(x - y)^2 = 0$, and thus

$$P(x = y) = 1.$$

All equalities between random variables occurring in the sequel should be understood in this sense, so that the two members are equal with probability one.

Convergence in norm in the Hilbert space H is identical with convergence in quadratic mean (q.m.) in probabilistic terminology.

A stochastic process is a family of real-valued random variables

$x(t) = x(t, \omega)$. We shall not be concerned in this paper with the more general cases of complex or vector-valued processes. The parameter t, which will be interpreted as time, is supposed to be confined to a finite or infinite real interval $T = (A, B)$, where A may be finite or $-\infty$, while B may be finite or $+\infty$. We shall always suppose that for every $t \in T$ we have

$$Ex(t) = 0, \qquad Ex^2(t) < \infty, \tag{2.1}$$

so that $x(t)$ is a point in the Hilbert space H, while the family of all $x(t)$ may be regarded as defining a *curve* in H.

We shall further also impose the important condition that the q.m. limits

$$x(t-0) = \lim_{h \uparrow 0} x(t+h), \qquad x(t+0) = \lim_{h \downarrow 0} x(t+h) \tag{2.2a}$$

exist for all $t \in T$, and that we have

$$x(t-0) = x(t), \tag{2.2b}$$

so that $x(t)$ is everywhere continuous to the left in q.m.

When a stochastic process satisfying (2.1) and (2.2) is given, we define the *Hilbert space* $H(x)$ *of the process* as the subspace of H spanned by the random variables $x(t)$ for all $t \in T$, i.e., the closure in norm of all finite linear combinations

$$c_1x(t_1) + c_2x(t_2) + \cdots + c_nx(t_n), \tag{2.3}$$

where the c_r and t_r are real constants, and $t_r \in T$. It can be shown [3, p. 253] that under the conditions (2.1) and (2.2) the Hilbert space $H(x)$ is *separable.*

We further define a family of subspaces $H(x, t)$ of $H(x)$, such that $H(x, t)$ is the closure in norm of all those linear combinations (2.3), where the t_r are restricted by the conditions $t_r \in T$ and $t_r \leqq t$. If the lower extreme A of T is finite, we define $H(x, t)$ for $t \leqq A$ as the space containing only the zero element, i.e., the random variable that is almost everywhere equal to zero.

Obviously the space $H(x, t)$ will never decrease as t increases. Consequently the limiting space $H(x, -\infty)$ will exist, whether A is finite or not, and will be identical with the intersection of all the $H(x, t)$. We then have for $t < u$

6

$$H(x, -\infty) \subset H(x, t) \subset H(x, u) \subset H(x).$$

Using the terminology of Wiener and Masani [18], we may say that $H(x, t)$ represents the *past and present* of the $x(t)$ process, from the point of view of the instant t, while $H(x, -\infty)$ represents the *remote past* of the process. With respect to $H(x, -\infty)$ there are two extreme particular cases that present a special interest.

If $H(x, -\infty) = H(x)$, we may say that already the remote past contains all available information concerning the process. As soon as the remote past is known, the future development of the process can be accurately predicted. In this case we say that the $x(t)$ process is *deterministic.*

The opposite extreme case occurs when $H(x, -\infty)$ contains only the zero element of $H(x)$, i.e., the random variable that is almost everywhere equal to zero. We have already noted above that this case is always present when A is finite. The remote past is in this case useless for prediction purposes. Speaking somewhat loosely, we may say that any piece of information present in the process, at the instant t for example, must have entered as a new impulse at some definite instant in the past. The $x(t)$ process is then said to be *purely nondeterministic* (some authors use the term *linearly regular*).

Any $x(t)$ satisfying (2.1) and (2.2) can be built up in a simple way by components belonging to these two extreme types. In fact, $x(t)$ can be uniquely represented in the form

$$x(t) = u(t) + v(t),$$

where $u(t)$ and $v(t)$ both belong to $H(x, t)$, while $u(t)$ is deterministic and $v(t)$ purely nondeterministic. Moreover, for any s, t the elements $u(s)$ and $v(t)$ are orthogonal or, in probabilistic terminology, the random variables $u(s)$ and $v(t)$ are uncorrelated.

From the point of view of the applications, deterministic processes do not seem to have very great interest, and we shall in the sequel be concerned only with the opposite extreme case of purely nondeterministic processes.

Consider now the never decreasing family $H(x, t)$ of subspaces of $H(x)$, and let P_t denote the projection operator on $H(x)$ with range $H(x, t)$. The family P_t of projections, where $t \in T$, then has the following properties

$$P_t \leqq P_u \qquad \text{for } t < u,$$
$$P_{t-0} = P_t,$$
$$P_A = 0, \qquad P_B = 1,$$

where 0 and 1 denote respectively the zero and the identity operator in $H(x)$. It follows that, in Hilbert space terminology, the P_t form a *resolution of the identity* or a *spectral family of projections.** We note that the P_t are uniquely determined by the given $x(t)$ process.

For any random variable z in $H(x)$, we now define a stochastic process $z(t)$ by writing

$$z(t) = P_t z. \tag{2.4}$$

It then follows from the properties of the P_t family that $z(t)$ will be a *process of orthogonal increments* defined for $t \in T$, and such that

$$E\, dz(t) = 0, \qquad E(dz(t))^2 = dF(t),$$

where $F(t)$ is a never decreasing function that is everywhere continuous to the left.

The subspaces $H(z)$ and $H(z, t)$ are, by definition, spanned by the random variables $z(t)$ in the same way as $H(x)$ and $H(x,t)$ are spanned by the $x(t)$. The space $H(z)$ is known as a *cyclic* subspace of $H(x)$, and consists of all random variables of the form

$$\int_A^B h(u)\, dz(u), \tag{2.5}$$

where $h(u)$ is a nonrandom F-measurable function such that

$$\int_A^B h^2(u)\, dF(u) < \infty,$$

the stochastic integral (2.5) being defined as an integral in q.m. Similarly, $H(z, t)$ is the set of all random variables obtained by replacing in (2.5) the upper limit by t, with a corresponding integrability condition over (A, t).

Consider now the class of all real-valued and never decreasing, not identically constant functions $F(t)$ defined for $t \in T$ and everywhere continuous to the left. We introduce a partial ordering in this

* For the elementary theory of Hilbert space, with applications to random variables, we may refer to [5, Ch. 5]. Those parts of the advanced theory that will be used here are developed in [17, Chs. 5–7], and in [15]. With respect to the condition $P_{t-0} = P_t$, which is usually replaced by $P_{t+0} = P_t$, see a remark in [12, p. 19].

class by saying that F_1 is *superior* to F_2, and writing $F_1 > F_2$, whenever F_2 is absolutely continuous with respect to F_1. If F_1 and F_2 are mutually absolutely continuous, we say that they are *equivalent*. The set of all F equivalent to a given F_0 forms the *equivalence class* of F_0. In the set of all equivalence classes C, a partial ordering is introduced in the obvious way by writing $C_1 > C_2$ when the corresponding relation holds for any $F_1 \in C_1$ and $F_2 \in C_2$. Evidently the relation $C_1 > C_2$ does not exclude that the classes may be identical.

We can now proceed to the formal statement of the representation theorem that will provide the basis of this paper. It has been shown above how a given $x(t)$ process uniquely determines the corresponding family of subspaces $H(x, t)$ and the associated spectral family of projections P_t. According to the theory of self-adjoint linear operators in a separable Hilbert space, the projections P_t, in their turn, uniquely determine a number N, which may be a positive integer, or equal to infinity, such that the following theorem holds.

Theorem 2.1. Let $x(t)$ be a purely nondeterministic stochastic process defined for $t \in T = (A, B)$, and satisfying the conditions (2.1) *and* (2.2). *Then there exists a number N uniquely determined by the $x(t)$ process, which will be called the multiplicity of the process and may have one of the values* $1, 2, \ldots, \infty$. *The random variable $x(t)$ can then be represented by the expansion*

$$x(t) = \sum_{1}^{N} \int_{A}^{t} g_n(t, u)\, dz_n(u), \tag{2.6}$$

which holds for all $t \in T$ and satisfies the following conditions $Q_1, \ldots, Q_4$:

(Q_1) *The $z_n(u)$ are mutually orthogonal processes of orthogonal increments such that*

$$E\, dz_n(u) = 0, \qquad E(dz_n(u))^2 = dF_n(u),$$

where F_n is never decreasing and everywhere continuous to the left.

(Q_2) *The $g_n(t, u)$ are nonrandom functions such that*

$$Ex^2(t) = \sum_{1}^{N} \int_{A}^{t} g_n^2(t, u)\, dF_n(u) < \infty.$$

(Q_3) $F_1 > F_2 > \cdots > F_N$.

9

(Q_4) *For any $t \in T$ the Hilbert space $H(x, t)$ is the vector sum of the mutually orthogonal corresponding subspaces of the z_n processes:*

$$H(x, t) = H(z_1, t) \oplus \cdots \oplus H(z_N, t).$$

The expansion (2.6) *is a canonical representation of $x(t)$ in the sense that no expansion of the same form, satisfying $Q_1, \ldots, Q_4$, exists for any smaller value of N.*

The expansion (2.6) gives a linear representation of $x(t)$ in terms of past and present N-dimensional *innovation elements* $[dz_1(u), \ldots, dz_N(u)]$. The *response function* $g_n(t, u)$ corresponds to the component $dz_n(u)$ of the innovation element. The quantity $g_n(t, t)$ is the corresponding *instantaneous response function.*

It should be observed that, while the multiplicity N is uniquely determined by the $x(t)$ process, the innovation processes $z_n(u)$ and the response functions $g_n(t, u)$ are *not* uniquely determined. It is, e.g., obvious that a positive and continuous factor $h(u)$ may be transferred from $g_n(t, u)$ to $dz_n(u)$ or vice versa. This remark will be used later.

However, the sequence of equivalence classes $C_1, \ldots, C_N$ of the functions $F_1, \ldots, F_N$ appearing in (Q_3) is, in fact, uniquely determined by the $x(t)$ process. Thus any expansion of $x(t)$ in the form (2.6), satisfying $Q_1, \ldots, Q_4$, will have functions $F_1, \ldots, F_N$ belonging to the same equivalence classes.

The representation (2.6) gives immediately an explicit expression for the best *linear least squares prediction* of $x(t)$ in terms of the development of the x process up to the instant $s < t$. In fact, this best prediction is

$$P_s x(t) = \sum_1^N \int_A^s g_n(t, u)\, dz_n(u),$$

the error of prediction being

$$x(t) - P_s x(t) = \sum_1^N \int_s^t g_n(t, u)\, dz_n(u). \tag{2.7}$$

The canonical representation (2.6) was first given by Hida [10] for the case of a normal process, and by Cramér [2, 3, 4] for the general case considered here, and also for finite-dimensional vector processes. Generalizations to higher vector processes have been investigated by Kallianpur and Mandrekar, e.g. in [11].

3. The Multiplicity Problem

Any $x(t)$ process considered in the sequel will be supposed to be defined for $t \in T = (A, B)$, to be purely nondeterministic and satisfy (2.1) and (2.2). Generally this will not be explicitly mentioned, but should always be understood.

For such a process we have a canonical representation of the form (2.6), and a problem that now immediately arises is *how to determine the multiplicity N corresponding to a given* $x(t)$. No complete solution of this *multiplicity problem* is so far known. In this and the two following sections, we propose to discuss the problem and present some preliminary results.

The corresponding problem for a process with discrete time is easily solved. In this case there is always a representation corresponding to (2.6) with $N = 1$, the integral being replaced by a sum of mutually orthogonal terms [2]. However, as soon as we pass to the case of continuous time, the situation is entirely different.

For the important case of a *stationary* process $x(t)$, defined over $T = (-\infty, \infty)$, we still have $N = 1$. In fact, it is well known that such a process admits a representation

$$x(t) = \int_{-\infty}^{t} g(t-u)\, dz(u), \tag{3.1}$$

where $z(u)$ is a process with orthogonal increments such that $E(dz(u))^2 = du$, while $H(x, t) = H(z, t)$. Obviously this is a canonical representation with $N = 1$ and $F(u) = u$.

On the other hand it is known [4, p. 174] that, in the more general class of *harmonizable* processes, it is possible to construct explicit examples of processes with any given finite or infinite multiplicity N, and even with any given sequence of equivalence classes $C_1, \ldots, C_N$ such that $C_1 > \cdots > C_N$.

From the point of view of statistical applications, it seems important to be able to define some fairly general class of $x(t)$ processes having multiplicity $N = 1$, since in this case the expression (2.6) will evidently be easier to deal with than in a case with $N > 1$. In this connection it should be observed that every example so far explicitly given of a process with $N > 1$ contains response functions $g_n(t, u)$

showing an extremely complicated behavior. Thus we might hope that, by imposing some reasonable regularity conditions on the $g_n(t, u)$ and the $z_n(u)$ occurring in (2.6), we should obtain a class of processes with $N = 1$.

This can in fact be done, as will be shown in Section 5. In the present section, we shall prove a lemma on multiplicity that will be used in the sequel.

Suppose that a stochastic process $x(t)$ is given by the expression

$$x(t) = \sum_1^N \int_A^t g_n(t, u)\, dz_n(u), \tag{3.2}$$

where the g_n and the z_n are known. This expression has the same form as (2.6), and we shall want to know if it may serve as a canonical representation of $x(t)$, in the sense of Theorem 2.1. Usually it will then be comparatively easy to find out whether the conditions Q_1, Q_2 and Q_3 of Theorem 2.1 are satisfied or not. Suppose now that these conditions are in fact satisfied. In order that (3.2) should be a canonical representation, it is then necessary and sufficient that the remaining condition Q_4 should also be satisfied. In the lemma given below, this condition will be expressed in a form that, in some cases, permits an easy application.

By (3.2) $x(t)$ is expressed as a sum of N terms, the nth term being an element of the Hilbert space $H(z_n, t)$. It follows that we have for all $t \in T$

$$H(x, t) \subset H(z_1, t) \oplus \cdots \oplus H(z_N, t).$$

In order that the two members of this relation should be identical, as required by the condition Q_4, it is then necessary and sufficient that the space in the second member should not contain any element orthogonal to $H(x, t)$. Now, by the remarks in connection with (2.5), any element y belonging to the vector sum of the mutually orthogonal $H(z_n, t)$ spaces has the form

$$y = \sum_1^N \int_A^t h_n(u)\, dz_n(u),$$

where $A < t < B$. If y is not almost everywhere equal to zero, we have

$$0 < Ey^2 = \sum_1^N \int_A^t h_n^2(u)\, dF_n(u) < \infty. \tag{3.3}$$

Further, if y is orthogonal to $x(s)$ for $A < s \leqq t$,

$$Eyx(s) = \sum_{1}^{N} \int_{A}^{s} h_n(u) g_n(s, u)\, dF_n(u) = 0. \tag{3.4}$$

Now the space $H(x, t)$ is spanned by the $x(s)$ for all s such that $A < s \leqq t$, and so we obtain the following lemma.

Lemma 3.1. Suppose that the expansion (3.2) *is known to satisfy conditions* Q_1, Q_2, *and* Q_3 *of Theorem* 2.1. *Then* (3.2) *is a canonical representation of* $x(t)$ *if and only if, for every* $t \in T$, *it is impossible to find functions* $h_1(u), \ldots, h_N(u)$ *satisfying* (3.3), *and such that* (3.4) *holds for all* s *in* $(A, t]$.

4. Regularity Conditions

Let us consider one single term in the expansion (3.2), say

$$\int_{A}^{t} g(t, u)\, dz(u),$$

where $z(u)$ is a process of orthogonal increments, satisfying the conditions Q_1 and Q_2 of Theorem 2.1. We shall say briefly that this term satisfies the *regularity conditions* R, if

(R_1) $g(t, u)$ and $\dfrac{\partial g(t, u)}{\partial t}$ are bounded and continuous for $u, t \in T$ and $u \leqq t$.

(R_2) $g(t, t) = 1$ for $t \in T$.

(R_3) $F(u)$ as defined by $E(dz(u))^2 = dF(u)$ is absolutely continuous and not identically constant, with a derivative $f(u) = F'(u)$ having at most a finite number of discontinuity points in any finite subinterval of T.

It will be convenient to add a few remarks with respect to the condition R_2. As pointed out in connection with Theorem 2.1, the canonical expansion (2.6) represents $x(t)$ as the accumulated effect of a stream of N-dimensional random impulses or innovations $[dz_1(u), \ldots, dz_N(u)]$ acting throughout the past of the process. The response function $g_n(t, u)$ measures the effect at time t of a unit impulse in the nth component acting at time u, and $g_n(t, t)$ is the corresponding instantaneous response function associated with the instant t. In an applied problem it may often seem natural to assume that we always have $g_n(t, t) > 0$. If this is so, the canonical representation (2.6) may be transformed by writing

$$\overline{g_n}(t, u) = \frac{g_n(t, u)}{g_n(u, u)},$$

$$d\overline{z_n}(u) = g_n(u, u)\, dz_n(u),$$

so that we obtain

$$x(t) = \sum_1^N \int_A^t \overline{g_n}(t, u)\, d\overline{z_n}(u),$$

which is obviously a canonical representation of $x(t)$ with $\overline{g_n}(t, t) = 1$. Thus it will be seen that, if we are willing to assume $g_n(t, t) > 0$ for all n and t, the condition R_2 will imply no further restriction of generality.

5. Classes of Processes with Multiplicity $N = 1$

We now proceed to prove the following theorem.

Theorem 5.1. Let X be the class of all $x(t)$ processes admitting a canonical representation (2.6) *such that each term in the second member satisfies the conditions R of Section* 4. *Then every $x(t) \in X$ has multiplicity $N = 1$.*

Let $x(t)$ be a process belonging to the class X, and suppose that $x(t)$ has multiplicity $N > 1$. We have to show that this is impossible.

By hypothesis, $x(t)$ has a canonical representation with $N > 1$, every term of which satisfies the regularity conditions R. From the regularity condition R_3 and the condition Q_3 of Theorem 2.1, it easily follows that we can find a finite subinterval of T, say $T_1 = (A_1, B_1)$, such that the derivatives $f_1(u)$ and $f_2(u)$ are bounded away from zero for all $u \in T_1$. Take any point t in T_1. It then follows from Lemma 3.1 that our theorem will be proved if we can find functions $h_1(u)$, $\ldots$, $h_N(u)$ satisfying (3.3), and such that (3.4) is satisfied for $A < s \leqq t$.

We now take $h_n(u) = 0$ for all u when $n > 2$, and $h_1(u) = h_2(u) = 0$ for $u \leqq A_1$ and $u \geqq B_1$. The relation (3.4) is then certainly satisfied for $A < s \leqq A_1$, while for $A_1 < s \leqq t$ it becomes

$$\sum_1^2 \int_{A_1}^s h_n(u) g_n(s, u) f_n(u)\, du = 0. \tag{5.1}$$

14

By the regularity conditions R_1 and R_2, this relation may be differentiated with respect to s, and gives

$$\sum_1^2 \left[h_n(s) f_n(s) + \int_{A_1}^{s} h_n(u) f_n(u) \frac{\partial g_n(s, u)}{\partial s} \, du \right] = 0. \tag{5.2}$$

Consider now the two integral equations of Volterra type obtained by equating the nth term of the first member of (5.2) to $(-1)^n$, where $n = 1$ and 2. Each of these can be solved by the classical iteration method, regarding $h_n(s) f_n(s)$ as the unknown function. The uniquely determined solutions are bounded and continuous for $A_1 < s \leqq t$, and are not almost everywhere equal to zero. These solutions satisfy (5.2), and, by integration, we obtain (5.1). Since by hypothesis $f_1(u)$ and $f_2(u)$ are bounded away from zero for $A_1 < u < B_1$, it follows that (3.3) is also satisfied, and the proof is completed.

Consider now the case of an $x(t)$ process given by the expression (3.2) with $N = 1$:

$$x(t) = \int_A^t g(t, u) \, dz(u), \tag{5.3}$$

where the second member is known to satisfy the regularity conditions R. It cannot be directly inferred from the preceding theorem that (5.3) is a canonical representation of $x(t)$ in the sense of Theorem 2.1. In fact, it might be possible that $x(t)$ would have muliplicity $N > 1$, with a canonical representation the terms of which do not satisfy the conditions R.

However, in the case when the lower extreme A of T is finite, we can actually prove that (5.3) is a canonical representation. On the other hand, when $A = -\infty$, we are only able to prove this under an additional condition, as shown by the following theorem.

Theorem 5.2. *Let* $x(t)$ *be given for* $t \in T = (A, B)$ *by* (5.3), *with a second member satisfying the conditions* Q_1 *and* Q_2 *of Theorem* 2.1, *as well as the regularity conditions* R *of Section* 4. *If* A *is finite,* $x(t)$ *has multiplicity* $N = 1$, *and* (5.3) *is a canonical representation. When* $A = -\infty$, *the same conclusion holds under the additional condition that*

$$\int_{-\infty}^{t} \left| \frac{\partial g(t, u)}{\partial t} \right| du < \infty \tag{5.4}$$

for all $t \in T$.

By Lemma 3.1 this theorem will be proved if we can show that, for any $t \in T$, it is impossible to find a function $h(u)$ such that

$$0 < \int_A^t h^2(u)f(u)\,du < \infty \tag{5.5}$$

and

$$\int_A^s h(u)g(s,u)f(u)\,du = 0 \tag{5.6}$$

for all s in $(A, t]$. As in the preceding proof, (5.6) may be differentiated, and gives

$$h(s)f(s) + \int_A^s h(u)f(u)\,\frac{\partial g(s,u)}{\partial s}\,du = 0.$$

This is a homogeneous integral equation of the Volterra type, and it follows from the classical theory that, in the case when A is finite, the only solution is $h(s)f(s) = 0$ for all s in $(A, t]$. If $A = -\infty$, it is easily shown that, under the assumption (5.4), the classical proof will hold without modification also in this case. Obviously the solution $h(u)f(u) = 0$ cannot satisfy (5.5), and so the proof is completed.

REMARK. We observe that Theorem 5.2 does not hold without the regularity conditions R. Consider, in fact, the case when $T = (A, B)$ is a finite interval, and $x(t)$ is given by (5.3), with a $z(u)$ satisfying R_3, and

$$g(t,u) = \frac{2(u-A)}{t-A} - 1.$$

Then R_2 is satisfied, but not R_1, as $\frac{\partial g}{\partial t} = -\frac{2(u-A)}{(t-A)^2}$ is not bounded as $t \to A$ and $A < u < t$. Taking $h(u)f(u) = 1$, and assuming that $f(u)$ is bounded away from zero over (A, B), it will be seen that (5.5) and (5.6) are both satisfied. Thus, in this case, (5.3) is not a canonical representation, and the multiplicity of $x(t)$ remains unknown.

6. Some Properties of Processes with $N = 1$

Let $x(t)$ be a process of multiplicity $N = 1$, with a canonical representation

$$x(t) = \int_A^t g(t,u)\,dz(u) \tag{6.1}$$

satisfying the conditions of Theorem 5.2. We shall discuss some properties of $x(t)$.

UNIQUENESS. It will be shown that, under the conditions stated, the representation (6.1) of $x(t)$ is unique in the sense that, for any two representations of $x(t)$ satisfying the same conditions, and characterized by the subscripts 1 and 2 respectively, we have

$$g_1(t, u) = g_2(t, u),$$

$$z_1(t) - z_1(u) = z_2(t) - z_2(u),$$

for any $t, u \in T$ and $u \leqq t$. We note that the last relation signifies that the random variables in the two members are equivalent, i.e., equal with probability one.

For the proof of this assertion we use the expression (2.7) for the error involved in a least squares prediction. Since this error depends only on the process $x(t)$ itself, it must be the same for both representations, so that we have

$$(6.2) \quad x(t+h) - P_t x(t+h) = z_n(t+h) - z_n(t) - \int_t^{t+h} [1 - g_n(t+h, u)]\, dz_n(u)$$

for $n = 1$ and 2, and for any $h > 0$. Now the variance of $z_n(t+h) - z_n(t)$ is $F_n(t+h) - F_n(t)$ while, by conditions R_1 and R_2, the variance of the last term in (6.2) is for small h small relative to $F_n(t+h) - F_n(t)$. Allowing h to tend to zero it then follows that we have $f_1(t) = f_2(t)$. For any $u < t$ we now divide the interval $[u, t)$ in n subintervals of equal length, and apply (6.2) to each subinterval. Summing over the subintervals and allowing n to tend to ∞, we then obtain by the same argument as before $z_1(t) - z_1(u) = z_2(t) - z_2(u)$. Finally, for $u < u + h < t$ we have by (2.7) for $n = 1$ and 2

$$P_{u+h}x(t) - P_u x(t) = g_n(t, u+h)[z_n(u+h) - z_n(u)] - \int_u^{u+h} [g_n(t, u+h) - g_n(t, v)]\, dz_n(v),$$

and an argument of the same type as above will then show that $g_1(t, u) = g_2(t, u)$.

THE COVARIANCE FUNCTION. Since $Ex(t) = 0$ we have for the covariance function

$$(6.3) \qquad r(s, t) = Ex(s)x(t) = \int_A^{\min(s,t)} g(s, u)g(t, u)f(u)\, du,$$

which shows that $r(s, t)$ is everywhere continuous in the interval $T \times T$ in the (s, t)-plane. Moreover, by the regularity conditions R_1 and R_2, it follows from (6.3) that $r(s, t)$ has partial derivatives $\frac{\partial r}{\partial s}$ and $\frac{\partial r}{\partial t}$, which are continuous at every point with $s \neq t$. On the diagonal $s = t$ these derivatives are discontinuous; we have, e.g.,

$$\lim_{s \uparrow t} \frac{r(s, t) - r(t, t)}{s - t} = \int_A^t g(t, u) \frac{\partial g(t, u)}{\partial t} f(u)\, du + f(t),$$

(6.4)

$$\lim_{s \downarrow t} \frac{r(s, t) - r(t, t)}{s - t} = \int_A^t g(t, u) \frac{\partial g(t, u)}{\partial t} f(u)\, du.$$

Thus at the point $s = t$, there is a jump of the height $f(t)$ in the partial derivatives. We shall see later that this fact has interesting applications in the case of normal processes.

SOME PARTICULAR CASES. In the case when $g(t, u) = p(t)q(u)$ is the product of one function of t and one function of u, it follows from the condition R_2 that $p(t)q(t) = 1$, so that

$$g(t, u) = \frac{p(t)}{p(u)}.$$

For the covariance function we then have

$$r(s, t) = p(s)p(t) \int_A^{\min(s,t)} \frac{f(u)}{p^2(u)}\, du.$$

Hence for $s < t < u$ the covariance function satisfies the relation

$$r(s, u)r(t, t) = r(s, t)r(t, u).$$

For the correlation coefficient

$$\rho(s, t) = \frac{r(s, t)}{[r(s, s)r(t, t)]^{1/2}}$$

the corresponding relation is

$$\rho(s, u) = \rho(s, t)\rho(t, u).$$

This is the characteristic relation of a *Markov process in the wide sense* according to the terminology of Doob [6, p. 233]. If, in addition, $x(t)$ is a normal process, it is even *strict sense Markov*. In the particular case when $A = -\infty$ and

$$p(t) = e^{-ct}, \qquad f(u) = 2c,$$

we have

$$r(s, t) = \rho(s, t) = e^{-c|s-t|},$$

so that $x(t)$ is the well-known stationary Markov process.

When $g(t, u) = 1$ for all t and u, and $z(A) = 0$, (6.1) gives $x(t) = z(t)$, so that $x(t)$ is now a *process of orthogonal increments*. In the special case when $x(t) = z(t)$ is normal, A is finite, and $f(u) = 1$, we have the Wiener (or Brownian movement) process.

Finally, when $g(t, u) = g(t - u)$, $f(u) = 1$, and $A = -\infty$, we have the case of a *stationary process*.

7. Normal Processes, Equivalence

From now on we shall restrict ourselves to *normal* (or Gaussian) processes, i.e., processes such that all their finite-dimensional distributions are normal. As far as statistical applications are concerned, the assumption of normality will often imply an oversimplification. Nevertheless it has been found useful and, in any case, the investigation of normal processes should be regarded as an important starting point for more realistic research work.

Consider a normal process $x(t)$ given by the canonical representation (6.1), and thus having multiplicity $N = 1$. As before, we assume that the second member of (6.1) satisfies the conditions of Theorem 5.2. Any finite number of elements of the Hilbert space $H(x)$ then have a normal joint distribution, and since $H(x) = H(z)$ it follows that the $z(u)$ process is also normal. Conversely, if $z(u)$ is known to be normal, (6.1) shows that $x(t)$ is also normal.

We shall begin by choosing our basic probability space in a way that will be convenient for the sequel. Consider the measurable space $(W, \mathbf{B})$, where W is the space consisting of all finite and real-valued functions $w(\cdot)$ defined on the real line, while $\mathbf{B}$ is the smallest σ-field of sets in W that includes every set formed by all functions $w(\cdot)$ satisfying a finite number of inequalities of the form

$$a_r < w(t_r) \leqq b_r$$

for $r = 1, 2, \ldots, n$.

Let P be a probability measure defined on the sets of $\mathbf{B}$, and consider the probability space $(W, \mathbf{B}, P)$. A point ω of the space W is a function $\omega = w(\cdot)$, and a random variable on $(W, \mathbf{B}, P)$ is a measurable functional of $w(\cdot)$.

Consider the family $x(t) = x(t, \omega)$ of random variables on this probability space defined by the relation

$$x(t, \omega) = x(t, w(\cdot)) = w(t).$$

This $x(t)$ family defines a stochastic process, any finite-dimensional distribution of which, say for the parameter values $t = t_1, t_2, \ldots, t_n$, is given by probabilities of the form

$$P(a_r < x(t_r, \omega) \leqq b_r) = P(a_r < w(t_r) \leqq b_r)$$

for $r = 1, 2, \ldots, n$. If P is such that all these distributions are normal, $x(t)$ is a normal stochastic process, and we say that P is a *normal probability measure* on $(W, \mathbf{B})$.

Any two probability measures P_0 and P_1 on $(W, \mathbf{B})$ are said to be *perpendicular,* if there exists a set of functions $w(\cdot)$ belonging to $\mathbf{B}$, say the set $S \in \mathbf{B}$, such that

$$P_0(S) = 0, \qquad P_1(S) = 1.$$

Thus, in this case, P_0 and P_1 have their total probability masses concentrated on two disjoint sets. It is then possible to test the hypothesis that the $x(t)$ distribution is given by P_0 against the alternative hypothesis P_1 with complete certainty by choosing S as our critical region. In practical applications this should be regarded as a singular limiting case that will hardly ever be more than approximately realized.

On the other hand, P_0 and P_1 are called *equivalent,* and we write $P_0 \sim P_1$, if the two measures are mutually absolutely continuous so that, for any set $S \in \mathbf{B}$, the relations $P_0(S) = 0$ and $P_1(S) = 0$ always imply one another. In this case the so called *Radon-Nikodym derivative*

$$p(\omega) = \frac{dP_1(\omega)}{dP_0(\omega)}$$

exists, and we have for any set $S \in \mathbf{B}$

$$P_1(S) = \int_S p(\omega)\, dP_0(\omega).$$

The derivative $p(\omega)$ corresponds to the *likelihood ratio* in the classical theory of statistical inference, and can be used, in the same way as the likelihood ratio, for the construction of statistical tests by means of the basic *Neyman-Pearson Lemma.*

Hajek [9] and Feldman [7] proved the important theorem that *two normal probability measures are either perpendicular or equivalent.* No intermediate case can occur. A fundamental problem in the theory of normal stochastic processes will thus be to find out when two given normal probability measures are equivalent, and when they are perpendicular. During recent years a large number of works dealing with this problem have been published. We refer, e.g., to papers by Parzen [13, 14], Yaglom [19], and Rozanov [16]. In our subsequent discussion of equivalence problems for normal processes of multiplicity $N = 1$, we shall follow mainly the last-mentioned author.

The finite-dimensional distributions of a normal process $x(t)$ are uniquely determined by the *mean* and *covariance functions*

$$m(t) = Ex(t), \qquad r(s, t) = E(x(s) - m(s))(x(t) - m(t)).$$

If $P(m, r)$ denotes the normal probability measure associated with $m(t)$ and $r(s, t)$, it can be proved without difficulty that we have

$$P(m_0, r_0) \sim P(m_1, r_1)$$

when and only when

$$P(m_0, r_0) \sim P(m_1, r_0) \sim P(m_1, r_1).$$

In order to find conditions for the equivalence of two normal probability measures it is thus sufficient to consider two particular cases, viz.: (a) the case of different means but the same covariance; and (b) the case of the same mean but different covariances. We shall consider these two cases separately.

CASE A. DIFFERENT MEANS, BUT THE SAME COVARIANCE. As we can always add a nonrandom function to the given stochastic $x(t)$ process, we may consider two normal probability measures P_0 and P_1 such that, denoting by E_0 and E_1 the corresponding expectations,

$$E_0x(t) = 0, \qquad E_0x(s)x(t) = r(s, t),$$

$$E_1x(t) = m(t), \qquad E_1(x(s) - m(s))(x(t) - m(t)) = r(s, t).$$

Let $H_0(x)$ be the Hilbert space spanned by the random variables $x(t)$, with the inner product

$$(y, z) = E_0 yz.$$

The mean

$$E_1 y = M(y)$$

is finite for all finite linear combinations y of the $x(t_i)$, and is a linear functional in $H_0(x)$ such that

$$M(x(t)) = m(t).$$

A necessary and sufficient condition for the equivalence of P_0 and P_1 is that $M(y)$ is a continuous linear functional in $H_0(x)$ [16, p. 31].

If this condition is satisfied, it is known from Hilbert space theory that there is a uniquely determined element $y \in H_0(x)$ such that

$$M(v) = (v, y) = E_0 yv \tag{7.1}$$

for all $v \in H_0(x)$. In particular, taking $v = x(t)$, we obtain

$$m(t) = E_0 yx(t). \tag{7.2}$$

Since the variables $x(t)$ span $H_0(x)$, the relation (7.2) implies (7.1), so that (7.2) is a necessary and sufficient equivalence condition.

We now apply this equivalence criterion to the case when P_0 is the probability measure corresponding to the normal $x(t)$ process given by the canonical expression (6.1):

$$x(t) = \int_A^t g(t, u)\, dz(u),$$

while P_1 is associated with the process $m(t) + x(t)$. Since in this case $H_0(x) = H_0(z)$, a necessary and sufficient condition for the equivalence of P_0 and P_1 is that $m(t)$ can be represented in the form

$$m(t) = E_0 yx(t) = \int_A^t h(u)\, g(t, u)\, f(u)\, du \tag{7.3}$$

for some $y \in H_0(z)$ given by

$$y = \int_A^B h(u)\, dz(u). \tag{7.4}$$

Then y is a normally distributed random variable with $E_0 y = 0$ and

22

(7.5) $$E_0 y^2 = \int_A^B h^2(u)\, f(u)\, du < \infty.$$

The set of all functions $m(t)$ such that P_0 and P_1 are equivalent is thus identical with the set of all functions represented by (7.3) for some $y \in H_0(x) = H_0(z)$, i.e., for some $h(u)$ satisfying (7.5). This set of $m(t)$ functions, with an inner product defined by $(m_1, m_2) = (y_1, y_2) = E_0 y_1 y_2$, is known as the *reproducing kernel Hilbert space* of the $x(t)$ process. The usefulness of this concept has been brought out by the works of Parzen quoted above.

The Radon-Nikodym derivative is in this case

$$p(\omega) = \frac{dP_1(\omega)}{dP_0(\omega)} = e^{y - 1/2 E_0 y^2},$$

where y is given by (7.4). The most powerful test of the hypothesis P_0 against the alternative P_1 is obtained by taking as critical region the set of all $\omega = w(\cdot)$ such that $p(\omega) > c$, the constant c being determined by

$$P_0(p(\omega) > c) = P_0(y > {}^1\!/_2 E_0 y^2 + \log c) = \alpha,$$

y being normally distributed with zero mean and variance given by (7.5), while α is the desired level of the test.

As before we assume that the canonical expression (6.1) satisfies the conditions of Theorem 5.2, so that we may differentiate (7.3) and obtain a Volterra integral equation

$$m'(t) = h(t)\, f(t) + \int_A^t h(u)\, f(u) \frac{\partial g(t, u)}{\partial t}\, du.$$

From this equation an explicit expression for $h(t)$ may be obtained by the iteration method.

In the particular case of a *normal Markov process* with $g(t, u) = p(t)/p(u)$ discussed in Section 6, the expression (7.3) for $m(t)$ becomes

$$m(t) = p(t) \int_A^t \frac{h(u)\, f(u)}{p(u)}\, du,$$

which gives

(7.6) $$h(t) = \frac{p(t)}{f(t)} \cdot \frac{d}{dt}\left(\frac{m(t)}{p(t)}\right).$$

It is thus necessary and sufficient for equivalence that the function $h(t)$ given by (7.6) satisfies the convergence condition (7.5).

For $p(t) = 1$ we have $g(t, u) = 1$, and $x(t)$ becomes a normal process with independent increments. Suppose that we observe this in the interval $0 \leqq t \leqq T$. From (7.6) we then obtain $h(t) = m'(t)/f(t)$, and the condition (7.5) now becomes

$$\int_0^T \frac{(m'(t))^2}{f(t)}\, dt < \infty.$$

Taking here $f(t) = 1$, we have the particular case of the Wiener process.

CASE B. THE SAME MEAN, BUT DIFFERENT COVARIANCES. We now consider two probability measures P_0 and P_1, and a normal process $x(t)$ with the canonical representation

$$x(t) = \int_A^t g(t, u)\, dz(u),$$

which, as before, is assumed to satisfy the conditions of Theorem 5.2. We suppose that, according to the probability measure P_i for $i = 0$ and 1,

$$g(t, u) = g_i(t, u),$$

$$E_i x(t) = E_i z(t) = 0,$$

$$E_i x(s)x(t) = r_i(s, t) = \int_A^{\min(s,t)} g_i(s, u) g_i(t, u) f_i(u)\, du.$$

We denote by $H_i(x) = H_i(z)$ the Hilbert space spanned by the random variables $x(t)$, with the inner product

$$(y, z)_i = E_i yz.$$

The direct product space

$$H_0(x) \otimes H_1(x) = H_0(z) \otimes H_1(z)$$

consists of all functions

$$y(\omega_0, \omega_1) = \int_A^B \int_A^B h(u, v)\, dz(u, \omega_0)\, dz(v, \omega_1),$$

where ω_i is a point in the probability space $(W, \mathbf{B}, P_i)$, and $h(u, v)$ satisfies

$$\int_A^B \int_A^B h^2(u, v) f_0(u) f_1(v)\, du\, dv < \infty. \tag{7.7}$$

Then [16, p. 54] *P_0 and P_1 are equivalent if and only if we have, for all s, t and for some $y \in H_0(z) \otimes H_1(z)$,*

$$\begin{aligned} r_1(s, t) - r_0(s, t) &= E_0 E_1 \,[y(\omega_0, \omega_1)\, x(s, \omega_0) x(t, \omega_1)] \\ &= \int_A^s \int_A^t h(u, v) g_0(s, u) g_1(t, v) f_0(u) f_1(v)\, du\, dv. \end{aligned} \tag{7.8}$$

Under the present conditions it follows from (7.8) that the difference $r_1(s, t) - r_0(s, t)$ has everywhere continuous partial derivatives of the first order. Hence the discontinuities of the partial derivatives of r_1 and r_0 at the diagonal $s = t$ must cancel in the difference, so that by (6.4) we obtain as a necessary condition for equivalence that

$$f_1(t) = f_0(t) \tag{7.9}$$

for all t.

In the case when P_0 and P_1 are both associated with normal Markov processes such that $g_i(t, u) = p_i(t)/p_i(u)$, the necessary and sufficient condition for equivalence becomes by (7.8) and (7.9)

$$r_1(s, t) - r_0(s, t) = p_0(s) p_1(t) \int_A^s \int_A^t h(u, v) \frac{f(u) f(v)}{p_0(u) p_1(v)}\, du\, dv,$$

where $f = f_0 = f_1$. This gives

$$h(s, t) = \frac{p_0(s) p_1(t)}{f(s) f(t)} \cdot \frac{\partial^2}{\partial s \partial t} \left(\frac{r_1(s, t) - r_0(s, t)}{p_0(s) p_1(t)} \right). \tag{7.10}$$

Thus P_0 and P_1 are equivalent if and only if $h(s, t)$ as given by (7.10) satisfies the convergence condition (7.7).

Consider the particular case when P_0 is associated with a Wiener process with $f_0(t) = 1$ observed in the interval $0 \leqq t \leqq T$, while P_1 corresponds to a normal Markov process with $g_1(t, u) = p_1(t)/p_1 u)$ observed in the same interval. By (7.9) a necessary condition for equivalence is that we have $f_1(t) = f_0(t) = 1$. If this condition is satisfied, (7.10) gives, after some calculation,

$$h(s, t) = \frac{p_1'(\max(s, t))}{p_1(t)}, \tag{7.11}$$

25

Now it follows from the regularity condition R_1 that $h(s, t)$ as given by (7.11) is bounded for all $s, t \in (0, T)$. Thus the convergence condition (7.7) is satisfied, and the condition $f_1(t) = f_0(t) = 1$ is in this case both necessary and sufficient for equivalence.

In this last case the Radon-Nikodym derivative is

$$p(\omega) = \frac{dP_1(\omega)}{dP_0(\omega)} = Ke^{-1/2y(\omega,\omega)},$$

where ω is a point in the probability space $(W, \mathbf{B}, P_0)$, while

$$y(\omega, \omega) = \int_0^T \int_0^T h(u, v)\, dz(u, \omega)\, dz(v, \omega),$$

$h(u, v)$ is given by (7.11), and K is a constant determined so as to render the integral of $p(\omega)\, dP_0(\omega)$ over the whole space equal to 1.

8: Normal Processes, Estimation

In this final section we shall briefly discuss some problems of statistical estimation for a normal process given by the canonical representation (6.1), subject to the same conditions as before. By (6.4) the diagonal $s = t$ of the (s, t)-plane is a line of discontinuity for the first order partial derivatives of the covariance function $r(s, t)$, the jump at the point $s = t$ being $f(t)$, in the sense specified by (6.4).

Let now $a < b$ be fixed constants, and consider the random variable S_n given by the expression

$$S_n = \sum_{a\cdot 2^n}^{b\cdot 2^n} [x(k/2^n) - x((k-1)/2^n)]^2.$$

For any given n an observed value of S_n can be computed as soon as a realization of the $x(t)$ process is known for $a \leqq t \leqq b$. It follows from a theorem due to Baxter [1] that, as $n \to \infty$,

$$S_n \to \int_a^b f(t)\, dt,$$

where the convergence takes place both in q.m. and with probability one.

Thus from one single observed realization of $x(t)$, we may compute the quadratic variation S_n for some large value of n, and the value obtained will give us an estimate of $\int_a^b f(t)\, dt$ with a variance tending to zero as $n \to \infty$.

In order to find an estimate of a parametric function connected with $g(t, u)$, we take $s < t$ and write for $k = 0, 1, 2, \ldots$

$$s_{k,n} = s - k \cdot 2^{-n}.$$

Consider the linear regression of the random variable $x(t)$ on the variables $x(s_{k,n})$ for $k = 0, 1, \ldots, m$, where $m = n \cdot 2^n$. The regression polynomial, say

$$x_{s,n}(t) = c_{0,n}x(s_{0,n}) + \cdots + c_{m,n}x(s_{m,n})$$

is the projection of $x(t)$ on the finite-dimensional Hilbert space spanned by the random variables $x(s_{0,n}), \ldots, x(s_{m,n})$. It is easily proved that, under the present conditions, $x_{s,n}(t)$ converges in q.m. as $n \to \infty$ to the best linear prediction

$$x_s(t) = P_s x(t)$$

of $x(t)$ in terms of the whole development of the process up to the instant s. Thus the residual

$$x(t) - x_{s,n}(t)$$

converges in q.m. to the prediction error

$$x(t) - P_s x(t)$$

given by (2.7). Hence we obtain from (2.7), as $n \to \infty$,

$$E[x(t) - x_{s,n}(t)]^2 \to \int_s^t g^2(t, u) f(u)\, du. \tag{8.1}$$

For a given large value of n, a sample value of the residual variance in the first member of (8.1) can be computed if a sufficient number of realizations of the $x(t)$ process are available. This will then serve as an estimate of the parametric function in the second member.

REFERENCES

1. Baxter, G., A strong limit theorem for Gaussian processes, *Proc. Am. Math. Soc.,* 7 (1956), 522–527.
2. Cramér, H., On some classes of non-stationary stochastic processes, *Proc. 4th Berkeley Symp. Math. Stat. and Prob.,* II (1961), 57–77.
3. ———, On the structure of purely non-deterministic stochastic processes, *Arkiv Mat.,* 4 (1961), 249–266.
4. ———, Stochastic processes as curves in Hilbert space, *Teoriya Veroj. i ee Primen.,* 9 (1964), 169–179.
5. ———, and Leadbetter, M. R., *Stationary and related stochastic processes,* New York: Wiley, 1967.
6. Doob, J. L., *Stochastic processes,* New York: Wiley, 1953.
7. Feldman, J., Equivalence and perpendicularity of Gaussian processes, *Pacif. J. Math.,* 8 (1958), 699–708.
8. Grenander, U., Stochastic processes and statistical inference, *Arkiv Mat.,* 1 (1950), 195–277.
9. Hajek, J., On a property of normal distributions of any stochastic process, *Czech. Math. J.,* 8 (1958), 610–618, and *Selected Transl. Math. Stat. Prob., 1* (1961), 245-252.
10. Hida, T., Canonical representations of Gaussian processes and their applications, *Mem. Coll. Sc. Kyoto Ser. A,* 32 (1960), 109–155.
11. Kallianpur, G., and Mandrekar, V., Multiplicity and representation theory of purely non-deterministic stochastic processes, *Teorija Veroj. i ee Primen.,* 10 (1965), 553–581.
12. Nagy, B. v. Sz., Spektraldarstellung linearer Transformationen des Hilbertschen Raumes. *Ergeb. Mathematik,* 5, Berlin, 1942.
13. Parzen, E., An approach to time series analysis, *Ann. Math. Stat.,* 32 (1961), 951–989.
14. ———, Statistical inference on time series by RKHS methods, *Techn. Report 14,* Stanford Univ., Statistics Dept., 1970.
15. Plessner, A. I., and Rokhlin, V. A., Spectral theory of linear operators, *Uspekhi Mat. Nauk,* 1 (1946), 71–191 (Russian).

16. Rozanov, Yu. A., Infinite-dimensional Gaussian distributions, *Trudy Ord. Lenina, Mat. Inst. V. A. Steklova,* 108 (1968), 1–136 (Russian).
17. Stone, M. H., Linear transformations in Hilbert space and their applications to analysis, *Am. Math. Soc. Colloquium Publications,* 25, New York, 1932.
18. Wiener, N., and Masani, P., The prediction theory of multivariate stochastic processes, *Acta Math.,* 98 (1957), 111–150, and 99 (1958), 93–137.
19. Yaglom, A. M., On the equivalence and perpendicularity of two Gaussian probability measures in function space, *Proc. Symp. Time Series Analysis,* New York: Wiley, 1963, 327–346.

30

Editors' Comments on Papers 18 and 19

These two papers, by Kadota and Shepp, both with the Bell Telephone Laboratories at Murray Hill, study the problem of nonsingular detection in the context of the equivalence between Gaussian probability measures. The existence of the likelihood ratio, as the Radon–Nikodym derivative of the respective measures, is related to the equivalence of the observation process to its innovation.

These two papers complement each other in many ways and were selected after careful sampling of the rich literature on the subject. The second, in particular, gives necessary and sufficient conditions for a Gaussian measure to be equivalent to a Wiener measure. These conditions are expressed in terms of the autocovariance functions of these measures.

These papers are reprinted with the permission of the authors, of the Director of Editorial Services of the Institute of Electrical and Electronics Engineers for the first paper, and of the Institute of Mathematical Statistics for the second. Dr. Shepp has kindly provided a list of errata for his paper, which has been placed at the end of the paper.

Reprinted from *IEEE Transactions on Information Theory,* **IT-16**, 291–298 (1970)

18

Nonsingular Detection and Likelihood Ratio for Random Signals in White Gaussian Noise

T. T. KADOTA

***Abstract*—This paper is concerned with the mathematical aspect of a detection problem (a random signal in white Gaussian noise). Specifically, we obtain a sufficient condition for nonsingular detection and derive a likelihood-ratio expression in terms of least-mean-square estimates. The problem itself is old, and the likelihood-ratio expression is also well known. The contribution of this paper is a relatively elementary and self-contained derivation of the likelihood-ratio expression as well as the nonsingularity condition.**

I. Introduction

CONSIDER a problem of optimally detecting a random signal in white Gaussian noise. One mathematical treatment of such a problem is to interpret it in terms of the integrated signal and noise. Thus, the signal portion is a time integral of the given signal process, and the noise a standard Wiener process. We regard the detection problem as one of discrimination between the signal-plus-noise and the noise processes. When optimality of this discrimination is defined in the sense of the Neyman–Pearson hypothesis test, the solution of the problem consists of obtaining a sufficient condition for nonsingularity of the two processes and expressing their likelihood ratio in terms of the observable.

It was conjectured that when the signal-plus-noise measure is absolutely continuous with respect to the noise measure, the likelihood ratio takes the same form as the one in the sure signal-in-noise problem, except for the fact that the signal is replaced by its least-mean-square estimate. Inasmuch as our interest is in the mathematical proof, we refer to Kailath [1] for the historical account and physical interpretation of this likelihood-ratio expression. In a special case where the signal is a diffusion process, the conjecture was proved by Duncan [2]. Kailath [1] and Wong [3] gave more general proofs under the condition that the signal has finite average energy and is independent of the noise. Kailath [4] later replaced the independence part of the condition by a weaker condition that the signal-plus-noise and the noise measures are equivalent. In applications, finiteness of average signal energy is not restrictive, but mutual independence between the signal and the noise is a serious restriction. For example, in communication systems with feedback the signal is a function of the past noise; thus, it necessarily depends on the noise. Unfortunately, this dependence makes the two measures no longer equivalent. The purpose of this paper is to establish the likelihood-ratio expression under a much weaker independence condition.

We prove that if the signal has finite energy with probability 1 and if it is independent of future increments of a delayed version of the noise (delay can be arbitrarily small), then the signal-plus-noise measure is absolutely continuous with respect to the noise measure; and if, in addition the expectation of the signal energy is finite, the likelihood ratio is given effectively by the same expression. We remark that in general neither of the first two conditions is necessary for absolute continuity. For example, if the signal is Gaussian, there is Shepp's necessary and sufficient condition [5], which is weaker than our two. In fact, we explicitly show that our first two conditions imply his, and our third coincides with the first in the case of Gaussian signals. In applications, however, it is inconceivable that a signal should have infinite average energy in a finite time interval. Also, dependence of a signal on additive noise is typically through feedback, which necessarily introduces some delay. Thus, though mathematically restrictive, these conditions seem physically acceptable.

Since this paper was submitted, our second condition has been relaxed by eliminating the delay [6]. Furthermore, assuming absolute continuity rather than equivalence, Kailath and Zakai (private communication) have verified

Manuscript received May 21, 1969; revised October 13, 1969.
The author is with Bell Telephone Laboratories, Inc., Murray Hill, N. J. 07971.

the likelihood-ratio expression. We nevertheless feel that a certain unique feature of our proof should justify its presentation. Unlike other proofs mentioned, ours is relatively elementary and self-contained, except for the use of Itô's differential rule [7]. With the independence condition and the condition of finite average signal energy, we calculate two probability densities of n time samples under two hypotheses (the signal-plus-noise and the noise-alone) and form their ratio. By using Itô's differential rule, this ratio is rewritten to conform with the proposed likelihood-ratio expression. We then show that as sampling becomes dense ($n \to \infty$), the logarithm of the ratio converges in the mean to the proposed expression with respect to the signal-plus-noise measure. From this, the absolute continuity is easily deduced, and it follows immediately that the actual likelihood ratio is equal to either the proposed expression or zero. Finally, through a simple contradiction argument, we relax the condition of finite *average* signal energy to that of finite signal energy with probability 1. The exact result is stated as a theorem in the Section II with pertinent remarks. The detailed proof is given in Section III and the Appendix.

II. Theorem and Discussions

Let $w(t)$ be a standard Wiener process and $y(t)$ a measurable and integrable process defined on a probability-measure space $(\Omega, \mathfrak{F}, P)$. Define a process $z(t)$ by

$$z(t) = \int_0^t y(s)\, ds + w(t), \tag{1}$$

and denote $\{z(s), 0 \le s \le t\}$ by z_t, namely, the "whole past" of z up to t. Let $x(t)$ be a measurable process with two associated measures P_0 and P_1, which are defined by

$$\begin{aligned} P_0\{x(t_1) < a_1, \cdots, x(t_n) < a_n\} &= P\{w(t_1) < a_1, \cdots, w(t_n) < a_n\}, \\ P_1\{x(t_1) < a_1, \cdots, x(t_n) < a_n\} &= P\{z(t_1) < a_1, \cdots, z(t_n) < a_n\}, \end{aligned} \tag{2}$$

where $0 \le t_1 < \cdots < t_n \le T$ and $a_1, \cdots, a_n$ and n are arbitrary.

Lemma 1

If

$$E \int_0^T |y(t)|\, dt < \infty,$$

then $E\{y(t) \mid z_t\}$, the conditional expectation of $y(t)$ given z_t, can be defined a.e. t (for almost every $t \in [0, T]$) so that it is a measurable functional $f(t, z_t)$ of z_t and is equal to a measurable process a.e. t.

Theorem

If

$$\int_0^T y^2(t)\, dt < \infty \quad \text{a.s.}$$

(almost surely with respect to P) and if $w(t) - w(s)$ is independent of

$$\{w(t_i) - w(s_i), y(u_i + \epsilon), i = 1, \cdots, n\}$$

for some $\epsilon > 0$ and arbitrary $s_i < t_i \le s < t$, $u_i \le s$ and n, then P_1 is absolutely continuous with respect to P_0. If in addition

$$E \int_0^T y^2(t)\, dt < \infty,$$

then the Radon–Nikodym derivative of P_1 with respect to P_0 can be given by

$$\frac{dP_1}{dP_0} = \begin{cases} \exp\left[\int_0^T f(t, x_t)\, dx(t) - \frac{1}{2}\int_0^T f^2(t, x_t)\, dt\right] & \text{a.s. } [P_1] \\ 0 & \text{otherwise.} \end{cases} \tag{3}$$

Remarks

Measurability and integrability of $y(t)$ are assumed in order to define $z(t)$, but they are obviously implied by $\int_0^T y^2(t) dt < \infty$ a.s., the first condition of the theorem. Processes $w(t)$, $y(t)$, and $z(t)$ are introduced mainly to define the measures P_0 and P_1 and the functional $f(t, z_t)$. The central process is $x(t)$, whose sample function is the "observable" in detection problems.

That $E\{y(t) \mid z_t\}$ can be written as a measurable functional of z_t follows from its measurability with respect to the σ field generated by z_t, one of the defining properties of the conditional expectation. However, its equality a.e. t to a measurable process is not immediate. Without this, the symbol $\int_0^T E\{y(t) \mid z_t\}\, dt$ would be vacuous and the expression for dP_1/dP_0 would become meaningless.

The second condition of the theorem states that an increment of $w(t)$ be independent of all its past increments and all past values of $y(t + \epsilon)$ for some positive ϵ simultaneously. We have referred to this condition as that of the signal being independent of future increments of a delayed version of the noise, since delaying the noise is equivalent to advancing the signal. The third condition $E \int_0^T y^2(t) dt < \infty$ is introduced solely to assure

$$\int_0^T |E\{y(t) \mid z_t\}|^2\, dt < \infty \quad \text{a.s.},$$

which enables us to express dP_1/dP_0 by the well-known formula. Even without this condition, dP_1/dP_0, of course, exists and can be expressed as a limit of a likelihood ratio of finite samples of $x(t)$ as seen in the proof of this theorem, see (4).

In the expression for dP_1/dP_0, $f(t, x_t)$ is a conditional expectation with respect to P_1 only. Thus, $\int_0^T f^2(t, x_t) dt$ is defined a.s. $[P_1]$; hence, the stochastic integral

$$\int_0^T f(t, x_t)\, dx(t)$$

is defined a.s. $[P_1]$ also. From the definition of dP_1/dP_0, however, it is with respect to P_0 that dP_1/dP_0 must be defined a.s., and in general P_0 may be nonzero on a set of P_1 measure zero. Since the equality (3) holds a.s. $[P_1]$ and dP_1/dP_0 must vanish on a set of P_1 measure zero, we have explicitly specified dP_1/dP_0 to be zero on the set where the equality (3) does not hold. It is worth noting that dP_1/dP_0 is expressed as a measurable functional of $x(t)$. In detection problems where $x(t)$, $0 \leq t \leq T$ is the observable and dP_1/dP_0 is the best test statistic, this means that (3) specifies in principle how the observed waveform is to be processed to form the test statistic.

Finally if $y(t)$ is Gaussian, P_0 and P_1 must be either equivalent or singular, and there is the following set of necessary and sufficient conditions for equivalence by Shepp.

1) $\int_0^T E^2 y(t)dt < \infty$ and the covariance of $z(t)$ can be written as $\min(t, s) + \int_0^t \int_0^s K(v, u)du\, dv$, where K is square integrable, namely, $\int_0^T \int_0^T K^2(t, s)ds\, dt < \infty$.
2) -1 is not in the spectrum of the integral operator with the kernel K.

Thus, our conditions for absolute continuity must imply those above. We verify this fact by first establishing

$$\int_0^T y^2(t)\, dt < \infty \quad \text{a.s.} \quad \Rightarrow E \int_0^T y^2(t)\, dt < \infty$$

(Lemma 4), then showing that

$$E \int_0^T y^2(t)\, dt < \infty$$

together with the independence condition (even with $\epsilon = 0$) imply Shepp's two conditions given above.

III. Proofs

Proof of Lemma 1

Let $([0, T], \mathcal{A}, \mu)$ be a Lebesgue measure space and

$$0 = t_0^{(n)} < t_1^{(n)} < \cdots < t_n^{(n)} = T$$

an n partition of $[0, T]$. Let $\mathcal{A}_i$ be the Lebesgue field of sets in $(t_{i-1}^{(n)}, t_i^{(n)}]$ and $\mathcal{B}_i$ the σ field generated by $z(t_1^{(n)}), \cdots, z(t_i^{(n)})$. Then, define a measure $\mathcal{Q}_i$ on $\mathcal{A}_i \times \mathcal{B}_i$ by

$$\mathcal{Q}_i(A_i \times B_i) = \int_{A_i \times B_i} y(t, \omega)\, d(\mu \times P),$$

where $A_i \times B_i$ is a rectangle in $\mathcal{A}_i \times \mathcal{B}_i$. Obviously, $\mathcal{Q}_i$ is totally finite and absolutely continuous with respect to $\mu \times P$. Hence, according to the Radon–Nikodym theorem [8, p. 128], there exists an $\mathcal{A}_i \times \mathcal{B}_i$ measurable and $\mu \times P$ integrable function $h_i^{(n)}(t, \omega)$ such that

$$\mathcal{Q}_i(A_i \times B_i) = \int_{A_i \times B_i} h_i^{(n)}(t, \omega)\, d(\mu \times P).$$

Then, from Fubini's theorem [8, p. 148],

$$\int_{A_i \times B_i} h_i^{(n)}(t, \omega)\, d(\mu \times P) = \int_{A_i} \int_{B_i} h_i^{(n)}(t, \omega)\, dP\, d\mu.$$

Hence

$$\int_{B_i} h_i^{(n)}(t, \omega)\, dP = \int_{B_i} y(t, \omega)\, dP \quad \text{a.e.} \quad t \in (t_{i-1}^{(n)}, t_i^{(n)}].$$

Thus, by definition, a conditional expectation of $y(t)$ relative to $\mathcal{B}_i$, $E\{y(t) \mid \mathcal{B}_i\}$, is

$$h_i^{(n)}(t, \omega) \quad \text{a.e.} \quad t \in (t_{i-1}^{(n)}, t_i^{(n)}].$$

Next, define

$$\left.\begin{aligned} h_t^{(n)}(t, \omega) &= h_i^{(n)}(t, \omega) \\ \mathcal{B}_t^{(n)} &= \mathcal{B}_i \end{aligned}\right\} \quad t \in (t_{i-1}^{(n)}, t_i^{(n)}], \quad i = 1, \cdots, n,$$

and

$$E\{y(t) \mid \mathcal{B}_t^{(n)}\} = f_t^{(n)}(t, \omega) \quad \text{a.e.} \quad t.$$

Note $f_t^{(n)}(t, \omega)$ is measurable with respect to $\mathcal{A} \times \mathcal{F}$. Now, from a martingale convergence theorem [9, p. 319] and continuity of $z(t)$,

$$\lim_{n \to \infty} E\{y(t) \mid \mathcal{B}_t^{(n)}\} = E\{y(t) \mid \mathcal{B}_t\} \quad \text{a.e.} \quad t,$$

where $\mathcal{B}_t$ is the σ field generated by z_t. Thus, $E\{y(t) \mid \mathcal{B}_t\}$, which is defined a.e. t, is equal to $\lim n \to \infty$ $f_t^{(n)}(t, \omega)$ a.e. ω. Hence $E\{y(t) \mid \mathcal{B}_t\}$ is equal to a $\mathcal{A} \times \mathcal{F}$ measurable function a.e. (t, ω).

Proof of Theorem

1) Assume $E \int_0^T y^2(t)dt < \infty$ and the independence condition of the theorem. Let

$$0 = t_0^{(n)} < t_1^{(n)} < \cdots < t_{n-1}^{(n)} < t_n^{(n)} = T$$

be a partition with

$$\max_{1 \leq i \leq n} (t_i^{(n)} - t_{i-1}^{(n)}) \leq \epsilon.$$

For convenience, we henceforth omit the superscript n. Put

$$\Delta t_i = t_i - t_{i-1}, x_i = x(t_i),$$

and

$$\Delta Y_i = \int_{t_{i-1}}^{t_i} y(s)\, ds, \qquad i = 1, \cdots, n.$$

Then, with the use of

$$P(\cdot \mid x_1, \cdots, x_n) = P(\cdot \mid x_1 - x_0, \cdots, x_n - x_{n-1}),$$

$$\begin{aligned} \Delta\, dP(x_1 - x_0 < \xi_1, \cdots, x_n - x_{n-1} < \xi_n) \\ = dP(x_1 - x_0 < \xi_1)\, dP(x_2 - x_1 < \xi_2 \mid x_1 - x_0) \cdots \\ \cdot dP(x_n - x_{n-1} < \xi_n \mid x_{n-1} - x_{n-2}, \cdots, x_1 - x_0), \end{aligned}$$

where $\xi_1, \cdots, \xi_n$ are real numbers, $x_0 = 0$ a.s. $[P_0, P_1]$, and $P(x_i - x_{i-1} < \xi_i \mid x_{i-1} - x_{i-2}, \cdots, x_1 - x_0)$ is a conditional probability measure of $x_i - x_{i-1} < \xi_i$ given $x_{i-1} - x_{i-2}, \cdots, x_1 - x_0$, two probability densities p_0 and p_1 of $x_1, \cdots, x_n$ are obtained as follows.

$$p_0(\xi_1, \cdots, \xi_n) = \prod_{i=1}^{n} (2\pi\Delta t_i)^{-1/2} \exp\left[-(\xi_i - \xi_{i-1})^2/2\Delta t_i\right],$$

$$p_1(\xi_1, \cdots, \xi_n) = \prod_{i=1}^{n} (2\pi\Delta t_i)^{-1/2} \int_{-\infty}^{\infty} \exp\left[-(\xi_i - \xi_{i-1} - \eta)^2/2\Delta t_i\right] \cdot dP(\Delta Y_i < \eta \mid z(t_1) = \xi_1, \cdots, z(t_{i-1}) = \xi_{i-1}).$$

Thus, the likelihood ratio of $x_1, \cdots, x_n$ is given by

$$\frac{p_1(x_1, \cdots, x_n)}{p_0(x_1, \cdots, x_n)} = \prod_{i=1}^{n} \int_{-\infty}^{\infty} \exp\left[\frac{\eta}{\Delta t_i}(x_i - x_{i-1}) - \frac{1}{2}\left(\frac{\eta}{\Delta t_i}\right)^2 \Delta t_i\right] \cdot dP_1(\Delta Y_i < \eta \mid x_1, \cdots, x_{i-1}), \tag{4}$$

where

$$P_1(\cdot \mid x_1, \cdots, x_{i-1}) \equiv P(\cdot \mid z(t_1) = x_1, \cdots, z(t_{i-1}) = x_{i-1}).$$

Now put

$$\zeta_i(\xi, t, \eta) = \frac{\eta}{\Delta t_i}(\xi - x_{i-1}) - \frac{1}{2}\left(\frac{\eta}{\Delta t_i}\right)^2 (t - t_{i-1}) \qquad t_{i-1} \le t \le t_i,$$

$$\psi_i(\xi, t) = \int_{-\infty}^{\infty} \exp\left[\zeta_i(\xi, t, \eta)\right] dP_1(\Delta Y_i < \eta \mid x_1, \cdots, x_{i-1}),$$

$$g_i(\xi, t) = \log \psi_i(\xi, t).$$

Then

$$g_i(x_{i-1}, t_{i-1}) = 0,$$

$$\frac{\partial}{\partial \xi} g_i(\xi, t) = \frac{1}{\psi_i}\int_{-\infty}^{\infty} \frac{\eta}{\Delta t_i} \cdot \exp(\zeta_i)\, dP_1(\Delta Y_i < \eta \mid x_1, \cdots, x_{i-1}),$$

$$\frac{\partial^2}{\partial \xi^2} g_i(\xi, t) = \frac{1}{\psi_i}\int_{-\infty}^{\infty} \left(\frac{\eta}{\Delta t_i}\right)^2 \cdot \exp(\zeta_i)\, dP_1(\Delta Y_i < \eta \mid x_1, \cdots, x_{i-1}) - \left(\frac{\partial}{\partial \xi} g_i\right)^2,$$

$$\frac{\partial}{\partial t} g_i(\xi, t) = -\frac{1}{2}\left[\frac{\partial^2}{\partial \xi^2} g_i + \left(\frac{\partial}{\partial \xi} g_i\right)^2\right], \tag{5}$$

where we have differentiated under the integral sign. This is valid since the derivatives are integrable with respect to $dP_1(\Delta Y_i < \eta \mid x_1, \cdots, x_{i-1})dt$. Note these derivatives are continuous in (ξ, t). Hence, according to Itô's differential rule [9, p. 194],

$$g_i(x_i, t) = \int_{t_{i-1}}^{t_i} \frac{\partial}{\partial \xi} g_i(\xi, t)\Big|_{\xi = x(t)} dx(t) - \int_{t_{i-1}}^{t_i} \left(\frac{1}{2}\frac{\partial^2}{\partial \xi^2} + \frac{\partial}{\partial t}\right) g_i(\xi, t)\Big|_{\xi = x(t)} dt,$$

where $dx(t) = y(t)dt + dw(t)$ with respect to P_1. Then, by substituting (5) into (4) and using an identity in the Appendix (Lemma 2, corollary):

$$E_1\{\Delta Y_i \mid x_1, \cdots, x_{i-1}, x(t)\} = \frac{1}{\psi_i}\int_{-\infty}^{\infty} \eta \exp\left[\zeta_i(x(t), t, \eta)\right] \cdot dP_1(\Delta Y_i < \eta \mid x_1, \cdots, x_{i-1}),$$

where

$$E_1\{\cdot \mid x_1, \cdots, x_{i-1}, x(t)\} \equiv E\{\cdot \mid z(t_1) = x_1, \cdots, z(t_{i-1}) = x_{i-1}, z(t) = x(t)\},$$

we obtain

$$\log \frac{p_1(x_1, \cdots, x_n)}{p_0(x_1, \cdots, x_n)} = \sum_{i=1}^{n}\left[\int_{t_{i-1}}^{t_i} E_1\left\{\frac{\Delta Y_i}{\Delta t_i} \,\Big|\, x_1, \cdots, x_{i-1}, x(t)\right\} dx(t) - \frac{1}{2}\int_{t_{i-1}}^{t_i} E_1^2\left\{\frac{\Delta Y_i}{\Delta t_i} \,\Big|\, x_1, \cdots, x_{i-1}, x(t)\right\} dt\right].$$

Now for any t, $0 \le t \le T$, define $j = j(t)$ so that $t_{j-1} < t \le t_j$, and define $y_n(t) = \Delta Y_j/\Delta t_j$ where the subscript n signifies n partition of $[0, T]$. Then

$$\log \frac{p_1(x_1, \cdots, x_n)}{p_0(x_1, \cdots, x_n)} = \int_0^T E_1\{y_n(t) \mid x_1, \cdots, x_{j-1}, x(t)\}\, dx(t) - \frac{1}{2}\int_0^T E_1^2\{y_n(t) \mid x_1, \cdots, x_{j-1}, x(t)\}\, dt. \tag{6}$$

2) Next, note $\int_0^T E^2\{y(t) \mid z_t\} dt < \infty$ a.s. since $\{E\{y(t) \mid z_t\}, 0 \le t \le T\}$ is equal to a measurable process for a.e. t according to Lemma 1 and

$$\int_0^T EE^2\{y(t) \mid z_t\}\, dt \le \int_0^T EE\{y^2(t) \mid z_t\}\, dt = E\int_0^T y^2(t)\, dt < \infty$$

where Jensen's inequality has been used. Now denote by $\mathcal{B}_t^{(n)}$ the σ field generated by $z(t_1^{(n)}), \cdots, z(t_{j-1}^{(n)}), z(t)$, and consider

$$E\int_0^T |E\{y(t) \mid z_t\} - E\{y_n(t) \mid \mathcal{B}_t^{(n)}\}|^2\, dt \le \int_0^T E\,|E\{y(t) \mid z_t\} - E\{y(t) \mid \mathcal{B}_t^{(n)}\}|^2\, dt + \int_0^T E\,|E\{y(t) - y_n(t) \mid \mathcal{B}_t^{(n)}\}|^2\, dt.$$

From a martingale convergence theorem and continuity of $z(t)$, the integrand of the first integral on the right-hand side vanishes a.e. t as $n \to \infty$, in such a way that

$$\max_{1 \le i \le n} (t_i^{(n)} - t_{i-1}^{(n)}) \to 0.$$

It is also bounded by $4Ey^2(t)$, which is integrable. Hence, from the dominated convergence theorem [8, p. 110], the first integral vanishes as $n \to \infty$. The second integral is bounded by $E \int_0^T |y(t) - y_n(t)|^2 dt$, which vanishes as $n \to \infty$ according to Lemma 3 (corollary) in the Appendix. Hence, we have

$$\lim_{n\to\infty} E \int_0^T |E\{y(t) \mid z_t\} - E\{y_n(t) \mid \mathcal{B}_t^{(n)}\}|^2 \, dt = 0. \tag{7}$$

Now

$$\begin{aligned} E_1 \Bigg| \int_0^T & E_1\{y(t) \mid x_t\} \, dx(t) \\ & - \int_0^T E_1\{y_n(t) \mid x_1, \cdots, x_{j-1}, x(t)\} \, dx(t) \Bigg|^2 \\ \le E \Bigg| & \int_0^T [E\{y(t) \mid z_t\} - E\{y_n(t) \mid \mathcal{B}_t^{(n)}\}] y(t) \, dt \Bigg|^2 \\ & + E \Bigg| \int_0^T [E\{y(t) \mid z_t\} - E\{y_n(t) \mid \mathcal{B}_t^{(n)}\}] \, dw(t) \Bigg|^2 \\ \le E & \int_0^T y^2(t) \, dt E \int_0^T |E\{y(t) \mid z_t\} - E\{y_n(t) \mid \mathcal{B}_t^{(n)}\}|^2 \, dt \\ & + E \int_0^T |E\{y(t) \mid z_t\} - E\{y_n(t) \mid \mathcal{B}_t^{(n)}\}|^2 \, dt \end{aligned}$$

which vanishes as $n \to \infty$ because of (7), and

$$\begin{aligned} E_1 \Bigg| \int_0^T & E_1^2\{y(t) \mid x_t\} \, dt \\ & - \int_0^T E_1^2\{y_n(t) \mid x_1, \cdots, x_{j-1}, x(t)\} \, dt \Bigg| \\ & \le E \int_0^T |E^2\{y(t) \mid z_t\} - E^2\{y_n(t) \mid \mathcal{B}_t^{(n)}\}| \, dt \end{aligned}$$

which also vanishes as $n \to \infty$ by virtue of (7). Therefore,

$$\begin{aligned} \lim_{n\to\infty} E_1 \Bigg| & \int_0^T E_1\{y(t) \mid z_t\} \, dx(t) \\ & - \frac{1}{2} \int_0^T E_1^2\{y(t) \mid z_t\} \, dt - \log \frac{p_1(x_1, \cdots, x_n)}{p_0(x_1, \cdots, x_n)} \Bigg| = 0. \end{aligned} \tag{8}$$

3) Let $\mathcal{Q}$ be a probability measure such that $P_0 < \mathcal{Q}$ (P_0 is absolutely continuous with respect to $\mathcal{Q}$) and $P_1 < \mathcal{Q}$. Denote by $\mathfrak{F}_n$ the σ field generated by $x(t_1^{(n)}), \cdots, x(t_n^{(n)})$ and by $P_0^{(n)}$ and $P_1^{(n)}$ the restrictions, respectively, of P_0 and P_1 on $\mathfrak{F}_n$. Then $dP_0/d\mathcal{Q}$, $dP_1/d\mathcal{Q}$, $dP_0^{(n)}/d\mathcal{Q}$, and $dP_1^{(n)}/d\mathcal{Q}$ exist, and $\lim dP_0^{(n)}/d\mathcal{Q} = dP_0/d\mathcal{Q}$ and $\lim dP_1^{(n)}/d\mathcal{Q} = dP_1/d\mathcal{Q}$ a.s. $[\mathcal{Q}]$. Thus

$$\begin{aligned} \lim_{n\to\infty} & \log (dP_1^{(n)}/d\mathcal{Q})/(dP_0^{(n)}/d\mathcal{Q}) \\ & = \log (dP_1/d\mathcal{Q})/(dP_0/d\mathcal{Q}) \quad \text{a.s.} \quad [\mathcal{Q}]. \end{aligned}$$

Now

$$\frac{dP_1^{(n)}/d\mathcal{Q}}{dP_0^{(n)}/d\mathcal{Q}} = \frac{dP_1^{(n)}}{dP_0^{(n)}} = \frac{p_1(x_1, \cdots, x_n)}{p_0(x_1, \cdots, x_n)} \quad \text{a.s.} \quad [\mathcal{Q}],$$

and from (8)

$$\begin{aligned} \infty > \lim_{n\to\infty} E_1 & \left| \log \frac{p_1(x_1, \cdots, x_n)}{p_0(x_1, \cdots, x_n)} \right| \\ & = \lim_{n\to\infty} \int_\Omega \left| \log \frac{dP_1^{(n)}/d\mathcal{Q}}{dP_0^{(n)}/d\mathcal{Q}} \right| \frac{dP_1}{d\mathcal{Q}} \, d\mathcal{Q}. \end{aligned}$$

Hence, from Fatou's lemma [8, p. 113],

$$\int_\Omega \left| \log \frac{dP_1/d\mathcal{Q}}{dP_0/d\mathcal{Q}} \right| \frac{dP_1}{d\mathcal{Q}} \, d\mathcal{Q} < \infty,$$

which implies that the integrand is finite a.s. $[\mathcal{Q}]$. Therefore, $dP_1/d\mathcal{Q} = 0$ whenever $dP_0/d\mathcal{Q} = 0$, implying $P_1 < P_0$. Thus,

$$dP_1/dP_0 = \lim p_1(x_1, \cdots, x_n)/p_0(x_1, \cdots, x_n) \quad \text{a.s.} \ [P_0],$$

hence, a.s. $[P_1]$ also. On the other hand, (8) implies

$$\begin{aligned} \lim_{k\to\infty} & \frac{p_1(x_1, \cdots, x_{n_k})}{p_0(x_1, \cdots, x_{n_k})} \\ & = \exp \left[\int_0^T E_1\{y(t) \mid x_t\} \, dx(t) - \frac{1}{2} \int_0^T E_1^2\{y(t) \mid x_t\} \, dt \right] \\ & \qquad \text{a.s.} \quad [P_1] \end{aligned}$$

for some subsequence $\{n_k\}$ of $\{n\}$. Hence (3) holds a.s. $[P_1]$.

4) Assume only $\int_0^T y^2(t) dt < \infty$ a.s., and define

$$y_M(t) = \begin{cases} y(t) & \int_0^t y^2(s) \, ds < M, \\ 0 & \text{otherwise.} \end{cases}$$

Let P_{1M} be another measure associated with the process $x(t)$, which is defined by $z_M(t) = \int_0^T y_M(s) ds + w(t)$ as P_1 is defined by $z(t)$ through (2). Then, according to what we have just established, $P_{1M} < P_0$ since

$$E \int_0^T y_M^2(t) \, dt < M < \infty.$$

Suppose $P_1 < P_0$ does not hold, namely, there exists a set A such that $P_0(A) = 0$, but $P_1(A) > a$ for some $a > 0$. Then, for a sufficiently large M, $P_0(A) = 0$, but $P_{1M}(A) > a_1$ for some $a_1 > 0$, since $\lim z_M(t) = z(t)$; thus, $\lim P_{1M}(A) = P_1(A)$. Hence, $P_{1M} < P_0$ does not hold, which is a contradiction. Therefore, $P_1 < P_0$ under $\int_0^T y^2(t) dt < \infty$ a.s. if $P_1 < P_0$ under $E \int_0^T y^2(t) dt < \infty$.

Compatibility With Shepp's Conditions

We first establish in the Appendix that $\int_0^T y^2(t) dt < \infty$ implies $E \int_0^T y^2(t) dt < \infty$ Δ (Lemma 4), and that under the second condition of the theorem $y(t)$ has an orthogonal decomposition;

$$y(t) = \tilde{y}(t) + \int_0^T L(t, s) \, dw(s),$$

where $\tilde{y}$ is a Gaussian process independent of the process $w(t)$ and $L(t, s)$ is a square-integrable Volterra kernel (Lemma 5). Then, from (1), the covariance of $z(t)$ becomes

$$E\left\{\int_0^t \left[\tilde{y}(u) - E\tilde{y}(u) + \int_0^T L(u, \tau)\, dw(\tau)\right] du + w(t)\right\}$$
$$\cdot\left\{\int_0^s \left[\tilde{y}(v) - E\tilde{y}(v) + \int_0^T L(v, \tau)\, dw(\tau)\right] dv + w(s)\right\}$$
$$= \int_0^t \int_0^s \left[\tilde{R}(u, v) + \int_0^T L(u, \tau)L(v, \tau)\, d\tau + L(u, v) + L(v, u)\right] du\, dv + \min(t, s),$$

where $\tilde{R}(t, s) = E[\tilde{y}(t) - E\tilde{y}(t)][\tilde{y}(s) - E\tilde{y}(s)]$. Note that

$$E\int_0^T\int_0^T L(t, u)L(s, v)\, dw(u)\, dw(v) = \int_0^T L(t, u)L(s, u)\, du$$
$$E\int_0^T L(t, u)\, dw(u)w(s) = \int_0^s L(t, u)\, du,$$

which follow from the definition of Wiener integrals ([9], pp. 426–433).

Now, the first half of 1) follows trivially from

$$\int_0^T y^2(t)\, dt < \infty \quad \text{a.s.}$$

and Lemma 4 through Schwarz's inequality. Namely,

$$\int_0^T E^2y(t)\, dt \le \int_0^T Ey^2(t)\, dt < \infty.$$

The second half follows by observing that

$$K(t, s) = \tilde{R}(t, s) + L(t, s) + L(s, t) + \int_0^T L(t, u)L(s, u)\, du,$$

where $L(t, s)$ and $\int_0^T L(t, u)L(s, u)du$ are square integrable because of Lemma 5, as is $\tilde{R}(t, s)$ since

$$\int_0^T\int_0^T E^2[\tilde{y}(t) - E\tilde{y}(t)][\tilde{y}(s) - E\tilde{y}(s)]\, ds\, dt \le \left(\int_0^T E\tilde{y}^2(t)\, dt\right)^2 \le \left(\int_0^T Ey^2(t)\, dt\right)^2.$$

To prove 2), we note

$$I + K = \tilde{R} + (I + L)(I + L^*),$$

where I is an identity operator, $\tilde{R}$ and L are integral operators with the kernels $\tilde{R}(t, s)$ and $L(t, s)$, and L^* is the adjoint of L. Observe that both $\tilde{R}$ and $(I + L)(I + L^*)$ are nonnegative definite, and furthermore $I + L$ has a bounded inverse because $L(t, s)$ is a square-integrable Volterra kernel [10, p. 34]. Hence, the spectrum of $I + K$ is bounded away from zero, which implies 2).

Appendix

Lemma 2

Let $y, z_1, z_2, \cdots$ be real random variables, and $p(\cdot \mid y)$ and $p(\cdot \mid y, z_1, \cdots, z_n)$ be conditional density functions of z_1 given y and z_{n+1} given $y, z_1, \cdots, z_n$, respectively. Assume both $p(\xi \mid y)$ and $p(\xi \mid y, z_1, \cdots, z_n)$ are measurable functions of $(\xi, y, z_1, \cdots, z_n)$. If $f(\cdot)$ is a Baire function with $E|f(y)| < \infty$, then

$$E\{f(y) \mid z_1, \cdots, z_{n+1}\} = \frac{\int_{-\infty}^{\infty} f(\eta)p(z_{n+1} \mid y = \eta, z_1, \cdots, z_n)\, dP(y < \eta \mid z_1, \cdots, z_n)}{\int_{-\infty}^{\infty} p(z_{n+1} \mid y = \eta, z_1, \cdots, z_n)\, dP(y < \eta \mid z_1, \cdots, z_n)}. \tag{9}$$

Proof: Let A and B_i, $i = 1, 2, \cdots$, be the set defined by

$$A = \{y < \alpha\},$$
$$B_i = \{z_i < \beta_i\}.$$

Then, from the definition of a conditional density,

$$P(AB_1) = \int_A \int_{-\infty}^{\beta_1} p(\xi \mid y)\, d\xi\, dP = \int_{-\infty}^{\beta_1}\int_{-\infty}^{\alpha} p(\xi \mid y = \eta)\, dP(y < \eta)\, d\xi,$$

where measurability of $p(\xi \mid y)$ is used to interchange the order of integrations. Hence, the density $p(\xi)$ of z_1 exists and

$$p(\xi) = \int_{-\infty}^{\infty} p(\xi \mid y = \eta)\, dP(y < \eta).$$

Suppose the joint density $p(\xi_1, \cdots, \xi_n)$ of $z_1, \cdots, z_n$ exists. Then, again from the definition of conditional probability and density,

$$P(AB_1 \cdots B_n) = \int_{-\infty}^{\beta_n} \cdots \int_{-\infty}^{\beta_1}\int_{-\infty}^{\alpha} dP(y < \eta \mid z_1 = \xi_1, \cdots, z_n = \xi_n) \cdot p(\xi_1, \cdots, \xi_n)\, d\xi_1 \cdots d\xi_n, \tag{10}$$

$$P(AB_1 \cdots B_{n+1}) = \int_{AB_1\cdots B_n}\int_{-\infty}^{\beta_{n+1}} p(\xi \mid y, z_1, \cdots, z_n)\, d\xi\, dP$$
$$= \int_{-\infty}^{\beta_{n+1}} \cdots \int_{-\infty}^{\beta_1}\int_{-\infty}^{\alpha} p(\xi_{n+1} \mid y = \eta, z_1 = \xi_1, \cdots, z_n = \xi_n) \cdot dP(y < \eta \mid z_1 = \xi_1, \cdots, z_n = \xi_n) \cdot p(\xi_1, \cdots, \xi_n)\, d\xi_1 \cdots d\xi_{n+1}, \tag{11}$$

where (10) is used in (11) and measurability of the conditional density is again used to interchange the order of integrations. Thus, the density of $z_1, \cdots, z_{n+1}$ also exists and the conditional density of z_{n+1} given $z_1, \cdots, z_n$ is

$$p(\xi \mid z_1, \cdots, z_n) = \int_{-\infty}^{\infty} p(\xi \mid y = \eta, z_1, \cdots, z_n)\, dP(y < \eta \mid z_1, \cdots, z_n). \tag{12}$$

Hence, through the usual induction argument, the density of $z_1, \cdots, z_n$ exists for all n and (12) is valid for all n. Then, by using (12) and the equality

$$p(\xi_1, \cdots, \xi_{n+1}) = p(\xi_{n+1} \mid z_1 = \xi_1, \cdots, z_n = \xi_n)p(\xi_1, \cdots, \xi_n)$$

and (11), the integral of the right-hand side of (9) on $B_1 \cdots B_{n+1}$ becomes

$$\int_{-\infty}^{\beta_{n+1}} \cdots \int_{-\infty}^{\beta_1} \int_{-\infty}^{\infty} f(\eta) p(\xi_{n+1} \mid y = \eta, z_1 = \xi_1, \cdots, z_n = \xi_n) \cdot dP(y < \eta \mid z_1 = \xi_1, \cdots, z_n = \xi_n) \cdot p(\xi_1, \cdots, \xi_n)\, d\xi_1 \cdots d\xi_{n+1} = \int_{B_1 \cdots B_n} f(y)\, dP.$$

This proves the lemma.

Corollary

Let $z_n = y_n + w_n$, $n = 1, 2, \cdots$, where w_n is a zero-mean Gaussian variable with variance τ_n. Assume w_n is independent of $w_1, \cdots, w_{n-1}, y_1, \cdots, y_n$. Then

$$\begin{aligned} p(\xi \mid y_n, z_1, \cdots, z_{n-1})\, d\xi &= dP(y_n + w_n < \xi \mid y_n, z_1, \cdots, z_{n-1}) \\ &= dP(w_n < \xi - y_n \mid y_n, z_1, \cdots, z_{n-1}) \\ &= dP(w_n < \xi - y_n) = (2\pi\tau_n)^{1/2} \exp[-(\xi - y_n)^2/2\tau_n]\, d\xi. \end{aligned}$$

Hence, from (9),

$$E\{f(y_n) \mid z_1, \cdots, z_n\} = \frac{\int_{-\infty}^{\infty} f(\eta) \exp\left[\frac{\eta_n}{\tau_n} z_n - \frac{1}{2}\left(\frac{\eta_n}{\tau_n}\right)^2 \tau_n\right] dP(y_n < \eta \mid z_1, \cdots, z_{n-1})}{\int_{-\infty}^{\infty} \exp\left[\frac{\eta_n}{\tau_n} z_n - \frac{1}{2}\left(\frac{\eta_n}{\tau_n}\right)^2 \tau_n\right] dP(y_n < \eta \mid z_1, \cdots, z_{n-1})}.$$

Lemma 3

Define an operator I_n on $\mathfrak{L}_2[0, T]$, the space of square-integrable functions on $[0, T]$, by

$$(I_n f)(t) = \frac{1}{t_j^{(n)} - t_{j-1}^{(n)}} \int_{t_{j-1}^{(n)}}^{t_j^{(n)}} f(s)\, ds$$

where $t_j^{(n)}$ and $t_{j-1}^{(n)}$ are two neighboring n-partition elements of t. Then

$$||I_n f|| \leq ||f||$$

$$\lim_{n\to\infty} ||f - I_n f|| = 0.$$

Proof:

$$\begin{aligned} ||I_n f||^2 &= \int_0^T \left[\frac{1}{t_j - t_{j-1}} \int_{t_{j-1}}^{t_j} f(s)\, ds\right]^2 dt \\ &= \sum_{i=1}^{n} \frac{1}{t_i - t_{i-1}} \left(\int_{t_{i-1}}^{t_i} f(s)\, ds\right)^2 \leq \sum_{i=1}^{n} \int_{t_{i-1}}^{t_i} f^2(s)\, ds \\ &= ||f||^2, \end{aligned}$$

where Jensen's inequality is used. If $g(t)$ is continuous,

$$\lim_{n\to\infty} |g(t) - (I_n g)(t)| = 0$$

and $g(t) - (I_n g)(t)$ is bounded (boundedness of $I_n g$ follows from the mean-value theorem). Hence, from the dominated convergence theorem,

$$\lim_{n\to\infty} ||g - I_n g|| = 0.$$

Now $C[0, T]$, the space of continuous functions on $[0, T]$, is dense in $\mathfrak{L}_2[0, T]$. Namely, for any $f \in \mathfrak{L}_2[0, T]$ and $\epsilon > 0$, there exists g such that $||f - g|| < \epsilon$. Then,

$$\overline{\lim_{n\to\infty}} ||f - I_n f|| \leq ||f - g|| + \overline{\lim_{n\to\infty}} ||g - I_n g|| + \overline{\lim_{n\to\infty}} ||I_n(g - f)|| = 2\epsilon,$$

implying

$$\lim_{n\to\infty} ||f - I_n f|| = 0.$$

Corollary

If $E \int_0^T y^2(t) dt < \infty$, then $\lim_{n\to\infty} E||y - I_n y||^2 = 0$.

Proof: From Lemma 3,

$$\lim_{n\to\infty} ||y - I_n y|| = 0 \quad \text{a.s.}$$

Note

$$||y - I_n y||^2 \leq (||y|| + ||I_n y||)^2 \leq 4\, ||y||^2$$

and $E||y||^2 < \infty$. Hence the dominated convergence theorem asserts

$$\lim_{n\to\infty} E\, ||y - I_n y||^2 = 0.$$

Lemma 4

If $y(t)$ is a measurable Gaussian process, then

$$\int_0^T y^2(t)\, dt < \infty \qquad \text{a.s.}$$

implies

$$E \int_0^T y^2(t)\, dt < \infty.$$

Proof: Without loss of generality, we take $Ey(t) = 0$. Assume $y(t)$ is continouous in quadratic mean. Then, the Karhunen–Loève expansion of $y(t)$ exists,

$$y(t) = \sum \eta_i \varphi_i(t) \qquad \eta_i = (y, \varphi_i),$$

$$\int_0^T Ey(t)y(s)\varphi_i(s)\, ds = \lambda_i \varphi_i(t),$$

$$E\eta_i\eta_j = \lambda_i\, \delta_{ij}.$$

Put $Y = \int_0^T y^2(t)dt$. Then

$$EY = \sum E\eta_i^2 = \sum \lambda_i < \infty,$$

$$E \exp(-Y) = \prod E \exp(-\eta_i^2) = \prod (1 + 2\lambda_i)^{-1/2}.$$

Hence

$$EY \leq [E \exp(-Y)]^{-2}.$$

Assume $y(t)$ is no longer continuous but only $\int_0^T y^2(t)dt < \infty$ a.s., and let $\{f_i\}$ be a complete orthonormal set of continuous functions in $\mathfrak{L}_2[0, T]$. Then

$$y_n(t) = \sum_{i=1}^{n} y_i f_i(t) \qquad y_i = (y, f_i),$$

is continuous in quadratic mean and

$$Y_n = \int_0^T y_n^2(t)\, dt = \sum_{i=1}^n y_i^2$$

converges a.s. to Y. Then, using the inequality just proved,

$$EY = \lim_{n\to\infty} EY_n \le \lim_{n\to\infty} (Ee^{-Y_n})^{-2} = (Ee^{-Y})^{-2}.$$

Namely, the inequality is valid with $\int_0^T y^2(t)dt < \infty$ a.s. only. This immediately establishes the assertion.

Lemma 5

Let $w(t)$ be a standard Wiener process and $y(t)$ a Gaussian process with $\int_0^T y^2(t)dt < \infty$ a.s. If

$$E[w(t) - w(s)][y(u) - Ey(u)] = 0 \qquad u \le s < t,$$

then $y(t)$ has an orthogonal decomposition

$$y(t) = \tilde{y}(t) + \int_0^T L(t, s)\, dw(s) \qquad 0 \le t \le T,$$

where $\tilde{y}(t)$ and $w(t)$ are mutually independent processes and

$$L(t, s) = 0 \qquad s > t$$

and

$$\int_0^T \int_0^T L^2(t, s)\, ds\, dt < \infty.$$

Proof: Note $Ey^2(t) < \infty$ and $E \int_0^T y^2(t)dt < \infty$ since $y(t)$ is a measurable Gaussian process. Define

$$\tilde{y}(t) = y(t) - E\{y(t) - Ey(t) \mid w_T\}$$

where w_T denotes $\{w(s), 0 \le s \le T\}$. Then, for any s and t in $[0, T]$,

$$Ew(s)\tilde{y}(t) = Ew(s)y(t) - EE\{w(s)[y(t) - Ey(t)] \mid w_T\} = 0.$$

Hence $w(t)$ and $\tilde{y}(t)$ are mutually independent Gaussian processes.

Let $\{f_i\}$ be an orthonormal basis of $\mathcal{L}_2[0, T]$. Then

$$w(t) = \sum w_i \int_0^t f_i(s)\, ds \quad \text{a.s.}$$

where $w_i = \int_0^T f_i(t)dw(t)$ and $Ew_iw_j = \delta_{ij}$ [5, p. 324]. Hence

$$E\{y(t) - Ey(t) \mid w_T\} = \lim_{n\to\infty} E\{y(t) - Ey(t) \mid w_1, \cdots, w_n\}$$

$$= \lim_{n\to\infty} \sum_{i=1}^n \alpha_i(t) w_i \quad \text{a.s.}$$

for some $\alpha(t)$ [9, p. 76], and the series converges in quadratic mean also [9, p. 319]. Thus, $\sum \alpha_i^2(t) < \infty$. Hence, there exists a kernel $\hat{L}(t, s)$ such that

$$\hat{L}(t, s) = \lim_{n\to\infty} \sum_{i=1}^n \alpha_i(t) f_i(s)$$

for every t in $[0, T]$ [11, p. 59]. Furthermore, $\hat{L}(t, s)$ is square integrable since

$$\int_0^T \int_0^T \hat{L}^2(t, s)\, ds\, dt = \int_0^T \sum \alpha_i^2(t)\, dt$$

$$= \int_0^T EE^2\{y(t) \mid w_T\}\, dt \le E \int_0^T y^2(t)\, dt.$$

Then, from the definition of w_i,

$$E\{y(t) - Ey(t) \mid w_T\} = \int_0^T \hat{L}(t, s)\, dw(s) \quad \text{a.s.}$$

Next note

$$E\{E\{y(t) - Ey(t) \mid w_T\} \mid w_t\} = E\{y(t) - Ey(t) \mid w_t\},$$

$$E\left\{\int_0^T \hat{L}(t, s)\, dw(s) \mid w_t\right\} = E\left\{\int_0^t \hat{L}(t, s)\, dw(s) \mid w_t\right\} + E\left\{\int_t^T \hat{L}(t, s)\, dw(s) \mid w_t\right\} = \int_0^t \hat{L}(t, s)\, dw(s).$$

Hence, by defining

$$L(t, s) = \begin{cases} \hat{L}(t, s) & s \le t, \\ 0 & s > t, \end{cases}$$

we have

$$E\{y(t) - Ey(t) \mid w_t\} = \int_0^T L(t, s)\, dw(s).$$

This completes the proof.

ACKNOWLEDGMENT

The proof 4) of the theorem and the proof of Lemmas 3 and 4 were provided by L. A. Shepp. The proof of Lemma 1 is due to J. M. C. Clark. The author is specially grateful for generous consultation provided by L. A. Shepp and V. E. Beneš. He is also indebted to T. Kailath, E. Wong, M. Zakai, and the reviewers for pointing out errors in the original manuscript.

REFERENCES

[1] T. Kailath, "A general likelihood-ratio formula for random signals in Gaussian noise," *IEEE Trans. Information Theory*, vol. IT-15, pp. 350–361, May 1969.
[2] T. E. Duncan, "Probability densities for diffusion processes with applications to nonlinear filtering theory and detection theory," Ph.D. dissertation, Stanford University, Stanford, Calif., 1967.
[3] E. Wong, "Likelihood ratio for random signals in Gaussian white noise," Intern. Memo., Bell Telephone Laboratories, August, 1968.
[4] T. Kailath, "A further note on a general likelihood formula for random signals in Gaussian noise," *IEEE Trans. Information Theory*, (to be published).
[5] L. A. Shepp, "Radon–Nikodym derivatives for Gaussian measures," *Ann. Math. Stat.*, vol. 37, pp. 321–354, 1966.
[6] T. T. Kadota and L. A. Shepp, "Conditions for absolute continuity between a certain pair of probability measures," (to be published in *Z. Wahrscheinlichkeitstheorie und Verwandte Gebiete*).
[7] W. M. Wonham, "Lecture notes on stochastic control," pt. I, Center for Dynamical Sys., Div. of Appl. Math., Brown University, Providence, R. I., 1967.
[8] P. R. Halmos, *Measure Theory*. New York: Van Nostrand, 1950.
[9] J. L. Doob, *Stochastic Processes*. New York: Wiley, 1953.
[10] F. Smithies, *Integral Equations*. London: Cambridge University Press, 1958.
[11] F. Riesz and B. Sz-Nagy, *Functional Analysis*. New York: Frederick Ungar, 1955.

Reprinted from THE ANNALS OF MATHEMATICAL STATISTICS
Vol. 37, No. 2, April, 1966
Printed in U.S.A.

RADON-NIKODYM DERIVATIVES OF GAUSSIAN MEASURES

19

BY L. A. SHEPP

Bell Telephone Laboratories, Inc., Murray Hill, New Jersey

I. SUMMARY

We give simple necessary and sufficient conditions on the mean and covariance for a Gaussian measure to be equivalent to Wiener measure. This was formerly an unsolved problem [26].

Another unsolved problem is to obtain the Radon-Nikodym derivative $d\mu/d\nu$ where μ and ν are equivalent Gaussian measures [28]. We solve this problem for many cases of μ and ν, by writing $d\mu/d\nu$ in terms of Fredholm determinants and resolvents. The problem is thereby reduced to the calculation of these classical quantities, and explicit formulas can often be given.

Our method uses Wiener measure μ_W as a catalyst; that is, we compute derivatives with respect to μ_W and then use the chain rule: $d\mu/d\nu = (d\mu/d\mu_W)/(d\nu/d\mu_W)$. Wiener measure is singled out because it has a simple distinctive property—the Wiener process has a random Fourier-type expansion in the integrals of any complete orthonormal system.

We show that any process equivalent to the Wiener process W can be realized by a linear transformation of W. This transformation necessarily involves stochastic integration and generalizes earlier nonstochastic transformations studied by Segal [21] and others [4], [27].

New variants of the Wiener process are introduced, both conditioned Wiener processes and free n-fold integrated Wiener processes. We give necessary and sufficient conditions for a Gaussian process to be equivalent to any one of the variants and also give the corresponding Radon-Nikodym (R-N) derivative.

Last, some novel uses of R-N derivatives are given. We calculate explicitly: (i) the probability that W cross a slanted line in a finite time, (ii) the first passage probability for the process $W(t + 1) - W(t)$, and (iii) a class of function space integrals. Using (iii) we prove a zero-one law for convergence of certain integrals on Wiener paths.

TABLE OF CONTENTS

Received 19 November 1965.

II. INTRODUCTION

All measures considered in this paper are *Gaussian* and are considered to be defined on the space of continuous functions $X = X(t), 0 \leqq t \leqq T$ with $T \leqq \infty$. Such a measure μ is determined by its mean m and covariance (cov) R

$$m(t) = \int X(t)\, d\mu(X),$$
$$R(s, t) = \int (X(s) - m(s))(X(t) - m(t))\, d\mu(X).$$

Wiener measure μ_W is the measure with mean zero and cov $= \min(s, t)$. Two measures are equivalent (denoted $\sim$) when they have the same sets of measure zero.

1. A necessary and sufficient condition that $\mu \sim \mu_W$. We denote by L^2 and $\mathbf{L}^2$ the space of square-integrable functions on $[0, T]$ and $[0, T] \times [0, T]$ respectively; two functions are considered equal if they coincide almost everywhere.

Suppose μ is a measure with mean m and cov R.

THEOREM 1. *$\mu \sim \mu_W$ if and only if there exists a kernel $K \in \mathbf{L}^2$ for which*

$$R(s, t) = \min(s, t) - \int_0^s \int_0^t K(u, v)\, du\, dv \tag{1.1}$$

and

$$1 \notin \sigma(K) \tag{1.2}$$

and a function $k \in L^2$ for which

$$m(t) = \int_0^t k(u)\, du. \tag{1.3}$$

The kernel K is unique and symmetric and is given by $K(s, t) = -(\partial/\partial s)(\partial/\partial t) \cdot R(s, t)$ for almost every (s, t). The function k is unique and is given by $k(t) = m'(t)$ for almost every t.

As usual, $\sigma(K) = \{\lambda: K\varphi = \lambda\varphi, \varphi \neq 0\}$ is the spectrum, or the set of eigenvalues, of the Hilbert-Schmidt operator K. Here $K\varphi = \lambda\varphi$ means $\varphi \varepsilon L^2$ and

$$\int_0^T K(t, u)\varphi(u)\, du = \lambda\varphi(t).$$

Since K is symmetric and in $\mathbf{L}^2$ the eigenvalues $\lambda_1, \lambda_2, \cdots$ are real and $\sum \lambda_j^2 < \infty$. We shall show in Section 11 that if R is given by (1.1) with $K \varepsilon \mathbf{L}^2$, then R is nonnegative-definite if and only if $\lambda_j \leqq 1$ for all j. The condition (1.2) therefore says that $\lambda_j < 1$ for all j and should thus be interpreted as a statement of *strict* positive-definiteness for R. Note that λ_j may assume negative values.

The condition (1.1) has a simpler restatement in case $R_1(s, t) = (\partial/\partial s)R(s, t)$ is continuous for $s \neq t$. In this case (1.1) becomes (see Section 11 for the proof)

$$R_1(s, s+) - R_1(s, s-) \equiv 1, \qquad 0 < s < T. \tag{1.4}$$

This means that R must have a fixed discontinuity of unit size in its derivative along $s = t$, the same discontinuity that min (s, t) has in its derivative. A theorem of G. Baxter [2] implies the necessity of (1.4). We note that $R(0, \cdot) \equiv 0$ because $X(0) = 0$ under μ_W.

2. The R-N derivative $d\mu/d\mu_W$. Whenever $\mu \sim \mu_W$ the R-N derivative $d\mu/d\mu_W$ exists. We will show that $d\mu/d\mu_W$ can be written in terms of the Fredholm determinant and resolvent of the unique kernel K appearing in (1.1).

When $\sum |\lambda_j| < \infty$, K is said to be of *trace class* and the Fredholm determinant is

$$d(\lambda) = \prod_j (1 - \lambda\lambda_j). \tag{2.1}$$

For the general K, $\sum |\lambda_j|$ may not exist. The modified, or Carleman-Fredholm, determinant of K is

$$\delta(\lambda) = \prod_j (1 - \lambda\lambda_j)e^{\lambda\lambda_j} \tag{2.2}$$

which converges for all λ because $\sum \lambda_j^2 < \infty$. For each value of λ for which $\lambda^{-1} \not\varepsilon$ $\sigma(K)$ there is a unique kernel $H_\lambda \varepsilon \mathbf{L}^2$ called the Fredholm resolvent of K at λ. The resolvent equation

$$H_\lambda - K = \lambda H_\lambda K = \lambda K H_\lambda \tag{2.3}$$

determines H_λ uniquely. We denote H_1 by H for simplicity and note that H is defined because $1 \not\varepsilon \sigma(K)$ by (1.2). The kernel H is symmetric and is continuous when K is continuous. There are known expansions of d, δ and H in powers of λ, cf. [7], pp. 1081–1086.

Let μ be a measure with mean m and cov R for which $\mu \sim \mu_W$. Let K be given by (1.1): $K(s, t) = -(\partial/\partial s)(\partial/\partial t)R(s, t)$ for almost every s and t.

THEOREM 2. *If K is continuous and of trace class then $d\mu/d\mu_W$ is given by*

$$d\mu/d\mu_W(X + m) = [d(1)]^{-\frac{1}{2}} \tag{2.4}$$

$$\cdot \exp\left[-\tfrac{1}{2}\int_0^T \int_0^T H(s, t)\, dX(s)\, dX(t) + \int_0^T k(u)\, dX(u) + \tfrac{1}{2}\int_0^T k^2(u)\, du\right]$$

Here $k(t) = m'(t)$. When $m = 0$, (2.4) simplifies. The integral $\int_0^T k(u)\, dX(u)$

is the Wiener integral evaluated at the point X. It exists because $k \varepsilon L^2$. The integral $I(X) = \int_0^T \int_0^T H(s, t)\, dX(s)\, dX(t)$ is the double Wiener integral evaluated at X and is a (non-Gaussian) random variable with mean value $\int_0^T H(s, s)\, ds$. In our case the integral I can be defined because H is continuous. If H is also of bounded variation, we may integrate by parts to obtain an ordinary integral. Both H and the constant $d(1)$ may sometimes be obtainable in closed form even when $\sigma(K)$ is not so obtainable, as we shall see in Section 15.

In order to give a formula for $d\mu/d\mu_W$ valid for all K, the notion of a double Wiener integral must be extended slightly. We will use the *centered double Wiener integral* denoted

$$J(X) = \int_0^T c\int_0^T H(s, t)\, dX(s)\, dX(t).$$

J is introduced in Section 9. When the mean of I exists, J is the ordinary double Wiener integral I minus its mean. By this simple trick of subtracting off the mean, J can be defined for all $H \varepsilon \mathbf{L}^2$. K. Ito [10] was the first to consider the centered multiple Wiener integral. He obtained it in an equivalent way, by ignoring the values of H on the diagonal.

THEOREM 3. *If* $\mu \sim \mu_W$ *then*

$$(2.5)\quad (d\mu/d\mu_W)(X + m) = (\delta(1) \exp \operatorname{tr} (HK))^{-\frac{1}{2}} \cdot \exp [-\tfrac{1}{2}J(X) + \int_0^T k(u)\, dX(u) + \tfrac{1}{2}\int_0^T k^2(u)\, du].$$

The trace of a product always exists and we have

$$(2.6)\qquad \operatorname{tr} (HK) = \int_0^T \int_0^T H(s, t)K(s, t)\, ds\, dt.$$

When the hypothesis of Theorem 2 holds, (2.4) and (2.5) agree. Of course (2.4) is simpler. (2.5) has the advantage of being valid in general.

3. A representation for W. Let $\eta_1, \eta_2, \cdots$ be a sequence of independent standard normal variables (mean zero and variance one). Let $\varphi_1, \varphi_2, \cdots$ be an *arbitrary* complete orthonormal sequence in $L^2[0, T]$ and set

$$(3.1)\qquad \Phi_j(t) = \int_0^t \varphi_j(u)\, du, \qquad j = 1, 2, \cdots.$$

THEOREM 4. *For each* t, $0 \leqq t \leqq T$, *the series*

$$(3.2)\qquad \sum_{j=1}^{\infty} \eta_j \Phi_j(t) = W(t)$$

converges almost surely. The sum is the Wiener process on $[0, T]$.

This result appears to be new except for two special cases due to Wiener and Lévy. To prove the theorem we observe

$$(3.3)\qquad \Phi_j(t) = (\varphi_j, 1_t)$$

where 1_t is the indicator of the interval $[0, t]$ and $(f, g) = \int_0^T f(u)g(u)\, du$. The *completeness* implies

$$(3.4)\qquad \sum_j (\varphi_j, 1_s)(\varphi_j, 1_t) = (1_s, 1_t) = \min (s, t)$$

and (3.2) follows immediately from the 3-series theorem since the sum is Gaussian and has the required mean and covariance.

Wiener himself studied the special case of Fourier series:

$$\Phi_j(t) = 2^{\frac{1}{2}}(\sin (j - \tfrac{1}{2})\pi t)/(j - \tfrac{1}{2})\pi, \qquad 0 \leqq t \leqq 1. \tag{3.5}$$

In this case Φ as well as φ are orthogonal and this property characterizes the Fourier case. In the Fourier case, (3.2) converges uniformly in t with probability one. However, for some other choices of the sequence φ the convergence of (3.2) is even better. In fact, there does not seem to be any particular advantage to (3.5). To prove this, let us take φ to be the Haar sequence. Using double indices for convenience, set

$$\begin{aligned} \varphi_{-1} = 1, \qquad \varphi_{n,j}(t) &= 2^{n/2}, && 2^n t \,\varepsilon\, (j, j + \tfrac{1}{2}), \\ &= -2^{n/2}, && 2^n t \,\varepsilon\, (j + \tfrac{1}{2}, j + 1), \\ &= 0, && \text{otherwise} \end{aligned} \tag{3.6}$$

for $j = 0, 1, \cdots, 2^n - 1; n = 0, 1, 2, \cdots$. The φ's are complete and orthonormal. The representation (3.2) in this case is due to Lévy, cf. [11], p. 19, and takes the form

$$W(t) = \eta t + \sum_{n=0}^{\infty} \sum_{j=0}^{2n-1} \eta_{n,j}\Delta_{n,j}(t) \tag{3.7}$$

where $\Delta_{n,j}(t) = \Phi_{n,j}(t) = \int_0^t \varphi_{n,j}$ is an isosceles triangle with base 2^{-n} centered at $(j + \frac{1}{2})2^{-n}$, and height, $\frac{1}{2} \cdot 2^{-n/2}$. This shows that the Wiener process is a random sum of triangles.

In Wiener's case, the series (3.2) fails to converge absolutely. By contrast (3.7) converges *absolutely* uniformly with probability one, as was pointed out by Z. Ciesielski. We follow [11], p. 19. Let

$$f_n(t) = \sum_{j=0}^{2n-1} \eta_{n,j}\Delta_{n,j}(t).$$

We have

$$|f_n(t)| \leqq 2^{-n/2} \max_j |\eta_{n,j}|, \qquad 0 \leqq t \leqq 1, \tag{3.8}$$

because $\sum_j \Delta_{n,j}(t) \leqq 2^{-n/2}$ uniformly in t. Now $M_n = \max_j |\eta_{n,j}|$ is the maximum of 2^n independent standard normal variables and so

$$P\{M_n \leqq a_n\} = (\Phi(a_n) - \Phi(-a_n))^{2^n}. \tag{3.9}$$

Choosing $a_n = 2n^{\frac{1}{2}}$ it is easy to check that

$$\sum_{n=1}^{\infty} P\{M_n > a_n\} < \infty. \tag{3.10}$$

By the Borel-Cantelli lemma we see that $M_n < a_n$ eventually and so, a.s.

$$\sum_1^{\infty} 2^{-n/2} M_n < \infty. \tag{3.11}$$

It follows that (3.7) converges absolutely uniformly.

H. P. McKean has informed me that by using a recent theorem of J. Delporte

[30], p. 201, Corollary 6.4B, it can be shown that (3.2) converges *uniformly* a.s. We will not use this strong type of convergence and so we omit the proof.[1]

4. Simultaneous representation of μ and μ_W in terms of independent random variables. Let μ be a measure with mean m and cov R satisfying (1.1)–(1.3). Let $\varphi_1, \varphi_2, \cdots$ be the eigenfunctions of K; these are orthonormal and complete. The representation (3.2) gives

$$W(t) = \textstyle\sum_j \eta_j \Phi_j(t), \qquad 0 \leqq t \leqq T. \tag{4.1}$$

We shall define a Gaussian process Y *on the same space* as $\eta_1, \eta_2, \cdots$ with mean m and cov R. Write

$$k = m' = \textstyle\sum_j k_j \varphi_j, \qquad k_j = (k, \varphi_j), \tag{4.2}$$

and define

$$Y(t) = \textstyle\sum_j (\eta_j(1 - \lambda_j)^{\frac{1}{2}} + k_j)\Phi_j(t). \tag{4.3}$$

Y is clearly Gaussian and has mean

$$EY(t) = \textstyle\sum_j k_j(\varphi_j, 1_t) = (k, 1_t) = m(t) \tag{4.4}$$

and covariance

$$\textstyle\sum (1 - \lambda_j)\Phi_j(s)\Phi_j(t) = \min(s, t) - \int_0^s \int_0^t K(u, v)\, du\, dv = R(s, t). \tag{4.5}$$

In (4.5) we have used the $\mathbf{L}^2$ expansion of K

$$K(s, t) = \textstyle\sum \lambda_j \varphi_j(s)\varphi_j(t). \tag{4.6}$$

We have proved the following theorem.

THEOREM 5. *The processes* (4.1) *and* (4.3) *give a simultaneous representation of W and Y in terms of sums of independent variables.*

It is now possible to give a formal expression for $d\mu/d\mu_W(X)$. We expand a path X as

$$X(t) = \textstyle\sum_j X_j \Phi_j(t), \qquad X_j = \int_0^T \varphi_j(t)\, dX(t) \tag{4.7}$$

where $X_j = X_j(X)$ is the Wiener integral evaluated at X. $d\mu/d\mu_W(X)$ is the relative likelihood of X under μ and μ_W. Under μ_W, $X_j = \eta_j$, independent random variables. Under μ, $X_j = \eta_j(1 - \lambda_j)^{\frac{1}{2}} + k_j$ *also independent.* Because of independence, the probabilities multiply and we get

$$(d\mu/d\mu_W)(X) = \textstyle\prod_{j=1}^{\infty} (1 - \lambda_j)^{-\frac{1}{2}} \cdot \exp[-\frac{1}{2}(X_j - k_j)^2/(1 - \lambda_j)]/\exp[-\frac{1}{2} X_j^2]. \tag{4.8}$$

The product (4.8) always converges and represents $d\mu/d\mu_W$. A rigorous proof is given in Section 10. In Section 11 we show how (4.8) reduces to (2.4). The reduction to (2.4) is important because (2.4) is in terms of classical quantities and, in addition, does not explicitly involve the eigenvalues or eigenvectors, which are usually difficult to find.

[1] John Walsh has found a shorter proof, based on an abstract martingale convergence theorem.

5. Calculating $d\mu/d\nu$**.** Suppose that μ and ν are measures, both equivalent to μ_W. Then $\mu \sim \nu$ and by the chain rule

$$(d\mu/d\nu)(X) = (d\mu/d\mu_W)(X)/(d\nu/d\mu_W)(X). \tag{5.1}$$

Applying (2.5) or (2.4) to obtain $d\mu/d\mu_W$ and $d\nu/d\mu_W$ we get an explicit formula for $d\mu/d\nu$, but only in the special case when $\mu \sim \mu_W$ and $\nu \sim \mu_W$. To get more general results we will study certain variants of W.

Let for $n = 0, 1, 2, \cdots$,

$$W_n(t) = \int_0^t [(t-u)^n/n!]\, dW(u), \qquad 0 \leqq t \leqq T, \tag{5.2}$$

denote the n-fold integrated Wiener process. We have

$$W_0(t) = W(t), \qquad W_n(t) = \int_0^t W_{n-1}(u)\, du, \qquad n = 1, 2, \cdots. \tag{5.3}$$

The processes W_n satisfy $W_n^{(j)}(0) = 0, j = 0, 1, \cdots, n$.

Suppose μ is a measure with mean m and cov R whose sample paths Y are n-times differentiable a.s. Let $Y^{(n)}$ denote the nth derivative of Y. The process $Y^{(n)}$ is Gaussian with mean $m^{(n)}(t)$ and covariance

$$D_1^n D_2^n R(s, t) = (\partial^n/\partial s^n)(\partial^n/\partial t^n)R(s, t).$$

Let $\mu^{(n)}$ be the measure induced by $Y^{(n)}$; $\mu^{(n)}$ has the same mean and covariance as $Y^{(n)}$.

THEOREM 6. *Suppose* $\mu \sim \mu_{W_n}$. *Then* $\mu^{(n)} \sim \mu_W$ *and*

$$(d\mu/d\mu_{W_n})(X) = (d\mu^{(n)}/d\mu_W)(X^{(n)}). \tag{5.4}$$

The mapping $X \to X^{(n)}$ is 1-1 on the set of n-times differentiable functions X for which $X^{(j)}(0) = 0, j = 0, 1, \cdots, n$ and standard arguments [9], pp. 163–164, give (5.4). The righthand side of (5.4) is given by (2.5).

Using (5.4), we can obtain $d\mu/d\nu$ explicitly whenever $\mu \sim \nu \sim \mu_{W_n}$ for some n. This generalizes (5.1) and, further, one may drop the assumption that n in (5.2) is an integer. However, the condition $\mu \sim \nu \sim \mu_{W_n}$ is still too restrictive. Excluded are *stationary measures* μ and ν because their sample paths do not vanish at zero. In order to remedy this lack we must unpin the process W_n at zero.

Let $\mathbf{W}_n$ be the *free* Wiener process

$$\mathbf{W}_n(t) = \textstyle\sum_{j=0}^n \xi_j t^j/j! + W_n(t), \qquad 0 \leqq t \leqq T, \tag{5.5}$$

where $\xi_0, \cdots, \xi_n$ are independent, standard normal variables. Suppose Y is a process for which $Y \sim \mathbf{W}_n$. The paths $Y(t)$ are then exactly n-times differentiable, and the derivatives of Y at $t = 0$ are *nonzero* random variables. The class of processes $Y \sim \mathbf{W}_n$ includes many processes of interest. We will see that stationary processes with rational spectral density are included as a special case. In the latter case, $d\mu/d\nu$ has already been found by Gelfand and Yaglom in a different way [28].

We shall give the conditions on m and R so that $Y \sim \mathbf{W}_n$. The condition on m is that $m^{(n+1)} \in L^2$, or

(5.6) $$m(t) = \sum_{j=0}^{n} [m^{(j)}(0)/j!]t^j + \int_0^t [(t-u)^n/n!]k(u)\,du,$$

where $k \varepsilon L^2$. In order to find the conditions on R, we first obtain a certain decomposition of an n-times differentiable covariance. The remainder of Section 5 will be needed only for Section 13 et seq.

An n-times differentiable process Y is called *nondegenerate at zero* when the random variables $Y(0), Y'(0), \cdots, Y^{(n)}(0)$ are linearly independent.

THEOREM 7. *The covariance R of an n-times differentiable process nondegenerate at zero may be written uniquely as*

(5.7) $$R(s, t) = \sum_{i=0}^{n} A_i(s)A_i(t) + R^*(s, t)$$

where

(5.8) $$\begin{array}{ll}\text{(i)} & R^* \text{ is a covariance,}\\ \text{(ii)} & D_1{}^iD_2{}^iR^*(0, 0) = 0, \qquad i = 0, 1, \cdots, n,\\ \text{(iii)} & A_j{}^{(i)}(0) = 0, \quad i < j; \qquad A_i{}^{(i)}(0) > 0, \quad i = 0, \cdots, n.\end{array}$$

Condition (ii) means that if Z is a process with cov R^* then

(5.9) $$Z^{(j)}(0) = 0, \qquad j = 0, 1, \cdots, n.$$

We will have $Z \sim W_n$. The decomposition (5.7) is designed to reduce the problem to Theorem 6.

There are elegant formulas for $A_0, \cdots, A_n$ closely related to the decomposition formulas in Gauss's elimination method. To obtain them suppose R is a covariance that is n-times differentiable in each argument. Define

(5.10) $$R_{ij}(s, t) = D_1{}^iD_2{}^jR(s, t), \qquad R_{ij} = R_{ij}(0, 0).$$

Let $\alpha_{-1} = 1$ and for $i \geqq 0$ let

(5.11) $$\alpha_i = \begin{vmatrix} R_{00} & \cdots & R_{0i} \\ \vdots & & \vdots \\ R_{i0} & \cdots & R_{ii} \end{vmatrix}.$$

If $R(s, t) = EY(s)Y(t)$ where Y is n-times differentiable then α_i is the Grammian of $Y(0), \cdots, Y^{(i)}(0)$ and α_i is strictly positive when Y is nondegenerate at zero. Whenever $\alpha_i > 0$, $i = 0, \cdots, n$ the functions $A_0, \cdots, A_n$ in (5.7) are unique and are given by

(5.12) $$A_i(t) = (\alpha_i\,\alpha_{i-1})^{-\frac{1}{2}} \begin{vmatrix} R_{00} & \cdots & R_{0i-1} & R_{00}(0, t) \\ & & R_{1i-1} & R_{10}(0, t) \\ \vdots & & \vdots & \vdots \\ R_{i0} & \cdots & R_{ii-1} & R_{i0}(0, t) \end{vmatrix},$$

$i = 0, 1, \cdots, n$. In particular, $A_0(t) = R(0, t)/(R(0, 0))^{\frac{1}{2}}$. Note that A_i is independent of n for $n \geqq i$.

Define for $i = 0, \cdots, n, j = 0, \cdots, n$,

(5.13) $$A_{ji} = A_i{}^{(j)}(0) = (\alpha_i\,\alpha_{i-1})^{-\frac{1}{2}} \begin{vmatrix} R_{00} & \cdots & R_{0i-1} & R_{0j} \\ \vdots & & \vdots & \vdots \\ R_{i0} & \cdots & R_{ii-1} & R_{ij} \end{vmatrix}.$$

The $(n + 1) \times (n + 1)$ matrix $A = A_{ij}$ is lower semidiagonal and has an inverse. The inverse matrix is denoted by C and can be written explicitly (13.6).

The next theorem gives the conditions for a measure μ to be equivalent to μ_{W_n} as well as a formula for $d\mu/d\mu_{W_n}$ whenever it exists. It is clear that whenever $\mu \sim \mu_{W_n}$ the path functions Y must be n-times differentiable and nondegenerate at zero. In this case the covariance R of μ satisfies the hypothesis of Theorem 7, and R^* and $A_0, \cdots, A_n$ are uniquely defined.

THEOREM 8. *Let μ be a measure with mean m and cov R. $\mu \sim \mu_{W_n}$ if and only if: m satisfies* (5.6), *R has a unique decomposition* (5.7), *and*

$$D_1^n D_2^n R^*(s, t) = \min(s, t) - \int_0^s \int_0^t K(u, v)\, du\, dv \tag{5.14}$$

for a (unique, symmetric) kernel $K \varepsilon \mathbf{L}^2$ with $1 \not\varepsilon \sigma(K)$ and

$$A_i^{(n)}(t) = \int_0^t a_i(u)\, du + A_i^{(n)}(0) \tag{5.15}$$

for (unique) $a_0, \cdots, a_n$ in L^2.

When $\mu \sim \mu_{W_n}$ the R-N *derivative is given by* (5.16) *provided that K satisfies the conditions of Theorem* 2:

$$\begin{aligned}(d\mu/d\mu_{W_n})(X + m) = [d(1)\alpha_n]^{-\frac{1}{2}} \exp [& -\tfrac{1}{2}\textstyle\int_0^T \int_0^T H(s, t)\, dX^{(n)}(s)\, dX^{(n)}(t) \\ & + \textstyle\int_0^T k(u)\, dX^{(n)}(u) + \tfrac{1}{2}(k, k) \\ & - \tfrac{1}{2}\textstyle\sum_{j=0}^{n} (e_j^2 - (X^{(j)}(0) + m^{(j)}(0))^2) \\ & - \tfrac{1}{2}\textstyle\sum_{i=0}^{n} \sum_{j=0}^{n} e_i e_j((I + H)a_i, a_j) \\ & + \textstyle\sum_{i=0}^{n} e_i \int_0^T ((I + H)a_i(t))\, dX^{(n)}(t)].\end{aligned} \tag{5.16}$$

Here I is the identity on L^2, H is the resolvent of K at $\lambda = 1$, $d(\cdot)$ is the determinant of K, and

$$e_j = \textstyle\sum_{i=0}^{n} C_{ji}(X^{(i)}(0) - m^{(i)}(0)), \qquad j = 0, 1, \cdots, n, \tag{5.17}$$

where $C = A^{-1}$, the inverse matrix of A in (5.13). *For general K we must replace the double stochastic integral in* (5.16) *by the centered integral and replace $d(1)$ by* $\delta(1) \exp \operatorname{tr}(HK)$. *This modification is completely analogous to that of Theorem* 3.

(5.16) is cumbersome. Its importance lies in its generality rather than in its simplicity; in many cases $d\mu/d\mu_{W_n}$ can be obtained more simply by other means.

6. Conditioned Wiener processes. We next consider sub-Wiener processes, obtained from W by linear conditioning. The Wiener integral

$$\eta = \eta(\psi) = \int_0^T \psi(t)\, dW(t) \tag{6.1}$$

is defined for $\psi \varepsilon L^2$; η is normal with mean zero and variance $(\psi, \psi) = \int_0^T \psi^2(u)\, du$.

Let V be any subspace of $L^2 = L^2[0, T]$ and let μ^V be the (Gaussian) measure obtained by conditioning μ_W so that $\eta(\psi) = 0$ for $\psi \varepsilon V$. Then μ^V has mean zero and cov

$$R^V(s, t) = \min(s, t) - \textstyle\sum_j \Psi_j(s)\Psi_j(t) \tag{6.2}$$

where

$$\Psi_j(t) = \int_0^t \psi_j(u)\,du \tag{6.3}$$

and $\psi_1, \psi_2, \cdots$ is any orthonormal basis for V, contained in V.

We may realize a process with measure μ^V as follows:

$$W^V(t) = W(t) - \sum_j \Psi_j(t) W_j, \qquad W_j = \eta(\psi_j). \tag{6.4}$$

It is easy to check that W^V has mean zero and cov R^V. Alternatively, let $\varphi_1, \varphi_2, \cdots$ be an orthonormal basis of the orthogonal complement of V in L^2. Of course, $\varphi_1, \varphi_2, \cdots$ are not complete. Let $\eta_1, \eta_2, \cdots$ be independent standard normal variables and define Φ_j as in (3.1). Then another realization of W^V is

$$W^V(t) = \sum_j \eta_j \Phi_j(t). \tag{6.5}$$

The interest in (6.4) is that it is a realization on the same space as the original process W.

As an example, take V to be the space generated by $\psi = \psi_1 = 1$. Then with $T = 1$, $\Psi(t) = t$ and (6.2) is

$$R^V(s, t) = \min(s, t) - st. \tag{6.6}$$

Now $W_1 = \eta(\psi) = \int_0^1 dW(t) = W(1)$ and (6.4) gives

$$W^V(t) \doteq W(t) - tW(1), \qquad 0 \leqq t \leqq 1. \tag{6.7}$$

The process (6.7) is called the pinned Wiener process [5].

The processes (6.4) are mutually singular for different subspaces V_1 and V_2. Indeed, if $\psi \varepsilon V_1$ but $\psi \not\varepsilon V_2$ then $\eta(\psi) = 0$ for μ^{V_1} but is normal with nonzero variance for μ^{V_2} so that $\mu^{V_1} \perp \mu^{V_2}$. What is the condition on μ so that $\mu \sim \mu^V$? The answer is given by the next theorem.

THEOREM 9. $\mu \sim \mu^V$ *if and only if*

$$R(s, t) = R^V(s, t) - \int_0^s \int_0^t K(u, v)\,du\,dv \tag{6.8}$$

where $K \varepsilon \mathbf{L}^2$ *and in addition*

$$1 \not\varepsilon \sigma(K) \quad \text{and} \quad K\psi = 0 \quad \text{for} \quad \psi \varepsilon V. \tag{6.9}$$

The mean must satisfy (1.3) *and in addition*

$$(k, \psi) = 0, \qquad \psi \varepsilon V. \tag{6.10}$$

In case (6.8)–(6.10) *hold the* R-N *derivative is given by* (2.5) *where K is the unique kernel satisfying* (6.8). *When K satisfies the conditions of Theorem* 2, (2.4) *is also valid.*

7. Stochastic linear transformations of W. Interesting classes of processes can be obtained by various linear transformations of W. Segal and others consider some nonstochastic transformations and with them realize *some* processes equivalent to W [21], p. 464. By means of a transformation depending on a *stochastic integral* we will realize *any* process equivalent to W.

Suppose $M \varepsilon \mathbf{L}^2$. The stochastic integral

$$Z(s) = \int_0^T M(s, u)\, dW(u), \qquad 0 \leqq s \leqq T, \tag{7.1}$$

can be defined for each s in such a way that Z is a.s. measurable [6], p. 430. Assuming Z so defined, we set

$$Y(t) = W(t) - \int_0^t Z(s)\, ds \tag{7.2}$$

and call Y the *affine transformation of* W *with kernel* M. Of course when M is of bounded variation in the second argument, we may write Y as an ordinary non-stochastic transformation by integrating by parts. The process Y is Gaussian and has mean zero and covariance

$$R(s, t) = \min (s, t) - \int_0^s \int_0^t K(u, v)\, du\, dv \tag{7.3}$$

where

$$K = M + M^* - MM^*. \tag{7.4}$$

As usual, $M^*(u, v) = M(v, u)$ and, of course,

$$MM^*(u, v) = \int_0^T M(u, y)M(v, y)\, dy.$$

Let I denote the identity. Then by (7.4), $I - K = (I - M)(I - M^*)$. Since $\sigma(M) = \sigma(M^*)$,

$$1 \not\varepsilon\, \sigma(K) \Leftrightarrow 1 \not\varepsilon\, \sigma(M). \tag{7.5}$$

By Theorem 1 *we obtain:* $Y \sim W$ *if and only if*

$$1 \not\varepsilon\, \sigma(M). \tag{7.6}$$

Suppose R is any covariance satisfying (1.1) and (1.2). It is simple to show that there is an M satisfying (7.4) and (7.6). Indeed, the kernel K of R has the Mercer expansion in $\mathbf{L}^2$,

$$K(s, t) = \sum \lambda_j \varphi_j(s)\varphi_j(t) \tag{7.7}$$

where $\lambda_j < 1$ for all j. We may take M to be

$$M(s, t) = \sum_j [1 \pm (1 - \lambda_j)^{\frac{1}{2}}]\varphi_j(s)\varphi_j(t). \tag{7.8}$$

When all but a finite number of the signs in (7.8) are negative we have $M \varepsilon \mathbf{L}^2$ and it is easy to see that $M = M^*$ and (7.4) and (7.6) hold.

We have proved the following theorem. Let μ be Gaussian with mean m and cov R.

THEOREM 10. *Suppose* $\mu \sim \mu_W$. *There is an* M *for which the process* $Y + m$, *where* Y *is given by* (7.2), *is a realization of* μ. *In other words, the measure* μ_{Y+m} *induced by* $Y + m$ *satisfies*

$$\mu_{Y+m} = \mu. \tag{7.9}$$

M *is not unique.*

There are additional conditions one could put on M in order to make it unique in Theorem 8. However, none seems natural.

When M is Volterra

$$M(s, t) = 0 \quad \text{for} \quad s \leqq t \tag{7.10}$$

the process Y is *causal*: $Y(t)$ depends only on values $W(\tau)$ for $\tau \leqq t$. But we could not solve the problem of the existence and uniqueness of kernels M of Volterra type in (7.4).

8. A discussion of previous work. I. Segal considered a nonstochastic transformation S of W [20], p. 22; [21], p. 464. He defined

$$S(t) = W(t) + \int_0^T N(t, u)W(u)\,du, \qquad 0 \leqq t \leqq T. \tag{8.1}$$

N is assumed continuous and $N_t \in \mathbf{L}^2$. The transformation (8.1) generalizes an earlier one of Cameron and Martin [8]. It is easily seen that (8.1) is a special case of (7.2). Segal asserts that $S \sim W$ under the stated conditions on N. However, note that a spectral condition analogous to (7.6) is needed. Even if this correction is made, (8.1) is not "best possible" as Segal claims. The stochastic transformation (7.2) is better because it gives the most general transformation equivalent to W (Theorem 10).

In the case $\mu = \mu_{W+m}$, a translate of μ_W, the condition (1.3) as well as the formula for $d\mu/d\mu_W$ was known and is due to Cameron and Martin [6], and in the general case to Segal [21], p. 462.

D. E. Varberg [26] gave a formula for $d\mu/d\mu_W$ when R admits a certain factorization. His formula is based on a transformation of Woodward [27], which is similar to (8.1). Varberg requires many complicated additional assumptions. These complications arise because: (a) his approach is based on a transformation which is not general enough and (b) $d\mu/d\mu_W$ itself depends on the transformation only through the covariance of μ—many transformations give the same covariance as is shown by the manifold nonuniqueness of M in Theorem 10. For this reason it is better to work with the covariance directly. It has been brought to my attention by R. H. Cameron that D. E. Varberg in an unpublished manuscript has independently obtained an equivalent form of the sufficiency half of Theorem 1.

We should mention that a condition similar to (1.1) and (5.14) appears in work of Yu. A. Rozanov [19], p. 455.

We acknowledge with pleasure an informative private lecture given by J. Feldman and L. Gross and many profitable discussions with S. P. Lloyd.[2]

9. Double Wiener integrals. We will now define the *centered double Wiener integral*

$$J(X) = \int_0^T c\int_0^T H(s, t)\,dX(s)\,dX(t) \tag{9.1}$$

for $H \in \mathbf{L}^2$ and for almost every function X.

A simple function H has the representation

$$H(s, t) = \sum_1^n \sum_1^n a_{jk}\chi_{jk}(s, t) \tag{9.2}$$

[3] Note added in proof: A. M. Yaglom communicated that I. M. Golosoy recently obtained results, announced in *Dokl. Akad. Nayk CCCP* **166** (1966) 263–266, which should be compared with ours. He obtains, among other things, an equivalent version of our Theorem 1

where

$$\chi_{jk}(s, t) = 1 \qquad (s, t) \in (t_{j-1}, t_j) \times (t_{k-1}, t_k) \tag{9.3}$$
$$= 0 \qquad \text{otherwise}$$

for some partition, $0 = t_0 < t_1 < \cdots < t_n = T$. When H is simple define

$$I(X) = \sum_1^n \sum_1^n a_{jk}(X(t_j) - X(t_{j-1}))(X(t_k) - X(t_{k-1})) \tag{9.4}$$

and

$$J(X) = I(X) - \int_0^T H(s, s)\, ds. \tag{9.5}$$

It is important and also easy to check that $J = J_H$ does not depend on the partition used to define it. J is not normally distributed, but has mean zero and variance

$$\int J^2\, d\mu_W = 2\int_0^T \int_0^T H^2(s, t)\, ds\, dt. \tag{9.6}$$

Now suppose that H is any element of $\mathbf{L}^2$ and that H_n is a sequence of simple functions for which $H_n \to H$ in $\mathbf{L}^2$. By (9.6), J_{H_n} is a Cauchy sequence in $L^2(\mu_W)$. We define $J_H = \lim J_{H_n}$. It is easy to check that J_H does not depend on the sequence H_n used to define it.

We call J the centered integral because

$$E_W J = \int J\, d\mu_W = 0. \tag{9.7}$$

The c between the integral signs in (9.1) calls attention to (9.7). For later use we point out that for $H(s, t) = \varphi(s)\varphi(t)$, a degenerate kernel, we get

$$\int_0^T c\int_0^T \varphi(s)\varphi(t)\, dX(s)\, dX(t) = \eta^2(\varphi) - (\varphi, \varphi) \tag{9.8}$$

where $\eta(\varphi)$ is the single Wiener integral

$$\eta(\varphi) = \int_0^T \varphi(t)\, dX(t). \tag{9.9}$$

The *(uncentered) double Wiener integral*

$$I(X) = \int_0^T \int_0^T H(s, t)\, dX(s)\, dX(t) \tag{9.10}$$

is now easy to define. We define $I = I_H$ for *continuous* H by

$$I(X) = J(X) + \int_0^T H(s, s)\, ds. \tag{9.11}$$

There is an important formula for I obtained by two integrations by parts when H is of bounded variation. We say H is of *bounded variation* when

$$\operatorname{Var}(H) = \sup \sum_1^n \sum_1^n |H(t_j, t_k) - H(t_{j-1}, t_k) - H(t_j, t_{k-1}) + H(t_{j-1}, t_{k-1})| \tag{9.12}$$

is finite, the sup being taken over all partitions. In this case we have

$$I(X) = H(T, T)X^2(T) - X(T)\int_0^T X(s)\, d_sH(s, T) - X(T)\int_0^T X(t)\, d_tH(T, t) + \int_0^T \int_0^T X(s)X(t)\, d_s\, d_tH(s, t). \tag{9.13}$$

---

by a rather different method. He then applies the theorem to obtain conditions for equivalence and singularity for an arbitrary Gaussian measure and a Gauss-Markov measure.

One proves (9.13) first for simple functions and then for general H of bounded variation by passage to the limit. The advantage of (9.13) is that it does not involve stochastic integration and so is defined for *every* continuous function X.

III. PROOFS OF THE MAIN RESULTS

We first obtained Theorems 1–3 and 9 heuristically, using a direct evaluation of $d\mu/d\mu_W$ as a limit of finite-dimensional densities,

$$(d\mu/d\mu_W)(X) = \lim [p_\mu(x_1, \cdots, x_n)/p_{\mu_W}(x_1, \cdots, x_n)], \quad x_i = X(t_i),$$

where

$$p_\mu(\bar{x}) = (2\pi)^{-n/2}|R|^{-\frac{1}{2}} \exp (R^{-1}(\bar{x} - \bar{m}), (\bar{x} - \bar{m})).$$

The limit is taken as the partition $0 = t_0 < t_1 < \cdots < t_n = T$ becomes dense. While a rigorous proof along these lines is difficult and involves many details, it can be given when the kernel K of R is smooth. We proceed by the easier but indirect method, via (3.2) and the simultaneous representation.

The advantage of using μ_W and the related measures as the catalysts stems from the properties of *white noise*, formally $W'(t)$. Many have tried to calculate $d\mu/d\nu$ directly by getting a simultaneous representation of μ and ν in terms of independent variables. What would be needed is a set of simultaneous eigenfunctions. However, these *do not exist* in general (note that the claim in [28], p. 334, about the existence of generalized eigenfunctions is not correct). In case $\nu = \mu_W$ such eigenfunctions do exist and are in L^2 as we have seen. Basic is the fact that the covariance of white noise is formally the δ-function and δ can be expanded in any complete set

$$\delta(s, t) = \textstyle\sum_j \varphi_j(s)\varphi_j(t).$$

Of course, this expansion is only formal, but upon integration it becomes precise.

10. Proof of Theorem 1: Sufficiency. Suppose μ is a measure with mean m and cov R that satisfy (1.1)–(1.3). We will prove that $\mu \sim \mu_W$ by actually giving $d\mu/d\mu_W$. Let $\varphi_1, \varphi_2, \cdots$ denote the eigenfunctions of the kernel K and define the Wiener integral

$$X_j = X_j(X) = \textstyle\int_0^T \varphi_j(t)\, dX(t), \qquad j = 1, 2, \cdots. \tag{10.1}$$

Define

$$F_j(X) = (1 - \lambda_j)^{-\frac{1}{2}}\{\exp [-\tfrac{1}{2}(X_j - k_j)^2/(1 - \lambda_j)]/\exp [-\tfrac{1}{2}X_j^2]\} \tag{10.2}$$

where $k_j = (k, \varphi_j)$, $k = m'$.

LEMMA 1. *The product*

$$F(X) = \textstyle\prod_{j=1}^\infty F_j(X) \tag{10.3}$$

converges a.e. (μ_W) *and is integrable* μ_W.

PROOF. First, $\prod_j (1 - \lambda_j)e^{\lambda_j}$ converges because $\sum \lambda_j^2 < \infty$. Using $\sum k_j^2 < \infty$ (by (1.3)) and the 3-series theorem, it is an easy exercise to prove that

$$\textstyle\sum_{j=1}^\infty (X_j - k_j)^2/(1 - \lambda_j) - X_j^2 - \lambda_j$$

converges for a.e. X. Thus (10.3) converges and F is a random variable.

To prove that $F \varepsilon L^1(\mu_W)$ we use an idea due to Kakutani [14]. Let

$$G_N = \prod_{j=1}^{N} (F_j)^{\frac{1}{2}}.$$

We already know that $G_N(X) \to (F(X))^{\frac{1}{2}}$ for a.e. X. We will show that G_N is a Cauchy sequence in $L^2(\mu_W)$. The limit of the Cauchy sequence must also be $(F)^{\frac{1}{2}}$. But L^2 is complete and so $(F)^{\frac{1}{2}} \varepsilon L^2(\mu_W)$, or $F \varepsilon L^1(\mu_W)$.

To prove that G_N is a Cauchy sequence observe that

$$E_W F_j = 1, \qquad E_W^{\,2}(F_j)^{\frac{1}{2}} = \beta_j^2 = \exp\,[-\tfrac{1}{2}k_j^2/(2 - \lambda_j)](1 - \lambda_j)^{\frac{1}{2}}/(1 - \lambda_j/2).$$

Now $X_1, X_2, \cdots$ are independent (μ_W) and hence so are $F_1, F_2, \cdots$. For $M < N$ we have by direct calculation

$$(10.4) \quad E_W(G_N - G_M)^2 = 2 - 2\prod_{M+1}^{N} E_W(F_j)^{\frac{1}{2}} = 2(1 - \prod_{M+1}^{N} \beta_j).$$

Again using $\sum \lambda_j^2 < \infty$ and $\sum k_j^2 < \infty$ it follows that $\prod \beta_j$ converges. The tail of the product tends to unity and so G_N is a Cauchy sequence. The lemma is proved.

LEMMA 2. *F is the* R-N *derivative*; $F = d\mu/d\mu_W$.

PROOF. What we must show is that

$$(10.5) \qquad \mu(A) = \int_A F(X)\, d\mu_W(X),$$

where A is any measurable set of functions X. We prove (10.5) first for sets A of the form

$$(10.6) \qquad A = \{X_1 < a_1, X_2 < a_2, \cdots, X_n < a_n\}.$$

Once again, $X_1, X_2, \cdots$ are independent (μ_W) and hence so are $F_1(X), \cdots, F_n(X)$. The expectation of a product of independent variables is the product of the expectations and by direct calculation

$$(10.7) \qquad \int_A F(X)\, d\mu_W(X) = \prod_{j=1}^{n} \Phi((a_j - k_j)/(1 - \lambda_j))^{\frac{1}{2}}$$

where Φ is the standard normal df.

Relative to μ the random variable X_j is Gaussian and has mean

$$(10.8) \qquad E_\mu \int_0^T \varphi_j(t)\, dX(t) = (\varphi_j, k) = k_j.$$

The covariance of $X_1, X_2, \cdots$ is

$$(10.9) \quad E_\mu(X_i - k_i)(X_j - k_j) = \int_0^T \int_0^T \varphi_i(s)\varphi_j(t)\, d_s\, d_t R(s, t).$$

Applying (1.1), we obtain

$$(10.10) \quad \int_0^T \int_0^T \varphi_i(s)\varphi_j(t)\, d_s\, d_t R(s, t) = (\varphi_i, \varphi_j) - (K\varphi_i, \varphi_j) = (1 - \lambda_i)\delta_{ij}.$$

Thus $X_1, X_2, \cdots$ are also independent with respect to μ. It is now easy to check that $\mu(A)$ agrees with (10.7) and hence (10.5) follows, at least for sets of the form (10.6).

In order to prove (10.5) for any measurable set A we find by direct calculation that

$$(10.11) \qquad E_W(\sum_j X_j \Phi_j(t) - X(t))^2 = 0$$

for each fixed t. It follows that almost surely (μ_W) we have

$$X(t) = \sum_j X_j \Phi_j(t) \tag{10.12}$$

and so the σ-field generated by $\{X_j\}$ and sets of probability zero (μ_W) includes the measurable sets (the σ-field generated by $X(t)$, for $0 \leqq t \leqq T$). We have already proved (10.5) for the σ-field generated by $\{X_j\}$ and Lemma 2 follows.

We have shown that when (1.1)–(1.3) hold, the R-N derivative $d\mu/d\mu_W$ exists. This derivative is a.s. positive because the infinite product converges and it follows that $\mu \sim \mu_W$. Without using the positivity, the equivalence would also follow from the dichotomy theorem of Feldman and Hájek. We turn to the other half of Theorem 1.

11. Proof of Theorem 1: Necessity. At this point we use an important theorem of Segal [21], p. 463. The following proof of the necessity was suggested by J. Feldman.

Let μ be a measure with mean m and cov R and suppose that $\mu \sim \mu_W$. Consider the Hilbert space $\mathbf{H} = L^2[0, T]$ and define the bilinear form $\mathbf{B}$: For $\varphi \varepsilon \mathbf{H}$ and $\psi \varepsilon \mathbf{H}$ define the random variables $\eta(\varphi)$ and $\eta(\psi)$ by the Wiener integral (9.9) and set

$$\mathbf{B}(\varphi, \psi) = \int \eta(\varphi)\eta(\psi)\, d\mu - \left(\int \eta(\varphi)\, d\mu\right)\left(\int \eta(\psi)\, d\mu\right). \tag{11.1}$$

The form $\mathbf{B}$ is positive, $\mathbf{B}(\varphi, \varphi) \geqq 0$, and is bounded,

$$\mathbf{B}(\varphi, \varphi) \leqq (\varphi, \varphi) \times \text{constant}$$

(see Lemma 1 of [15]). Consequently, there is a linear transformation B [18], p. 202, with

$$(B\varphi, \psi) = \mathbf{B}(\varphi, \psi). \tag{11.2}$$

Since B is a positive linear transformation it has a squareroot T [18], p. 265.

Segal's theorem says that if $\mu \sim \mu_W$ (in his terminology $n_T \sim n$) we must have

$$T^*T = B = I - K \tag{11.3}$$

where I is the identity and K is Hilbert-Schmidt, $K \varepsilon \mathbf{L}^2$. For all $\varphi \varepsilon \mathbf{H}$, $\psi \varepsilon \mathbf{H}$ we get by (11.3),

$$(B\varphi, \psi) = (\varphi, \psi) - (K\varphi, \psi) = \mathbf{B}(\varphi, \psi). \tag{11.4}$$

Now choose $\varphi = 1_s$, $\psi = 1_t$. We have $\eta(1_t) = \int_0^t dX(u) = X(t)$, since $X(0) = 0$ a.s. μ_W. We get by (11.1)

$$\mathbf{B}(1_s, 1_t) = R(s, t). \tag{11.5}$$

By (11.4)

$$R(s, t) = \min(s, t) - \int_0^s \int_0^t K(u, v)\, du\, dv. \tag{11.6}$$

This proves (1.1) necessary.

Next we prove the spectral condition on K, $\lambda_j < 1$ for all j. By (11.4) we get

$$(11.7) \qquad (\varphi, \varphi) \geqq (K\varphi, \varphi)$$

since $(B\varphi, \varphi) \geqq 0$. The largest eigenvalue of K is given by $\lambda_{\max} = \sup_\varphi (K\varphi, \varphi)/(\varphi, \varphi)$ and so $\lambda_j \leqq 1$. To show that $1 \varepsilon \sigma(K)$ we may argue as follows. Suppose $K\varphi = \varphi$, $(\varphi, \varphi) = 1$. We see that $\eta(\varphi)$ is an a.s. constant with respect to μ since it has variance $(B\varphi, \varphi) = 0$ by (11.4). On the other hand, with respect to μ_W, $\eta(\varphi)$ has variance one. This means $\mu \perp \mu_W$ and we get a contradiction. This proves (1.2) necessary.

(11.7) shows that if R has the form (1.1) *and R is a covariance* then $\sigma(K) \subseteq (-\infty, 1]$. Next we show that if R has the form (1.1) and $\text{sp}(K) \subseteq (-\infty, 1]$ then R is a covariance. This will prove the remark made below Theorem 1, R is nonnegative definite if and only if $\lambda_j \leqq 1$ for all j.

What we must show is that

$$(11.8) \qquad (R\varphi, \varphi) \geqq 0$$

for all $\varphi \varepsilon L^2$. Let

$$\Phi(t) = -\int_t^T \varphi(u)\, du.$$

Integration by parts applied twice gives

$$(11.9) \qquad (R\varphi, \varphi) = \int_0^T \Phi(s)\Phi(t)\, d_s\, d_t R(s, t) = (\Phi, \Phi) - (K\Phi, \Phi).$$

But $\sigma(K) \subseteq (-\infty, 1]$ and so $(\Phi, \Phi) - (K\Phi, \Phi) \geqq 0$. We get (11.8) immediately. We remark that (1.3) cannot be replaced by the condition: $(R\varphi, \varphi) > 0$ for nonzero $\varphi \varepsilon L^2$.

Next we prove (1.3). Let $\nu(R, m)$ denote the Gaussian measure with mean m and cov R so $\mu = \nu(R, m)$. We have

$$(11.10) \qquad \mu = \nu(R, m) \sim \nu(R_W, 0) = \mu_W$$

where $R_W(s, t) = \min(s, t)$. By a theorem of C. R. Rao and V. S. Varadarajan [17], p. 308, the measures must also be equivalent when the means are ignored:

$$(11.11) \qquad \nu(R, 0) \sim \nu(R_W, 0).$$

By considering the 1-1 path transformation $X(t) \to X(t) + m(t)$ we see that (11.11) gives

$$(11.12) \qquad \nu(R, m) \sim \nu(R_W, m) = \mu_{W+m}.$$

Comparing (11.10) and (11.12) and using the fact that $\sim$ is an equivalence relation we get

$$(11.13) \qquad \mu_W \sim \mu_{W+m}.$$

The condition on m in order that $\mu_W \sim \mu_{W+m}$ was found by Segal [21], p. 462. It is precisely (1.3). This completes the proof of Theorem 1.

In order to prove (1.4) we observe that whenever (1.1) holds K is given by

$$K(s, t) = -(\partial/\partial s)(\partial/\partial t)R(s, t), \qquad s \neq t. \tag{11.14}$$

Putting (11.14) back into (1.1) we get for $s < t$,

$$R(s, t) = s + \int_0^s \left(\int_0^u R_{12}(u, v)\, dv\right) du + \int_0^s \left(\int_u^t R_{12}(u, v)\, dv\right) du. \tag{11.15}$$

Integrating on v and using $R(0, t) = 0$ we get

$$R(s, t) = s + \int_0^s R_1(u, u-)\, du + \int_0^s (R_1(u, t) - R_1(u, u+))\, du. \tag{11.16}$$

This gives immediately

$$s = \int_0^s (R_1(u, u+) - R_1(u, u-))\, du \tag{11.17}$$

which is the integrated version of (1.4).

12. Proof of Theorems 2 and 3. We will now show that the product formula (10.3) for $F = d\mu/d\mu_W$ can be expressed in terms of the Fredholm quantities.

Let K denote, as usual, the kernel in (1.1) of the covariance of $\mu \sim \mu_W$. The eigenfunctions of the resolvent H of K are $\varphi_1, \varphi_2, \cdots$, the same as those of K, and the eigenvalues γ of H satisfy $\gamma_j = \lambda_j/(1 - \lambda_j)$, $j = 1, 2, \cdots$. We note that $\sum \gamma_j^2 < \infty$ and so H has the $\mathbf{L}^2$ expansion

$$H(s, t) = \sum_j \gamma_j \varphi_j(s)\varphi_j(t). \tag{12.1}$$

With $X_j = \int_0^T \varphi_j(t)\, dX(t)$ as in (10.1) we get formally from (12.1),

$$\sum_{j=1}^\infty X_j^2 \gamma_j = \int_0^T \int_0^T H(s, t)\, dX(s)\, dX(t). \tag{12.2}$$

Since $k = \sum k_j\varphi_j$ and $Hk = \sum k_j\gamma_j\varphi_j$ we get formally

$$\sum_{j=1}^\infty X_j k_j/(1 - \lambda_j) = \int_0^T k(t)\, dX(t) + \int_0^T \int_0^T H(s, t)k(s)\, ds\, dX(t) \tag{12.3}$$

and

$$\sum_{j=1}^\infty k_j^2/(1 - \lambda_j) = (k, k) + (Hk, k). \tag{12.4}$$

Substituting (12.2)–(12.4) into (10.3) we get

$$\begin{aligned} F(X) = \prod_{j=1}^\infty F_j(X) = (d(1))^{-\frac{1}{2}} \exp\Big[-\tfrac{1}{2} \int_0^T \int_0^T H(s,t)\, dX(s)\, dX(t) \\ + \int_0^T k(u)\, dX(u) + \int_0^T Hk(t)\, dX(t) - \tfrac{1}{2}(k, k) - \tfrac{1}{2}(Hk, k)\Big]. \end{aligned} \tag{12.5}$$

Replacing X by $X + m$ gives (2.4).

To prove (12.2) rigorously we observe that by (9.8) the partial sum

$$\sum_{j=1}^N X_j^2\gamma_j = \sum_{j=1}^N \gamma_j + \int_0^T c \int_0^T H_N(s, t)\, dX(s)\, dX(t), \tag{12.6}$$

where

$$H_N(s, t) = \sum_{j=1}^N \gamma_j\varphi_j(s)\varphi_j(t), \qquad N = 1, 2, \cdots.$$

Since $H_N \to H$ in $\mathbf{L}^2$ we have in $L^2(\mu_W)$

$$\int_0^T c \int_0^T H_N(s, t)\, dX(s)\, dX(t) \to \int_0^T c \int_0^T H(s, t)\, dX(s)\, dX(t). \tag{12.7}$$

Now suppose that K satisfies the assumptions of Theorem 2 so that K is continuous and of trace class. It follows that the resolvent H is continuous and of trace class. H is also symmetric, of course, and it is *known* that these assumptions on H imply that

$$\text{tr}(H) = \textstyle\sum_{j=1}^{\infty} \gamma_j = \int_0^T H(s, s)\, ds. \tag{12.8}$$

Although this fact is well known in the theory of integral equations, apparently no published reference exists. Passing to the limit in (12.6) we have

$$\textstyle\sum_{j=1}^{\infty} X_j^2 \gamma_j = \sum_{j=1}^{\infty} \gamma_j + \int_0^T c \int_0^T H(s, t)\, dX(s)\, dX(t). \tag{12.9}$$

Using (12.8) and

$$\textstyle\int_0^T c \int_0^T H(s,t)\, dX(s)\, dX(t) = \int_0^T \int_0^T H(s,t)\, dX(s)\, dX(t) - \int_0^T H(s, s)\, ds,$$

we obtain (12.2).

To prove (12.3) rigorously we observe that

$$\textstyle\sum_{j=1}^{N} X_j k_j/(1 - \lambda_j) = \int_0^T (k_N(t) + Hk_N(t))\, dX(t), \tag{12.10}$$

where $k_N = \sum_{j=1}^{N} k_j \varphi_j$. As $N \to \infty$, $k_N \to k$ in L^2. By the convergence properties of Wiener integrals we get (12.3).

Since the rigorous justification of (12.4) is straightforward we have proved Theorem 2. Theorem 3 can be proved similarly. Instead of (12.2) we must use

$$\textstyle\sum_{j=1}^{\infty} (X_j^2 - 1)\gamma_j = \int_0^T c \int_0^T H(s, t)\, dX(s)\, dX(t), \tag{12.11}$$

which follows from (12.6), letting $N \to \infty$. Of course, now $\sum_j \gamma_j$ does not necessarily converge.

The trace of HK is

$$\text{tr}(HK) = \textstyle\sum_j \gamma_j \lambda_j \tag{12.12}$$

and the proof of (2.5) is completed by a short calculation.

We omit the proof of Theorem 9 because it proceeds along the same lines as those of Theorems 1–3.

13. Proof of Theorem 7. Let R be the covariance of an n-times differentiable process Y nondegenerate at zero. Let $A_0, \cdots, A_n$ be defined by (5.12) and R^* be defined by (5.7). We will prove that R^* is a covariance by showing that it is the covariance of the process Z,

$$Z(t) = Y(t) - \textstyle\sum_{j=0}^{n} Y^{(j)}(0) B_j(t), \tag{13.1}$$

where

$$B_j(t) = \textstyle\sum_{i=0}^{n} B_{ij} R_{i0}(0, t), \qquad j = 0, \cdots, n, \tag{13.2}$$

and $B_{ij} = R_{ij}^{-1}$ is the inverse matrix of R in (5.10). The matrix $R = R_{ij}$ is positive-definite and so has a lower semidiagonal (lsd) square root which is unique and is given by Gauss's formula [8], p. 37. Comparing Gauss's formula with (5.13) we see that A is the lsd square root, that is

$$R_{ij} = \textstyle\sum_{k=0}^{n} A_{ik} A_{jk}, \qquad A_{ji} = A_i^{(j)}(0), \tag{13.3}$$

$i = 0, \cdots, n$ and $j = 0, \cdots, n$; in symbolic notation $R = AA^T$.

The covariance of Z is by direct calculation

$$R_Z(s, t) = R(s, t) - \sum_{j=0}^{n} \sum_{k=0}^{n} B_{jk} R_{j0}(0, s) R_{k0}(0, t). \tag{13.4}$$

We will now show that the second term on the right,

$$\sum_{j=0}^{n} \sum_{k=0}^{n} B_{jk} R_{j0}(0, s) R_{k0}(0, t) = \sum_{k=0}^{n} A_k(s) A_k(t), \tag{13.5}$$

which will prove that $R^*(s, t) = R_Z(s, t)$, the covariance of Z. Define the matrix $C = C_{ij}$, $i = 0, \cdots, n$ and $j = 0, \cdots, n$ by

$$C_{ij} = (\alpha_i \, \alpha_{i-1})^{-\frac{1}{2}} \begin{vmatrix} R_{00} & \cdots & R_{0i-1} & 0 \\ & & \vdots & \vdots \\ \vdots & & R_{ji-1} & 1 \\ & & \vdots & \vdots \\ R_{i0} & \cdots & R_{ii-1} & 0 \end{vmatrix}, \tag{13.6}$$

where the last column is zero except for the element in the jth row, $j \leqq i$, which is unity. We have $C_{ij} = 0$ if $i < j$ and so C is lsd. Multiplying (13.3) on the right by R_{jl}^{-1} and adding, $j = 0, \cdots, n$, we get

$$\delta_{il} = \sum_{k=0}^{n} A_{ik} \sum_{j=0}^{n} A_{jk} R_{jl}^{-1}. \tag{13.7}$$

But using the formula (5.13) for A_{jk} we get

$$\sum_{j=0}^{n} A_{jk} \, R_{jl}^{-1} = (\alpha_k \, \alpha_{k-1})^{-\frac{1}{2}} \begin{vmatrix} R_{00} & \cdots & R_{0k-1} & \sum_{j=0}^{n} R_{0j} \, R_{jl}^{-1} \\ \vdots & & \vdots & \vdots \\ R_{k0} & & R_{kk-1} & \sum_{j=0}^{n} R_{kj} \, R_{jl}^{-1} \end{vmatrix} = C_{kl}. \tag{13.8}$$

Putting (13.8) into (13.7) we get $\delta_{il} = \sum_{k=0}^{n} A_{ik} C_{kl}$ and so $C = A^{-1}$.

We see that $R^{-1} = (AA^T)^{-1} = C^T C$ and so

$$\sum_{j=0}^{n} \sum_{k=0}^{n} R_{jk}^{-1} \, u_j \, v_k = \sum_{l=0}^{n} (\alpha_l \, \alpha_{l-1})^{-1} \begin{vmatrix} R_{00} & \cdots & R_{0l-1} & u_0 \\ \vdots & & \vdots & \vdots \\ R_{l0} & \cdots & R_{ll-1} & u_l \end{vmatrix} \cdot \begin{vmatrix} R_{00} & \cdots & R_{0l-1} & v_0 \\ \vdots & & \vdots & \vdots \\ R_{l0} & \cdots & R_{ll-1} & v_l \end{vmatrix}, \tag{13.9}$$

where $u_0, \cdots, u_n, v_0, \cdots, v_n$ are variables. Setting $u_j = R_{j0}(0, s), j = 0, \cdots, n$ and $v_k = R_{k0}(0, t)$, $k = 0, \cdots, n$ we obtain (13.5).

We have proved that R^* is a covariance. In fact it is the covariance of Z. (5.8) (i) is proved. To prove (5.8) (ii) we observe that $D_1{}^i D_2{}^i R^*$ is the covariance of $Z^{(i)}$ and by (13.1)

$$Z^{(i)}(0) = Y^{(i)}(0) - \sum_{j=0}^{n} Y^{(j)}(0) B_j{}^{(i)}(0). \tag{13.10}$$

But $B_j^{(i)}(0) = \sum_{k=0}^n B_{kj}R_{ki} = \delta_{ij}$ by (13.2) since $B = R^{-1}$. Hence (5.9) holds and so (5.8) (ii) is proved. Now (5.8) (iii) follows immediately from the definition (5.12) of $A_0, \cdots, A_n$. Note that

$$(13.11)\quad A_i^{(i)}(0) = (\alpha_i\,\alpha_{i-1})^{-\frac{1}{2}}\begin{vmatrix} R_{00} & \cdots & R_{0i-1} & R_{0i} \\ & & \vdots & \vdots \\ R_{i0} & \cdots & R_{ii-1} & R_{ii}\end{vmatrix} = (\alpha_i/\alpha_{i-1})^{\frac{1}{2}}.$$

It remains only to prove the uniqueness of (5.7). Suppose that (5.7) holds where $A_0, \cdots, A_n$ and R^* satisfy (5.8). Differentiating i times on s and j times on t and setting $s = t = 0$ we get for $i = 0, \cdots, n$ and $j = 0, \cdots, n$,

$$(13.12)\qquad R_{ij} = \sum_{k=0}^n A_k^{(i)}(0)A_k^{(j)}(0).$$

We have used the fact that $D_1^iD_2^jR^*(0, 0) = 0$ which follows from (i) and (ii) of (5.8). Using (5.8) (iii) we see that $A = A_k^{(i)}(0)$ is an lsd square root of R which we know to be unique. Now we differentiate (5.7) j times on s only and then set $s = 0$. Again, $D_1^jR^*(0, t) = 0$ and we get

$$(13.13)\qquad R_{j0}(0, t) = \sum_{i=0}^n A_i^{(j)}(0)A_i(t), \qquad j = 0, \cdots, n.$$

The matrix $A = A_i^{(j)}(0)$ is nonsingular because of the nondegeneracy at zero and so the linear equations (13.13) have a unique solution $A_0(t), \cdots, A_n(t)$ for each fixed t. We have proved Theorem 7.

14. Proof of Theorem 8. Let μ be a measure with mean m and cov R. The proof of Theorem 8 will be based on the decomposition (5.7). Given a process Z with covariance R^* we may realize a process Y with covariance R by setting

$$(14.1)\qquad Y(t) = \sum_{j=0}^n \xi_jA_j(t) + Z(t)$$

where $\xi_0, \cdots, \xi_n$ are standard normal variates independent of Z. Using (14.1) we will obtain the simultaneous representation of μ and $\mu_{\mathbf{W}_n}$.

Suppose that m and R satisfy the hypothesis of Theorem 8. Let $\varphi_1, \varphi_2, \cdots$ be the complete o.n. system of eigenfunctions of K in (5.14) and set

$$(14.2)\qquad \Phi_{j,n}(t) = \int_0^t [(t-u)^n/n!]\varphi_j(u)\,du.$$

The process $\mathbf{W}_n$ has the representation

$$(14.3)\qquad \mathbf{W}_n(t) = \sum_{j=0}^n \xi_jt^j/j! + \sum_{j=1}^\infty \eta_j\Phi_{j,n}(t),$$

where $\xi_0, \cdots, \xi_n, \eta_1, \cdots$ are independent standard normal variates on some space. The process Y defined on the same space as the ξ's and η's by

$$(14.4)\quad Y(t) = \sum_{j=0}^n \xi_jA_j(t) + \sum_{j=1}^\infty [k_j + (1-\lambda_j)^{\frac{1}{2}}\eta_j]\Phi_{j,n}(t) + \sum_{j=0}^n m^{(j)}(0)t^j/j!$$

has mean

$$EY(t) = \sum_{k=1}^\infty k_j\Phi_{j,n}(t) + \sum_{j=0}^n m^{(j)}(0)t^j/j! = m(t)$$

by (5.6) and covariance R in (5.7) as a short calculation shows. Now it is easy to obtain a formal expression for $d\mu/d\mu_{\mathbf{W}_n}$. We compute the formal likelihood ratio under the two measures as follows:

We expand (formally)

$$(14.5)\quad X(t) = \sum_{i=0}^{n} X^{(i)}(0)t^i/i! + \sum_{j=1}^{\infty} X_j\Phi_{j,n}(t), \qquad X_j = \int_0^T \varphi_j(u)\, dX^{(n)}(u)$$

and denote $X^i = X^{(i)}(0)$. Under $\mu_{\mathbf{W}_n}$ we have $X^i = \xi_i$, $X_j = \eta_j$. Under μ we get with $m^i = m^i(0)$ from (14.4),

$$(14.6)\qquad X^i = m^i + \sum_{j=0}^{n} \xi_j A_j^{(i)}(0),$$

$$(14.7)\qquad X_j = k_j + \eta_j(1 - \lambda_j)^{\frac{1}{2}} + \sum_{i=0}^{n} \xi_i a_{ij},$$

where $a_{ij} = (a_i, \varphi_j)$. Inverting (14.6) with $C = A^{-1}$ given by (13.7), we get

$$(14.8)\qquad \xi_j = e_j = \sum_{i=0}^{n} C_{ji}(X^i - m^i).$$

Therefore,

$$(14.9)\qquad \eta_j = (X_j - k_j - u_j)/(1 - \lambda_j)^{\frac{1}{2}}, \qquad u_j = \sum_{i=0}^{n} e_i a_{ij}.$$

The likelihood ratio becomes

$$(14.10)\quad d\mu(X)/d\mu_{\mathbf{W}_n}(X) = \exp\left[-\tfrac{1}{2}\sum_{j=0}^{n}(e_j^2 - (X^j)^2) - \tfrac{1}{2}\sum_{j=1}^{\infty}(\eta_j^2 - X_j^2)\right]|J|$$

where $|J|$ is the Jacobian of the transformation (14.6), (14.7) of $X^j, X_j \to \xi_j, \eta_j$. Now $|\partial\xi_j/\partial X^i| = |C_{ji}| = C_{00}\cdots C_{nn}$ because C is lsd. Now $C_{ii} = 1/A_{ii}$ and $A_{ii} = (\alpha_i/\alpha_{i-1})^{\frac{1}{2}}$ by (5.13). We get $C_{00}\cdots C_{nn} = (\alpha_n)^{-\frac{1}{2}}$. We have $|\partial\eta_i/\partial X_j| = [\prod_j (1 - \lambda_j)^{\frac{1}{2}}]^{-1} = [d(1)]^{-\frac{1}{2}}$ and so $|J| = (d(1)\alpha_n)^{-\frac{1}{2}}$. Now,

$$\sum_{j=1}^{\infty}(\eta_j^2 - X_j^2) = \sum_{j=1}^{\infty}[(X_j - k_j)^2/(1 - \lambda_j) - X_j^2] - 2\sum (X_j - k_j)u_j/(1 - \lambda_j) + \sum u_j^2/(1 - \lambda_j).$$

For the first term on the right see (12.2)–(12.4). For the others we obtain

$$(14.11)\qquad \sum_{j=1}^{\infty}(X_j - k_j)u_j = \sum_{i=0}^{n} e_i \int_0^T a_i(t)\, d\bar{X}^{(n)}(t),$$

$$\sum_{j=1}^{\infty}(X_j - k_j)u_j\lambda_j/(1 - \lambda_j) = \sum_{i=0}^{n} e_i \int_0^T\int_0^T H(s, t)a_i(s)\, ds\, d\bar{X}^{(n)}(t),$$

where $\bar{X}(t) = X(t) - m(t)$ and also

$$(14.12)\qquad \sum_{j=1}^{\infty} u_j^2 = \sum_{i=0}^{n}\sum_{k=0}^{n} e_i e_k(a_i, a_k),$$

$$\sum_{j=1}^{\infty} u_j^2\lambda_j/(1 - \lambda_j) = \sum_{i=0}^{n}\sum_{k=0}^{n} e_i e_k(Ha_i, a_k).$$

Now (14.11) and (14.12) give (5.16). The formal calculations can be made precise when the hypothesis—(5.6), (5.14) and (5.15)—of Theorem 8 holds. This procedure is an imitation of the sufficiency proof for Theorems 2 and 3 and is omitted.

The necessity of (5.14) is easy because if X is a process with $X(t) \sim \mathbf{W}_n(t)$, it follows from Theorem 6 that $X^{(n)}(t) - X^{(n)}(0) \sim W(t)$, and so (5.14) follows from (1.1).

To prove the necessity of (5.6), suppose X is a process with mean m and $X \sim \mathbf{W}_n$. It follows, see Section 12, that $\mathbf{W}_n \sim \mathbf{W}_n + m$. Using the transformation $Y(t) \to Y^{(n)}(t) - Y^{(n)}(0)$ on both sides of the latter equivalence we obtain

$$(14.13) \quad \mathbf{W}_n^{(n)}(t) - \mathbf{W}_n^{(n)}(0) \sim \mathbf{W}_n^{(n)}(t) - \mathbf{W}_n^{(n)}(0) + m^{(n)}(t) - m^{(n)}(0).$$

But $W(t) = \mathbf{W}_n^{(n)}(t) - \mathbf{W}_n^{(n)}(0)$ and so $W \sim W + m^{(n)}(t) - m^{(n)}(0)$. Now (1.1) gives (5.6) immediately.

The necessity of (5.15) is formally obvious but apparently is difficult to prove. We proceed as follows: Suppose X is a process and $X \sim \mathbf{W}_n$. Using the simultaneous representation of X and $\mathbf{W}_n$, (14.3) and (14.4), we see that the sequences of random variables $S_j^1 = \eta_j$, $j = 1, 2, \cdots$, and $S_j^2 = (1 - \lambda_j)^{\frac{1}{2}}\eta_j + k_j + \sum_{i=0}^{n} a_{ij}\xi_i$, $j = 1, 2, \cdots$, are equivalent. Here $\xi_0, \cdots, \xi_n, \eta_1, \cdots$ are independent, standard normal variates. In order to prove (5.15) we must show that

$$(14.14) \qquad \sum_{j=1}^{\infty} a_{ij}^2 < \infty, \qquad i = 0, 1, \cdots, n.$$

Let μ_1 and μ_2 be the measures induced by S^1 and S^2 respectively on the space of infinite sequences. Then $\mu_1 \sim \mu_2$ and so the Hellinger integral [14]

$$(14.15) \qquad \int (d\mu_1\, d\mu_2)^{\frac{1}{2}} > 0.$$

The integral can be evaluated explicitly. Given $\xi_0, \cdots, \xi_n$ both S^1 and S^2 are sequences of independent random variables and so

$$(14.16) \quad \int (d\mu_1\, d\mu_2)^{\frac{1}{2}} = \int_{-\infty}^{\infty} \cdots \int_{-\infty}^{\infty} \exp\left(-\tfrac{1}{2}(\xi_0^2 + \cdots + \xi_n^2)\right) \cdot \prod_{j=1}^{\infty} r_j(\xi_0, \cdots, \xi_n)\, d\xi_0 \cdots d\xi_n,$$

where

$$(14.17) \quad r_j = (2\pi)^{-\frac{1}{2}} \int_{-\infty}^{\infty} \exp\left(-\tfrac{1}{4}x^2\right) \exp\left(-\tfrac{1}{4}(x - b_j)^2/(1 - \lambda_j)\right) dx/(1 - \lambda_j)^{\frac{1}{4}},$$

and $b_j = k_j + \sum_{i=0}^{n} a_{ij}\xi_i$, $j = 1, 2, \cdots$. Evaluating the integral (14.17), (14.16) becomes

$$(14.18) \quad \Lambda \cdot \int_{-\infty}^{\infty} \cdots \int_{-\infty}^{\infty} \exp\left(-\tfrac{1}{2}(\xi_0^2 + \cdots + \xi_n^2)\right) \cdot \prod_{j=1}^{\infty} \exp\left(-b_j^2/4(2 - \lambda_j)\right) d\xi_0 \cdots d\xi_n,$$

where

$$(14.19) \qquad \Lambda = \prod_{j=1}^{\infty} (1 - \lambda_j)^{\frac{1}{4}}/(1 - \lambda_j/2)^{\frac{1}{2}}.$$

The product (14.19) converges because $\sum \lambda_j^2 < \infty$. Now (14.18) is positive by (14.15) and it follows that

$$(14.20) \qquad \sum_{j=1}^{\infty} b_j^2 < \infty$$

for a set of $\xi_0, \cdots, \xi_n$ of positive measure in Euclidean $n + 1$ space. We have already proved the necessity of (5.6) and so $\sum k_j^2 < \infty$. Thus

$$(14.21) \qquad \sum_{j=1}^{\infty} \left(\sum_{i=0}^{n} a_{ij}\xi_i\right)^2 < \infty$$

for a set of $\xi_0, \cdots, \xi_n$ of positive measure. It follows readily that (14.14) must hold and so (5.15) is proved. This completes the proof of Theorem 8.

IV. EXAMPLES OF R-N DERIVATIVES

The examples given here will be applied in the next section in order to obtain new information about W and other processes. They will also serve to illustrate how Theorems 2 and 8 are applied.

15. Scale changes of W. Let h be an absolutely continuous and *increasing* function on $[0, T]$ with $h(0) = 0$. Define the process

$$Z(t) = [h'(t)]^{-\frac{1}{2}}W(h(t)), \qquad 0 \leqq t \leqq T. \tag{15.1}$$

As can be seen, the differential increments of Z have the same variance as those of W. Under very general conditions on h, which are given precisely by the following theorem, $Z \sim W$.

THEOREM 11. *$Z \sim W$ if and only if $h' = g^{-2}$ where g is absolutely continuous and $g' = \gamma \varepsilon L^2$.*

For smooth h, say $h \varepsilon C^3$ the R-N *derivative is*

$$(d\mu_Z/d\mu_W)(X) = (h'(T)/h'(0))^{\frac{1}{4}} \cdot \exp\{-[X^2(T)/4]h''(T)/h'(T) - \tfrac{1}{2}\textstyle\int_0^T f(t)X^2(t)\,dt\} \tag{15.2}$$

where $f = -\frac{1}{2}((h''/h')' - \frac{1}{2}(h''/h')^2)$, *the Schwarzian derivative of h.*

Doob [5] considered the more general scale change $Z(t) = v(t)W(u(t)/v(t))$ with covariance

$$R(s, t) = u(\min(s, t))v(\max(s, t)). \tag{15.3}$$

Such covariances were called *triangular* by Varberg [25], who calculated the R-N derivative for equivalent triangular processes by evaluating directly the limit of the finite dimensional densities. The densities can be written explicitly because Z is a Markov process.

In order to prove the theorem we find from (15.3) that

$$\begin{aligned} K(s, t) &= -u'(s)v'(t), \qquad s \leqq t, \\ &= -u'(t)v'(s), \qquad t \leqq s, \end{aligned} \tag{15.4}$$

where $u = h/(h')^{\frac{1}{2}}$, $v = (h')^{-\frac{1}{2}}$. Now if $\gamma \varepsilon L^2$ then $v' = \gamma$, $u' = h\gamma + |g|^{-1}$, are both in L^2 and so $K \varepsilon \mathbf{L}^2$. Conversely if $K \varepsilon \mathbf{L}^2$ then v' and hence $\gamma \varepsilon L^2$. We will see later that $1 \not\varepsilon \sigma(K)$ is automatically true and so by Theorem 1 we have proved that $\gamma \varepsilon L^2$ is necessary and sufficient.

Now assume that h is smooth, say $h \varepsilon C^3$. We make this assumption in order to use integration by parts at one step to express $d\mu_Z/d\mu_W$ in as simple a form as possible.

We will calculate the Fredholm determinant and resolvent of K at $\lambda = 1$. The eigenvalue problem $K\varphi = \lambda\varphi$ can be written

$$-v'\textstyle\int_0^t (u'\varphi) - u'\int_t^T (v'\varphi) = \lambda\varphi \tag{15.5}$$

where the arguments have been suppressed. Now it is easy to check that

$$u'' = fu, \qquad v'' = fv, \tag{15.6}$$

$$u(0) = 0, \qquad v'(T) = -\tfrac{1}{2}v(T)h''(T)/h'(T) \tag{15.7}$$

where

$$f = -\tfrac{1}{2}[(h''/h')' - \tfrac{1}{2}(h''/h')^2] \tag{15.8}$$

and so R is in fact a Green function. Letting $y(t) = \int_0^t \varphi$, a direct calculation shows that the eigenvalues λ of (15.5) are those λ for which

$$y'' = (1 - \lambda^{-1})fy, \quad y(0) = 0; \quad y'(T)/y(T) = (1 - \lambda^{-1})v'(T)/v(T) \tag{15.9}$$

has a nontrivial solution y. Now it is clear that $\lambda = 1 \notin \mathrm{sp}\,(K)$ because if $K\varphi = \varphi$ then by (15.9) $y = 0$ and so $\varphi = 0$. To handle nonsmooth h, we would replace (15.6) and (15.9) by integrated versions, integral equations. The details are straightforward.

We want $d(1) = \prod (1 - \lambda_j)$ where λ_j satisfy (15.9). Define for any complex β the unique solution y_β to

$$y_\beta'' = \beta f y_\beta, \qquad y_\beta(0) = 0, \qquad y_\beta'(0) = 1. \tag{15.10}$$

Set

$$D(\beta) = y_\beta'(T) - \beta y_\beta(T)v'(T)/v(T). \tag{15.11}$$

It is known from the general theory of differential equations that $y_\beta(T)$ and $y_\beta'(T)$ depend analytically on β. Furthermore it is not difficult to show that $y_\beta(T)$ is entire and of order $\frac{1}{2}$. Also $y_\beta'(T)$ is entire and of order $\frac{1}{2}$. It follows that $D(\beta)$ is entire and of order $\frac{1}{2}$—hence $D(\beta)$ is its own canonical product:

$$D(\beta) = \prod (1 - \beta/\beta_j), \tag{15.12}$$

where $\beta_1, \beta_2, \cdots$ are the zeros of D. Now, we have

$$\beta_j = 1 - \lambda_j^{-1} \tag{15.13}$$

because $D(\beta) = 0$ if and only $\beta = 1 - \lambda^{-1}$ where λ satisfies (15.9) for some nonzero $y = y_\beta$ (note that y_β is nonzero because $y_\beta'(0) = 1$). Therefore

$$d(1) = \prod (1 - \lambda_j) = \left(\prod (1 - \beta_j^{-1})\right)^{-1} = (D(1))^{-1}.$$

What is $D(1)$? By (15.11),

$$D(1) = [y_1'(T)v(T) - y_1(T)v'(T)]/v(T). \tag{15.14}$$

For $\beta = 1$ (and only for this value),

$$y_\beta'(t)v(t) - \beta y_\beta(t)v'(t) = \text{constant} \tag{15.15}$$

being the Wronskian. Putting $t = 0$ to evaluate the constant we get

$$D(1) = v(0)/v(T) = (h'(T)/h'(0))^{\frac{1}{2}}. \tag{15.16}$$

Finally,

(15.17) $$d(1) = (h'(0)/h'(T))^{\frac{1}{2}}.$$

Next we calculate the resolvent kernel H. It will turn out that

(15.18) $$H(s, t) = \theta(\max(s, t))$$

where

(15.19) $$\theta'(t) = -f(t), \qquad \theta(T) = \tfrac{1}{2}h''(T)/h'(T)$$

H is the unique continuous solution to

(15.20) $$H - K = HK = KH.$$

It is straightforward to check that (15.18) satisfies (15.20) and so (15.18) is proved since H is unique in (15.20).

By (2.4) we have since $m = 0$,

(15.21) $$(d\mu_Z/d\mu_W)(X) = [d(1)]^{-\frac{1}{2}} \exp[-\tfrac{1}{2} \int_0^T \int_0^T H(s,t)\, dX(s)\, dX(t).$$

Applying (9.13) or proceeding formally (note that $X(0) = 0$) we have

(15.22)
$$\begin{aligned} &\int_0^T \int_0^T H(s, t)\, dX(s)\, dX(t) \\ &\quad = 2 \int_0^T (\int_0^t H(s, t)\, dX(s))\, dX(t) \\ &\quad = 2 \int_0^T (\int_0^t \theta(t)\, dX(s))\, dX(t) = \int_0^T \theta(t)\, dX^2(t) \\ &\quad = X^2(T)\theta(T) + \int_0^T X^2(u) f(u)\, du. \end{aligned}$$

We obtain (15.2) immediately and Theorem 11 is proved.

Let W^* be the pinned Wiener process,

(15.23) $$W^*(t) = W(t) - tW(1), \qquad 0 \leqq t \leqq 1.$$

The scale change

(15.24) $$Z(t) = [h'(t)]^{-\frac{1}{2}} W^*(h(t))$$

will have $Z \sim W^*$ when h satisfies the hypotheses of Theorem 11 and in addition

(15.25) $$h(1) = 1.$$

By identical techniques, this time using Theorem 9, one shows that the R-N derivative is

(15.26) $$(d\mu_Z/d\mu_{W^*})(X) = (h'(0)h'(1))^{-\frac{1}{4}} \exp[-\tfrac{1}{2} \int_0^1 f(t) X^2(t)\, dt]$$

where f is again the Schwarzian derivative of h. (15.26) will be applied in Section 18 in order to evaluate certain integrals.

16. The linear covariance. Consider the measure μ with $m = 0$ and covariance

(16.1) $$R(s, t) = \tfrac{1}{2}(1 - |t - s|)$$

with $T \leqq 2$. Now $\mu \sim \mu_{W_0}$ and we will use the decomposition method of Theorems 7 and 8 in order to obtain $d\mu/d\mu_{W_0}$. Recall that $\mathbf{W}_0(t) = \xi_0 + W(t)$.

For $n = 0$ and $m = 0$, (5.16) becomes

$$(16.2)\quad \begin{aligned} (d\mu/d\mu_{W_0})(X) = [d(1)\, R(0,0)]^{-\frac{1}{2}} \exp [&- \tfrac{1}{2} \textstyle\int_0^T\int_0^T H(s,t)\, dX(s)\, dX(t) \\ &+ \tfrac{1}{2}X^2(0)(1 - [R(0,0)]^{-1}((I+H)a_0, a_0)) \\ &+ X(0)[R(0,0)]^{-\frac{1}{2}} \textstyle\int_0^T (I+H)a_0(t)\, dX(t)]. \end{aligned}$$

In the case (16.1) we have $A_0(t) = (1-t)/2^{\frac{1}{2}}$, $a_0(t) = -1/2^{\frac{1}{2}}$. We get in turn

$$(16.3)\quad \begin{aligned} R^*(s,t) &= s - st/2; \qquad s < t, \\ K(s,t) &= \tfrac{1}{2}, \\ H(s,t) &= 1/(2-T), \end{aligned}$$

so that K and H are constants. The spectrum of K is the single element $\lambda = T/2$. We must exclude $T = 2$ by (1.2) and indeed for $T = 2$ it is clear that $\mu \perp \mu_{W_0}$ since $Z(2) = -Z(0)$ a.s. for $\mu = \mu_Z$. We obtain $d(1) = 1 - T/2$.

We obtain further

$$(16.4)\quad \begin{aligned} (I+H)a_0(t) &= a_0(t) + \textstyle\int_0^T H(t,s)a_0(s)\, ds \\ &= -2^{\frac{1}{2}}/(2-T), \\ ((I+H)a_0, a_0) &= T/(2-T), \\ \textstyle\int_0^T\int_0^T H(s,t)\, dX(s)\, dX(t) &= (X(t) - X(0))^2/(2-T). \end{aligned}$$

We put all the above ingredients into (16.2) and we get

$$(16.5)\quad \begin{aligned} &(d\mu/d\mu_{W_0})(X) \\ &\quad = [2/(2-T)^{\frac{1}{2}}] \exp [X^2(0)/2 - (X(0) + X(T))^2/2(2-T)]. \end{aligned}$$

(16.5) can also be obtained by a direct passage to the limit from the finite dimensional densities of μ, which have been found explicitly by Slepian [23]. The results agree.

We draw attention to the normalization inherent in considering the Wiener process with *unit* diffusion constant. A process Z may be equivalent to σW, for some $\sigma \neq 1$, rather than to W itself. This is only a scale factor and gives no trouble. We get $Z/\sigma \sim W$ and

$$(16.6)\quad (d\mu_Z/d\mu_{\sigma W})(X) = (d\mu_{Z/\sigma}/d\mu_W)(\sigma^{-1}X).$$

Sometimes $\sigma = 1$ is not the most convenient normalization. Instead of (16.1) it is more usual to consider

$$(16.7)\quad \bar{R}(s,t) = 1 - |t-s|.$$

Let us denote by $\bar{\mu}$ the measure with cov $\bar{R}$ and mean 0. Let $\bar{\mu}_0$ denote the measure $\bar{\mu}$ conditioned so that $X(0) = x_0$ given. The mean of $\bar{\mu}_0$ is

$$(16.8)\quad \bar{m}_0(t) = x_0\bar{R}(0,t)/\bar{R}(0,0)$$

and the cov of $\bar{\mu}_0$ is

$$(16.9) \qquad \bar{R}_0(s, t) = \bar{R}(s, t) - \bar{R}(0, s)\bar{R}(0, t)/\bar{R}(0, 0).$$

$\bar{\mu}_0$ is a Gaussian measure and is equivalent to the Wiener-type measure corresponding to the process

$$(16.10) \qquad \omega(t) = x_0 + 2^{\frac{1}{2}}W(t), \qquad 0 \leqq t \leqq T.$$

The R-N derivative is easy to obtain and is another form of (16.5); it can be computed also from (2.4). We have

$$(16.11) \quad (d\bar{\mu}_0/d\mu_\omega)(X)$$
$$= [2/(2 - T)]^{\frac{1}{2}} \exp\left(\tfrac{1}{2}x_0^2\right) \exp\left[-\tfrac{1}{2}(x_0 + X(T))^2/2(2 - T)\right].$$

We will apply (16.11) in Section 17 to solve a certain first passage problem.

The example of this section was chosen for its simplicity. We can obtain with our methods all known examples of R-N derivatives in the Gaussian case as well as many new examples. The calculations, although straightforward, are sometimes quite tedious.

We have given illustrations by example of all the main theorems. The reader may find it instructive to obtain the R-N derivative in the Ornstein-Uhlenbeck case (Example 3 of [25]) by our methods.

V. SOME APPLICATIONS OF R-N DERIVATIVES

The R-N derivative has found its main application in statistics as the likelihood ratio. However, it can also be used as a theoretical tool, as Skorokhod [22], p. 408, pointed out. We will use it both to calculate probabilities and to evaluate function space integrals.

17. Calculating probabilities. We shall prove that if $b > 0$,

$$(17.1) \quad \Pr\{W(t) < at + b, 0 \leqq t \leqq T\}$$
$$= \Phi((aT + b)/T^{\frac{1}{2}}) - e^{-2ab}\Phi((aT - b)/T^{\frac{1}{2}}).$$

We see that (17.1) is the probability that a gambler with income does not go broke in time $T \leqq \infty$ if b represents his initial capital, a his fixed rate of income (or outgo), and W his losses due to chance fluctuations. (17.1) was found by Doob [5] for $T = \infty$ and was given in general by Malmquist in a different form and by a different method [16].

To prove (17.1) we observe that by definition of $d\mu/d\mu_W$,

$$(17.2) \qquad \mu(A) = \int_A (d\mu/d\mu_W)(X)\, d\mu_W$$

for any event A. Let $\mu = \mu_{W+m}$ where $m(t) = -at$ and take $A = \{X: X(t) < b, 0 \leqq t \leqq T\}$. By (2.4) we obtain since $k(t) = -a$,

$$(17.3) \qquad F(X) = (d\mu/d\mu_W)(X) = e^{-aX(T)}e^{-\frac{1}{2}a^2T}.$$

Now by (17.2),

$$(17.4) \qquad \mu(A) = e^{-\frac{1}{2}a^2T}\int_A e^{-aX(T)}\, d\mu_W(X)$$
$$= e^{-\frac{1}{2}a^2T}\int_{-\infty}^{b} e^{-ax}\,\Pr\{M(T) < b;\, W(T) = x\}\, dx$$

where $M(T) = \max_{0\leq t\leq T} W(t)$. But the reflection principle gives for $x < b$,

$$(17.5) \quad \Pr\{M(T) < b;\, W(T) = x\}$$
$$= \Pr\{W(T) = x\} - \Pr\{W(T) = 2b - x\}.$$

Using

$$\Pr\{W(T) = x\} = (2\pi T)^{-\frac{1}{2}}e^{-\frac{1}{2}x^2/T}$$

and $\mu(A) = \Pr\{W(t) < at + b,\, 0 \leq t \leq T\}$ we obtain (17.1) after a simple calculation. The point of using (17.2) is that the reflection principle fails for slanted lines $at + b$, while (17.2) enables us to reduce the problem to the horizontal line.

As a second example of the same technique, we solve a first passage problem originally given by Slepian [23] in a different form. Let $S = S(t)$, $0 \leq t \leq T$ be the process with mean zero and covariance

$$(17.6) \qquad R(s, t) = 1 - |t - s|, \qquad |t - s| \leq 1,$$
$$= 0, \qquad |t - s| > 1.$$

Suppose $T \leq 1$. Let

$$(17.7) \qquad P_a^+(T \mid x_0) = \Pr\{S(t) < a,\, 0 \leq t \leq T \mid S(0) = x_0\}, \qquad x_0 < a,$$
$$P_a^-(T \mid x_0) = \Pr\{S(t) > a,\, 0 \leq t \leq T \mid S(0) = x_0\}, \qquad x_0 > a,$$

be the first passage probabilities.

We shall prove

$$(17.8) \quad P_a^\pm(T \mid x_0) = \Phi(\pm[a - x_0(1 - T)]/[T(2 - T)]^{\frac{1}{2}})$$
$$- e^{-\frac{1}{2}(a^2 - x_0^2)}\Phi(\pm[x_0 - a(1 - T)]/[T(2 - T)]^{\frac{1}{2}}).$$

The formulas break down for $T > 1$ and in this case the problem remains unsolved.

Let S_0 denote the process S conditioned to pass through x_0 at $t = 0$. We have seen that $S_0 \sim \omega = x_0 + 2^{\frac{1}{2}}W$ and the R-N derivative is given by (16.11). Let

$$A = \{X : X(t) \leq a,\, 0 \leq t \leq T\}.$$

Applying (17.2) with $\mu = \mu_{S_0}$ and μ_W replaced by μ_ω we get

$$(17.9) \quad P_a^+(T \mid x_0)$$
$$= [2/(2 - T)]^{\frac{1}{2}}e^{\frac{1}{4}x_0^2}\int_A \exp\,[-\tfrac{1}{2}(x_0 + X(T))^2/2(2 - T)]\, d\mu_\omega(X).$$

Using the reflection principle to evaluate the integral, we obtain (17.8). P^- is obtained in a similar way. In comparing our form of the answer with Slepian's note that his $Q_a(T \mid x_0)$ is the derivative on T of our $P_a(T \mid x_0)$.

In principle one could use the method of this section to solve other first passage problems. However, these are the only cases in which we could evaluate the integrals explicitly.

18. Evaluating some special integrals. The integral of the R-N derivative over all space is unity. This fact will permit us to evaluate the Wiener integral of any functional that is the exponential of a quadratic form. Some special cases of this result are due to Kac and Siegert [12] who obtained them by other methods. Their results were later simplified by Anderson and Darling [1]. We will then apply the evaluations to prove a dichotomy theorem about Wiener paths.

We begin with a general identity. *Let L be any symmetric continuous kernel of trace class. Then*

$$(18.1) \qquad E_W \exp\left[\tfrac{1}{2}\lambda \int_0^T \int_0^T L(s, t)\, dX(s)\, dX(t)\right] = (D^L(\lambda))^{-\frac{1}{2}}$$

where E_W is the Wiener integral, d^L is the Fredholm determinant of L, and λ satisfies

$$(18.2) \qquad 1 - \lambda\lambda_j > 0, \qquad j = 1, 2, \cdots,$$

where $\lambda_1, \lambda_2, \cdots$ are the eigenvalues of L. When (18.2) fails the integral is $+\infty$.

To prove (18.1) note that by (2.4) we have

$$(18.3) \qquad 1 = E_W[d(1)]^{-\frac{1}{2}} \exp\left[-\tfrac{1}{2}\int_0^T \int_0^T H(s, t)\, dX(s)\, dX(t)\right]$$

where $d(1) = d^K(1)$. Now if γ denotes eigenvalues of H then $\gamma - \lambda = \gamma\lambda$ and so

$$(18.4) \qquad d^K(1) = \textstyle\prod_\lambda (1 - \lambda) = \prod_\gamma (1 + \gamma)^{-1} = d^H(-1) = d^{-H}(1).$$

Let now $L = -\lambda H$ and we obtain (18.1). When (18.2) fails the quadratic form

$$(18.5) \qquad (\varphi, \varphi) - \lambda(L\varphi, \varphi)$$

is not positive-definite and it is at least formally clear that (18.1) should diverge. The proof is omitted. It is amusing to expand both sides of (18.1) in powers of λ and check coefficients of λ^n. The general equality seems to be difficult to prove directly.

Next we evaluate

$$(18.6) \qquad A(f) = E_W \exp\left[-\tfrac{1}{2}\int_0^T f(t) X^2(t)\, dt\right]$$

for any, say continuous, f. *Solve the one-point problem*

$$(18.7) \qquad g'' = fg, \qquad g'(T) = 0, \qquad g > 0 \text{ on } [0, T).$$

Then

$$(18.8) \qquad A(f) = (g(T)/g(0))^{\frac{1}{2}}.$$

When (18.7) *fails to have a solution positive on the half-open interval* [0, T) *then*

$A(F) = +\infty$. The result is also valid for $T = \infty$ provided $\lim g(x) = g(\infty) > 0$ and $\lim g'(x) = g'(\infty) = 0$.

Suppose now that the basic process is W^*, the pinned Wiener process, instead of W. *We have*

$$E_{W^*} \exp\left[-\tfrac{1}{2}\int_0^1 f(t)X^2(t)\,dt\right] = (l(1))^{-\frac{1}{2}} \tag{18.9}$$

where l is the unique solution to

$$l'' = fl, \qquad l(0) = 0, \qquad l'(0) = 1, \tag{18.10}$$

provided $l > 0$ in (0, 1]. *When the solution l of* (18.10) *has a zero in* (0, 1] *then* (18.9) *is* $+\infty$.

For positive f, (18.9) is finite and was obtained by Anderson and Darling [1] who simplified a previous formula due to Kac and Siegert [13].

To prove (18.9) we proceed as follows. By (15.26) we have for any h with $h(0) = 0$, $h(1) = 1$ and $h' > 0$,

$$E_{W^*} \exp\left[-\tfrac{1}{2}\int_0^1 \bar{f}(t)X^2(t)\,dt\right] = (h'(0)h'(1))^{\frac{1}{4}} \tag{18.11}$$

where $\bar{f} = -\frac{1}{2}((h''/h')' - \frac{1}{2}(h''/h')^2)$. Now choose h to satisfy

$$h(0) = 0, \qquad h' = g^{-2}/\int_0^1 g^{-2} \tag{18.12}$$

where g is any *positive* solution of

$$g'' = fg. \tag{18.13}$$

We have $h' > 0$, $h(1) = 1$ and the Schwarzian of h is f,

$$\bar{f} = -\tfrac{1}{2}((h''/h')^1 - \tfrac{1}{2}(h''/h')^2) = g''/g = f. \tag{18.14}$$

By (18.11) we have

$$E_{W^*} \exp\left[-\tfrac{1}{2}\int_0^1 f(t)X^2(t)\,dt\right] = \left(g(0)g(1)\int_0^1 g^{-2}\right)^{-\frac{1}{2}}. \tag{18.15}$$

Let

$$l(t) = g(0)g(t)\int_0^t g^{-2}. \tag{18.16}$$

It is easy to check that l satisfies (18.10) and this gives (18.9). Now it is known from Sturm-Liouville theory that a positive solution g exists to (18.13) if and only if the solution l of (18.10) is positive in (0, 1]. It is easy to see that when l has a zero then (18.9) is actually $+\infty$. A similar proof can be given for (18.8). The formula (18.9) for W^* is more symmetric than (18.8) for W as is evident by (18.15). This is because W^* is time-reversible.

We have assumed f to be continuous but this is not necessary. In general, one expresses the differential equation $g'' = fg$, $g'(T) = 0$ in the form of an integral equation

$$g(t) = g(T) + \int_t^T (u - t)g(u)f(u)\,du. \tag{18.17}$$

One can now allow f to be simply measurable. Further, f can be replaced by a measure: $fdu = dm$.

As mentioned above $A(f)$ has been calculated for continuous $f > 0$ by Kac and others by using various techniques. However, the most direct way to obtain $A(f)$ in this case was found by Gel'fand and Yaglom [29]:

By Riemann integration and bounded convergence we have

$$A(f) = \lim_{n\to\infty} E_W \exp\left[-\tfrac{1}{2}\textstyle\sum_1^n f(jT/n)X^2(jT/n)T/n\right]. \tag{18.18}$$

Now the expected value on the right is an n-dimensional integral that can be evaluated explicitly. Let $f_j = f(jT/n)$, $x_j = X(jT/n)$. The integral in (18.18) is

$$\int_{-\infty}^{\infty} \cdots \int_{-\infty}^{\infty} \exp\left[-\tfrac{1}{2}\textstyle\sum f_j X_j^2 T/n - \tfrac{1}{2}\sum (x_j - x_{j-1})^2 n/T\right] dx_1 \cdots dx_n/(2\pi T/n)^{n/2} = (|b_{ij}^{(n)}|_1^n)^{-\frac{1}{2}} \tag{18.19}$$

where

$$\begin{aligned} b_{ij}^{(n)} &= 2 + f_j T/n^2, & i &= j = 1, \cdots, n-1, \\ &= -1, & |i-j| &= 1, \\ &= 1 + f_n T^2/n^2, & i &= j = n, \\ &= 0, & |i-j| &> 1. \end{aligned}$$

Define

$$g_k = g_k^{(n)} = |b_{ij}^{(n)}|_k^n, \qquad k = 1, 2, \cdots, n. \tag{18.20}$$

We see that $g_n = 1 + f_n T^2/n^2$, $g_n - g_{n-1} = O(T^2/n^2)$ and

$$\Delta^2 g_k = f_{k+1} g_k, \qquad k = 1, \cdots, n-1. \tag{18.21}$$

We compare (18.21) with the differential equation

$$g'' = fg, \qquad g(T) = 1, \qquad g'(T) = 0, \tag{18.22}$$

with $g(kT/n) \leftrightarrow g_k$. Standard techniques of comparison of difference and differential equations give

$$A(f) = \lim (g_1^{(n)})^{-\frac{1}{2}} = (g(0))^{-\frac{1}{2}} \tag{18.23}$$

which agrees with (18.8) since $g(T) = 1$.

If μ and ν are Gaussian then $d\mu/d\nu(X)$ is always the exponential of a quadratic form in X. With $\nu = \mu_W$, *the quadratic form is diagonal,*

$$\int_0^T X^2(u) f(u)\, du, \tag{18.24}$$

if and only if $\mu = \mu_Z$ *where* Z *is a scale change as in* (15.1). Indeed, as we have seen, (18.24) holds for $\mu = \mu_Z$. Conversely, we can obtain (18.24) for any f by means of a properly chosen scale change. Thus (18.24) characterizes scale change processes.

19. A dichotomy theorem for Wiener paths. Suppose $f \geqq 0$ on $[0, T]$. *The integral*

$$\int_0^T f(u) W^2(u)\, du \tag{19.1}$$

either converges a.s. *or diverges* a.s. *according as*

$$\int_0^T uf(u)\,du \tag{19.2}$$

converges or diverges. The convergence is trivial because if (19.2) is finite then

$$E_W\left(\int_0^T f(u)W^2(u)\,du\right) = \int_0^T uf(u)\,du < \infty \tag{19.3}$$

and so (19.1) must be finite a.s.

Now suppose $\int_0^T uf(u)\,du = \infty$. We will prove that

$$E_W \exp\left[-\int_0^T f(u)X^2(u)\,du\right] = 0 \tag{19.4}$$

and so $\int_0^T f(u)W^2(u) = \infty$ a.s. will follow. In order to avoid details we assume that f is continuous away from zero. By (18.8) it is enough to prove the following simple lemma.

LEMMA. *Suppose f is continuous on* $(0, T]$ *and* $\int_0^T uf(u)\,du = \infty$. *Then*

$$g'' = fg, \qquad g'(T) = 0, \qquad g(T) = 1, \tag{19.5}$$

has $g(0) = \infty$.

PROOF. Since g is convex and $g'(T) = 0$, g is monotonically decreasing. Suppose $g(0) = M < \infty$. Then $0 < -\epsilon g'(\epsilon) \leqq g(0) - g(\epsilon) \leqq M$ by convexity. We have, since $g \geqq 1$, for any $\epsilon > 0$

$$\begin{aligned} \int_\epsilon^T uf(u)\,du &< \int_\epsilon^T uf(u)g(u)\,du \\ &= \int_\epsilon^T ug''(u)\,du = Tg'(T) - \epsilon g'(\epsilon) - g(T) + g(\epsilon) \leqq 2M. \end{aligned} \tag{19.6}$$

This contradicts $\int_0^T uf(u)\,du = \infty$ and proves the lemma.

AN EXAMPLE. Let $f(u) = 2/(\epsilon + u)^2$, $0 < \epsilon$, $T < \infty$. The solution to (18.7) is

$$g(u) = A(\epsilon + u)^{-1} + B(\epsilon + u)^2, \qquad A = 2B(\epsilon + T)^3. \tag{19.7}$$

We obtain

$$E_W \exp\left[-\int_0^T (\epsilon + u)^{-2}W^2(u)\,du\right] = (3\epsilon/2(\epsilon + T)(1 + (\epsilon/\epsilon + T)^3/2))^{\frac{1}{2}}. \tag{19.8}$$

For other examples where (18.7) has an explicit solution see [1]. Letting $\epsilon \to 0$ we have by monotone convergence

$$E_W \exp\left[-\int_0^T u^{-2}W^2(u)\,du\right] = 0 \tag{19.9}$$

and so

$$\int_0^T u^{-2}W^2(u)\,du = \infty \tag{19.10}$$

for almost every path W. (19.10) also follows from a form of the iterated logarithm theorem recently found by V. Strassen [24]. However, it does not seem possible to obtain the general case of (19.2) in this way.

REFERENCES

[1] ANDERSON, T. W. and DARLING, D. A. (1952). Asymptotic theory of certain "goodness of fit" criteria based on stochastic processes. *Ann. Math. Statist.* **23** 193–212.

[2] BAXTER, G. (1956). A strong limit theorem for Gaussian processes. *Proc. Amer. Math. Soc.* **7** 522–527.

[3] CAMERON, R. H. and MARTIN, W. T. (1944). Transformations of Wiener integrals under translations. *Ann. Math.* **45** 386–396.

[4] CAMERON, R. H. and MARTIN, W. T. (1945). Transformations of Wiener integrals under a general class of linear transformations. *Trans. Amer. Math. Soc.* **58** 184–219.

[5] DOOB, J. L. (1949). Heuristic approach to the Kolmogorov-Smirnov theorems. *Ann. Math. Statist.* **20** 393–403.

[6] DOOB, J. L. (1953). *Stochastic Processes.* Wiley, New York.

[7] DUNFORD, N. and SCHWARZ, J. T. (1963). *Linear Operators Part II.* Interscience, New York.

[8] GANTMACHER, F. R. (1959). *The Theory of Matrices.* **1** Chelsea, New York.

[9] HALMOS, P. R. (1950). *Measure Theory.* Van Nostrand, New York.

[10] ITO, K. (1951). Multiple Wiener integral. *J. Math. Soc. Japan* **3** 157–169.

[11] ITO, K. and MCKEAN, H. P. (1965). *Diffusion Processes and Their Sample Paths.* Academic Press, New York.

[12] KAC, M. (1951). On some connections between probability theory and differential and integral equations. *Proc. Second Berkeley Symp. Math. Statist. Prob.* Univ. of California Press.

[13] KAC, M. and SIEGERT, A. J. F. (1947). On the theory of noise in radio receivers with square law detectors. *J. Appl. Phys.* **18** 383–397.

[14] KAKUTANI, S. (1948). On equivalence of infinite product measures. *Ann. Math.* **49** 214–224.

[15] KALLIANPUR, G. and OODAIRA, H. (1962). The equivalence and singularity of Gaussian measures, Chap. 19. *Proc. Symp. Time Series Analysis,* (M. Rosenblatt, ed.). Wiley, New York.

[16] MALMQUIST, S. (1954). On certain confidence contours for distribution functions. *Ann. Math. Statist.* **25** 523–533.

[17] RAO, C. R. and VARADARAJAN, V. S. (1963). Discrimination of Gaussian processes. *Sankhyā Ser. A* **25** 303–330.

[18] RIESZ, F. and SZ-NAGY, B. (1955). *Functional Analysis.* Ungar, New York.

[19] ROZANOV, YU. A. (1964). On probability measures in function spaces, corresponding to Gaussian stationary processes (in Russian). *Teor. Verojatnost. i Primenen.* **9** 448–465.

[20] SEGAL, I. E. (1958). Distributions in Hilbert space and canonical systems of operators. *Trans. Amer. Math. Soc.* **88** 12–41.

[21] SEGAL, I. E. (1965). Algebraic integration theory. *Bull. Amer. Math. Soc.* **71** 419–489.

[22] SKOROKHOD, A. V. (1957). On the differentiability of measures which correspond to stochastic processes. *Theor. Prob. Appl.* **2** 407–432.

[23] SLEPIAN, D. (1961). First passage time for a particular Gaussian process. *Ann. Math. Statist.* **32** 610–612.

[24] STRASSEN, V. (1964). An invariance principle for the law of the iterated logarithm. *Z. Wahrscheinlichkeitstheorie und Verw. Gebiets* **3** 211–226.

[25] VARBERG, D. E. (1961). On equivalence of Gaussian measures. *Pacific J. Math.* **11** 751–762.

[26] VARBERG, D. E. (1964). On Gaussian measures equivalent to Wiener measure. *Trans. Amer. Math. Soc.* **113** 262–273.

[27] WOODWARD, D. A. (1961). A general class of linear transformations of Wiener integrals. *Trans. Amer. Math. Soc.* **100** 459–480.

[28] YAGLOM, A. M. (1963). On the equivalence and perpendicularity of two Gaussian probability measures in function space, Chapter 22. *Proc. Symp. Time Series Analysis* (M. Rosenblatt, ed.). Wiley, New York.

[29] GEL'FAND, I. M. and YAGLOM, A. M. (1956). Integration in function spaces (in Russian). *Uspekhi Mat. Nauk.* **11** 77–114.

[30] DELPORTE, J. (1964). Fonctions aléatoires presque sûrement continues sur un intervalle fermé. *Ann. Inst. Henri Poincaré Sec. B* **1** 111–215.

Errata for Paper 19:

p. 332, footnote: for	Golosoy	substitute	Golosov.
p. 352, line 8, (18.19):	X_j	substitute	x_j.
line 11: for	$2+f_jT/n^2$	substitute	$2+f_jT^2/n^2$.
line 18, (18.21): for	$\Delta^2 g_k = f_{k+1} g_k$	substitute	$\Delta^2 g_k = T^2 f_k g_{k+1}/n^2$.

Editors' Comments on Paper 20

At present, knowledge of the Cramér–Hida original representation is widespread. Its theoretical consequences and its applied ramifications have been studied extensively. Its relationship to earlier landmark ideas by Wiener and others has placed it in historical perspective.

The following paper, by Z. Ivkovich and Prof. Yu. A. Rozanov of the Steklov Mathematical Institute of the U.S.S.R., the latter widely known for his contributions to many areas of mathematical statistics and applied probability, renews interest in the subject by uncovering a number of new aspects.

As this volume went to print, further results by Professor Rozanov had been presented but were not yet available for publication.

This paper is reprinted with the permission of the authors and the Society for Industrial and Applied Mathematics.

Reprinted from *Theory of Probability and Its Applications*, **16**, 348–353 (1971)

20

ON THE CANONICAL HIDA-CRAMÉR REPRESENTATION FOR RANDOM PROCESSES

Z. IVKOVICH AND YU. A. ROZANOV

(*Translated by K. Durr*)

This work contains a series of remarks on the so-called canonical Hida–Cramér representation: directly on the representation itself (see below (1) and (2)), on the notion of regularity of nonstationary processes, on the change of rank (multiplicity) under certain transformations (in particular, under projection), etc.

1°. Let $\xi = \xi(t)$ be a random process of second order on a closed or open interval $[a, b]$, where possibly $a = -\infty$ and $b = \infty$, with correlation function $B(s, t) = \mathbf{E}\xi(s)\xi(t)$. We shall consider $\xi = \xi(t)$ as a curve in the Hilbert space of all random variables η, $\mathbf{E}\eta^2 < \infty$, with scalar product

$$\langle \eta_1, \eta_2 \rangle = \mathbf{E}\eta_1\eta_2,$$

assuming the function $\xi = \xi(t)$ to be left continuous:

$$\xi(t-0) = \lim_{s \to t-0} \xi(s) = \xi(t).$$

Let $H_t(\xi)$ denote the closed linear hull of the variables $\xi(s), s \leqq t$. We set

$$H_{t+0}(\xi) = \bigcap_{h>0} H_{t+h}(\xi)$$

and denote by E_t the operator of projection onto $H_{t+0}(\xi)$.

The operator function $E_t, a \leqq t \leqq b$, is monotonically nondecreasing and right continuous. Considering E_t only on the separable space $H = H_b(\xi)$, we set $E_{a-0} = 0$ and $E_b = I$.

For any variables $\eta_1, \cdots, \eta_N$, the N-dimensional random process $\eta(t) = \{\eta_i(t)\}_{i=1,\cdots,N}$, considered as a vector column with components $\eta_i(t) = E_t\eta_i, a \leqq t \leqq b$, is a process with orthogonal increments: for any $s_1 \leqq s_2 \leqq t_1 \leqq t_2$,

$$\langle \eta_i(s_2) - \eta_i(s_1), \eta_j(t_2) - \eta_j(t_1) \rangle = 0, \qquad i, j = 1, \cdots, N.$$

Here the number N can also be infinite. Choosing $\eta_1, \cdots, \eta_N$ so that $H_b(\eta) = H_b(\xi)$ we have, for all t,

$$H_t(\eta) = H_{t+0}(\xi);$$

it is also clear that, for all $t > a$, $H_{t-0}(\eta) = H_t(\xi)$, where $H_{t-0}(\eta)$ denotes the closure of the subspace $\bigcup_{h<0} H_{t-h}(\eta)$. For $H_b(\eta) = H_b(\xi)$ we call $\eta(t) = \{\eta_i(t)\}_{i=1,\cdots,N}$ *the renewal process for* $\xi = \xi(t)$.

Let

$$F_{ij}(t) = \langle \eta_i(t), h_j(t) \rangle, \qquad i, j = 1, \cdots, N.$$

As is known, for a process with orthogonal increments $\eta(t) = \{\eta_i(t)\}_{i=i,\cdots,N}$ the matrix function $F(t) = \{F_{ij}(t)\}_{j=1,\cdots,N}^{i=1,\cdots,N}$ is nondecreasing and right continuous; we call $F(t)$ *the structure matrix* of the process $\eta(t)$.

Let $L^2(F)$ denote the Hilbert space of all vector functions $c(t) = \{c_i(t)\}^{i=1,\cdots,N}$ on the interval $[a, b]$ satisfying the condition[1]

$$\int_{a-0}^{b} c(t)\, dF(t) c(t)^* < \infty,$$

with scalar product

$$\langle c_1, c_2 \rangle = \int_{a-0}^{b} c_1(t)\, dF(t) c_2(t)^*.$$

Every element $\xi \in H_b(\eta)$ can be written in the form

$$\xi = \int_{a-0}^{b} c(t)\, d\eta(t) = \int_{a-0}^{b} \sum_{i=1}^{N} c_i(t)\, d\eta_i(t).$$

In particular, if $\eta(t) = \{\eta_i(t)\}_{i=1,\cdots,N}$ is the renewal process for $\xi = \xi(t)$, then

$$\xi(t) = \int_{a-0}^{t} c(t, s)\, d\eta(s), \tag{1}$$

where $c(t, s)$, as a function of s, belongs to $L^2(F)$ with $c(t, s) = 0$ for $s > t$.

In considering the process $\xi = \xi(t)$, it is natural to turn to its renewal process $\eta(t) = \{\eta_i(t)\}_{i=i,\cdots,N}$ of MINIMAL DIMENSION N; such an N will be called the *rank* of the process $\xi = \xi(t)$, and the corresponding representation of this process in the form of a stochastic integral (1) will be called its *canonical Hida–Cramér representation*.

A representation of this type was first proposed by Wold [1] in the special case of a regular stationary process $\xi = \xi(t)$ with discrete time t, $-\infty < t < \infty$, and by Cramér [2] for an arbitrary regular process; here regularity means that $\bigcap_t H_t(\xi) = 0$. In the Gaussian case such a representation was considered by Hida [3].

2°. As is known, for a regular stationary process, the rank N is always equal to 1. Considering various processes with known rank N, it is natural to ask how the rank is affected by various simple transformations of the initial process $\xi = \xi(t)$. Let us say that the rank is not affected by multiplication of $\xi(t)$ by a nonzero scalar function, by a monotonic and continuous substitution of time t, and by many other transformations. Let us dwell in more detail on projection of the

---

[1] The function $c(t) = \{c_i(t)\}^{i-1,\cdots,N}$ is a vector row; the function $c^*(t) = \{c_j(t)\}_{j=1,\cdots,N}$ is the vector column conjugate to it.

process $\xi(t)$—on a transformation of the form

$$\tilde{\xi}(t) = P\xi(t), \qquad a \leqq t \leqq b,$$

where P is projection onto some subspace $L \subseteq H$.

It originally seemed to us that, under such a transformation, the rank of the process would be preserved or diminished. However, this turned out to be false.

Example. Let $\xi(t)$ be a process with orthogonal increments on the interval $[0, 1]$, $\xi(0) = 0$. One can extend it formally by setting $\xi(t) = 0$ for $t < 0$ and $\xi(t) = \xi(1)$ for $t > 1$. Clearly, $\xi = \xi(t)$ is a regular process having rank $N = 1$. For definiteness, let its correlation function be $B(s, t) = \min(s, t)$ for $0 \leqq s, t \leqq 1$. Set

$$\eta_k = \int_0^1 \xi(t)\varphi_k(t)\,dt, \qquad k = 0, 1, \cdots,$$

where

$$\varphi_k(t) = \sin(k + \tfrac{1}{2})\pi t, \qquad 0 \leqq t \leqq 1,$$

are the eigenfunctions of the kernel of $B(s, t)$, and let us consider the subspace L generated by the variables $\eta_0, \eta_1, \cdots, \eta_n$.

As is known, the process $\xi(t)$ can be written in the form of a series in the eigenfunctions $\varphi_k(t)$:

$$\xi(t) = \sum_{k=0}^{\infty} \eta_k\varphi_k(t), \qquad 0 \leqq t \leqq 1,$$

where $\eta_0, \eta_1, \cdots,$ is an orthogonal basis in H. We shall show that the *regular continuous process*

$$\tilde{\xi}(t) = P\xi(t) = \sum_{k=0}^{n} \eta_k\varphi_k(t), \qquad 0 \leqq t \leqq 1,$$

$$(\tilde{\xi}(t) = 0 \text{ for } t < 0, \quad \tilde{\xi}(t) = \tilde{\xi}(1) \text{ for } t > 1)$$

has rank $N = n + 1$.

In fact, for any $t > 0$, there are $t_0, t_1, \cdots, t_n$, for which the matrix $\{\varphi_k(t_j)\}$ will be nonsingular, so that from the equations

$$\sum_{k=0}^{n} \eta_k\varphi_k(t_j) = \tilde{\xi}(t_j), \qquad j = 0, 1, \cdots, n,$$

we obtain

$$\eta_k = \sum_{j=0}^{n} c_{kj}\tilde{\xi}(t_j), \qquad k = 0, 1, \cdots, n;$$

thus $H_{+0}(\tilde{\xi}) = H_t(\tilde{\xi}) = H_\infty(\tilde{\xi})$, while $H_0(\tilde{\xi}) = 0$, and the rank $\tilde{N}$ of the process $\tilde{\xi}(t)$ is equal to $n + 1$, the dimension of the subspace

$$H_{+0}(\tilde{\xi}) \ominus H_0(\tilde{\xi}).$$

3°. Note that in the context of the general theory, the notion of regularity formally means that $H_{a+0} = H_a(= 0)$ and characterizes the "renewal" of the process $\xi = \xi(t)$ at the initial point $t = a$. Hence, the question of regularity is a part of the question concerning the presence of jumps in the renewal process $\eta(t) = \{\eta_i(t)\}_{i=1,\cdots,N}$. In other words, it is a part of the question of the existence of points t at which the process is "renewed by a jump":

$$\Lambda_t(\xi) = H_{t+0}(\xi) \ominus H_t(\xi) \neq 0.$$

Clearly, this condition is equivalent to the fact that the structure matrix $F(t) = \{F_{ij}(t)\}_{j=1,\cdots,N}^{i=1,\cdots,N}$ of the renewal process $\eta(t)$ has a jump at the corresponding point t.

Let us give an example of a *regular continuous* process $\xi = \xi(t)$ completely renewed by a jump and having rank $N = \infty$. Namely, let $\xi_0(u)$, $-\infty < u < \infty$, be a stationary singular process for which $\bigcap H_u(\xi_0) = H_\infty(\xi_0)$, and

$$\xi(t) = \begin{cases} 0 & \text{for } -\infty < t < -\dfrac{\pi}{2}, \\ \alpha(t)\xi_0(t) & \text{for } -\dfrac{\pi}{2} \leqq t \leqq 0, \\ \xi_0(1) & \text{for } 0 < t < \infty, \end{cases}$$

where $\alpha(t)$ is a positive continuous o function on the interval $(-\pi/2, 0)$ such that $\alpha(-\pi/2) = 0$, $\alpha(0) = 1$. Clearly, $H_{-\pi/2}(\xi) = 0, H_{-\pi/2+0}(\xi) = H_\infty(\xi_0) = H_\infty(\xi)$, and the dimension N of every renewal process $\eta(t) = \{\eta_i(t)\}_{i=1,\cdots,N}$ is equal to the dimension of the space

$$H_\infty(\xi_0) = H_{-\pi/2+0}(\xi) \ominus H_{-\pi/2}(\xi).$$

We modify the notion of regularity as follows: the process $\xi = \xi(t), a \leqq t \leqq b$, is said to be *regular at the point* t if $\Delta_t(\xi) = 0$, and *regular everywhere* if it is regular at every point t.

It is easy to see that every process $\xi(t)$ can be represented as a sum of an everywhere regular process $\xi_1(t)$ and a process $\xi_2(t)$ orthogonal to $\xi_1(t)$ which is renewed only by jumps[2] (see [5]):

$$\xi(t) = \xi_1(t) \oplus \xi_2(t),$$

where

$$\xi_1(t) = \int_a^t \chi_{\Delta_1}(s) c(t, s)\, d\eta(s), \; \xi_2(t) = \int_a^t \chi_{\Delta_2}(s) c(t, s)\, d\eta(s)$$

(Δ_1 denotes the set of jumps of the structure matrix $F(t) = \{F_{ij}(t)\}_{j=1,\cdots,N}^{i=1,\cdots,N}$ of the renewal process $\eta(t) = \{\eta_i(t)\}_{i=1,\cdots,N}$ in the canonical representation (1), Δ_2 is the complementary set on the interval $[a, b]$, $\chi_{\Delta_1}(s)$ and $\chi_{\Delta_2}(s)$ are the indicators of these sets) and, for all t,

$$H_{t+0}(\xi) = H_{t+0}(\xi_1) \oplus H_{t+0}(\xi_2).$$

Note that for the process $\xi = \xi(t)$, renewed only by jumps, the following assertion is true: under any transformation of the form $\tilde{\xi} = A\xi(t)$, where A is a bounded operator on the Hilbert space H, the rank does not increase ($\tilde{N} \leqq N$).

In fact, in our case, representation (1) has the form

$$\xi(t) = \sum_{t_k < t} c(t, t_k)\, d\eta(t_k) = \sum_{t_k < t} \sum_{i=1}^{N} c_i(t, t_k)\, d\eta_i(t_k),$$

and it is easy to see that

$$\tilde{\xi}(t) = \sum_{t_k < t} \sum_{i=1}^{N} c_i(t, t_k)[A\, d\eta_i(t_k)] = \sum_{t_k < t} \sum_{i=1}^{N} \tilde{c}_i(t, t_k)\, d\tilde{\eta}_i(t_k),$$

where $A\, d\eta_i(t_k) \in H_{\tilde{\xi}}(t_k + 0)$ and $d\tilde{\eta}_i(t_k), i = 1, \cdots, N$, is an orthogonal basis in the subspace formed by the projections of the variables $A\, d\eta_i(t_k), i = 1, \cdots, N$, onto $H_{\tilde{\xi}}(t_k)$.

4°. The role played in the theory of prognosis and filtration by the canonical representation (1) is well known; however, up till now it was studied only for the case of regular stationary processes.

It seems that one of the general approaches to the study of the structure of random processes of second order from the point of view of this canonical representation, can be taken to be the following generalization of a theorem of Karhunen [6]:

The random process $\xi = \xi(t), a \leqq t \leqq b$, can be written in the form (1) (*where $d\eta(t)$ is some renewal process with orthogonal increments having structure matrix $dF(t)$*) *if and only if its correlation function $B(s, t)$ admits a representation*

$$B(s, t) = \int_a^b c(s, u)\, dF(u) c(t, u)^* = \int_a^{\min(s,t)} c(s, u)\, dF(u) c(t, u)^*, \tag{2}$$

in which, for each t_0, the system of vector functions $c(v, u)$, $a \leqq u \leqq b$, ($c(v, u) = 0$ for $u > v$, the parameter v ranges over the interval $[a, t_0)$) is complete in the corresponding subspace $L^2_{t_0}(F)$ of all functions in $L^2(F)$ vanishing outside the interval $[a, t_0)$.

Note that it is convenient to extend the notion of a renewal process $d\eta(t)$ by considering it as a measure; an example is the known representation of a regular stationary process in the form

$$\xi(t) = \int_{-\infty}^{t} c(t - s)\, d\eta(s), \qquad -\infty < t < \infty, \tag{3}$$

where $d\eta(s)$ is the so-called fundamental measure, $\mathbf{E}|d\eta(s)|^2 = ds$. In this case, in order to find the canonical representation (3), one must find the corresponding function $c(t)$, $c(t) = 0$ for $t < 0$, in

[2] For stationary processes the indicated representation coincides with the well-known decomposition into regular and singular components (see, for example, [4]).

the usual space L^2 on the line, such that all its shifts $c(v-u)$, $-\infty < u < \infty$, (the parameter v ranges from $-\infty$ to 0) form a complete system in the subspace L_0^2 of all the functions in L^2, vanishing on the positive half-axis, which is related to the correlation function $B(s, t) = B(t-s)$ by the formula

$$B(t-s) = \int_{-\infty}^{\infty} c(s-u)c(t-u)\,du = \int_{-\infty}^{\infty} c(t-s+u)c(u)\,du.$$

As is known, this is equivalent to the fact that the Fourier transform $\varphi(\lambda) = \frac{1}{2\pi}\int_0^{\infty} e^{-i\lambda t}c(t)\,dt$ is a maximal analytic function of the Hardy class H_2 in the lower half-plane, satisfying the condition

$$|\varphi(\lambda)|^2 = \frac{1}{2\pi}f(\lambda),$$

where $f(\lambda)$ is the spectral density of the stationary process:

$$B(t) = \int_{-\infty}^{\infty} e^{i\lambda t}f(\lambda)\,d\lambda$$

(see, for example, [1]).

5°. In conclusion, let us dwell on the simplest type of random processes of second order, the so-called *wide sense Markov processes*, defined by the condition that, for any $s \leqq t$, the projection $\hat{\xi}(t, s)$ of $\xi(t)$ onto the subspace $H_s(\xi)$ coincides with the projection onto $\xi(s)$. More precisely,

(4) $$|\hat{\xi}(t, s) = \frac{B(t, s)}{B(s, s)}\xi(s)$$

(recall that $B(s, t) = \langle \xi(s), \xi(t) \rangle$).

We shall assume that $\xi(t) \neq 0$ for all t, $a < t \leqq b$. In this connection, note that for $\xi(t_0) = 0$, the process $\xi(t)$, $a \leqq t \leqq b$, decomposes into two independent Markov processes $\xi(t)$, $a \leqq t \leqq t_0$, and $\xi(t)$, $t_0 < t \leqq b$, which are *mutually orthogonal* (the projection of each variable $\xi(t)$, $t > t_0$, onto the subspace $H_{t_0}(\xi)$, coinciding by hypothesis with the projection onto $\xi(t_0) = 0$, is equal to 0).

Clearly, the process obtained after multiplication of $\xi(t)$ by the scalar function $B(t, t)^{1/2}$ is Markovian. We shall assume that $B(t, t) = 1$, $a < t \leqq b$, and moreover, that the extreme value $\xi(b)$ is known (this assumption is not essential).

Let us set $\eta(t) = \hat{\xi}(b, t)$. Clearly the projection of $\xi(b)$ onto the subspace $H_s(\xi)$ coincides, for $s \leqq t$, with the projection of $\eta(t)$ onto the same subspace, and is equal to $\eta(s)$. We see that the difference $\eta(t) - \eta(s)$ is orthogonal to $H_s(\xi)$. But $\eta(t)$ differs from $\xi(t)$ only by a scalar function: $\eta(t) = B(t, b)\xi(t)$. Here $F(t) = |B(t, b)|^2$, $a \leqq t \leqq b$, is a monotonically nondecreasing function, and if $F(t_0) = 0$ at some point $t_0 > a$, then this implies the orthogonality of $\xi(t)$, $a < t \leqq t_0$, and $\xi(t)$, $t_0 < t \leqq b$. As already noted, such processes can be considered separately and, without loss of generality, one can suppose that $F(t) \neq 0$ for all t, $a < t \leqq b$. Then, clearly, $H_t(\xi) = H_t(\eta)$, and the orthogonality of the difference $\eta(t) - \eta(s)$ to the subspace $H_s(\eta)$ implies that $\eta = \eta(t)$, $a \leqq t \leqq b$, is a renewal process with orthogonal increments for $\xi = \xi(t)$. Hence, in this case, the canonical representation (1) has the form[3] $\xi(t) = c(t)\eta(t)$, where $c(t) = B(t, b)^{-1}$. The corresponding representation (2) for the correlation function is

$$B(s, t) = c(s)F(\min(s, t))c(t),$$

which yields the following formula for $B(s, t) = B(t, s)$:

(5) $$B(s, t) = c(t)/c(s) \text{ for } s \leqq t.$$

Note that the regularity of a wide sense Markov process $\xi = \xi(t)$ at the point t_0, equivalent, in the general case, to the continuity at t_0 of the structure function $F(t)$ of the renewal process, is equivalent to the continuity of the process $\xi(t)$ at $t = t_0$ (to the continuity of its correlation function $B(s, t)$ as $s = t_0$, $t = t_0$). Regularity at the point $a = -\infty$ for $\xi(t)$, $-\infty < t \leqq b$, is equivalent to the fact that $B(t, b) \to 0$ as $t \to -\infty$ (cf. [7]).

Received by the editors
October 30, 1970

[3] In formula (1) we must set $c(t, u) = c(t)$ for $u \leqq t$, and $c(t, u) = 0$ for $u > t$.

REFERENCES

[1] H. Wold, *A Study in the Analysis of Stationary Time Series*, 2nd Ed., Amgrist and Wiksells, Uppsala, 1954.

[2] H. Cramér, *Stochastic processes as curves in Hilbert space*, Theory Prob. Applications, Vol. 9 (1964), pp. 169–179.

[3] T. Hida, *Canonical representation of Gaussian processes and their applications*, Mem. Coll. Sci. Univ. Kyoto, Ser. A, 33 (1960), pp. 109–155.

[4] Yu. A. Rozanov, *Stationary Random Processes*, Holden-Day, San Francisco, 1967.

[5] G. Kallianpur and U. Mandrekar, *Multiplicity and representation theory of purely non-deterministic stochastic processes*, Theory Prob. Applications, 10 (1965), pp. 553–581.

[6] K. Karhunen, *Über lineare Methoden in der Wahrscheinlichkeitsrechnung*, Ann. Acad. Sci. Fennicae, Ser. A., 37 (1947), pp. 3–79.

[7] U. Mandrekar, *On multivariate wide-sense Markov processes*, Nagoja Math. J., 33 (1968), pp. 7–19.

Author Citation Index

Subject Index